Analysis of Seawater

Analysis of Seawater

T.R. Crompton BSc (Hons), MSc
Northwest Water Authority,
Dawson House, Liverpool Road,
Great Sankey, Warrington

Butterworths
London Boston Singapore Sydney Toronto Wellington

 PART OF REED INTERNATIONAL P.L.C.

First published 1989

© **Butterworths & Co (Publishers) Ltd. 1989**

British Library Cataloguing in Publication Data

Crompton, T.R. (Thomas Roy)
  Analysis of seawater.
  1. Seawater. Chemical analysis
  I. Title
  551.46'01

  ISBN 0–407–01610–4

Library of Congress Cataloging–in–Publication Data

Crompton, T.R. (Tomas Roy)

  Analysis of seawater/T.R. Crompton.
    p.   cm.
  ISBN 0–407–01610–4 :
  1. Sea-water--Analysis.   I. Title.
  GC101.2.C76   1989
  551.46'01--dc19

Typeset by Scribe Design, Gillingham, Kent
Printed and bound by Hartnolls Ltd, Bodmin, Cornwall

Preface

This book covers all aspects of the analysis of seawater using both classical and the most advanced recently introduced physical techniques.

Until fairly recently, the analysis of seawater was limited to a number of major constituents such as chloride and alkalinity.

It is now generally agreed that any determinations of trace metals carried out on seawater prior to about 1975 are questionable, principally due to the adverse effects of contamination during sampling, which were then little understood and lead to artificially high results. It is only in the last few years that methods of adequate sensitivity have become available for true ultra-trace metal determinations in water.

Similar comments apply in the case of organics in seawater because it has become possible to resolve the complex mixtures of organics in seawater and achieve the required very low detection limits only since the advent of sample preconcentration coupled with gas chromatography and high-performance liquid chromatography, and possibly derivitization of the original sample constituents to convert them into a form suitable for chromatography.

Fortunately, our interest in micro-constituents in seawater both from the environmental and the nutrient balance points of view has coincided with the availability of advanced instrumentation capable of meeting the analytical needs.

Chapter 1 discusses a very important aspect of seawater analysis, namely sampling. If the sample is not taken correctly, the final result is invalidated, no matter how sophisticated the final analytical procedure. Recent important work on sampling is discussed in detail.

Chapter 2 discusses the determination of anions. Direct application of many of the classical procedures for anions fail for seawater owing to interfering effects of the sample matrix. Suitable modifications are discussed that are amenable to seawater. Dissolved gases in seawater are of interest in certain contexts and their determination is discussed in Chapter 3.

Chapter 4 discusses the application of new techniques such as atomic absorption spectrometry with and without graphite furnace and Zeeman background correction, inductively coupled plasma emission spectrography, X-ray fluorescence spectroscopy, neutron activation analysis, voltammetric techniques and others. In the first part of the chapter elements are discussed singly in alphabetical order, then as groups of elements because the newer techniques often cover ranges of elements. Finally, there is a section on metal preconcentration techniques. By concentrating all the metals present in a large volume of sample into a few millilitres, dramatic improvements in detection limit can be achieved, and it is this that enables the techniques to be applicable at the very low basal concentrations at which many metals exist in seawater. Increasingly, owing to fall-out and the introduction of nuclear power stations and processing plants, it is necessary to monitor the levels of radioactive elements in seawater, and this is discussed in Chapter 5. Chapters 6 and 7 cover the determination of a wide range of organics in seawater whilst Chapter 8 covers organometallic compounds. An increasingly long list of organometallics can be detected in seawater, many of these being produced by biologically induced metal methylation processes occurring in sediments and fish tissues.

Finally, in Chapter 9, is discussed the present state of knowledge on the determination of oxygen demand parameters in seawater. Amongst others these include total, dissolved and volatile organic carbon and total inorganic carbon as well as recent work on the older oxygen demand parameters, such as chemical oxygen demand and biochemical oxygen demand. The confusion that formerly existed regarding these methodologies is now being resolved to the point that meaningful measurements can now be reported. The determination of other non-metallic elements is also discussed in this chapter.

Whilst the book will be of obvious interest to anyone concerned with seawater environmental protection, it is believed that it will also be of interest to other groups of workers including River Authorities who have to implement legal requirements regarding seawater pollution, oceanographers, fisheries experts and biologists concerned with nutrient balances and toxicity effects in seawater and also politicians who create and implement environmental policies and the news media who are responsible for making the general public aware of environmental matters. The book will also be of interest to practising analysts and, not least, to the scientists and environmentalists of the future who are currently passing through the university system and on whom, more than even previously, will rest the responsibility of ensuring that our oceans are protected in the future.

Contents

Chapter 1

Sample handling and storage

1.1 Sampling

Common to all analytical methods is the need for correct sampling. It is still the most critical stage with respect to risks to accuracy in aquatic trace metal chemistry, owing to the potential introduction of contamination. Systematic errors introduced here will make the whole analysis unreliable. Very severe errors were commonly made during sampling by most laboratories until about a decade ago, owing to ignorance or at least underestimation of the problems connected with sampling. This is the principal reason why nearly all trace metal data before about 1975 for the sea and many fresh water systems are to be regarded as inaccurate or at least doubtful.

Surface-water samples are usually collected manually in precleaned polyethylene bottles (from a rubber or plastic boat) from the sea, lakes and rivers. Sample collection is performed in front of the bow of the boat, against the wind. In the sea, or in larger inland lakes, sufficient distance (about 500 m) in an appropriate wind direction has to be kept between the boat and the research vessel to avoid contamination. The collection of surface water samples from the vessel itself is impossible, considering the heavy metal contamination plume surrounding each ship. Surface water samples are usually taken at 0.3–1 m depth, in order to be representative and to avoid interference by the air/water interfacial layer in which organics and consequently bound heavy metals accumulate. Usually, sample volumes between 0.5 and 2 litres are collected. Substantially larger volumes could not be handled in a sufficiently contamination free manner in the subsequent sample pretreatment steps.

Reliable deep-water sampling is a special and demanding art. It usually has to be done from the research vessel. Special devices and techniques have been developed to provide reliable samples.

Samples for mercury analysis should preferably be taken in pre-cleaned glass flasks. If, as required for the other ecotoxic heavy metals, polyethylene flasks are commonly used for sampling, then an aliquot of the collected water sample for the mercury determination has to be transferred as soon as possible into a glass bottle, because mercury losses with time are to be expected in polyethylene bottles.

1

1.2 Sampling devices

The job of the analyst begins with the taking of the sample. The choice of sampling gear can often determine the validity of the sample taken; if contamination is introduced in the sampling process itself, no amount of care in the analysis can save the results. Sampling devices and sample handling for hydrocarbon analysis have been exhaustively reviewed by Green [1].

The highest concentrations of dissolved and particulate organic matter in the oceans are normally found in the surface films. When organic pollutants are present, they, too, tend to accumulate in this surface film, particularly if they are either non-polar or surface active. Much of the available information on these surface films is reviewed by Wangersky [2,3].

A major problem in the sampling of surface films is the inclusion of water in the film. In the ideal sampler, only the film of organic molecules, perhaps a few molecular layers in thickness, floating on the water surface, would be removed; the analytical results could then be expressed either in terms of volume taken or of surface area sampled.

In practice, all of the samplers in common use at this time collect some portion of the surface layer of water. Each sampler collects a different slice of the surface layer; thus results expressed in weights per unit volume are only comparable when the samples were taken with the same sampler. Of the surface film samplers described in the literature, the 'slurp' bottle used by Gordon and Keizer [4] takes perhaps the deepest cut, sampling as far as 3 mm into the water. The sampling apparatus consists simply of an evacuated container from which extend floating tubes with holes. The main advantages of the sampler are its simplicity and low cost. Its disadvantages are the thickness of the water layer sampled and the inherent inability to translate the results of the analysis into units of weight per surface area swept.

A thinner and more uniform slice of the surface can be collected with the Harvey [5] rotating drum sampler. This consists of a floating drum which is rotated by an electric motor, the film adhering to the drum being removed by a windshield wiper blade. The thickness of the layer sampled depends upon the coating on the drum, the speed with which it rotates, and the water temperature. The slice taken will usually run between 60 and 100 μm. This method has the advantage of sampling a known area of ocean surface, and of collecting a large sample very quickly. However, like most of the other surface film samplers, it must be operated from a small boat and can therefore only be used in calm weather.

The collector most commonly used for surface films is the Garrett screen [6], a monel or stainless steel screen which is dipped vertically below the surface and then raised in a horizontal position through the surface film. The material clinging to the screen is then drained into a collection bottle. This sampler collects a slice of the surface somewhere between 150 and 450 μm thick. A relatively small sample, about 20 ml, is collected on each dip, so that the collection of a sample of a reasonable size is very time consuming. This method is also limited to calm weather, since the sampling must be conducted from a small boat. While the Garrett screen has been adopted by many investigators because of its simplicity and low cost, as well as its relative freedom from contamination, it is a far from satisfactory solution to the problem of surface-film sampling. The small size of the sample taken greatly restricts the kinds of information which can be extracted from the sample. Also, the uncertainty as to the thickness of the film samples makes comparison between samplers difficult.

Only one actual comparison of these two techniques is available in the literature. Daumas *et al.* [7] compared the Harvey drum sampler to the Garrett screen, and found greater organic enrichment in both dissolved and particulate matter in the drum samples. The size of the difference between the two samplers suggests that the Garrett screen included 2–3 times as much water as did the Harvey drum.

Several samplers have been constructed on the principle of the use of a specially treated surface to collect surface-active materials. Harvey and Burzell [8] used a glass plate, both inserted and removed vertically with the material adhering to the plate then collected with a wiper. This method of sampling still includes some water; if a surface which preferentially adsorbs hydrophobic or surface-active materials is used, as for instance, Teflon, either normal or specially treated, only the organic materials will be removed, and the water will drain away. Such samplers have been described by Garrett and Burger [9], Larsson, Odham and Sodergren [10] and Miget *et al.* [11], among others. Anufrieva *et al.* [12] used polyurethane sheets, rather than Teflon, as the adsorbent. While this sort of sampler seems, at least theoretically, to have many advantages, since the surface swept is easily defined and the underlying water is excluded, the sample taken is very small; and it must be removed from the sampler by elution with an organic solvent. Thus the chances of contamination from the reagents used are sharply increased.

An interesting method which combines sampling and analysis in one step has been described by Baier [13]. A germanium probe is dipped into the water and carefully withdrawn, bringing with it a layer of surface-active material. This layer is then analysed directly by internal reflectance infra-red spectroscopy. Since there is no handling of the sample, contamination is reduced to a minimum. However, only infra-red spectral analysis is possible with this system; since the material adsorbed on the germanium prism is always a mixture of compounds, and since the spectrophotometer used for the production of the spectra is not a high-precision unit, the information coming from this technique is limited. While identification of specific compounds is not usually possible, changes in spectra–which can be related to time of day, season, or to singular events–can be observed.

Taken in all, there is still no really satisfactory method for sampling the surface film. Of the methods now in use, the method favoured by most workers is favoured for practical reasons, and not for any inherent superiority as a collector. Also, as long as we have no simple, accurate method for measuring total dissolved organic carbon, it will be difficult to estimate the efficiency of any of the surface film collectors.

Sampling the subsurface waters, although simpler than sampling the surface film, also presents some not completely obvious problems. For example, the material from which the sampler is constructed must not add any organic matter to the sample. To be completely safe, then the sampler should be constructed either of glass or of metal. All-glass samplers have been used successfully at shallow depths; these samplers are generally not commercially available [14,15]. To avoid contamination from material in the surface film, these samplers are often designed to be closed while they are lowered through the surface, and then opened at the depth of sampling. The pressure differential limits the depth of sampling to the upper 100 m; below this depth, implosion of the sampler becomes a problem.

Implosion at greater depths can be prevented either by strengthening the container or by supplying pressure compensation. The first solution has been applied in the Blumer sampler [16]. The glass container is actually a liner inside an aluminium pressure housing; the evacuated sampler is lowered to the required depth, where a rupture disc breaks, allowing the sampler to fill. Even with the aluminium pressure casing, however, the sampler cannot be used below a few thousand metres without damage to the glass liner.

Another approach to the construction of glass sampling containers involves equalization of pressure during the lowering of the sampler. Such a sampler has been described by Bertoni and Melchiorri-Santolini [17]. Gas pressure is supplied by a standard diver's gas cylinder, through an automatic delivery valve of the type used by SCUBA divers. When the sampler is opened to the water, the pressurizing gas is allowed to flow out as the water flows in. The sampler in its original form was designed for use in Lago Maggiore, where the maximum depth is about 200 m, but in principle it can be built to operate at any depth.

Stainless steel samplers have been devised, largely to prevent organic contamination. Some have been produced commercially. The Bodega–Bodman sampler and the stainless steel Niskin bottle, formerly manufactured by General Oceanics, Inc., are examples. These bottles are both heavy and expensive. The Bodega–Bodman bottle, designed to take very large samples, can only be attached to the bottom of the sampling wire; therefore, the number of samples taken on a single station is limited by the wire time available, and depth profiles require a great deal of station time.

The limitations of the glass and stainless steel samplers have led many workers to use the more readily available plastic samplers, sometimes with a full knowledge of the risks and sometimes with the pious hope that the effects resulting from the choice of sampler will be small compared with the amounts of organic matter present. The effects of the containers can be of three classes:

1. Organic materials may be contributed to the sample, usually from the plasticizers used in the manufacture of the samplers.
2. Organic materials, particularly hydrophobic compounds, may be adsorbed from solution on the walls of the sampler.
3. Organic materials may be adsorbed from the surface film or surface waters; then desorbed into the water samples at depth, thereby smearing the real vertical distributions.

The first case is most likely to be a problem with new plastic samplers. Although there is little in the literature to substantiate the belief, folklore has it that ageing most plastic samplers in seawater reduces the subsequent leaching of plasticizers markedly. The second case is known to be a problem; in fact, the effect is used in the various Teflon surface film samplers already mentioned. This problem alone would seem to militate against the use of Teflon for any sampling of organic materials, unless a solvent wash of the sampler is included routinely. With such a solvent wash, we introduce all of the problems of impurities in the reagents.

The third, and largely unexpected, case appeared as a problem in the analysis of petroleum hydrocarbons in seawater [18]. In this case, petroleum hydrocarbons picked up, presumably in the surface layers or surface film, were carried down by the sampling bottles and were measured as part of the pollutant load of the deeper waters. While the possibility of adsorption and subsequent release is obviously most acute with hydrophobic compounds and plastic samplers, it does raise a question as to whether any form of sampler which is open on its passage through the water column can be used for the collection of surface-active materials. The effect of such transfer of material may be unimportant in the analysis of total organic carbon, but could be a major factor in the analysis of single compounds or classes of compounds, where the total amount present in deeper water might be equal to or less than the amount carried down by the sampler surfaces.

Again, as in the case of the surface film samplers, information on the comparative merits of the various water samplers is largely anecdotal. Although such studies are not inspiring and require an inordinate amount of time, both on the hydrographic wire and in the laboratory, they are as necessary for the proper interpretation of data as are intercalibration studies of the analytical methods. The lack of comparison studies of the various samplers increases the probability of polemics in the literature.

Smith [19] has described a device for sampling immediately above the sediment water interface of the ocean. The device consists of a nozzle supported by a benthic sled, a hose and a centrifugal deck pump, and is operated from a floating platform. Water immediately above the sediment surface is drawn through the nozzle and pumped through the hose to the floating platform, where samples are taken. The benthic sled is manipulated by means of a hand winch and a hydrowire.

Intercomparison of seawater sampling devices for trace metals

Recently several round-robin intercalibrations for trace metals in seawater [20–24] have demonstrated a marked improvement in both analytical precision and numerical agreement of results among different laboratories. However, it has often been claimed that spurious results for the determination of metals in seawater can arise unless certain sampling devices and particular methods of sampler deployment are applied to the collection of seawater samples. It is, therefore, desirable that the biases arising through the use of different, commonly used, sampling techniques be assessed to decide upon the most appropriate technique(s) for both oceanic baseline and nearshore pollution studies.

Two international organizations, the International Council for the Exploration of the Sea (ICES) and the Intergovernmental Oceanographic Commission (IOC) have recently sponsored activities aimed at improving the determination of trace constituents in seawater through intercalibrations. Since 1975, ICES has conducted a series of trace metal intercalibrations, to assess the comparability of data from several tens of laboratories. These exercises have included the analyses of both standard solutions and real seawater samples [20–25]. The considerable improvement in the precisions and relative agreement between laboratories has been reflected in the results of these intercalibrations. By 1979 it had been concluded that sufficient laboratories were capable of conducting high-precision analyses of seawater for several metals to allow an examination of the differences between commonly used sampling techniques for seawater sample collection.

In early 1980, the IOC, with the support of the World Meteorological Organization (WMO) and the United Nations Environment Program (UNEP), organized a workshop on the intercalibration of sampling procedures at the Bermuda Biological Station during which the most commonly used sampling bottles and hydrowires were to be inter-compared. This exercise forms part of the IOC/WMO/UNEP Pilot Project on monitoring background levels of selected pollutants in open-ocean waters. Windom [26] had already conducted a survey of the seawater sampling and analytical techniques used by marine laboratories and the conclusions of this survey were largely used for the selection of sampling devices to be intercompared. The bottles selected for comparison in Bermuda were modified and unmodified GO-FLO® samplers, modified Niskin® bottles and unmodified Hydro-Bios® bottles. GO-FLO samplers are the most widely used sampling device for trace metals in seawater. The other two devices continue to be used by several marine laboratories. Windom's [26] 1979 survey established that the most common method of sampler deployment was on hydrowires, as opposed to the use of rosette systems. The hydrowires selected for intercomparison were Kevlar®, stainless steel and plastic coated steel. Kevlar and plastic-coated steel were selected because they are widely used in continental shelf and nearshire environments and are believed to be relatively 'clean'.

The method of intercomparison of the various devices was to deploy pairs of sampler types on different hydrowires to collect water samples from a homogeneous body of deep water at Ocean Station S ('Panulirus Station') near Bermuda (Figure 1.1). The water at this depth has characteristics of $3.97 \pm 0.05°C$ temperature and $35.01 \pm 0.02‰$ salinity for the month of January [27]. The restricted length of Kevlar hydrowire available necessitated the collection of samples in the lower thermocline at depths between 1150 and 1250 m.

Data analysis was reduced to a separate one-way analysis of variance on the data from individual laboratories to examine the differences between types of sampling

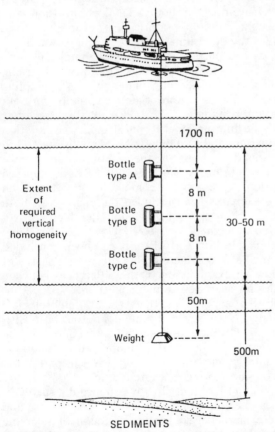

Figure 1.1 Sampling strategy. (Reproduced from Thiband, Y., *et al.* (1980) *Report CIEM 1979*, p. 1, Marine Chemistry Group, ICES, by courtesy of authors and publishers.)

bottle on a single (common) hydrowire, and to determine the influences of the three types of hydrowire using a single type of sampling bottle (modified GO-FLO). Samples were replicated so that there were, in all cases, two or more replicates to determine the lowest level and analytical error.

Replicate [23] unfiltered water samples were collected for each participant for the comparison of pairs of sampling bottles on different hydrowires. Modified GO-FLO bottles were employed on each of the three hydrowires and this permitted a comparison of the three types of hydrowire. Only in the cases of iron and manganese were there indications of inhomogeneity at levels that might invalidate the intercomparisons. This is assumed to be due to inhomogeneity in the distribution of suspended particulate material that will influence metals that have major fractions in the particulate phase.

The results obtained by the various calibrations in the determinations of nickel and copper are shown in Tables 1.1 and 1.2. Table 1.3 gives the differences between sampling devices for copper, as determined by each participant, when these are significant at the 95% and 90% levels of confidence. Only the results of participants that had acceptable analytical performance, as measured by precision and agreement with contemporary concensus values for deep North Atlantic waters (Table 1.4) were used for drawing conclusions.

Table 1.1 Numerical comparisons for nickel ($\mu g\,l^{-1}$)*

Wire:	PCS	PCS	SS mod	SS exw	KEV	KEV	PCS	SS	KEV
Bottle:	HB	MGF	GF	GF	NIS	MGF	MGF	MGF	MGF
Laboratory									
1 m 0.224	0.205	0.209	0.243	0.233	0.207	0.201	0.226	0.207
sd 0.015	0.020	0.023	0.013	0.023	0.025	0.029	0.025	0.025
2 m 0.298	0.278	0.235	0.235	0.218	0.150	0.223	0.235	0.231
sd 0.031	0.123	0.077	0.048	0.019	0.018	0.128	0.059	0.119
4 m 0.18	0.47	0.47	0.35			0.47	0.41	0.23
sd 0.07	–	–	0.17			–	0.12	–
5 m 0.478	0.221	0.240	0.235	0.273	0.237	0.193	0.238	0.237
sd 0.066	0.026	0.014	0.012	0.021	0.016	0.048	0.012	0.016
6 m 0.340	0.159	0.220	0.238	0.237	0.232	0.159	0.230	0.230
sd 0.052	0.035	0.030	0.021	0.023	0.026	0.035	0.027	0.024
7 m 1.93	1.63	1.60	1.75			1.63	1.68	1.72
sd 0.24	0.10	0.37	0.13			0.10	0.27	0.25
8A m 0.185	0.100	0.123	0.113	0.160	0.105	0.100	0.119	0.123
sd 0.041	–	0.005	0.012	0.054	0.010	–	0.009	0.041
10 m 0.737	0.357	0.634	0.353	0.385	0.461	0.413	0.493	0.462
sd 0.078	0.077	0.309	0.131	0.064	0.162	0.159	0.262	0.162
11 m 0.511	0.367	0.349	0.365	0.421	0.393	0.367	0.357	0.404
sd 0.034	0.008	0.034	0.009	0.009	0.027	0.008	0.025	0.037
12 m 0.230	0.165							
sd 0.024	0.036							
13 m 0.230	0.200	0.238	0.265	0.204	0.236	0.200	0.250	0.236
sd 0.036	0.020	0.045	0.007	0.056	0.007	0.020	0.035	0.007

*Numbers result from common computer analyses and not all such figures will be necessarily significant.
PCS = Plastic-coated steel hydrowire. SS = Stainless steel (type 302 unlubricated) hydrowire. KEV = Kevlar® hydrowire. HB = Hydro-Bios sampler. MGF = Modified GO-FLO sampler. mod GF = Modified GO-FLO sampler. exw GF = Unmodified GO-FLO sampler. NIS = Modified Niskin sampler. m = mean. sd = standard deviation.
Reproduced from Thibaud, Y. *et al.* (1980) *Report CIEM 1979*, by Marine Chemistry Group, ICES, by courtesy of authors and publishers.

Table 1.2 Statistical comparisons for nickel

Base comparison	PCS MGF/MGF	PCS HB/MGF	SS MGF/GF	KEV MGF/MGF	KEV NIS/MGF	MGF WIRES
Laboratory						
1	NS	Sig HB>MGF	Sig GF>MGF	NS	Sig NIS>MGF	Sig SS>KEV>PCS
2	NS	NS	NS	Sig	Sig NIS>MGF	NS
4		Sig MGF>HB	NS			NS
5		Sig HB>MGF	NS		90 NIS>MGF	Sig SS>KEV>PCS
6		Sig HB>MGF	Sig GF>MGF	NS	NS	Sig KEV>SS>PCS
8A		Sig HB>MGF				
10	NS	Sig HB>MGF	NS	Sig	NS	NS
11		Sig HB>MGF	NS	NS	NS	Sig KEV>PCS>SS
12	90	Sig HB>MGF				
13		NS	NS	NS	NS	90 SS>KEV>PCS

PCS = Plastic-coated steel hydrowire. SS = Stainless steel (type 302 unlubricated) hydrowire. KEV = Kevlar® hydrowire. HB = Hydro-Bios sampler. MGF = Modified GO-FLO sampler. GF = Unmodified GO-FLO sampler. NIS = Modified Niskin sampler. Sig = Difference is significant ($P<0.05$). 90 = Difference is significant ($P<0.1$). NS = Not significant ($P>0.1$).
From Thibaud, Y. *et al.* (1980) *Report CIEM, 1979*, Marine Chemistry Group, ICES, by courtesy of authors and publishers.

The experiment reveals that the differences between results obtained through the use of various combinations of hydrowires and samplers are not large and in no case can they account for the recent decline in the oceanic concentrations of trace metals reported in the literature. Nevertheless, for several metals, most notably copper, nickel and zinc, significant differences are evident between both bottles and hydrowires. For deep ocean studies the best combination of those tested is undoubtedly modified GO-FLO samplers and plastic-coated steel hydrowire. Except in the cases of mercury and manganese, Hydro-Bios samplers appear to yield higher metal values than modified GO-FLO samplers. In contrast, Niskin bottles, modified by the replacement of the internal spring by silicone tubing, are capable of collecting samples of comparable quality to those collected by modified GO-FLO sampler for all metals except zinc. Modification to factory supplied Teflon® coated GO-FLO bottles (i.e. replacement of 'O' rings with silicone equivalents and the substitution of all-Teflon drain cocks for those originally supplied), do appear to result in a significant reduction in the levels of most metals in seawater samples collected with them. Kevlar and stainless steel hydrowires generally yield measurably greater concentrations of most metals than does plastic-coated steel. These differences, however, are small enough to suggest that these hydrowires are still suitable for trace metal studies of all but the most metal-depleted waters if proper precautions are taken [28–31].

Table 1.3 Numerical comparisons for copper ($\mu g\,l^{-1}$)*

Wire:		PCS	PCS	SS mod	SS exw	KEV	KEV	PCS	SS	KEV
Bottle:		HB	MGF	GF	GF	NIS	MGF	MGF	MGF	MGF
Laboratory										
1	m	0.094	0.092	0.095	0.103	0.111	0.131	0.093	0.099	0.120
	sd	0.007	0.009	0.012	0.012	0.011	0.012	0.011	0.012	0.021
2	m	1.000	0.765	0.553	0.620	0.455	0.272	0.650	0.586	0.403
	sd	0.857	0.289	0.261	0.100	0.487	0.185	0.291	0.186	0.298
3	m	0.437	0.180	0.211	0.205	0.447	0.550	0.233	0.208	0.455
	sd	0.347	0.084	0.067	0.034	0.540	0.317	0.081	0.051	0.314
4	m	0.533	0.435	1.25	1.065			0.435	1.158	1.065
	sd	0.163	0.177	0.35	0.177			0.177	0.252	0.177
5	m	0.188	0.063	0.064	0.142	0.101	0.072	0.101	0.103	0.072
	sd	0.108	0.003	0.004	0.010	0.049	0.012	0.059	0.043	0.012
6	m	0.615†	0.070	0.074	0.083	0.121	0.120	0.070	0.079	0.120
	sd	0.560	0.005	0.003	0.004	0.039	0.022	0.005	0.006	0.026
7	m	0.35	0.27	0.71	0.28			0.27	0.50	0.32
	sd	0.29	0.12	0.59	0.23			0.12	0.47	0.37
8B	m	0.155	0.045	0.163	0.278	0.133	0.160	0.045	0.220	0.140
	sd	0.076	0.006	0.044	0.059	0.030	0.037	0.006	0.078	0.038
9	m	0.84	0.32			0.35	0.55	0.32		0.44
	sd	0.79	0.03			0.02	0.21	0.03		0.17
10	m	0.123	0.135	0.158	0.119	0.096	0.100	0.130	0.138	0.101
	sd	0.015	0.003	0.033	0.032	0.015	0.019	0.024	0.037	0.019
11	m	0.195	0.137	0.102	0.106	0.109	0.132	0.137	0.104	0.149
	sd	0.089	0.027	0.005	0.001	0.013	0.019	0.027	0.004	0.073
12	m	0.059†	0.172							
	sd	0.325	0.040							
13	m	0.168	0.101	0.105	0.292	0.133	0.121	0.101	0.186	0.121
	sd	0.063	0.028	0.013	0.006	0.009	0.020	0.028	0.100	0.200

*Numbers result from common computer analyses and not all such figures will be necessarily significant.
†Suspected contamination.
All other symbols are the same as those used in Table 1.1.
Reproduced from Thibaud, Y. et al. (1980) Report CIEM, 1979, Marine Chemistry Group, ICES, by courtesy of authors and publishers.

Table 1.4 Results of sampling bottle and hydrowire intercomparisons

Metal	Concentration*	No. of laboratories	Best combined sampling/ analytical precisions ($\mu g/l^{-1}$)	Comparisons	
				Hydrowires	Samplers
Cd	0.035±0.016	12	0.001	PCS<(KEV ≈ SS)	(MGF ≈ NIS)<HB<GF
Cu	0.13±0.04	6	0.010	PCS<(KEV ≈ SS)	(MGF ≈ NIS)<HB<GF
	0.51±0.28	6			
Ni	0.21±0.05	7	0.02	PCS<(KEV ≈ SS)	(MGF ≈ NIS ≈ GF)<HB
	0.42±0.11	3			
Zn	0.35±0.18	5	0.05	PCS<(KEV ≈ SS)	MGF<(NIS ≈ HB ≈ GF)
Fe	0.41±0.29	3	0.05	PCS<KEV<SS	(MGF ≈ NIS)<GF<HB
Mn	0.064±0.038	2	0.010	(PCS ≈ KEV ≈ SS)	(MGF ≈ NIS ≈ GF ≈ HB)
	0.012±0.006	1	0.003		
Hg	0.007±0.002	2	0.002	Insufficient comparisons	
	0.001	1			

*Mean (±SD), $\mu g\,l^{-1}$
Reproduced from Thibaud, Y. et al. (1980) Report CIEM 1979, Marine Chemistry Group, ICES, by courtesy of authors and publishers.

A major conclusion of the Bermuda experiment is that the use of differing sampling devices and hydrowires only accounts for a small portion of the differences between trace metals results from different laboratories. It appears that the major contributions to such differences are analytical artefacts. It is stressed that although the sampling tools available to marine geochemists appear adequate for the measurement of metal distributions in the ocean, the execution of co-operative monitoring programs for metals should be preceded by a mandatory intercomparison of sample storage and analytical procedures.

Intercomparison of sampling devices and analytical techniques using seawater from a Copex (controlled, Ecosystem Pollution Experiment) enclosure

Wong *et al.* [32] conducted an intercomparison of sampling devices using seawater at 9 m in a plastic enclosure of 65 m in Saanich Inlet, BC, Canada. The sampling methods were:

1. Peristaltic pumping with Teflon tubing.
2. Niskin PVC sampler.
3. Go-Flow sampler.
4. Close–open–close sampler.
5. Teflon-piston sampler.

Sampling was conducted for 4 days:

1. Day 1 (August 1978) for mercury.
2. Day 2 for lead, cadmium, copper, cobalt and nickel by Chelex extraction and differential pulse polarography, as well as manganese by Chelex and flameless atomic absorptiometry.
3. Day 3 for lead by isotope dilution.
4. Day 4 for cadmium, copper, iron, lead, nickel and zinc by Freon extraction and flameless atomic absorptiometry.

Samples were processed in clean rooms in the shore laboratory within 30 minutes of sampling. Results indicated the feasibility of inter-calibrating using the enclosure approach, the availability of chemical techniques of sufficient precision in the cases of copper, nickel, lead and cobalt for sampler intercomparison and storage tests, a problem in sub-sampling from the captured seawater in a sampler, and the difficulty of commonly used samplers to sample seawater in an uncontaminated way at the desired depth.

The Teflon tubing used in the pumping system, the Niskin sampler and the Go-Flow sampler were cleaned by immersion in 0.05% nitric acid for the tubing and by soaking the inside of the samplers in 0.05% nitric acid overnight, rinsing with distilled water and repeating the dilute acid/distilled water cycle. The close–open–close sampler was cleaned by 0.1 N nitric acid overnight, then rinsed with distilled water till the blank was acceptable. The Teflon–piston sampler was cleaned by sucking in 0.05% nitric acid and standing overnight (in the case of the poly bag liner used in the Teflon–piston sampler hydrochloric acid was used instead of nitric acid).

The storage bottles were cleaned as follows. The Pyrex bottles (2 litres) were used for mercury samples only. They were cleaned by filling with a solution of 0.1% $KMnO_4$, 0.1% $K_2S_2O_8$ and 2% nitric acid, heating to 80°C for 2 hours, and after cooling and rinsing, stored filled with 2% nitric acid containing 0.01%

$K_2Cr_2O_7$ until ready for use. Conventional polyethylene bottles of 1 or 2 litre sizes, were used for the other metal samples. They were cleaned by Patterson's method [33]. All bottles were stored inside two or three plastic bags to prevent contamination.

For the pumping system, seawater was pumped up from 9 m and collected in the appropriate bottles on the raft and returned to the shore clean laboratory for preservation and/or analysis. For the other four sampling devices, the sampler was lowered to 9 m, allowed to equilibrate for 10 minutes, closed by a triggering mechanism activated by the Teflon messenger, raised to the surface, transferred into the container, transported back by boat and trucked back to the shore clean laboratory, where the subsamples were drawn. The time between messenger activation and subsampling was about 30 minutes. For handling of the samples, messengers, Teflon tubing, vinyl-coated hydrowires and sampling devices, all personnel wore polyethylene gloves to avoid contamination.

The clean laboratory for trace metals was divided into three areas: entrance laboratory (with clothes changing annexe), instrument laboratory and the ultra-clean sample preparation laboratory, all under positive pressure with active charcoal filtered air. Personnel using the clean rooms were required to wear hair caps, polyethylene gloves, laboratory coats and designated shoes. These items are worn only in the clean rooms.

Mercury was determined after suitable digestion by the cold vapour atomic absorption method [34]. Lead was determined after digestion by a stable isotope dilution technique [35–37]. Copper, lead, cadmium, nickel and cobalt were determined by differential pulse polarography following concentration by Chelex 100 ion-exchange resin [38,39] and also by the Freon TF extraction technique [40]. Manganese was determined by flameless atomic absorption spectrometry.

The precision of the procedures under clean room conditions is shown in Table 1.5.

Table 1.5 Precision of the procedure for Cu, Ni, Cd, Fe and Pb as applied in ocean chemistry clean room

Metal	Seawater concentration (nmol kg^{-1})	Blank (nmol)	Relative standard deviation at test level (average of 10 analyses) (%)	Recovery (%)
Cd	1.16	0.009	6	83–98
Cu	13.0	0.08	2	95
Ni	14.8	0.12	2	95
Zn	32.3	0.69	2	95
Fe	7.8	0.54	3	90
Pb	0.10	0.02	30	85–100

Reproduced from Wong, C.S., *et al.* (1985), Ocean Chemistry Division, Institute of Ocean Sciences, Sydney.

The results in Figure 1.2 show values between 0.06 and 0.12 nmol kg^{-1}. The average mercury contents obtained by pumping, Niskin sampler, Go-Flow sampler and the close–open–close device are 0.09 ± 0.03, 0.08 ± 0.01, 0.08 ± 0.03 and 0.10 ± 0.02 nmol kg^{-1} respectively. The mercury values obtained by the Teflon–piston sampler were high at $0.21 \pm$ nmol kg^{-1} due to malfunction with incomplete filling and previous contamination as indicated by the very low salinity in this set. The values inside the bag were higher than those outside, measured about 1 month after the intercomparison to be 0.02, 0.03 and 0.04 nmol kg^{-1}. There was a

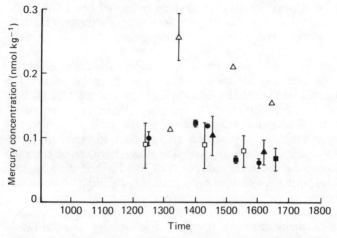

Figure 1.2 Comparison of mercury concentrations found in a CEPEX enclosure using five different sampling methods. ● = Pump; ▲ = Hydrobios; ■ = Go-Flow; □ = Niskin; △ = Seakern (Reproduced from Wong, C.S. *et al.* (1985), Ocean Chemistry Division, Institute of Ocean Sciences, Sydney.)

subsampling problem. The first and second draw of the sampling bottle usually showed a very wide spread in values, as much as 0.07 nmol kg^{-1}, e.g. between 0.05 and 0.12 nmol kg^{-1}. This difference was real since the technique of cold vapour atomic absorption should be capable of detecting difference in subsamples from the same digested sample in a Pyrex bottle. The peristaltic pumping method appears to yield the best agreement between subsamples: a difference of 0.02, 0.00 and 0.01 nmol kg^{-1} between subsamples from the three casts. The average Hg values for each sampler appeared to converge towards lower values on repeated casts within the same day. Further work is required to clarify contamination in mercury sampling.

Isotope dilution and mass spectrometry showed the lead values to be 0.73 ± 0.02, 0.72 ± 0.03, 0.75 ± 0.02, 0.78 ± 0.05 and 0.81 ± 0.03 nmol kg^{-1} for sampling by peristaltic pump, Niskin sampler, Go-Flow sampler, close–open–close sampler and Teflon–piston sampler, respectively (Figure 1.3). For the other two techniques, the Teflon–piston sampler showed considerable variability and statistically much higher values. The results were not used in the comparison. The Freon extraction and F.A.A. approach showed the same range of values as the isotope dilution approach, i.e. 0.71 ± 0.36, 0.76 ± 0.13, 0.73 ± 0.13 nmol kg^{-1} for the pumping, Niskin sampler and close–open–close sampler respectively, with the exception of the Go-Flow sampler with a low value of 0.58 ± 0.15 nmol kg^{-1}. However, the range of values was wide, e.g. for the peristaltic pumping, 1.12 nmol kg^{-1} for the first cast dropping to 0.46 nmol kg^{-1} for the third cast. Chelex extraction and differential pulse polarography showed an even larger spread from 0.38 ± 12 nmol kg^{-1} for the three casts with the Niskin sampler to 1.09 ± 0.26 nmol kg^{-1} for the close–open–close sampler.

Wong *et al.* concluded that:

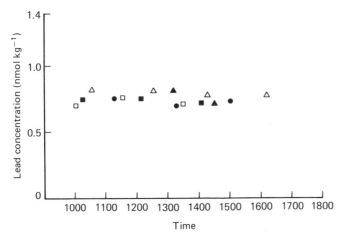

Figure 1.3 Comparison of lead concentrations found in a CEFEX enclosure using five different sampling methods and IDMS analysis. ● = Pump; ▲ = Hydrobios; ■ = Go-Flow; □ = Niskin; △ = Seakem. (Reproduced from Wong, C.S. *et al.* (1985), Ocean Chemistry Division, Institute of Ocean Sciences, Sydney.)

1. It is feasible to capture a large volume of sea water in the range of 65 000 litres by the CEPEX approach for the purpose of sampler intercomparison. It is possible by artificial stimulation of a plankton bloom and detritus removal to produce a reasonably homogeneous body of seawater for the study. Proximity of the *in-situ* enclosure for the experiment and shore clean laboratory facilities eliminate errors introduced by shipboard contamination under less than ideal conditions on cruises.
2. The following analytical techniques seem to be adequate for the concentrations under consideration: copper and nickel by Freon extraction and flame atomic absorption spectrometer, cobalt by Chelex extraction and differential pulse polarography, mercury by cold vapour absorptiometry, and lead by isotope dilution plus clean room manipulation and mass spectrometry. These techniques may be used to detect changes in the above elements for storage tests: Cu at $8\,nmol\,kg^{-1}$, Ni at $5\,nmol\,kg^{-1}$, Co at $0.5\,nmol\,kg^{-1}$, Hg at $0.1\,nmol\,kg^{-1}$ and Pb at $0.7\,nmol\,kg^{-1}$.
3. Salinity of seawater captured by various sampling devices in the CEPEX enclosure indicates problems not revealed in the usual oceanographic sampling situation. Relative to peristaltic pumping, all samplers exhibited some salinity anomalies. Inadequate flushing to rinse the sampler of any concentrated brine or entrapped seawater is thought to be the problem.
4. Logistics and cleaning procedures are important factors in successful sampler intercomparisons. It is not desirable or possible to endorse or to condemn the performance of a certain type of sampler or analytical technique based on results of one set of tests, especially if procedures are changed.
5. The problems of subsampling from the same seawater sample has to be studied in greater detail.

6. A long-term but sustained effort on sampler inter-comparisons would be advantageous in identifying problems.

1.3 Sample preservation and storage

If we were to choose the ideal method for the analysis of any component of seawater, it would naturally be an *in-situ* method. Where such a method is possible, the problems of sampling and sample handling are eliminated and in many cases we can obtain continuous profiles rather than a limited number of discrete samples. In the absence of an *in-situ* method, the next most acceptable alternative is analysis on board ship. A 'real-time' analysis not only permits us to choose our next sampling station on the basis of the results of the last station, it also avoids the problem of the storage of samples until the return to a shore laboratory.

While there are a few methods of this type for major constituents, and the advent of the automatic analyser has made possible the adaptation of some micro-methods to shipboard analysis, the majority of chemical analyses, particularly those using the newer, more sophisticated instruments, must still be run on shore. For such samples, the problems of storage and sample preservation become all-important, since the quantity we wish to measure is the *in-situ* value and not the amount remaining after some period of biological and chemical activity. Complete oxidation of the organic material to inorganic constituents takes longer than a year [41]; however, decomposition great enough to free most inorganic micronutrients takes place within the first 2 weeks. Important changes in micronutrient levels resulting from bacterial utilization of organic compounds can be seen after 1 day. Therefore, some method of preservation of the organic compounds must be sought if the samples are to be taken back to a shore laboratory.

Changes in the distribution of organic compounds in a seawater sample can be due to physical, chemical, or biological factors. As a physical factor, we might consider the adsorption of surface-active materials on the walls of the sample container. While this effect cannot be eliminated, it can be minimized by the use of the largest convenient sample bottle, and the avoidance of plastic (especially Teflon) containers. Another possible method of eliminating this source of error would be to draw the water sample directly into the container in which the analytical reaction is to be run.

If volatile organic materials are present in the sample, plastic sample containers cannot be used. Many small organic molecules are able to diffuse through plastics. If the sample container is not tightly sealed, the volatiles may escape into the environment. Even with a proper seal, an escape from solution into the head space will occur. The loss will be greater if the sample is allowed to warm to room temperature. To avoid such losses, sample bottles must be made of glass, with gas-tight seals. They should be filled to the top and stored at low temperature, but above freezing.

The most probable purely chemical reaction to be expected would be photolysis. This possibility can be prevented by storing the sample in the dark. Since most of these organic materials have been present together in seawater for some considerable time, any other reactions are unlikely to occur without an increase in temperature or the inadvertent addition of some catalyst. However, if biologically mediated reactions were to result in the formation of highly reactive species, such as enzymes, further chemical reactions might then occur. Changes in pH, brought about either by biological activity or by the addition of mineral acids as preservatives, might also promote reactions.

Most of the changes taking place in stored seawater are due to biological, and principally bacterial, reactions. In one sense, it might seem strange that removing a parcel of water from the ocean, complete with its normal bacterial complement, should result in any increase in bacterial activity. However, the factor controlling bacterial activity seems to be the available surface area, since free-floating bacteria are not usually growing actively [42,43]. Thus, enclosing a sample of sea water in a bottle serves to furnish the free-floating bacteria with a large surface to which they can cling and upon which they can multiply. This 'bottle effect' is well known, and must be prevented if the results of analysis are to have any meaning.

Table 1.6 Details of bottles and fixing reagents

Analysis	Bottle material	Bottle size	Reagents in bottle	Shelf life	Notes
Dissolved oxygen	Glass	125 ml	None. Fixing kit supplied	Unlimited	
Toxic metals	Polyethylene	250 ml 1 ml	10 ml 25% Analar or Aristar HNO_3 40 ml 25% HNO_3	1 Month	If soluble metals to be determined, filter sample before acid addition. Clean plastic Cassella by soaking in acid
Phenol	Glass	500 ml	1 ml H_3PO_4 and 1 g $CuSO_4$ $5 H_2O$	Made up as needed	
Silica	Polyethylene		None		
Mercury	Glass	125 ml	7.5 ml 5% $K_2Cr_2O_7$ plus 4 ml conc H_2SO_4	Made up as needed	Bottles filled with 2 N HNO_3 when not in use; rinsed out before adding reagents
Chlorophyll	Brown glass	40 oz	None	Unlimited	To be kept out of direct sunlight after filling with sample
Sulphide	Glass	(A) 4 oz.ctg.	5 ml 132 g l^{-1} Zn acetate and 1 ml acetic acid.		To bottle already containing 5 ml solution 'A' add sample as quickly as possible, avoiding entrainment of air bubbles. 5 ml of solution 'B' is then added to fill the bottle, which should be stoppered and thoroughly mixed
		(B) Reagent bottle ctg.	96 g l^{-1} sodium carbonate		
Pesticides	Glass	40 oz	None	Unlimited	Bottle chromic acid cleaned, ground-glass stoppered and wrapped in polyethylene bag
Oils	Glass	40 oz	None	Unlimited	Chromic acid cleaned. Separate bottle from main sample, bottle ⅔ filled with sample
Cyanide	Glass DO bottles		2 Na OH pellets	Made up as needed	

Table 1.7 Sample preservation

Parameter	Suggested methods	Remarks
Acidity/alkalinity	Refrigeration	Gains or losses of CO_2 affect the result. Microbial action can affect this, should be one of the first analyses of the sample
BOD	Refrigeration	Refrigeration is only partially effective. Analysis is best done immediately. Difficulties with composites upon sewage works. Large changes can occur over a few hours. Glass bottles preferred
COD	1. Refrigeration 2. Acidification with sulphuric acid to pH2	Changes can occur quite rapidly with some samples. Glass bottles preferred. DoE recommendation. Hydrolysis can occur over longer periods
Chlorine (residual)	Immediate analysis	No storage possible
Cyanide	Addition of ascorbic acid then render alkaline to pH 11–12 with sodium hydroxide	This parameter must be preserved immediately. Need for separate container. Any residual chlorine must be removed with ascorbic acid
Dissolved oxygen	Winkler reagents	Addition of reagents immediately after sampling. Fixed sample should be kept in the dark and analysed as soon as possible. Where measurements are needed on liquids containing high levels of easily oxidizable organic matter, a portable meter should be used, otherwise the copper sulphate/sulphamic acid mix should be used. DoE recommendation
Metals (excepting mercury)	*Dissolved* Filter on site into acid to pH 1–2 *Suspended* Filter on site *Total* Acidify to pH 1–2 with hydrochloric or nitric acid	This procedure should be used wherever possible at sewage works when compositing effluents. DoE recommendations vary between 2 and 20 Mls, depending on the element. Polythene bottle required For potable water, acidification must be carried out on receipt in the laboratory; acidified bottles should not be left with private householders
Mercury	Nitric acid (5 ml) + 5% potassium dichromate (5 ml) per litre	This treatment must be rendered immediately on sampling otherwise a large proportion can be lost within minutes. Glass bottle must be used (DoE recommendation.)
Nitrogen (ammonia, nitrate, organic)	1. Sulphuric acid 2. Chloroform 3. Refrigeration	Chloroform as used effectively by one division. Sulphuric acid is effective and is frequently recommended. Acidification may be recommended by DoE
Nitrite	Chloroform	Best analysed as soon as possible
pH	Refrigeration	Immediate analysis advisable, changes can occur quite rapidly, particularly if the container is opened. Well-buffered samples are less susceptible to changes
Phenolics	Copper sulphate 1 g l^{-1} + phosphoric acid to reduce pH to less than 4, refrigerated	

Table 1.7 (contd)

Parameter	Suggested methods	Remarks
Phosphate	1. Sulphuric acid conditioning 2. Use of prepared bottle/ refrigeration	Intensive investigation has been carried out on this subject. The concentration of the determinand may affect the method. The DoE recommendation will be for low levels, i.e. less than 25 μg l^{-1}, iodized plastic bottles for high levels, acid conditioned glass bottle
Sulphide	Zinc acetate/sodium carbonate, i.e. 5 ml zinc acetate (110 g zinc acetate + 1 ml acetic acid per litre) + 5 ml sodium carbonate (80 g l^{-1}) to 100 ml sample	Must be preserved immediately. Care needed for samples containing much suspended matter; may be DoE recommendation
Surfactants, anionic	Refrigeration and either: 1. Mercuric chloride 1% 2 ml l^{-1} 2. Chloroform 2 ml l^{-1}	These materials can be quite unstable over periods longer than a few hours Care will be needed where composites are taken at sewage works
Surfactants, non-ionic	Formaldehyde 5 ml l^{-1}	

The two most popular methods of sample preservation are quick-freezing and the addition of inorganic poisons. Samples that are frozen without the addition of preservatives are less likely to pick up contamination from the sample handling. If the samples are to be frozen quickly enough to prevent bacterial growth, the sample bottles must be immersed in a freezing bath, to facilitate heat exchange. Such freezing baths are usually organic, the sealing and handling of sample bottles must be carried out with extreme care, in order to prevent contamination.

While it is normally considered that both biological and chemical reactions will be essentially halted by freezing, this is not necessarily true. It has been shown that some reactions of considerable biochemical importance are in fact enhanced in the frozen state [44–46]. In any given case it cannot simply be taken for granted that freezing will be a sufficient preservative; the efficiency of the method must be tested for the compounds in question. The method is also limited to those analyses that can be performed on a small sample, perhaps 100–200 ml. Larger volumes of seawater take too long to freeze. If the next step in the sample preparation is to be freeze-drying, a considerable saving in time, as well as a decrease in possible contamination, can result from freezing the sample in the container to be used in the freeze-drying.

We are almost forced into hoping that freezing will prove satisfactory as a method of sample preservation, since none of the usual inorganic poisons works in every case. While mercuric chloride has been found to be effective with those marine organisms responsible for N_2O production [47], it has been found that added radioactive glycine, in the presence of mercuric chloride in seawater, was decomposed slowly in the dark in the apparent absence of bacteria, the radioactive label appearing in the vapour phase. Of the other two most favoured poisons, sodium azide proved to be ineffective in stopping bacterial activity at concentrations as high as 10 mg l^{-1}; and cyanide, although effective at levels between 1 and 10 mg l^{-1}, interfered with many of the chemical reactions.

Acidification with mineral acids is also often used as a method for preservation of organics. It would be expected that the distribution of organic compounds would be changed by such treatment, even if the total amount of dissolved organic carbon were not appreciably altered. Certainly the volatility of some of the organic molecules of low molecular weight is pH dependent; certain compounds normally retained at the slightly alkaline pH of seawater will be lost during freeze-drying or degassing of acidified samples.

Preservation of organic samples is thus still a major problem; there is no general, foolproof method applicable to all samples and all methods of analysis. The most generally accepted method of sample preservation is storage under refrigeration in the dark, with a preservative. This is another area that still needs extensive investigation.

Tables 1.6 and 1.7 show a selection of reagents used for preserving or fixing various determinants. These reagents are placed in the sample bottle before it is filled with sample; consequently the sample is 'protected' from the moment it is taken.

The importance that sample container materials can have on seawater sample composition is illustrated below by two examples: one concerning the storage of metal solutions in glass and plastic bottles, the other concerning the storage of solutions of phthalic acid esters and polychlorinated biphenyls in glass and plastic.

Losses of silver, arsenic, cadmium, selenium and zinc from seawater by sorption on various container surfaces [48]

The following container materials were studied: polyethylene, polytetrafluoroethylene and borosilicate glass. The effect was studied of varying the specific surface

Table 1.8 Sorption behaviour (as percentages) of silver, cadmium and zinc in artificial seawater

	Silver												Cadmium			
Material:	Polyethylene				Borosilicate glass				PTFE				Poiyethelene			
pH:	4		8.5		4		8.5		4		8.5		4		8.5	
R (cm^{-1}): Contact time	1.4	3.4	1.4	3.4	1.0	4.2	1.0	4.2	1.0	5.5	1.0	5.5	1.4	3.4	1.4	3.4
1 min	–	–	–	7	–	–	–	–	–	–	–	–	–	–	–	–
30 min	–	–	7	8	–	–	–	–	–	–	–	3	–	–	–	–
1 h	–	–	6	5	–	–	3	3	–	–	–	4	–	–	–	–
2 h	–	–	10	9	–	–	3	5	–	–	–	6	–	–	–	–
4 h	–	–	14	13	–	–	3	4	–	–	–	7	–	–	–	–
8 h	–	–	16	18	3	4	5	10	–	–	–	8	–	–	–	–
24 h	–	–	24	28	4	4	6	9	–	–	6	12	–	–	–	–
2 d	–	–	35	36	6	7	10	31	–	–	13	17	–	–	–	–
3 d	–	–	44	45	6	11	31	80	–	–	13	23	–	–	–	–
7 d	–	–	64	64	74	60	27	73	–	–	14	29	–	–	–	–
14 d	–	–	66	72	81	76	39	84	–	–	20	30	–	–	–	–
21 d	–	–	58	77	80	73	39	64	–	–	26	37	–	–	–	–
28 d	–	–	46	78	82	71	40	67	–	–	27	37	–	–	–	–

Reproduced from Wasse, R. and Maessen, J.J. (1981) *Analytical Chemistry*, 127, p. 181, American Chemical Society, by courtesy of authors and publishers.

R (cm^{-1}) (ratio of inner container surface in contact with the solution to volume of the solution) on adsorption of metals on the container surface. New bottles were used exclusively. The differences in R values were achieved by adding pieces of the material considered. To avoid the possibility of highly active sites for sorption arising from fresh fractures, the edges of the added pieces of borosilicate glass were sealed in a flame. Prior to the use of all materials, the surfaces were cleaned by shaking with 8 M nitric acid for at least 3 days and by washing five times with distilled water.

Working solutions (1 litre) which were 10^{-7} mol l^{-1} in one of the elements to be studied were prepared by appropriate addition of the radioactive stock solutions to pH-adjusted artificial seawater. After the pH had been checked, 100 ml portions were transferred to the bottles to be tested. The filled bottles were shaken continuously and gently in an upright position, at room temperature and in the dark. At certain time intervals, ranging from 1 min to 28 days, 0.1 ml aliquots were taken. These aliquots were counted in a 3 × 3 in NaI(TI) well-type scintillation detector, coupled to a single-channel analyser with a window setting corresponding to the rays to be measured.

The counting times were chosen in such a way that at least 15 000 pulses were counted. The sorption losses were calculated from the activities of the aliquots and the activity of the aliquot taken at time zero. Taking into account the various sources of errors, mainly counting statistics, the maximum imprecision is about 3%. Therefore, calculated sorption losses of 3% and lower are omitted from the listings as being not significant.

Table 1.8 shows the percentage loss as a function of time of, respectively, silver, cadmium, and zinc from artificial seawater stored in polyethylene, borosilicate glass, and PTFE at various pH and R values.

Table 1.8 (contd)

Cadmium								Zinc											
Borosilicate glass				PTFE				Polyethylene				Borosilicate glass				PTFE			
4		8.5		4		8.5		4		8.5		4		8.5		4		8.5	
1.0	4.2	1.0	4.2	1.0	5.5	1.0	5.5	1.4	3.4	1.4	3.4	1.0	4.2	1.0	4.2	1.0	5.5	1.0	5.5
—	—	—	—	—	—	—	—	—	—	—	—	—	—	—	—	—	—	—	—
—	—	—	—	—	—	—	—	—	—	—	—	—	—	12	31	—	—	—	—
—	—	—	—	—	—	—	—	—	—	—	—	—	—	9	31	—	—	—	—
—	—	—	—	—	—	—	—	—	—	—	—	—	—	10	29	5	—	—	—
—	—	—	—	—	—	—	—	—	—	—	—	—	—	9	30	4	—	—	—
—	—	—	—	—	—	—	—	—	—	—	—	—	—	9	28	5	—	—	—
—	—	—	—	—	—	—	—	—	—	—	—	—	—	5	26	4	—	—	—
—	—	—	—	—	—	—	—	—	—	—	—	—	—	4	21	4	—	—	—
5	—	—	—	—	—	—	—	—	—	—	—	—	—	4	18	5	—	—	—
14	40	—	—	—	—	—	—	10	4	—	—	—	—	3	9	4	—	—	—
13	43	—	—	—	—	—	—	27	19	—	—	—	—	—	10	5	—	—	—
15	41	—	—	—	—	—	—	25	17	—	—	—	—	3	9	5	—	—	—
14	36	—	—	—	—	—	—	20	19	—	—	—	—	4	9	5	—	—	—

For arsenic (added as sodium arsenate) and selenium (added as sodium selenite), losses were insignificant in all the container materials considered, irrespective of matrix composition.

The sorption behaviour of trace elements depends on a variety of factors which, taken together, make sorption losses rather difficult to predict. However, the data from this study and from the literature indicate for which elements sorption losses may be expected as a function of a number of factors, such as trace element concentration, container material, pH and salinity.

As is shown above, reduction of contact time and specific surface may be helpful in lowering sorption losses, and acidification with a strong acid will generally prevent the problems of losses by sorption. However, it must be emphasized that the use of acids may drastically change the initial composition of the aqueous sample, making unambiguous interpretation of the analytical results cumbersome or even impossible [49].

For cases of sample storage where losses cannot be excluded a priori, some sort of check is required. This should be done under conditions which are representative of the actual sampling, sample storage and sample analysis. As this study indicates, the use of radiotracers is helpful in making such checks.

The various factors in sorption losses may be classified into four categories. The first category is concerned with the analyte itself, especially chemical form and concentration. The second category includes the characteristics of the solution, such as the presence of acids (pH), dissolved material (e.g. salinity, hardness), complexing agents, dissolved gases (especially oxygen, which may influence the oxidation state), suspended matter (competitor in the sorption process) and micro-organisms (e.g. trace element take-up by algae). The third category comprises the properties of the container, such as its chemical composition, surface roughness, surface cleanliness and as this study demonstrates, the specific surface. Cleaning by prolonged soaking in 8 M nitric acid [50] is to be recommended. The history of the containers (e.g. age, method of cleaning, previous samples, exposure to heat) is important because it may be of direct influence on the type and number of active sites for sorption. Finally, the fourth category consists of external factors, such as temperature contact time, access of light, and occurrence of agitation. All of these factors must be considered in assessing the likelihood of sorption losses during a complete analysis.

Robertson [51] have measured the adsorption of zinc, caesium, strontium, antimony, indium, iron, silver, copper, cobalt, rubidium, scandium and uranium onto glass and polyethylene containers. Radioactive tracks of these elements were added to samples of seawater, the samples were adjusted to the original pH of 8.0, and aliquots were poured into polyethylene bottles, Pyrex-glass bottles and polyethylene bottles containing 1 ml concentrated hydrochloric acid to bring the pH to about 1.5. Adsorption on to the containers was observed for storage periods of up to 75 days with the use of a NaI(Tl) well crystal. Negligible adsorption on all containers was registered for zinc, caesium, strontium and antimony. Losses of indium, iron, silver, copper, rubidium, scandium and uranium occurred from water at pH 8.0 in polyethylene (excepting rubidium) or Pyrex glass (excepting silver). With indium, iron, silver and cobalt, acidification to pH 1.5 eliminated adsorption on polyethylene but this was only partly effective with scandium and uranium.

Pellenberg and Church [52] have discussed the storage and processing of estuarine water samples for analysis by atomic absorption spectrometry.

Losses of phthalic acid esters and polychlorinated biphenyls from seawater samples during storage

During the storage of the sample, loss of the analyte can occur via vaporization, degradation and/or adsorption. Adsorption of trace organic and inorganic species in seawater to container walls can severely affect the accuracy of their determination. The adsorption of dichlorodiphenyltrichloroethane [53] and hexachlorobiphenyl [54] onto glass containers has been observed.

Sullivan *et al.* [55] studied the loss of phthalic acid esters and chlorinated biphenyls from seawater whilst stored in glass containers. Equilibrium was essentially reached in 12 hours at 25°C. Labelled compounds were used in some of the studies. Table 1.9 shows that between 2.2 and 49.9% of the organic solutes were last from the spiked solutions.

Table 1.9 Distribution of solutes among original solutions and subsequent water and solvent rinses of containers

Solute	Initial aqueous concentration ($\mu g\ l^{-1}$)	Percentage of solute recovered		
		Original spiked water	Water rinses of test tubes	Acetonitrile rinses of test tubes
DBP	4420 ($\pm$140)	94.6	ND	5.4
		85.4	ND	14.6
^{14}C DBP	28.9 ($\pm$2.2)	95.8	4.1	<0.1
		97.0	2.9	<0.1
^{14}C DBP	19.2 ($\pm$2.0)	96.2	3.6	<0.1
		97.8	2.1	0.1
BEHP	407 ($\pm$9)	68.0	20.5	11.4
		72.1	19.5	8.4
		74.1	17.5	8.4
BEHP	229 ($\pm$7)	56.2	3.1	40.7
		73.0	2.0	25.0
^{14}C BEHP	15.5 ($\pm$0.3)	56.0	27.8	16.2
		50.1	31.2	18.7
		54.6	30.4	15.0
^{14}C PCB (alone)	6.8 ($\pm$0.5)	71.1	5.5	23.4
		81.1	0.6	18.3
^{14}C PCB (after BEHP)	6.8 ($\pm$0.5)	55.2	11.4	33.4
		55.3	4.4	40.3

DBP = Dibutyl phthalate; BEHP = bis(2-ethylhexyl) phthalate; PCB = polychlorinated biphenyls; ND = not done.
Reproduced from Sullivan, K.F., *et al.* (1981) *Analytical Chemistry*, **53**, p. 1718, American Chemical Society, by courtesy of authors and publishers.

The absorbed compounds were only partially recovered by subsequent water rinses. A solvent rinse of the containers recovered the remainder of the compounds adsorbed to the surface.

The amount of phthalate bound to the glass test tubes appears to be a function of the aqueous solubility of the phthalate. The solubilities of the phthalates have been reported to be 3.2 mg dibutyl phthalate and 1.2 mg bis (2-ethylhexyl) phthalate per litre of artificial seawater [56]. Table 1.9 shows that the more soluble dibutyl phthalate is absorbed far less than bis (2-ethyl-hexyl) phthalate.

The amount of solute also appears to be altered by the presence of other material already on the surface. Between 70% and 80% of the total polychlorinated biphenyls stayed in the aqueous phase when bis (2-ethylhexyl) phthalate was not present. When approximately 400 ng bis (2-ethylhexyl) phthalate was adsorbed to each test tube, only 55% of the total polychlorinated biphenyls stayed in the aqueous phase. The increased lipophilicity due to the presence of bis (2-ethylhexyl) phthalate apparently increased the adsorption of polychlorinated biphenyl by the glass test tubes.

The significant percentages of these organic pollutants in the unspiked water rinses indicates that their adsorption is reversible. Once the compounds are adsorbed to the glass, they can desorb into the next solution that is placed into the container, which is a concern during the repeated use of the same storage or sampling container. The cross-contamination of samples and the loss of the analyte can be reduced by rinsing the container with clean water or solvent which is then processed with the sample or by extracting the sample in the container.

1.4 Sample contamination during analysis

The environment in which samples are collected and processed during an oceanographic cruise would be considered impossible by any non-oceanographic microanalyst. On even the best-planned oceanographic vessels the spaces in which the samples are actually taken, the winch room and the wet lab, are normally awash in seawater, with a thin film of oil over most of the exposed surfaces. The worst case is to be found where the wet lab and winch room are combined, or where the wet lab is the natural passageway between important parts of the ship, such as the engine room and the galley. These circumstances are the rule rather than the exception on oceanographic vessels, even on those planned from scratch for oceanographic research. The reasons for these apparent flaws in planning are historical; chemical oceanographers were interested either in major components of seawater or in trace nutrients, and neither of these kinds of analyses would be seriously damaged by the contamination to be found in such wet labs.

With the recent emphasis upon the analysis of trace metals and or organic materials, particularly possible pollutants, it has become obvious that cleaner working areas are necessary. The winch room, with its assorted greases and oils, must be separated from the sampling room, and ideally the people working on the hydrographic wire, handling the samplers, should not also be drawing the samples. The samplers should come into the sampling room through a hatch which can be closed. The sampling room should also be a dead-end room, not a throughway, to discourage visitors. It would be unrealistic to expect a wet lab to be as free of contamination as a clean room, but it should approach the clean room in general arrangements. Even the air entering the sampling area should be cleaned of hydrocarbons, perhaps by filtration through charcoal; the all-pervasive smell of diesel fuel in most oceanographic vessels does not bode well for the accuracy of any analyses for petroleum hydrocarbons.

If the analyses are to be performed on board, the room in which the samples are prepared and analysed should in fact be built as a clean room. Many modern oceanographic vessels are constructed to accept modular laboratories, which can be removed between voyages. Clean room modules, complete with air conditioning and filtered air supply, have been built for several vessels. It is possible to perform accurate, precise microanalyses on board ship under less favourable conditions, but the analyst is really fighting the odds.

If the samples are to be brought back to a shore laboratory for analysis, contamination during analysis is more easily controlled. The analyst on shore must have confidence in the people taking the samples; with the pressure on berth space and wire time on board ship, too often the samples are taken on a 'while you're out there, take some for me' basis. Again, ideally, the analyst should at least oversee every part of the process, from the cleaning of the sampler to the final calculation of the amounts present. His confidence in the accuracy of the final calculation must decrease as he departs from the ideal arrangement.

In the shore laboratory, the samples must be handled with the care needed for any trace analysis. It must be remembered that the total amount of organic carbon in seawater is around 1 ppm; single compounds are likely to be present at ppb levels. In order to collect enough material even for positive identification of some of the compounds present, the materials must often be concentrated.

Analytical chemists have long been aware of the necessity for purification of any organic solvents used in trace analysis. The advent of the gas and liquid–liquid chromatographs has made plain just how many impurities can hide behind a 'high purity' label. Redistillation of organic solvents just before use is a commonplace in most analytical laboratories. What has not been so evident is the amount of organic material to be found in most inorganic reagents. The actual amount present, let us say, in reagent grade sodium chloride may be low enough so that it is not listed on the label, but still high enough to produce an artificial seawater containing more organic carbon than the real article. The presence of these compounds becomes serious when the analyst wishes to concoct an artificial seawater for standards and blanks. If the chemical in question can withstand oxidation, either at high temperature or in the presence of active oxygen, the organic material may be eliminated. However, many compounds used in routine analysis cannot be treated in this manner. If such chemicals must be used, the calculation of a true methods blank can become a major analytical problem.

Another problem, equally unrecognized, is the organic content of the distilled water. For most analytical procedures, simple distillation is sufficient treatment; perhaps in special cases, such extremes as distillation from permanganate or distillation in quartz is considered necessary. For the analysis of organic materials at the ppm level in aqueous solutions, these methods are far from sufficient. The experience of many workers has been that no form of chemical pretreatment will remove all of the organic material from distilled water, and that some form of high-temperature oxidation of the organic impurities in the water must be used [57–59]. Depending upon the original source of the water, normal distillation will leave between 0.25 and 0.6 mg C/l^{-1} in the distillate. The amounts and the kinds of compounds may vary with the seasons and with the dominant phytoplankton species in the reservoirs. Since ocean water taken from depths greater than 500 m will usually contain only 0.3–0.7 mg C/l^{-1}, it can be seen that the purity of the distilled water used to make up reagents can be quite important. Sub-boiling distillation of deep seawater might be an efficient starting point for the production of carbon-free blanks.

While at least partial solutions have been found to most of the problems of contamination, these solutions have largely been adopted piecemeal by the various laboratories engaged in research on organic materials in seawater. The reasons for the adoption of half-way measures are largely historical. The study of organic materials in seawater is relatively new, and the realization that Draconian measures are needed in the analysis is just becoming accepted.

The problems that can be encountered in sampling, sample preservation, and analysis have been discussed by King [60], Grice *et al.* [61] (sampling); Bridie *et al.* [62] (non-hydrocarbon interference); Farrington [63] (hydrocarbon analysis); Kaplin and Poskrebysheva [64] (organic impurities); Hume [65], Riley [66] (oceanographic analysis); Acheston *et al.* [67] (polynuclear hydrocarbons); Giam and Chan [68] (phthalates); Giam *et al.* [69] (organic pollutants) and Grasshoff [70,71] (phosphates).

Mart [72] has described a typical sample bottle cleaning routine for use when taking samples for very low level metal determinations.

Sampling bottles and plastic bags, both made of high-pressure polyethylene, were rinsed by the following procedure. First clean with detergent in a laboratory washing machine, rinse with deionized water, soak in hot (about 60°C) acid bath, beginning with 20% hydrochloric acid, reagent grade, followed by two further acid baths of lower concentration, the last being of Merck, Suprapur quality or equivalent. The bottles are then filled with dilute hydrochloric acid, Merck, Suprapur, this operation being carried out on a clean bench. They are soaked once more in dilute acid and heated up. Empty bottles under clean bench, rinse and fill them up with very pure water (pH 2). Bottles are wrapped into two polyethylene bags. For transport purposes, lots of ten bottles are enclosed hermetically into a larger bag.

The determination of traces of heavy metals in natural waters can be greatly affected by contamination (positive or negative) during filtration and storage of samples [73–78]. Until the recent work of Scarponi *et al.* [79], this problem, especially as regards filtration, has not been studied adequately and systematically with respect to the determination of cadmium, lead and copper in seawater. Frequently, in order to make the determination easier, synthetic matrix samples [78–80] and/or high metal concentrations [78–82] have been used; otherwise, in order to demonstrate possible filter contamination, washed and unwashed filters have been analysed after washing [83–89].

The same filter can release or adsorb trace metals depending on the metal concentration level and the main constituents of the sample [78,90] therefore, claimed results must be considered with caution in working with natural samples. To avoid contamination, the following procedures have often been used. Filters have been cleaned by soaking them in acids [74–76,86,89–93] or complexing agents [73,74,87] and/or conditioned either by soaking in seawater or a simulated seawater solution [89,90,94] or by passing a 0.2–2.1 sample before aliquots are taken for analysis [74,76,89,95,96]. Sometimes, however, the washing procedure has not been found to be fully satisfactory [73]. For example, it has been reported that strong adsorption of cadmium and lead occurs on purified unconditioned membrane filters when triple-distilled water is passed through the filter, whilst there is no change in the concentration with a river water sample after filtration of 500 ml [90]. Some investigators prefer to avoid filtration when the particulate matter does not interfere with the determination (in which case the analyses must be completed soon after sampling) [89,97] or when open seawater is analysed; in the latter case, filtered and unfiltered samples do not seem to differ significantly in measurable metal content [89,98,99].

The optimal conditions for uncontamination long-term storage of dilute heavy metal solutions, particularly seawater, are now a topic of great interest and contradictory results have been frequently reported. For example, with regard to the type of material to be used for the container, some workers have recommended the use of linear (high density) polyethylene instead of conventional (low density) polyethylene [74,76,100–102] while others have reported that linear polyethylene is totally unsuitable or inferior to low-density polyethylene [73,75,103–105]. Moreover, as a general rule, findings for particular conditions are not necessarily applicable to elements, concentrations, matrices, containers, or experimental conditions different from those tested. As the macro and micro constituents of natural waters can differ widely [106], extreme caution must be used in handling published results. Possibly, as recommended [74,76,107], it is best to ascertain the effectiveness of the storage system adopted in ones own laboratory.

Scarponi *et al.* [79] used anodic stripping voltammetry to investigate the contamination of seawater by cadmium, lead and copper during filtration and storage of samples collected near an industrial area. Filtrations were carried out under clean nitrogen to avoid sample contamination. Seawater leaches metals from uncleaned membrane filters but, after 1 litre of water has passed through, the contamination becomes negligible. Samples stored in conventional polyethylene containers (properly cleaned and conditioned with prefiltered seawater) at 4°C and natural pH remain uncontaminated for 3 months (5 months for cadmium); losses of lead and copper occur after 5 months storage. Reproducibility (95% confidence interval) was 8–10%, 3–8% and 5–6% at concentration levels of about 0.06, 2.5 and 6.0 $\mu g\,l^{-1}$, for cadmium, lead and copper, respectively.

The first aim of this work was to study the influence of an unwashed membrane filter on the cadmium, lead and copper concentrations of filtered seawater samples. It was also desirable to ascertain whether, after passage of a reasonable quantity of water, the filter itself could be assumed to be clean so that subsequent portions of filtrate would be uncontaminated. If this were the case, it should be possible to eliminate the cleaning procedure and its associated contamination risk. The second purpose of the work was to test the possibility of long-term storage of samples at their natural pH (about 8) at 4°C, kept in low-density polyethylene containers which had been cleaned with acid and conditioned with seawater.

Before use, new containers were cleaned by soaking in 2 M hydrochloric acid for 4 days and conditioned with prefiltered seawater for a week, all at room temperature [76,90]. Teflon-covered stirring bars (required for the voltammetric measurements) were introduced into the containers at the beginning of the cleaning procedure. The containers used in one procedure were rinsed and left filled with prefiltered seawater until re-use. Containers used in another procedure and in the study of long-term storage, could be regarded as having been conditioned for about 1 month and more than 2 months, respectively. Other plastic ware used in the sampling and filtration processes, and the components of the voltametric cell that came in contact with the sample solution, underwent the same cleaning procedure as the containers.

Figure 1.4 shows a typical curve demonstrating the dependence of concentrations of copper, lead and cadmium in the filtrate on the volume of seawater sampled. Metal levels become constant after 1–1.5 litres of sample have been filtered, and it can be concluded that at this point contamination of the sample by the filtration equipment is negligible.

Table 1.10 gives the results of analytical measurements on aliquots of a conditioning seawater stock, stored at about 4°C for 3 and 5 months in old low-density polyethylene containers (acid-washed for 4 days and conditioned for more

26

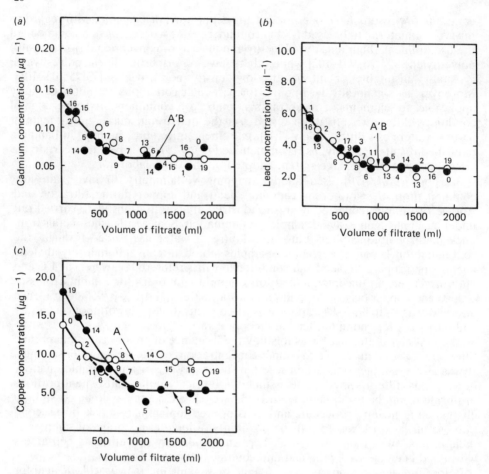

Figure 1.4 Concentration dependences on filtrate volume by procedure 1: (*a*) Cd; (*b*) Pb; (*c*) Cu. Numbers refer to storage time in days. (●) Measured in order of sampling: (○) measured in reverse order of sampling. (Reproduced from Scarponi, G., *et al.* (1982) *Analytica Chimica Acta*, **135**, 263, Elsevier Science Publishers, by courtesy of authors and publishers.)

Table 1.10 Results after long-term storage

Date (1979)	Storage time (months)	Metal concentration (µg l⁻¹)							
		Cd			*Pb*			*Cu*	
		Conc.	*Mean*	*Change (%)*	*Conc.*	*Mean*	*Change (%)*	*Conc.*	*Mean*	*Change (%)*
July 2	0	0.16, 0.19 0.17, 0.16	0.17		4.0, 4.3 4.3, 4.2	4.2		7.7, 8.0 8.4, 8.0	8.0	
Sept 30	3	0.15, 0.16 0.16, 0.18	0.16	−6	5.0, 3.6 4.2, 3.6	4.1	−2	7.8, 7.5 8.1	7.8	−2.5
Nov 28	5	0.18, 0.18 0.16	0.17	0	3.7, 3.1 2.6, 3.8	3.3	−21	5.9, 5.8 6.5, 6.0	6.05	−24

Reproduced from Scarponi, G., *et al.* (1982) *Analytica Chimica Acta*, **135**, p. 263, Elsevier Science Publishers B.V., by courtesy of authors and publishers.

than 2 months). Apart from the observation that the concentrations are generally higher than those measured previously, which indicates contamination during conditioning and manipulation, and the necessity of frequently renewing seawater for equilibration purposes, it can be seen that there are no changes in the metal concentrations for 3 months for lead and copper, or for 5 months for cadmium. Also, after 5 months storage, some loss of lead and copper (21% and 24%, respectively) can be observed, possibly because of the formation and slow adsorption on container surfaces of hydroxo- and carbonato-complexes [76,108]. Hence at 4°C in polyethylene containers no significant changes of heavy metal concentrations occur over a 3-month period [91,96,109].

Scarponi *et al.* [79] concluded that filtration of seawater through uncleaned membrane filters shows positive contamination by cadmium, lead and copper. In the first filtrate fractions, the trace metal concentration may be increased by a factor of two or three. During filtration, the soluble impurities are leached from the filter, which is progressively cleaned, and the metal concentration in the filtrate, after passage of 0.8–1 litres seawater, reaches a stable minimum value. Thus it is recommended that at least 1 litre seawater at natural pH be passed through uncleaned filters before aliquots for analysis are taken from subsequent filtrate. The same filter can be re-used several times, and then only the first 50–100 ml filtrate need be discarded. This system seems simpler and more reliable for avoiding contamination than that of washing and conditioning filters before use, especially since in the latter case it has been suggested that the first 0.2–2.0 litres of filtrate should also be discarded.

Low-density polyethylene containers are suitable for storing seawater samples at 4°C and natural pH, provided that they are thoroughly cleaned (in 2 M hydrochloric acid for at least a week) and adequately conditioned (with prefiltered seawater for at least 1–2 weeks). Storage can be prolonged for at least 3 months (or 5 months for cadmium) without significant concentration changes; for lead and copper, adsorption losses are observed after 5 months.

The use of a special device that allows filtration under nitrogen, the direct introduction of sample into containers for storage during filtration and the use of these containers as analysis cells are all improvements that minimize external sample contamination and improve between-sample reproducibility.

References

1. Green, D.R. *Committee of Marine Analytical Chemistry*, Canada Centre for Inland Waters Fisheries and Environment, Ottawa, Canada (1977)
2. Wangersky, P.J. *Annual Review of Ecological Systems*, **7**, 161 (1976)
3. Wangersky, P.J. *Deep Sea Research*, **23**, 457 (1976)
4. Gordon, D.C., Jr. and Keizer, P.D. Fisheries Research Board, Canada. *Technical Report No. 481* (1974)
5. Harvey, G.W. *Limnology and Oceanography*, **11**, 608 (1966)
6. Garrett, W.D. *Limnology and Oceanography*, **10**, 602 (1965)
7. Daumas, R.A., Laborde, P.L., Marty, J.C. and Saliot, A. *Limnology and Oceanography*, **319**, 319 (1976)
8. Harvey, G.W. and Burzell, L.A. *Limnology and Oceanography*, **19**, 156 (1972)
9. Garrett, W.D. and Burger, W.R. *Sampling and Determining the Concentration of Film-forming Organic Constituents of the Air–Water Interface.* US NRL Memo Report 2852 (1974)
10. Larsson, K., Odham, G. and Sodergren A. *Marine Chemistry*, **2**, 49 (1974)
11. Miget, R., Kator, H. and Oppenheimer, C. *Analytical Chemistry*, **45**, 1154 (1974)

12. Anufrieva, N.M., Gornitsky, A.B., Nesterova, M.P. and Nemirovskaya, I.A. *Okeanologiya*, **16**, 255 (1976)
13. Baier, R.E. *Journal of Geophysical Research*, **77**, 5062 (1972)
14. Gump, B.H., Hertz, H.A., May, W.E., Chesler, S.N., Dyszel, S.M. and Enagonio, D.P. *Analytical Chemistry*, **47**, 1223 (1975)
15. Keizer, P.D., Gordon, D.C., Jr. and Dale, J. *Journal of the Fisheries Research Board*, **34**, 347 (1977)
16. Clark, R.C. Jr., Blumer, M. and Raymond, S.O. *Deep Sea Research*, **14**, 125 (1967)
17. Bertoni, R. and Melchiorri-Santolini, U. *Mem. Inst. Ital. Hydrobiol.*, **29**, 97 (1972)
18. Gordon, D.C., Jr., Keizer, P.D. and Dale, J. *Marine Chemistry*, **251**, 251 (1974)
19. Smith, K.L. *Limnology and Oceanography*, **16**, 675 (1971)
20. Bewers, J.M., Dalziel, J., Yeats, P.A. and Barron, J.L. *Marine Chemistry*, **10**, 173 (1981)
21. Olafsson, J. *Marine Chemistry*, **87**, 87 (1978)
22. Olafsson, J. *A Preliminary Report on ICES Intercalibration of Mercury in Seawater for the Joint Monitoring Group of the Oslo and Paris Commissions*, submitted to the Marine Chemistry Working Group of OCES, Feb. 1980 (1980)
23. Thibaud, Y. *Exercise d'intercalibration CIEM, 1979, cadmium en eau de mer*. Report submitted to the Marine Chemistry Working Group of ICES, Feb. 1980
24. Jones, P.G.W. *A Preliminary Report on the ICES Intercalibration of Sea water Samples for the Analyses of Trace Metals*. ICES CM1977/E:16 (1977)
25. Jones, P.G.W. *An ICES Intercalibration Exercise for Trace Metal Standard Solutions*. ICES CM1979/E:15 (1976)
26. Windom, H.L. *Report on the Results of the ICES Questionnaire on Sampling and Analysis of Seawater for Trace Elements*. Submitted to the first meeting of the Marine Chemistry Working Group, Lisbon, May 1979 (1979)
27. Pocklington, R. *Variability of the Ocean off Bermuda*. Bedford Institute of Oceanography Report-Series BI-R-72-3 (1972)
28. Boyle, E.A., Sclater, F.R. and Edmond, J.M. *Science Letters*, **37**, 38 (1977)
29. Bruland, K.W., Knauer, G.A. and Martin, J.H. *Limnology and Oceanography*, **23**, 618 (1978)
30. Bruland, K.W., Knauer, G.A. and Martin, J.H. *Nature (London)*, **271**, 741 (1978)
31. Sclater, F.R., Boyle, E. and Edmond, J.M. Earch Planet. *Science Letters*, **31**, 119 (1976)
32. Wong, C.S., Kremling, K., Riley, J.P., *et al*. Ocean Chemistry Division, Institute of Ocean Sciences, PO Box 6000 Sidney, BC, V8L 4B2, Canada. Ocean Chemistry Division contract to SEAKEM Oceanography Ltd., Sidney, BC, Canada. Marine Chemistry Department, Institut fur Meereskunde und der Universitat, Kiel, Dunsternbrooker Weg 20, 2300 Kiel, FR Germany. Department of Oceanography, University of Liverpool, PO Box 147, Liverpool L69 3BX, UK (1985)
33. Patterson, C.C. and Settle, D.M. The reduction of orders of magnitude errors in lead analysis, In *Accuracy in Trace Analysis: Sampling, Sample Handling, Analysis*. (ed. P.D. La Fleur). NBS Special Publication, **422**, p.321 (1976)
34. Bothner, M.H. and Robertson, D.E. *Analytical Chemistry*, **47**, 592 (1975)
35. Participants in the IDOE interlaboratory analyses workshop, 1975. Comparison determinations of lead by investigators analyzing individual samples of sea water in both their home laboratory and in an isotope dilution standardization laboratory. *Marine Chemistry*, **4**, 389 (1975)
36. Stukas, V.J. and Wong, C.D. *Science*, **211**, 1424 (1976)
37. Wong, C.S., Kremling, K., Riley, J.P., *et al*. *Accurate Measurement of Trace Metals in Sea water: an Intercomparison of Sampling Devices and Analytical Techniques using CEPEX Enclosure of Sea water*. Unpublished manuscript report, NATO study funded by NATO Scientific Affairs Division (1979)
38. Abdullah, M.I., El-Rayis, O.A. and Riley, J.P. *Analytica Chimica Acta*, **84**, 363 (1976)
39. Abdullah, M.I. and Royale, L.G. *Analytica Chimica Acta*, **80**, 58 (1972)
40. Danielsson, L.G., Magnusson, B. and Westerlund, S. *Analytica Chimica Acta*, **98**, 47 (1978)
41. Otsuki, A. and Hanya, T. *Limnology and Oceanography*, **87**, 248 (1972)
42. Jannasch, N.W. and Pritchard, P.H. The role of inert particulate matter in the activity of aquatic microorganisms. *Mem. Inst. Ital. Hydrobiol.*, **24 Suppl.**, 289–306 (1972)
43. Wiebe, W.S. and Pomeroy, L.R. Microorganisms and their association with aggregates and detritus in the sea: A microscopic study. *Mem. Inst. Ital. Hydrobiol.*, **29 Suppl.**, 325–352 (1972)
44. Alburn, H.E. and Grant, N.H. *Journal of the American Chemical Society*, **87**, 4174 (1965)
45. Grant, N.H. and Alburn, H.E. *Biochemistry*, **4**, 1913 (1965)
46. Grant, N.H. and Alburn, H.E. *Archives of Biochemistry and Biophysics*, **118**, 292 (1967)
47. Yoshinari, I. *Marine Chemistry*, **4**, 189 (1976)
48. Massee, R. and Maessen, F.J.M.J. *Analytica Chimica Acta*, **127**, 181 (1981)

49. Florence, T.M. and Bailey, G.E. *C.R.C. Critical Reviews on Analytical Chemistry*, **August**, 219 (1980)
50. Karin, R.W., Buone, J.A. and Fashing, J.L. *Analytical Chemistry*, **47**, 2296 (1975)
51. Robertson, D.E. *Analytica Chimica Acta*, **42**, 533 (1968)
52. Pellenburg, R.E. and Church, T.M. *Analytica Chimica Acta*, **97**, 81 (1978)
53. Picer, M., Picer, N. and Strohal, P. *Science of the Total Environment*, **8**, 159 (1977)
54. Pepe, H.G. and Byrne, J.J. *Bulletin of Environmental Contamination and Toxicology*, **25**, 936 (1980)
55. Sullivan, K.R., Altas, E.L. and Giam, C.S. *Analytical Chemistry*, **53**, 718 (1981)
56. Kakreka, J.P.M.S. *Thesis*, Texas A M University, December (1974)
57. Wangersky, P.J. *American Science*, **53**, 358 (1965)
58. Hickman, K., White, I. and Stark, E. *Science*, **180**, 15 (1973)
59. Conway, B.E., Angerstein-Kozlowska, H. and Sharp, W.B.A. *Analyical Chemistry*, **45**, 1331 (1973)
60. King, D.L. and Ciaccio, L.L. (eds) *Sampling of Natural Waters and Waste Effluents*, Marcel Dekker, New York, pp. 451–481 (1971)
61. Grice, G.D., Harvey, G.R., Bown, V.T. and Backus, R.H. *Bulletin of Environmental Contamination and Toxicology*, **7**, 125 (1972)
62. Bridie, A.L., Box, J. and Herzberg, S. *Journal of the Institute of Petroleum, London*, **59**, 263 (1973)
63. Farrington, J.W. *A.D. Report No. 777695/GA*, U.S. National Technical Information Service (1974)
64. Kaplin, A.A. and Poskrebysheva, L.M. *Izuestia Tomak Politckh Inst.*, **233**, 91 (1974)
65. Hume, D.N. Fundamental problems in oceanographic analysis. In *Analytical Methods in Oceanography* (ed. R.P. Gibb, Jr.), American Chemical Society, Washington, pp. 1–8 (1975)
66. Riley, J.P. In *Chemical Oceanography*, Vol. 3, (eds J.P. Riley and G. Skirrow), Academic Press, London, pp. 193–514 (1975)
67. Acheson, M.A., Harrison, R.M., Perry, R. and Wellings, R.A. *Water Research*, **10**, 207 (1976)
68. Giam, C.S. and Chan, H.S. *Special Publication*, National Bureau Standards (US), No. 422, p. 761 (1976)
69. Giam, C.S., Chan, H.S. and Neff, G.S. In *Proceedings of International Conference Environmental Sensing Assessment*, American Chemical Society, Washington (1976)
70. Grasshoff, K. *Analytical Chemistry*, **220**, 89 (1966)
71. Grasshoff, K. *Verlag Chemie*, **1**, 50–70 (1976)
72. Mart, L. *Fresenius Zeitschrift fur Analytische Chemie*, **296**, 350 (1979)
73. Roberton, D.E. In *Ultrapurity, Methods and Techniques* (eds M. Zief and R. Speights), Dekker, New York, p. 207 (1972)
74. Riley, J.P., Robertson, D.E., Dutton, J.W.R., *et al.* In *Chemical Oceanography* (eds J.R. Riley and G. Skirrow), 2nd edn, Vol. 3, Academic Press, London, p. 193 (1975)
75. Zief, M. and Mitchell, J.W. *Contamination Control in Trace Element Analysis*, Wiley, New York (1976)
76. Batley, G.E. and Gardner, D. *Water Research*, **11**, 745 (1977)
77. Salim, R. and Cooksey, B.G. *Journal of Electroanalytical Chemistry*, **106**, 251 (1980)
78. Truitt, R.E. and Weber, J.H. *Analytical Chemistry*, **51**, 2057 (1979)
79. Scarponi, G., Capodaglio, G., Oescon, P., *et al. Analytica Chimica Acta*, **135**, 268 (1982)
80. Weber, J.H. and Truitt, R.E. *Research Report 21*, Water Resource Research Center, University of New Hampshire, Durham, N.H. (1979)
81. Marvin, K.T., Proctor, R.R. Jr. and Neal, R.A. *Limnology and Oceanography*, **15**, 320 (1970)
82. Gardiner, J. *Water Research*, **8**, 157 (1974)
83. Spencer, D.W. and Manheim, F.T. *U.S. Geological Survey Professional Paper*, 650-D, p. 288 (1969)
84. Spencer, D.W., Brewer, P.G. and Sachs, P.L. *Geochimica Cosmochimica Acta*, **36**, 71 (1972)
85. Dams, R., Rahn, K.A. and Winchester, J.W. *Environmental Science and Technology*, **6**, 441 (1972)
86. Wallace, G.T., Jr., Fletcher, I.S. and Duce, R.A. *Journal of Environmental Science and Health*, **A12**, 493 (1977)
87. Duychserts, G. and Gillain, G. In *Essays on Analytical Chemistry* (ed. E. Wanninen), Pergamon, Oxford, p. 417 (1977)
88. Smith, R.G. *Talanta*, **25**, 173 (1978)
89. Mart, L. and Fresenius, Z. *Analytical Chemistry*, **296**, 350 (1979)
90. Nurnberg, H.W., Valenta, P., Mart, L., Raspor, B. and Sipos, L. *Fresenius Zeitschrift fur Analytisch Chemie*, **282**, 357 (1976)

91. Batley, G.E. and Gardner, D. *Estuarine Coastal Marine Science*, **7**, 59 (1978)
92. Bruland, K.W., Franks, R.P. and Knauer, G.A. *Analytica Chimica Acta*, **105**, 233 (1979)
93. Burrell, D.C. *Marine Science Communication*, **5**, 283 (1979)
94. Figura, P. and McDuffie, B. *Analytical Chemistry*, **52**, 1433 (1980)
95. Burrell, D.C. and Lee, M.L. *Water Quality Parameters*, ASTM STP-573, American Society for Testing and Materials, p. 58 (1975)
96. Fukai, R. and Huynh-Ngoc, L. *Marine Pollution Bulletin*, **7**, 9 (1976)
97. DeForest, A., Pettis, R.W. and Fabris, G. *Australian Journal of Marine and Freshwater Research*, **29**, 193 (1978)
98. Fukai, R. and Huynh-Ngoc. *Analytica Chimica Acta*, **83**, 375 (1976)
99. Zirino, A., Lieberman, S.H. and Clavell, C. *Environmental Science and Technology*, **12**, 73 (1978)
100. Gardiner, J. and Stiff, M.J. *Water Research*, **9**, 517 (1975)
101. Bowen, V.T., Strohal, P., Saiki, M., *et al. Reference Methods for Marine Radioactivity Studies* (eds Y. Nishiwaki and R. Fukai), International Atomic Energy Agency, Vienna, pp. 12–14 (1970)
102. Bowditch, D.C., Edmond, C.R., Dunstan, P.J. and McGlynn, J. *Technical Paper No. 16*, Australian Water Resources Council, p. 22 (1976)
103. Tolg, G. *Talanta*, **19**, 1489 (1972)
104. Tolg, G. In *Comprehensive Analytical Chemistry* (ed. G. Svehia) Vol. 3, Elsevier, Amsterdam, p. 1 (1975)
105. Moody, J.R. and Lindstrom, R.M. *Analytical Chemistry*, **49**, 2264 (1977)
106. Davison, W. and Whitfield, M. *Journal of Electroanalytical Chemistry*, **75**, 763 (1977)
107. Scarponi, G., Miccoli, E. and Frache, R. In *Proceedings of the 3rd Congress of the Association of Italian Oceanology and Limnology*, Sorrento, Italy. Pergamon Press, Oxford, p. 433 (1978)
108. Subramanian, K.S., Chakrabati, C.L., Sheiras, J.E. and Maines, I.S. *Analytical Chemistry*, **50**, 444 (1978)
109. Carpenter, J.H., Bradford, W.L. and Grant, V. In *Estuarine Research* (ed. L.E. Cronin) Academic Press, New York, p. 188 (1975)

Chapter 2

Analysis for anions and cations

2.1 Chloride

Chloride, at the levels that occur in seawater, is usually determined by classical titration procedures using standard silver nitrate as the titrant and potassium chromate indicator, or alternatively by the mercuric thiocyanate procedure using dithizone as indicator. As large dilutions of the original sample are used in these analyses it is essential to use grade A glassware and take all other suitable precautions, such as temperature control.

Chloride can also be estimated by potentiometric titration using standard silver nitrate [1]. The results are recorded directly on punched tape and by teletyper and evaluated by means of a computer program based on the Gran extrapolation method. The determinations have a precision of $\pm$ 0.02% and since many samples can be titrated simultaneously, the time for a single determination including evaluation and editing of titration data can be reduced to less than 5 min.

Chronopotentiometry has also been used to determine chloride ion in seawater [2]. The chloride in the solution containing an inert electrolyte were deposited on a silver electrode (1.1 cm^2) by the passage of an anodic current. The cell comprised a silver disc as working electrode, a symmetrical platinum-disc counter-electrode, and a silver-AgCl reference-electrode to monitor the potential of the working electrode. This potential was displayed on one channel of a two-channel recorder, and its derivative was displayed on the other channel. The chronopotentiometric constant was determined over the chloride concentration range 0.5–10 mmol l^{-1} and the concentration of the unknown solution was determined by altering the value of the impressed current until the observed transition time was about equal to that used for the standard solution.

Noborn [3] has studied the dynamic properties of chloride selective electrodes and their application to seawater concentrates. An Orion 94-17 electrode was used in these studies.

In the presence of bromide ions the electrode was subject to a drop in potential (e.g. 1.5–5.7 mV at a molar ratio of Br$^-$ to Cl$^-$ of 2000 : 3) and to delayed response. A considerable hysteresis effect is also observed in concentrated solutions of chloride when the electrode is used in a 1 M chloride solution and then dipped in one that is 0.02 M in chloride. Equilibrium is reached only after 10 min. The junction potential is minimized by diluting the test solution with the salt-bridge solution (10% aq. KNO$_3$).

Grasshoff and Wenck [4] have described a modern version of the Mohr–Knudsen silver nitrate titration procedure for the determination of the chlorinity of seawater. In this method, which overcomes the disadvantages of conventional burettes, use is made of a motor-driven piston burette of 20 ml capacity, which is sufficient for the range of chlorinities 0–45 per thousand. The accuracy is the same as for conventional titration. The apparatus is compact and portable.

Several autoanalyser procedures are available for the determination of chloride.

2.2 Bromide

Bromide in seawater can be determined by the procedure described below which is capable of determining down to $0.1\,mg\,l^{-1}$ bromide.

The sample is acidified with sulphuric acid. The bromide content is then determined by the volumetric procedure described by Kolthoff and Yutzy [5]. In this procedure the buffered sample is treated with excess sodium hypochlorite to oxidize bromide to bromate. Excess hypochlorite is then destroyed by addition of sodium formate. Acidification of the test solution with sulphuric acid followed by addition of excess potassium iodide liberates an amount of iodine equivalent to the bromate (i.e. the original bromide) content of the sample. The liberated iodine is titrated with standard sodium thiosulphate.

Foti [6] has studied the feasibility of concentrating traces of radioactive bromide ions by passing the seawater sample through a column of inactive AgBr (to effect isotopic exchanges). The effects of column height and of flow rate, volume and/or residence time of the seawater on the extent of exchange were examined; each of these variables had a significant effect.

Rose and Cuttita [7] proposed a graphical method for evaluating the background when determining traces of bromide by X-ray emission spectrography (carried out on cellulose pellets containing the residue from evaporation of the sample in the presence of standard bromide solution). Based on this work Liebhafsky et al. [8] proposed a graphical method which provides a value (m_B) of the amount of bromide equivalent to the background, which is substituted into the equation: counts per sec = $k(m + m_s + m_B)$, where k is a factor depending on the sample, m is the Br^- content of the original solution and m_s is the amount of Br^- added as 'spike'. Results obtained by both methods agree satisfactorily but it is stressed that both methods are approximations.

Walters [9] has examined the effect of chloride on the use of bromide and iodide solid state membrane electrodes and he calculated selectivity constants. Multiple linear regression analysis was used to determine the concentrations of bromide, fluoride and iodide in geothermal brines and indicated high interferences at high salt concentrations. The standard curve method was preferred to the multiple standard addition method because of:

1. The deviation from linearity at high salt concentrations of the bromide electrode.
2. The loss of accuracy due to increase in sample volume by using volume increments.
3. The limitations in reading from the Orion meter (0.1 mV).
4. The fact that bromide, fluoride and iodide were present at relatively low concentrations (where the electrode exhibited non-linear behaviour).

2.3 Iodide

Iodine in seawater can be determined by the procedure described below which is capable of determining down to $0.1\,mg\,l^{-1}$ iodide [10]. In this method the sample is strongly acidified with hydrochloric acid titrated with $0.00125\,M$ potassium iodate solution which converts iodide via iodine into iodine monochloride:

$$KIO_3 + 2KI + 6HCl = 3KCl + 3ICl + 3H_2O$$

The end-point, which occurs with the complete conversion of iodide to iodine monochloride, is indicated by the disappearance of the violet iodine colour from a chloroform layer present in the titration flask.

Sugawara [11] has described a method for the determination of iodide in seawater. Various workers [12,13] have modified this procedure.

Matthews and Riley [12] preconcentrated iodide by coprecipitation with chloride ions. This is achieved by adding $0.23\,g$ silver nitrate per 500 ml seawater sample. Treatment of the precipitate with aqueous bromine and ultrasonic agitation promote recovery of iodide as iodate which is caused to react with excess of iodide under acid conditions, yielding I_3^-. This is determined either spectrophotometrically or by photometric titration with sodium thiosulphate. Photometric titration gave a recovery of $99.0 \pm 0.4\%$ and a coefficient of variation of $\pm 0.4\%$ compared with $98.5 \pm 0.6\%$ and $\pm 0.8\%$ respectively for the spectrophotometric procedure.

Shizuo [13] allowed the silver halide precipitate obtained in the coprecipitation process to stand in contact with the solution for more than 20 hours to ensure quantitative collection of iodide on the precipitate. They then evaporated the oxidized iodate solution to 5–10 ml and again allowed the solution to stand for more than 12 hours before the colorimetric determination. No interference occurred by bromine compounds. The errors were then within $\pm 3\%$.

2.4 Iodate

Truesdale [14] has described autoanalyser procedures for the determination of iodate and total-iodine in seawater. The total-iodine content of seawater (approximately $50–60\,\mu g\,l^{-1}$) is believed to be composed of iodate- $(30–60\,\mu g\,l^{-1}.I)$ and iodide-iodine $(0–20\,\mu g\,l^{-1})$ with perhaps a few $\mu g\,l^{-1}$ of organically bound iodine. In both of the methods, the iodine species of interest are first converted to iodate. Then the iodate is reacted with acid and excess iodide to give iodonium ions, I_3^-, which are detected spectrophotometrically at 350 nm. In the total-iodine procedure a preoxidation step is therefore required; here bromine-water was chosen. Interference from nitrite ions, which in acid also oxidize iodide to iodonium ions, is suppressed by sulphamic acid which destroys the nitrite ions.

Originally the objective of this work was only the automatic iodate-iodine procedure. This was needed to give the extra precision called for in an earlier study (Truesdale [15]) which suggested that most of the observed variation in iodate-iodine results for the deeper waters (>200 m) of the oceans is due to analytical imprecision. However, in addition to the original objective the procedure for total-iodine has developed. This development stemmed from the need to test the likelihood of iodonium ions, produced in the iodate procedure, being reduced by substances occurring naturally in seawater. This problem was defined when Truesdale [16] showed that molecular iodine is reduced rapidly in some seawaters;

he found that the oxidizing capacity of 260 µg l^{-1} of iodine-iodine added to a filtered coastal seawater disappeared within 30 min. The effect of such a process on the iodate method would be to lower the observed iodate concentrations. Further, the effect, if it occurred, would be expected to have its maximum effect in coastal and oceanic surface waters where primary productivity is highest; these waters also appear to contain the lowest recorded iodate concentrations. To test for the iodonium-ion reduction, a pre-oxidation step including iodine-iodine was incorporated in the analytical method for iodate-iodine. Having accomplished the iodine-water pre-oxidation step successfully, it was logical to attempt the bromine-water pre-oxidation and thereby produce the total-iodine method.

Methods for the following three determinations were described by Truesdale [14]:

1. *Determination of iodate without pre-oxidation*–Iodate in the buffered sample is reacted with sulphamic acid (to destroy nitrite) and potassium iodide to produce the iodonium ion I$_3^-$ which is determined spectrophotometrically at 350 nm.
2. *Determination of iodate with pre-oxidation*–Iodine water is added to an acetic acid sodium acetate buffered sample to reoxidize to iodate any iodine-containing substances produced by reduction of iodate by naturally occurring reducing substances present in the sample. Total iodate (i.e. iodate present in the original sample as iodate plus additional iodate produced by iodine water treatment) is then reacted with phenol solution at pH 5.4 to destroy excess free iodine and then with sulphamic acid and potassium iodide to produce the iodonium ion which is estimated spectrophotometrically at 350 nm.

   This determination will test for the presence of naturally occurring reducing agents in seawater which by their action on iodonium ions could lead to an underestimate in iodate concentration. (The use of the method on anoxic waters containing sulphide is a prime example of when this precaution should be taken.)
3. *Determination of total iodine with preoxidation*–This procedure is the same as that described under (2) except that the iodine water is replaced by bromine water, and the buffer contains added sodium bromide (i.e. sodium bromide–acetic acid–sodium acetate instead of the acetic acid–sodium acetate buffer used in method (1)).

In all three methods a blank is obtained. To ascertain the blank excess sodium thiosulphate is added to the potassium iodide reagent at a concentration of 4.0 × 10^{-14} mol l^{-1}. Samples were reanalysed and the appropriate blank subtracted from the sample signal.

Variations in the salinity of the sample have very little effect on the accuracy of the results obtained in all three methods provided that the difference in salinity between the samples and the standards does not exceed 2.5‰.

Determinations of iodate without preoxidation in Pacific seawater by the above method gave a mean result of 583 µg l^{-1} with a standard deviation of 0.23 µg l^{-1}. For samples containing between 40 and 60 µg l^{-1} standard deviations of 0.19 µg l^{-1} (iodate method with preoxidation), 0.12 µg l^{-1} (iodate method without pre-oxidation) and 0.43 µg l^{-1} (total iodine method) were obtained.

A set of Pacific open-ocean samples were analysed for iodate-iodine using both the procedure which incorporates pre-oxidation with iodine-water and that which does not. Also, in a similar exercise total-iodine was determined using both the

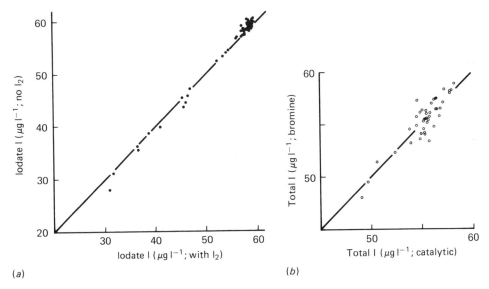

Figure 2.1 The distribution of sample concentrations (means) encountered in the comparison of methods exercise (*a*) for iodate–iodine with and without iodine–water pre-oxidation (*b*) for total-iodine using the bromination procedure and older catalytic procedure. The analysis of variance has shown that in both exercises there was no significant difference between methods; any apparent bias is not significant when the variance due to sample replication is included. For convenience the line of unit gradient is shown. Note that an expanded scale has been used in the plot for total-iodine results (Reproduced from Truesdale, V.W. (1978) *Marine Chemistry*, **6**, 253, Elsevier Science Publishers by courtesy of author and publishers.)

method that incorporates pre-oxidation with bromine water and the catalytic method using the reaction between Ce^{IV} and As^{III} (Truesdale and Smith [17]). Variance tests showed that differences between either replicates or methods was not significant (Figure 2.1).

Truesdale and Smith [18] have also carried out a comparative study of the determination of iodate in open ocean, inshore Irish seawaters and waters from the Menai Straits, using the spectrophotometric method (with and without preoxidation using iodine water) and also by a polarographic method [19].

Figure 2.2 shows the results obtained in a comparison of the four methods on a range of deep-sea and offshore samples. The line of gradient 1.0 on each diagram shows the result which would have occurred had agreement been obtained. The Students t-test (Table 2.1) showed that in both exercises the colorimetric method with iodine–water treatment yielded higher values than that without iodine–water treatment, and the polarographic method yielded, on average, a concentration lower than that obtained by the colorimetric procedure without iodine–water.

Schnepfe [20] has described yet another procedure for the determination of iodate and total iodine in seawater. To determine total iodine, 1 ml 1% aqueous sulphamic acid is added to 10 ml seawater which, if necessary, is filtered and then adjusted to a pH of less than 2.0. After 15 minutes, 1 ml 0.1 M sodium hydroxide and 0.5 ml 0.1 M potassium permanganate are added and the mixture heated on a steam bath for 1 hour. The cooled solution is filtered and the residue washed. The

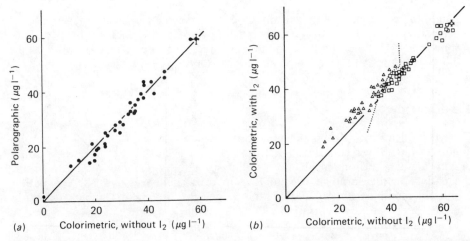

Figure 2.2 The distribution of sample concentrations encountered in the comparisons of the colorimetric method without iodine with (*a*) the polarographic method and (*b*) the colorimetric method with iodine water. For convenience the line of unit gradient is shown in both cases. In each case the precision (95% confidence level) for each measurement is as shown for the most concentrated sample. In (*b*) the broken line divides onshore ($\triangle$) and offshore ($\square$) samples (Reproduced from Truesdale, V.W. and Smith, J.S. (1979) *Marine Chemistry*, 7, 133, Elsevier Science Publishers, by courtesy of authors and publishers.)

Table 2.1 Results of testing the difference in the means obtained by the different iodate methods

Method	No. of samples	Mean concentration ($\mu g\ l^{-1}$)	Variance ($\mu g\ l^{-1})^2$	t
(a) Two colorimetric procedures:				
Without iodine	85	40.5	100.7	94.6
With iodine	85	42.4	131.9	
(b) Colorimetric/polarographic procedures:				
Colorimetric	36	31.1	150.5	10.7
Polarographic	36	30.3	161.5	

Reproduced from Truesdale, U.W. and Smith, J.S. (1979) *Marine Chemistry*, 7, 133, Elsevier Science Publishers, B.V. by courtesy of authors and publishers.

filtrate and washings are diluted to 16 ml and 1 ml of a 0.25 M phosphate solution added (containing 0.3 μg iodine as iodate per ml) at 0°C. Then 0.7 ml 0.1 M ferrous chloride in 0.2% v/v sulphuric acid, 5 ml 10% aqueous sulphuric acid phosphoric acid (1 : 1) are added at 0°C followed by 2 ml starch–cadmium iodide reagent. The solution is diluted to 25 ml and after 10–15 minutes the extinction of the starch–iodine complex is measured in a 5 cm cell. To determine iodate the same procedure is followed as is described above except that the oxidation stage with sodium hydroxide-potassium permanganate is omitted and only 0.2 ml ferrous chloride solution is added. A potassium iodate standard was used in both methods.

The total iodine procedure is claimed to be relatively free from interference by foreign ions. The iodate procedure is subject to interference by bromate and sulphite ions. This method is claimed to be capable of determining down to 0.1 μg iodine in the presence of 500 mg chloride ion and 5 mg bromide ion.

2.5 Fluoride

Fletsch and Richards [212] determined fluoride in seawater spectrophotometrically as the cerium alizarin complex. The cerium alizarin complexan chelate was formed in 20% aqueous acetone at pH 4.35 (sodium acetate buffer) and, after 20–60 min, the extinction measured at 625 nm (2.5 cm cell) against water. The calibration graph was rectilinear for 80–200 μg fluoride per litre; the mean standard deviation was ± 10 μg l^{-1} at a concentration of 1100 μg fluoride per litre.

Spectrophotometric procedures based on the lanthanum–alizarin complex are also described [22].

Photoactivation analysis has also been used to determine fluoride in seawater [23]. In this method a sample and simulated seawater standards containing known amounts of fluoride are freeze-dried, and then irradiated simultaneously and identically, for 20 min, with high-energy photons. The half-life of ^{18}F (110 min) allows sufficient time for radiochemical separation from the seawater matrix before counting. The specific activities of sample and standards being the same, the amount of fluoride in the unknown may be calculated. The limit of detection is 7 ng fluoride, and the precision is sufficient to permit detection of variations in the fluoride content of oceans. The method can be adapted for the simultaneous determination of fluorine, bromine and iodine.

Ion-selective electrodes are emerging as the method of preference for the determination of fluoride in seawater [24–29].

Anfalt and Jaguer [28] measured total fluoride ion concentration by means of a single-crystal fluoride selective electrode (Orion, model 94-09). Samples of seawater were adjusted to pH 6.6 with hydrochloric acid and were titrated with 0.01 M sodium fluoride with use of the semi-automatic titrator described by Jagner [30]. Equations for the graphical or computer treatment of the results are given. Calibration of the electrode for single-point potentiometric measurements at different seawater salinities is discussed.

Rix *et al.* [29] have commented that earlier published potentiometric methods for determining fluoride in seawater appear to be unnecessarily complicated in several respects. Reagents, such as a total ionic strength adjustment buffer, TISAB, are added to the sample at high concentrations to set the pH and ionic strength and to liberate the majority of fluoride bound in metal complexes. The possibility of introducing solution contamination when using high concentrations of TISAB is always a potential hazard and, if this step can be avoided, this will obviously be advantageous. Furthermore, a significant drawback with most earlier methods was the difficulty of establishing an acceptable or representative matrix in which to prepare standards, as the actual matrix is often unknown in natural waters. This aspect of fluoride determination in seawater has received considerable attention, and extensive and elaborate procedures have been used to match the sample and standard matrices [24].

The method of standard additions is frequently employed in environmental analysis for samples of variable matrix [31]. The method of data treatment used by

Table 2.2 Determination of fluoride in seawater under a variety of conditions to demonstrate the reliability of the method

Sample		$|F|_T$ found (mg l^{-1})	Conditions
Point Lonsdale, Victoria, Australia	1	$\begin{cases}1.40 \pm 0.05 \\ 1.35 \pm 0.05^*\end{cases}$	pH $\simeq$ 8
	2	1.30 ± 0.05	1.5 ppm spike added. $[F]_T$ determined = 2.80 ± 0.05 ppm
	3	1.40 ± 0.05	4.0 ppm spike added. $[F]_T$ determined = 5.40 ± 0.05 ppm
	4	1.30 ± 0.05	8.6 ppm spike added. $[F]_T$ determined = 9.90 ± 0.05 ppm
	5	1.28 ± 0.05	pH = 5.50 (HCl added)
	6	1.36 ± 0.05	5 mg potassium alum added/100 ml seawater $\simeq 2.7$ ppm Al^{3+}
	7	1.43 ± 0.10	50 mg potassium alum added/100 ml seawater $\simeq 27$ ppm Al^{3+}. Equilibration time of electrode very long
	8	$\begin{cases}1.39 \pm 0.05 \\ 1.37 \pm 0.05^*\end{cases}$	50 mg potassium alum and 3 g ammonium citrate added/100 ml seawater
	9	1.30 ± 0.05	37 mg ferric alum added to 100 ml seawater and pH adjusted to 3 by adding few drops of glacial acetic acid. Soln $\simeq$ 40 ppm Fe^{3+}
	10	1.37 ± 0.05	50 ml TISAB added/100 ml seawater
	11	1.28 ± 0.10	Calculated via addition of TISAB and calibration vs. synthetic seawater
Synthetic		0.0068 ± 0.0008	pH 5.8
		0.0053 ± 0.0008	pH 4.3

*This determination was performed approximately 3 weeks after the first.
Reproduced from Rix, C.J., *et al.* (1976) *Analytical Chemistry*, **48**, 1236, American Chemical Society, by courtesy of authors and publishers.

Rix *et al.* [29] is similar to that originally described by Gran [32] for the exact determination of potentiometric end-points. Brand and Rechnitz [33], Baumann [34] and Craggs *et al.* [35] have already described procedures employing the method of standard additions to selective ion-electrode potentiometry, but no data are available for natural waters using this direct method except for the work of Liberti and Mascini [36] who determined fluoride in mineral waters. The procedure used by Rix *et al.* [29] is exceedingly simple, and the method of data treatment provides a linear plot, the slope and intercept of which give independent measures of the original concentration of fluoride in the sample. Total fluoride concentration is determined despite the fact that the activity of the fluoride ion is the parameter monitored by the fluoride selective ion electrode.

The concentration of fluoride in seawater is approximately 1.4 ppm or 7×10^{-5} mol l^{-1} and the fluoride electrode has been shown to give a Nernstian response at this level and indeed three orders of magnitude lower [37].

It has been well established that the fluoride electrode is highly selective in its response to the activity of fluoride with the only common interferant being hydroxide [38]. Nevertheless, although seawater has a uniform pH of 8, possible hydroxide interference was shown not to be a problem as indicated by the results presented in Table 2.2.

Complexation of fluoride by metal ions in seawater has previously been overcome by the addition of TISAB solution. This reagent is presumed to release the bound fluoride by preferential complexation of the metal ions with EDTA type ligands present in the TISAB. Examination of the metal ions present in seawater [39,40] suggests that magnesium is the major species forming fluoride complexes. Theoretical calculations for the simple case when only $MgF»fa$, $Mg^2»fa$ and $F-$ are considered demonstrate that even this species is unlikely to interfere.

Results for seawater of 35.21 o/oo salinity are presented in Table 2.2, and the average value of $1.35 \pm 0.05\,mg\,l^{-1}$ is in excellent agreement with previously published data [41,24,25,42,39].

The first result in Table 2.2 is the value found directly. The next set of three results shows that recovery of fluoride after deliberate addition to the spiked samples is excellent. The fifth result is that obtained on an acidified sample. The sixth, seventh, eighth and ninth sets of data show that the expected fluoride concentration is still obtained after deliberate addition of aluminium or iron in the form of their alums. Aluminium (III) and iron (III) form very strong fluoride complexes [43] and, provided that sufficient time is allowed for equilibration (as also noted by Raumann [37] for very low fluoride concentrations), total fluoride could still be determined accurately by the method of standard additions even though the artificial concentration levels of aluminium and iron were three orders of magnitude greater than those normally occurring in seawater [39]. This observation is noteworthy, since the work of Liberti and Mascini [36] suggests that the direct method cannot be used for fluoride in the presence of high relative concentrations of aluminium (III) or iron (III) aquo species.

This complication is probably avoided in seawater (even with relatively high concentrations of Al^{III} and Fe^{III} by the large excess of chloride ions which partially complex with any aluminium (III) or iron (III) present to produce chloro-complexes of the metals, thus sequestering their influence on the fluoride concentration. Data obtained after addition of TISAB to release bound fluoride and using the calibration method further confirm the validity of the direct method. The data in Table 2.2 demonstrate the absence of hydroxide ion interference, the absence of interference caused by metal ion complexation and indicate that total rather than free fluoride is determined by the method of standard additions. Figure 2.3 (a) shows a plot of potential versus the logarithm of the concentration of added

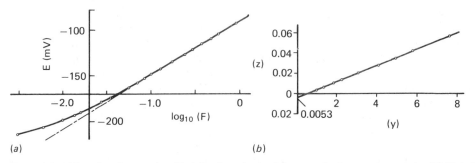

(a)   (b)

Figure 2.3(a) Plot of emf versus log (F−) for data obtained from synthetic seawater sample. **(b)** Plot of z versus y for data in (a) showing the linear extrapolation (Reproduced from Rix, C.J., *et al.* (1976) *Analytical Chemistry*, **48**, 1236, American Chemical Society, by courtesy of authors and publishers.)

fluoride in synthetic seawater. The slope of the linear portion of the graph was 58.8 mV per decade change in fluoride ion concentration, and this value was used in the calculations of fluoride concentration in seawater. At lower levels of fluoride ($<10^{-6}$ mol l^{-1}), it might be assumed that the electrode behaves in a non-Nernstian fashion. However, calculations of the kind suggested by Rix *et al.* [29] reveal that the deviation from linearity can satisfactorily be attributed to residual fluoride resulting either from reagent contamination or from the electrode itself at a concentration of about 3×10^{-7} mol l^{-1}.

The level of fluoride in seawater is at least two orders of magnitude higher than this residual level, hence contamination does not significantly influence the results.

When handling large numbers of samples, computerized rather than graphical analysis is considered to be substantially superior. The precision of the data reported in Table 2.2 is largely limited by the accuracy of the potential measurement ($\pm$ 1 mV). Using instrumentation capable of measuring 0.1 mV and with least-square fitting of data, average deviations on replicate determinations were routinely found to be 2%.

2.6 Ammonium

The determination of ammonia in seawater has long been recognized as an important measurement in environmental and ecological studies. The procedures used for such determination fall into three categories:

1. Based on the formation of bis-(3-methyl-1-phenyl-5-) pyrazolone, described by Prochazkova [44] and modified by Johnston [45] and Strickland and Austin [46].
2. Based on reaction with sulphanilamide coupled to N-(1-naphthyl) ethylene diamine [58,59].
3. Based on the Berthelot reaction between ammonia, phenol and hypochloride at alkaline pH to form the indophenol blue.

Although method (1) is more sensitive, it is generally multi-stage and often uses organic solvent extraction so that its automation is rendered less successful. Numerous procedures have been described for the production of the indophenol blue. Newell [47] employed chloramine-T instead of hypochlorite and subsequently extracted the indophenol blue using the method of Newell [47]. Matsunaga and Nishimura [57] investigated the determination of ammonia in seawater by extraction of indophenol. They modified the method by using thymol instead of phenol. The modified method showed no interference from nitrogen compounds including urea, glycine and glutamic acid and the calibration graph was rectilinear up to 5 μg N per litre. Emmet [48] using the original reaction, buffered the sample to pH 9.4 to obtain the colour. Attempts to increase the sensitivity and the speed of the reaction by Koroleff [49] using a more alkaline buffer together with sodium nitroprusside as a catalyst, produced a precipitation and serious interference when polluted waters were analysed. In order to avoid the precipitation of magnesium hydroxide at alkaline pH, Solorzano [50] developed the indophenol blue method (at pH 10.5) using citrate buffer.

The simplicity of the latter methods renders these most suitable for automation. Attempts made to automate Solarzano's method, however, appeared to be fraught with difficulties; for example, Head [51] encountered precipitation in the ethanol–phenol reagent (not met in the original manual method), while Grasshoff and

Johannsen [52] attributed some difficulties to the uncontrollable content of the hypochlorite reagent. Liddicoat *et al.* [53], however, suggested that in their manual procedure, controlled photo-oxidation generally produced a more reproducible colour. In almost all the published automatic procedures, heating was employed in the final indophenol blue colour development (70°C [52], 60°C [51], 90°C [54], 65°C [55]) with the result that serious interference from amino acids was unavoidable [52].

Harwood and Huyser's [56] study of the indophenol blue reaction for ammonia determination showed the effect of pH variation on the final colour and emphasized the necessity for the efficient buffering to obtain reproducible results. The optimum conditions found by the authors (pH 10.8 and using sodium nitroprusside) were similar to those used by Solorzano. Harwood and Kuhn [60] also replaced acetone with sodium nitroprusside.

Degobbis [61] recommended 25–45°C. Dal Pont *et al.* [62] increased Solorzano's [50] citrate and sodium hydroxide concentrations to work in seawater at the optimum pH (about 10.5). They used sodium dichloroisocyanurate as a convenient chlorine source, but needed a period at about 70°C to hasten colour development.

Berg and Abdullah [63] have described a spectrophotometric autoanalyser method based on phenol, sodium hypochlorite and sodium nitroprusside for the determination of ammonia in sea and estuarine water (i.e. the indophenol blue method) (Figure 2.4).

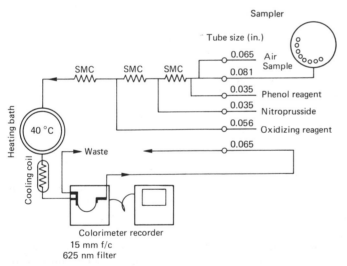

Figure 2.4 Schematic flow-diagram for the automatic analysis for ammonia (Reproduced from Beng, B.R. and Abdullah, M.I. (1977) *Water Research*, **11**, 637, Pergamon Publications, by courtesy of authors and publishers.)

The manifold design allows for the determination of ammonia concentration in the range of $0.2–20\,\mu g\,l^{-1}$ as NH_4 over a salinity range of 35–10‰ with negligible interference from amino acids.

The interference from amino acids was investigated and found to be negligible as reported by Solorzano [50] and Harwood and Huyser [56] who employed no heating for the indophenol blue colour development. Solutions containing

$50 \, \mu g \, N \, l^{-1}$ of urea, histidine, lycine, glycine and alanine were analysed. The NH_4-N detected ranged between 0.4% (for urea) and 2.2% (for alanine) of the nitrogen added.

Hampson [64] used ultraviolet photon activation energy of appropriate frequency and a ferrocyanide catalyst to activate selectively the reaction between ammonia, phenol and sodium hypochlorite. The reaction is carried out at an optimum pH of 10.5 since urea may break down at a pH over 11 and at a low temperature of $30 \pm 1°C$ to avoid alkaline hydrolysis of amino acids to ammonia, a process known to occur extensively at higher temperatures [65]. All conditions, including photon flux and ionic activities, are precisely controlled to give stability and reproductibility in a kinetic system. Serious colour suppression had been noted when indophenol blue type methods such as those of Koroleff and Solarzano [49,50] had been applied to certain types of samples. Hampson investigated the causes of such colour suppression.

In this method colour development is complete in 40 min and remains constant for many hours. About 3% precision is obtainable at ammonia concentrations above 0.1 ppm N. In routine daily measurements over a year in fish-rearing-tank waters precision was as in Table 2.3.

Table 2.3 Precision in routine use of IPB procedure for ammonia determination in fish-tank water

Type of fresh water or seawater	Range of ammonia concentration (ppm N)	Mean coefficient of variation (%)	Average no. of determinations on each sample	SEM of each days measurements
F W with much organic N and NO_2^-	0.75–5.4	18.5	4.7	8.5
F W with low organic N and NO_2^-	0.045–0.13	13.4	3.4	7.3
S W with much organic N and NO_2^-	0.25–0.98	9.2	3.5	4.9
S W with low organic N and NO_2^-	0.06–0.33	11.2	2.8	6.7

From Hampson, B.L. (1977) *Water Research*, **11**, 305, Pergamon Publications Ltd, by courtesy of author and publishers.

The absorbance–concentration relationship is linear with 1 absorbance unit from 0.1 ppm ammonia N using 10 cm cuvettes. Reagent blanks (untreated reagents) are equivalent to about 0.01 ppm ammonia N. The detection limit of ammonia in fresh- or seawater is about 0.001–0.002 ppm N.

The methyl and ethyl primary, secondary and tertiary amines were examined for possible interference in the ammonia–indophenol blue reaction by allowing the reaction to take place in pure seawater containing a known addition of ammonia convenient for measurement (0.40 ppm ammonia N), in parallel paired experiments with and without a known addition of each amine in turn. The amines were present at 100-fold excess over the ammonia, all concentrations being expressed on a N atom basis. All the amines suppressed the indophenol blue reaction with ammonia very strongly and in the general order: primary amines < secondary amines < tertiary amines.

The effects of varying concentrations of amine in the range 0–1000 ppm N on the indophenol blue response to ammonia was investigated using pure seawater containing 0.40 ppm ammonia N and expressed as percentage reduction in colour

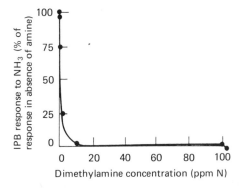

Figure 2.5 Suppression of indophenol blue response to ammonia by dimethylamine in seawater; 0.4 ppm ammonia N in seawater (Reproduced from Hampson, B.L. (1977) *Water Research*, **11**, 305, Pergamon Publications, by courtesy of author and publishers.)

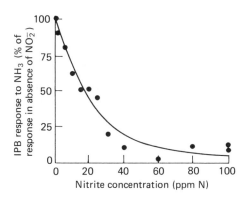

Figure 2.6 Suppression of indophenol blue response to ammonia by nitrite ion in seawater; 0.4 ppm ammonia N in seawater (Reproduced from Hampson, B.L. (1977) *Water Research*, **11**, 305, Pergamon Publications, by courtesy of author and publishers.)

(Figure 2.5). Very strong suppression is noted, even at low amine concentrations and with even the indophenol blue formation from the reagent blank suppressed at amine concentrations above 25 ppm N. Numerical analysis of the type of relationship between amine concentration and indophenol blue response to ammonia suggests a third, or possibly fourth power polynomial function (as distinct from a sample power or exponential relationship).

None of these amines gave any indophenol blue response when pure seawater (ammonia free) to which they had been added, was subjected to the standard analytical procedure.

To overcome the suppression effect of amines in the determination of ammonia, Hampson [64] investigated the effect of nitrite ions added either as nitrite or as nitrous acid. Figure 2.6 indicates that very considerable suppression by nitrite does occur, although it is not as strong as with any of the amines. Again, it is not great so long as the nitrite N concentrations is less than the ammonia N concentration but rapidly increases as the nitrite concentration exceeds the ammonia concentration. In fact the nitrite modified method was found to be satisfactory in open seawater samples and polluted estuary waters.

The determination of ammonia in non-saline waters does not present any analytical problems and, as seen above, reliable methods are now available for the determination of ammonia in seawaters. In the case of estuarine waters, however, new problems present themselves. This is because the chloride content of such waters can vary over a very wide range from almost nil in rivers entering the estuary to about $18\,g\,l^{-1}$ in the edges of the estuary where the water is virtually pure seawater.

Particularly in auto-analyser methods of analysis this wide variation in chloride content of the sample can lead to serious 'salt errors' and, indeed, in the extreme case, can lead to negative peaks in samples that are known to contain ammonia. Salt errors originate because of changes of pH, ionic strength and optical properties with salinity. This phenomenon is not limited to ammonia determinations by auto-analyser methods; it has, as will be discussed later, also been observed in the automated determination of phosphate in estuarine samples by molybdenum blue methods.

In a typical survey carried in an estuary, the analyst may be presented with several hundred samples with a wide range of chloride contents. Before starting any analysis, it is good practice to obtain the electrical conductivity data for such samples so that they can be grouped into increasing ranges of conductivity and each group analysed under the most appropriate conditions.

In this connection, Mantoura and Woodward [66] have described an indophenol blue method for the automated determination of ammonia in estuarine waters.

The reaction manifold describing the automated determination of ammonia is shown in Figure 2.7. Two alternative modes of sampling are shown: discrete and

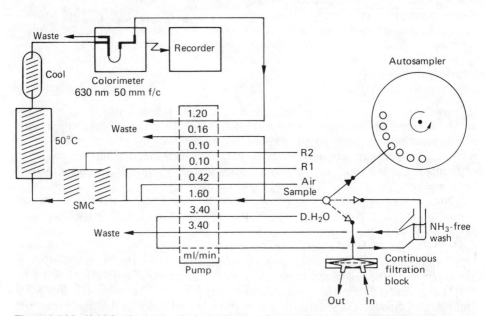

Figure 2.7 Manifold for the automatic determination of ammonia (# on pump tube R1 indicates 'Solvaflex' tubing) (Reproduced from Montoura, R.F.C. and Woodward, E.M.I. (1983) *Estuarine, Coastal and Shelf Science*, **17**, 219, Academic Press, by courtesy of authors and publishers.)

Table 2.4 Analytical performance of the automated NH$_3$ analyser

Linear detection range	0.2–18 μg-at N l^{-1}
Reproducibility (% SD of 10 replicates at 3 μg-at N l^{-1})	± 1.0%
Detection limit (S/N = 2)	0.02 μg-at N l^{-1}
Delay time	11.7 minutes
Response time (95%)	2.5 minutes
Sample/wash times	1.5 minutes
Sample throughput	20 h^{-1}

From Mantouri, R.F.C. and Woodward, E.M. (1983) *Estuarine, Coastal and Shelf Science*, **17**, 219, Academic Press Inc., by courtesy of authors and publishers.

continuous. Discrete 5 ml samples contained in ashed (450°C) glass vials are sampled from an autosampler (Hook and Tucker model A40-11; 1.5 min sample/ wash). For high-resolution work in the estuary, the continuous sampling mode is preferred. The indophenol blue complex was measured at 630 nm with a colorimeter and the absorbance recorded on a chart recorder. The analytical performance figures based on the colorimeter are summarized in Table 2.4.

Mantoura and Woodward [66] overcame the problem of magnesium precipitation by ensuring a stoichiometric excess of citrate (about 120%). These workers believe that 'salt errors' occurring with estuarine samples originate from poor pH buffering rather than ionic strength variations. They, in fact, used phenol at a concentration of 0.06 M to make the system self-buffering. Even in the presence of 1 mg NH$_3$-N per litre the indophenol blue reaction will consume only 3% of the phenol leaving most of the phenol to act as a pH buffer. Ethanol was used to solubilize the high concentration of phenol used in the system. The salt error of this method, as determined by standard addition of ammonia into waters of different salinities, is shown in Figure 2.8 (a). When compared with other methods, the method displays minimal salt error (about 8%) even though the final pH of the river water mixture (pH 10.9) was greater than seawater (pH 9.9).

In addition to the chemical effects of varying salinity, there are optical interferences in colorimetric analysis which are peculiar to estuarine samples. Saline waters and river waters have, in the absence of colorimetric reagents, an apparent absorbance arising from:

1. Refractive bending of light beams by sea salts–'refractive index blank' [67].
2. Background absorbance by dissolved organics of riverine origin.

The former is a function of the optical geometry of the light beam and the flow cell, and the latter is related to the organic loading of river water. Figure 2.8 (b) shows that both are linearly related to salinity, which makes optical blank corrections easy to apply to estuarine samples.

Other workers who have investigated automated methods for the determination of ammonia in seawater include Grasshoff and Johannsen [68], Berg and Abdullah [69], Truesdale [70], Le Corre and Treguer [71] and Matsumaga and Nishimura [72]. McLean *et al.* [73] have applied polarography to the determination of ammonia and other nitrogen compounds in brine samples, and Gilbert and Clay [74] have investigated the determination of ammonia in seawater using the ammonia electrode. These latter workers showed that down to 0.01 ppm ammonia can be determined using an electrode (Orion model 95-10) incorporating a hydrophobic membrane that separates the sample solution (adjusted to pH 11 with

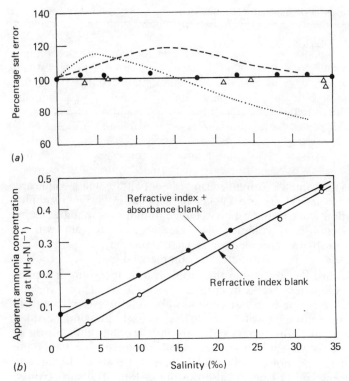

Figure 2.8(a) The effect of salinity on the sensitivity of standard additions of ammonia in laboratory mixed waters (●) and in waters from the Tamar estuary (▲) expressed as % of response in river water. For comparison, the salt error curves reported by Grasshoff and Johannsen [64] and Loder and Glibert [65] are also shown (... and ---, respectively). **(b)** Contribution of refractive index and organic absorbance to the optical blanks in the Chemlab Colorimeter. ● = River Water–seawater mixture. ○ = De-ionized water–seawater mixture (Reproduced from Mantoura, R.F.C. and Woodward, E.M.I. (1983) *Estuarine, Coastal and Shelf Science*, **17**, 219, Academic Press, by courtesy of authors and publishers.)

NaOH) from an internal solution 0.1 M in ammonium chloride. A glass pH-electrode and a silver-AgCl reference electrode are immersed in the aqueous ammonium chloride. The ammonia in the sample passes through the membrane and the change in pH in the internal solution is detected by the glass electrode. The behaviour and characteristics, including theoretical detection limits, of the system are discussed.

Le Corre and Treguer [71] developed an automated procedure based on oxidation of the ammonium ion to nitrite by hypochlorite in the presence of sodium bromide followed by spectrophotometric determination of the nitrite. The validity of automatic analysis of ammonium nitrogen in seawater was tested. The standard deviation on a set of samples containing $1 \mu g$ NH_4^--N per litre was 0.02. This method was compared with an automated method for the determination of ammonia as indophenol blue. The results from the two methods are in good agreement.

Urea and amino acids interfere in this procedure. Le Corre and Treguer [71] discuss the effect of salinity on the determination of ammonia and describe a suitable correction procedure.

Degobbis [75] studied the storage of seawater samples for ammonia determination. The effects of freezing, rate of freezing, filtration, addition of preservatives and type of container on the concentration of ammonium ion in samples stored for up to a few weeks were investigated. Both rapid and slow freezing were equally effective in stabilizing ammonium ion concentration and addition of phenol as preservative was effective in stabilizing non-frozen samples for up to 2 weeks.

Willason and Johnson [76] have described a modified flow-injection analysis procedure for ammonia in seawater. Ammonium ions in the sample were converted to ammonia which diffused across a hydrophobic membrane and reacted with an acid–base indicator. Change in light transmittance of the acceptor steam produced by the ammonia was measured by a light emitting diode photometer. The automated method had a detection limit of $0.05\ \mu\mathrm{mol\,l}^{-1}$ and a sampling rate of 60 or more measurement per h.

2.7 Nitrite

Spencer and Brewer [77] have reviewed methods for the determination of nitrite in seawater. Workers at the Water Research Centre, UK [78] have described an automated procedure for the determination of oxidized nitrogen and nitrite in estuarine waters. The procedure determines nitrite by reaction with N-1 naphthyl-ethylene diamine hydrochloride under acidic conditions to form an azo dye which is measured spectrophotometrically. The reliability and precision of the procedure were tested and found to be satisfactory for routine analyses provided that standards are prepared using water of an appropriate salinity. Samples taken at the mouth of an estuary require standards prepared in synthetic seawater whilst samples taken at the tidal limit of the estuary require standards prepared using deionized water. At sampling points between these two extremes there will be an error of up to 10% unless the salinity of the standards is adjusted accordingly. In a modification of the method nitrate is reduced to nitrite in a micro cadmium/copper reduction column and total nitrite estimated. The nitrate content is then obtained by difference.

Matsunaga et al. [79] have also described a similar procedure for the determination of nitrite in seawater. To 500 ml of sample is added 10 ml 1% solution of sulphanilamide in 2 M hydrochloric acid and 5 ml 0.1% aqueous N-1-naphthylethylene-diamine hydrochloride. After 10 min 5 ml 0.5% aq. dodecylbenzenesulphonate is added and the mixture extracted for 2 min with 50 ml carbon tetrachloride; 75 ml acetone and 10 ml 0.1 M hydrochloric acid are added to the separated organic layer and the azo-dye back-extracted into the acid. The extinction of the acid layer is measured at 543 nm in a 2 cm cell. The sensitivity for 0.001 extinction in 1 cm cells is 0.7 ng atom nitrite-N per litre.

Gianguzza and Orecchio [80] have carried out comparative trials of various methods for estimating nitrites in seawaters. These workers compared a method using sulphanilic acid/alpha-naphthylamine complexes, with a method using sulphanilamide/N(1-naphthyl) ethylenediamine complexes for the determination of nitrites in saline waters. The second method has the greater sensitivity and lower detection limits. The former method is subject to interference from chlorides and this interference can be completely eliminated by the coupling diazotization procedure of the latter method.

Anion exchange resins have been used to determine extremely low concentrations of nitrite down to nanomoles in seawater. Wada and Hattori [81] formed an azo dye from the nitrite using N-1 naphthylethylene diamine dihydrochloride and then adsorbed the dye in an anion exchange resin. The dye is then eluted from the column with 60% acetic acid and evaluated spectrophotometrically at 550 nm.

2.8 Nitrate

Spencer and Brewer [77] have reviewed methods for the determination of nitrate in seawater. Classical methods for determining low concentrations of nitrate in seawater use reduction to nitrite with cadmium/copper [72] (see section on nitrite) or zinc powder [82] followed by conversion to an azo dye using N-1-naphthylethylenediamine dihydrochloride and spectrophotometric evaluation. Malhotra and Zanoni [83] and Lambert and Du Bois [84] have discussed the interference by chloride in reduction–azo dye methods for the determination of nitrate.

Ultraviolet spectrometry has also been used to determine nitrates. In a method for determining high levels of nitrate described by Mertens and Massart [85] the sample, diluted to contain 0.5–1 ppm nitrate is acidified, filtered through a 0.5 μm filter and the extinction measured against a blank at 210–220 nm. The concentration of nitrate is obtained from a calibration graph. Interference from chloride, bromide, organic matter, carbonate, bicarbonate and nitrite is largely removed by using as blank a solution prepared by boiling 10 ml of the sample with 0.5 g Raney nickel for 30 min and then stirring for 90 min at 90°C, to reduce nitrate to ammonia. The ϵ value is 8500 at 210 nm and 4100 at 220 nm. Beer's law is valid up to 10 ppm at 210 nm and up to 15 ppm at 220 nm. The standard deviations for a sample containing 17.4 ppm of nitrate were 0.3 and 0.5 ppm at 210 and 220 nm respectively. No systematic errors were detected.

Previous ultraviolet methods for determining nitrate have attempted to allow for humic acid interference [86–89]. However, with the exception of Morries [89] these methods of allowance are inaccurate at humic acid concentrations above about 3.5 mg l^{-1}. Unfortunately, none of these previous methods have attempted to make any allowance for ultraviolet absorbing pollutant organic compounds or interfering inorganic ions. Thus their applications to water analyses other than for relatively unpolluted fresh waters is open to question.

Brown and Bellinger [90] have proposed an ultraviolet technique that is applicable to both polluted and unpolluted fresh and some estuarine waters. Himic acid and other organics are removed on an ion-exchange resin. Bromide interference in seawater samples can be minimized by suitable dilution of the sample but this raises the lower limit of detection such that only on relatively rich (0.5 mg l^{-1} NO$_3$N) estuarine and inshore waters could the method be used. Chloride at concentrations in excess of 10 000 mg l^{-1} do not interfere.

The method is either not affected by or can allow for interference from phosphate, sulphate, carbonate, bicarbonate, nitrite, coloured metal complexes, ammonia dyes, detergents, phenols and other ultraviolet absorbing substances. The method incorporates three features designed to reduce interferences:

1. Humic acid interference is reduced by carrying out measurements at 225 nm, a higher wavelength than that used by previous workers (210–220 nm).
2. Removal of inorganic interferences, particularly the removal of bromide interference in seawater by diluting the sample fivefold, the removal of nitrite by the addition of sulphamic acid and the removal of metals by passage through Amberlite IR 120 cation exchange resin.
3. Removal of ultraviolet absorbing organics by passage through a specific ion-exchange resin such as Amberlite XAD-2.

Table 2.5 compares nitrate determinations in the presence of various interfering substances for this method and for two alternate methods–phenoldisulphonic acid

Table 2.5 The effects of various interferences on the determination of nitrates by three methods. For all determinations (with the exception of natural river waters) the concentration of NO₃N was 1.125 mg l⁻¹. All concentrations given are in mg l⁻¹

Interference	Phenoldisulphonic acid		Method [90] New u.v. method		Selective ion electrode	
	Uncorrected	Corrected	Uncorrected	Corrected	Uncorrected	Corrected
Chloride:						
100	0.62	1.07*	1.08		1.08	1.55§
1000			1.05			
Bicarbonate–carbonate:						
100	1.13		1.06		7.87	1.24‖
200	1.23					
1000			1.13			
Bicarbonate–chloride:						
100 Cl:200 HCO₃						1.58¶
100 Cl:100 HCO₃						1.46¶
50 Cl:100 HCO₃						1.46¶
Nitrate:						
0.304	1.11		1.48	1.10†		
1.500						1.31¶
3.040	1.13			1.06†		
Ammonium:						
1.4	0.98					
14.0	0.66		1.08			1.13¶
1400.0			1.08			
Phenol:						
1.0	1.10		1.29	1.10‡		1.13¶
10.0			1.86	1.08‡		
Daz**:						
1.0			1.12			2.48¶
10.0			1.56	1.05‡		202.5¶
Unpolluted river water thumic acid–inorganic salts NO₃N by polarography 1.170	1.69		1.77	1.77‡		3.15¶

*Addition of exact amount of silver sulphate to precipitate chloride with no excess silver.
†Addition of sulphamic acid (final concentration, in sample approximately 20 mg l⁻¹).
‡Treatment with Amberlite XAD-2 resin.
§Addition of saturated silver sulphate (1–10 ml sample). Not recalibrated.
‖Addition of 2 M acetic acid to sample (0.09 ml per 30 ml sample). Not recalibrated.
¶Calibrate for presence of 3 ml saturated silver sulphate and 0.09 ml 2 M acetic acid in 30 ml sample.
**Daz–a commercially available household detergent.
Reproduced from Brown, L. and Bellinger, E.L., *Water Research*, **12**, 223, Pergamon Publications Ltd, by courtesy of authors and publishers.

and ion-selective electrode methods. In general, the method proposed by Brown and Bellinger [90] is less subject to interference.

The speed of the nitrate selective ion electrode makes its use potentially ideal for nitrate determinations on a large number of samples. However, the results from adding various interfering substances (Table 2.5) seem to cast some doubt upon the values obtained in the presence of chloride and bicarbonate for, although the results are precise, they are not accurate–approximately 20–30% high.

Chemiluminescent techniques have been used to determine nanomolar quantities of nitrate and nitrite in seawater [91,92]. This method depends on the selective reduction of these species to nitric oxide which is then determined by its chemiluminescent reaction with ozone, using a commercial nitrogen oxides analyser. The necessary equipment is compact and sufficiently sturdy to allow shipboard use. A precision of $\pm$ 2 nmol l^{-1} is claimed and an analytical range of 2 nm mol l^{-1} with analytical rates of 10–12 samples hourly.

In this method [91] nitrate, nitrate plus nitrite or nitrite alone are selectively reduced to nitric oxide which is swept from the sample in a helium carrier gas flow. Nitric oxide is allowed to react with ozone in a nitrogen oxides analyser where it forms excited nitrogen dioxide. The return of the nitrogen dioxide to the ground state is accompanied by release of a photon which is detected by a photomultiplier. The integrated output of the photomultiplier over the time that the nitric oxide is purged from the sample is proportional to the nitrate or nitrite content of the sample.

The apparatus is shown schematically in Figure 2.9. Helium is supplied to the system at a pressure of 20 psi (138 kpa) and the flow is regulated by means of a precision needle valve and flowmeter. The flow is adjusted to provide a small positive pressure at the input to the nitrogen oxides analyser. The helium flow is directed through an impinger consisting of a coarse porosity sintered glass frit (20–50 μm) in a 20 $\times$ 3.5 cm diameter glass sample tube fitted with a 24/40 ground-glass

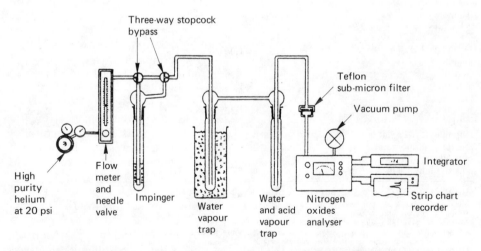

Figure 2.9 Schematic diagram of apparatus (Reproduced from Garside, C. (1982) *Marine Chemistry*, **11**, 159, Elsevier Science Publishers, by courtesy of author and publishers.)

joint. The sample tube can be bypassed from the helium flow by means of a pair of three-way stopcocks to allow changing of samples. Polyethylene 1/4 inch o.d. tubing and nylon Swagelock connectors are used to this point in the system. Comparable Teflon tubing and connectors are used downstream from the impinger.

The helium carrier gas and nitric oxide from the sample pass sequentially through a salt/ice cold trap to condense water vapour, an anhydrous sodium carbonate drying tube to remove remaining traces of water vapour and acid gases and a Teflon submicron filter to exclude particles. The gas stream then enters a Bendix Model 8101-C Nitrogen Oxides Analyser operated in the 'NO only' mode, where nitric oxide is determined chemiluminescently.

The instantaneous concentration in the gas flow is recorded on a strip chart recorder operated at an input sensitivity of 10 mV full scale, and a chart speed of 0.2 in min^{-1} (5 mm min^{-1}). The integrated area is recorded by using an Analog Devices AD537J V to f converter connected to the 1 V output of the nitrogen oxides analyser and the output of the V to f converter is totalled using a Heath Model IM 4100 frequency counter operated in the totalize mode. The V to f converter is mounted internally in the frequency counter and power is supplied from its logic circuits.

Linear calibration plots are obtained by this procedure covering the range up to 1 nmol l^{-1} for nitrate and for nitrite.

Flow injection analysis is another technique that has been applied to the determination of nitrate and nitrite in seawater. Anderson [93] used flow injection analysis to automate the determination of nitrate and nitrite in seawater. The detection limit of his method was 0.1 μmol l^{-1}. However, the sampling rate was only 30 per hour which is low for flow injection analysis. Reactions seldom go to completion in a determination by flow injection analysis [94,95] because of the short residence time of the sample in the reaction manifold. Anderson selected a relatively long residence time so that the extent of formation of the azo dye was adequate to give a detection limit of 0.1 μmol l^{-1}. This reduced the sampling rate because only one sample is present at a time in the postinjector column in flow injection analysis. Any increase in reaction time causes a corresponding increase in the time needed to analyse one sample.

Johnson and Petty [96] reduced nitrate to nitrite with copperized cadmium which was then determined as an azo dye, but the method is automated by means of flow injections analysis technique [97]. More than 75 determinations can be made per hour. The detection limit is 0.1 %.112mol l^{-1} and precision is better than 1% at concentrations greater than 10 μmol l^{-1}.

A block diagram of the equipment used is shown in Figure 2.10. The reaction manifold was made from 0.8 mm i.d. PTFE tubing. A four-channel Buchler peristaltic pump was used to propel the seawater and reagents. Tygon pump tubes (1.6 mm i.d.) were used to pump the sample (seawater or standard) and the ammonium chloride buffer. The sulphanilamide and diamine reagents were pumped with 0.8 mm i.d. Tygon tubing. The flow rates of these reagents were 4.5 times smaller than the sample and buffer flow rates. Altex low pressure fittings were used to make all of the connections between pump and manifold tubing and at the T fittings.

All of the tubing in which the sample was carried from the inlet, through the pump and reduction column to the confluence with the sulphanilamide line was made as short as possible.

52

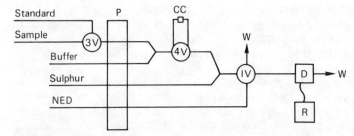

Figure 2.10 Block diagram of the apparatus used to determine nitrate and nitrite. P = Pump; CC = cadmium reduction column; 3V = three-way valve used to switch between the sample and the standard 4V = switching valve used to take the cadmium column out of line; IV = injection valve; D = detector; W = waste; R = recorder; Sulf. and NED are the sulphanilamide and N-(1-naphthyl)ethylenediamine dihydrochloride reagents (Reproduced from Johnson, K.S. and Pettry, R.L. (1983) *Limnology and Oceanography*, **28**, 1260, Springer Verlag, by courtesy of the author and publisher.)

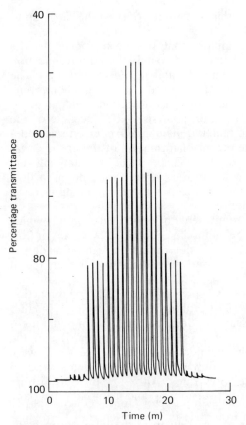

Figure 2.11 Detector response for a series of nitrate standards with concentrations of 0, 10, 20 and 40 μmol l^{-1} (Reproduced from Johnson, K.S. and Pettry, B.L. (1983) *Limnology and Oceanography*, **28**, 1260, Springer Verlag, by courtesy of authors and publishers.)

The reduction column consisted of a piece of copperized cadmium, 1 cm long in a piece of PTFE tubing 3 cm long and 3.2 mm i.d. (¼ in.). Glass wool plugs (3 mm long) were placed in each end to retain the cadmium powder and were in turn held in place by short pieces of PTFE tubing (1.6 mm i.d., 3.2 mm o.d.). Connections to the reaction manifold tubing (1.6 mm o.d.) were simply press fit into these adapters until the ends were flush.

The peaks obtained for a series of standard solutions are shown in Figure 2.11. A small amount of carryover between samples, amounting to about 1%, is evident. A calibration curve obtained from these standards is distinctly non-linear; errors in accuracy, under 1%, appear at concentrations above $10 \, \mu\text{mol} \, l^{-1}$ if a linear calibration is assumed.

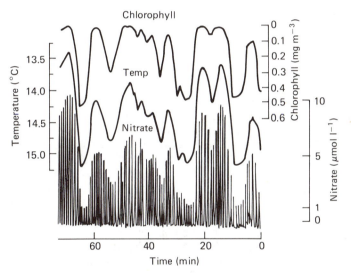

Figure 2.12 Nitrate, temperature and chlorophyll measured at 15 m in the Santa Barbara Channel (34°01′56″N, 119°19′30″W). Temperature and chlorophyll were determined with a thermistor and a Turner Designs fluorometer at the effluent of the debubbler (Reproduced from Johnson, K.S. and Pettry, B.L. (1983) *Limnology and Oceanography*, **28**, 1260, Springer Verlag, by courtesy of authors and publishers.)

Figure 2.12 shows distributions of nitrate determined by this method, chlorophyll and water temperature in a sea cruise over a thermal front. The concentrations of nitrate changed by nearly an order of magnitude within a few minutes; these changes are highly correlated with the temperature and with chlorophyll fluorescence.

Bajic and Jaselskis [98] described a spectrophotometric method for the determination of nitrate and nitrite in seawater. It included the reduction of nitrate and nitrite to hydroxylamine by the zinc amalgam reductor (Jones reductor) at pH 3.4 and reoxidation of the product with iron (III) in the presence of Ferrozine. Interference by high levels of nitrite could be eliminated with azide treatment. Levels of nitrate of 0.1 ppm could be detected with a precision of 3% in the presence of large amounts of nitrite and chloride.

Hydes and Hill [99] used the copper–cadmium reduction method to determine nitrate in seawater. The construction of a copper–cadmium (50–50 w : w) reductor column for use during continuous flow analysis at sea is described. A 100% yield could be obtained using a 20 cm × 3 mm column fitted with grains of copper–cadmium alloy between 500 and 350 μm in size. The column maintained its reactivity during 3 months storage prior to an oceanographic cruise, and during the 4-week cruise period. Its performance was similar to that of the Stainton-type cadmium wire column but it had the advantages of easier preparation and easier control of reductor volume.

Tyree and Bynum [100] describe a ion-chromatographic method for the determination of nitrate and phosphate in seawater. The pretreatment comprised vigorous mixing of the sample with a silver-based cation-exchange resin, followed by filtration to remove the precipitated silver salt.

2.9 Phosphate

Spencer and Brewer [77] have reviewed methods for the determination of phosphate in seawater. Earlier methods for the determination of phosphate in seawater are subject to interferences, particularly by nitrate. In one early method [101] the filtered seawater sample is acidified with nitric acid and perchloric acid in the presence of hydroxylammonium chloride and ammonium chloride and the phosphomolybdate complex extracted with methylisobutyl ketone. This extract is then reduced with acidic stannous chloride and ascorbic acid and the extinction measured at 725 nm. Hosokawa and Ohshima [102] heated the sample (50 ml) for 20 min on a boiling water bath with 2 ml Mo^{VI}–Mo^V reagent (hydrochloric acid added (10 ml) to $2 m$–Mo^{VI} (10 ml) then added zinc (0.3 g), and after the zinc dissolved added hydrochloric acid to 100 ml). The mixture was cooled and the extinction at 830 nm measured. The resulting blue complex remains stable for at least 2 months and has an ϵ value of 26 000. Nitrate interferes if present in concentrations more than $1 \mu g l^{-1}$ and should first be reduced to nitrite with zinc in acidic medium, AsO_4^{3-} interfere at $10 mg l^{-1}$. The salt error was about 5% at a chloride concentration of 1.9%.

Isaeva [103] described a phosphopolybdate method for the determination of phosphate in turbid seawater. Molybdenum titre methods are subject to extensive interferences and are not considered to be reliable when compared with more recently developed methods based on solvent extraction [104–109] such as the solvent extraction–spectrophotometric determination of phosphate using molybdate and malachite green [110]. In this method the ion-pair formed between malachite green and molybdophosphate is extracted from the seawater sample with an organic solvent. This extraction achieves a useful 20-fold increase in the concentration of the phosphate in the extract. The detection limit is about $0.1 ng ml^{-1}$, standard deviation $0.05 ng l^{-1}$ ($4.3 ng ml^{-1}$ in tap water) and relative standard deviation 1.1%. Most cations and anions found in natural waters do not interfere, but arsenic (V) causes large positive errors.

In earlier work [109] Motomizu et al. used ethyl violet as counter-ion for molybdophosphate. The spectrophotometric determination of phosphate by solvent extraction of molybdophosphate with ethyl violet is more sensitive than those previously reported [97–101], and less troublesome; a single extraction is adequate. However, in the determination of phosphorus at $ng ml^{-1}$ levels in waters it has certain disadvantages. First, the absorbance of the reagent blank becomes too large for the concentration effect achieved by the solvent extraction to be of much use; for example, when 20 ml of sample solution and 5 ml organic solvent were used, the absorbance of the reagent blank was 0.14. Secondly, the shaking time needed was long and the colour of the extract faded gradually if the shaking lasted more than 30 min.

However malachite green [110] gave a stable dark yellow species in 1.5 M sulphuric acid (probably a protonated one), whereas ethyl violet became colourless within 30 min even in only 0.5 M sulphuric acid.

In this procedure 10 ml of the sample solution containing up to 0.7 µg phosphorus as orthophosphate, was transferred into a 25 ml test tube. To this solution was added 1 ml each of 4.5 M sulphuric acid and the reagent solution. The solution was shaken with 5 ml of a 1:3 v/v mixture of toluene and 4-methylpentan-2-one for 5 min. After phase separation, the absorbance of the organic phase was measured at 630 nm against a reagent blank, in 1 cm cells.

The absorption spectra of the ion-pair and the reagent blank are shown in Figure 2.13. Linear calibration graphs are obtained even when the aqueous phase volume is increased to 100 ml (and 7 ml of extracting solvent are used). The results obtained

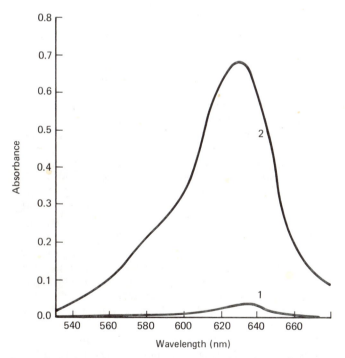

Figure 2.13 Absorption spectra in the organic phase (curve 1) reagent blank (reference, extracting solvent); (curve 2) phosphorus, 0.428 µg (reference, extracting solvent); aqueous phase, 12 ml, organic phase. 5 ml (Reproduced from Motanizi, S., *et al.* (1984) *Talanta*, **31**, 235, Pergamon Publications, by courtesy of authors and publishers.)

Table 2.6 Effect of volume of sample solution

Sample solution taken (ml)	Sulphuric acid added (ml)	Reagent solution added (ml)	Absorbance of reagent blank	Absorbance units for 1 ng ml^{-1} phosphorus	Molar absorptivity* $(10^5$ l mol^{-1} cm$^{-1})$
5	0.5	0.5	0.034	0.007	2.30
10	1.0	1.0	0.044	0.015	2.26
25	2.5	2.5	0.062	0.038	2.31
50	5.0	5.0	0.078	0.080	2.48
100†	10.0	10.0	0.122	0.124	2.69

Extracting solvent 5 ml of a 1:3 mixture of toluene and 4-methylpentan-2-one.
Phosphorus 0–0.8 µg.
*Organic phase volume was assumed to be 5 ml.
†7 ml of extracting solvent were used.
Reproduced from Motomizi, S., et al. (1984) *Talanta*, **31**, 235, Pergamon Publications Ltd, by courtesy of authors and publishers.

by varying the volume of sample solution are shown in Table 2.6, which shows that when 50 ml of sample solution are used, 0.1 ng ml^{-1} of phosphorus can be detected.

Arsenic (V) causes large positive errors; arsenic (V) at a concentration of 10 ng ml^{-1} produces an absorbance of 0.07 but can be masked with tartaric acid (added in the reagent solution). When arsenic (V) was present at concentrations of 50 ng ml^{-1} it was masked with 0.1 ml 10^{-4} M sodium thiosulphate added after the sulphuric acid.

For a 25 ml sample, almost all of the foreign ions tested, when present at the concentrations listed in Table 2.7, produced an absorbance of less than 0.005.

Table 2.8 shows the results for determination of phosphorus in waters. Less than 15 ml of seawater was taken as sample and was diluted to 25 ml with distilled water because of its high chloride content (>0.5 M).

Recovery tests were done by adding known amounts of phosphate. The results are also shown in Table 2.8. The recovery of phosphorus was good; 99–103%. The relative standard deviation for phosphorus was 0.6% for 21.0 ng ml^{-1} in seawater.

A commonly used procedure for the determination of phosphate in seawater and estuarine waters uses the formation of the molydenum blue complex at 35–40°C in an autoanalyser and spectrophotometric evaluation of the colour produced. Unfortunately, when applied to seawater samples, depending on the chloride content of the sample, peak distortion or even negative peaks occur which make it impossible to obtain reliable phosphate values (Figure 2.14). This effect can be overcome by the replacement of the distilled water wash solution used in such methods by a solution of sodium chloride of an appropriate concentration related to the chloride concentration of the sample. The chloride content of the wash solution need not be exactly equal to that of the sample. For chloride contents in the sample up to 18 000 mg l^{-1} (i.e. seawater) the chloride concentration in the wash should be within ± 15% of that in the sample (Table 2.9). The use of saline standards is optional but the use of saline control solutions is mandatory. Using good equipment, down to 0.02 mg l^{-1} phosphate can be determined by such procedures. For chloride contents above 18 000 mg l^{-1} the chloride content of the wash should be within ± 5% of that in the sample.

Recently, the technique of reversed flow injection analysis has been applied to the determination of phosphate in seawater. Reversed flow injection analysis

Table 2.7 Effect of other ions

Ion	Concentration		
	Added as	mol l^{-1}	Absorbance*
None			0.549
Na^+, Cl^-	NaCl	0.3	0.545
K^+	KCl	1×10^{-2}	0.548
SO_4^{2-}	Na_2SO_4	1×10^{-2}	0.552
Ca^{2+}	$CaCl_2$	1×10^{-2}	0.551
Mg^{2+}	$MgCl_2$	1×10^{-2}	0.548
HCO_3^-	$NaHCO_1$	1×10^{-2}	0.556
Al^{3+}	$KAl(SO_4)_2$	1×10^{-3}	0.550
Fe^{3+}	$FeNH_4(SO_4)_2$	1×10^{-3}	0.551
NH_4^+	NH_4Cl	1×10^{-3}	0.557
NO_3^-	$NaNO_3$	1×10^{-3}	0.548
Cr^{3+}	$CrCl_3$	1×10^{-3}	0.549
Mn^{2+}	$MnCl_2$	1×10^{-3}	0.549
Co^{2+}	$CoCl_2$	1×10^{-3}	0.549
Cd^{2+}	$CdCl_2$	1×10^{-3}	0.552
Ni^{2+}	$NiCl_2$	1×10^{-3}	0.549
Cu^{2+}	$CuCl_2$	1×10^{-3}	0.553
Zn^{2+}	$ZnCl_2$	1×10^{-3}	0.546
Br^-	NaBr	1×10^{-3}	0.546
B(III)	$Na_2B_4O_7$	1×10^{-3}	0.552
SiO_3^{2-}	Na_2SiO_3	5×10^{-4}†	0.553
VO_3^-	NH_4VO_3	1×10^{-5}	0.556
WO_4^{2-}	$(NH_4)_{10}W_{12}O_{41}$	1×10^{-5}	0.554
I^-	NaI	1×10^{-5}	0.550
Laurylsulphate	Na salt	5×10^{-6}†	0.549
Laurylbenzenesulphonate	Na salt	5×10^{-6}†	0.547
ClO_4^-	$NaClO_4$	1×10^{-6}†	0.557
Ti^{4+}	$TiCl_4$	1×10^{-6}	0.551

*Sample solution 10 ml; organic phase 5 ml; phosphorus 0.366 µg; measured against reagent blank.
†Tolerance limit.
Reproduced from Motomizi, S. et al. (1984) Talanta, **31**, 235, Pergamon Publications Ltd, by courtesy of authors and publishers.

Table 2.8 Determination of phosphorus in river and seawater and recovery test

Sample	Phosphorus found (ng/ml^{-1})	Recovery test				
		Sample taken (ml)	P in sample (ng)	P added (ng)	P found (ng)	Recovery (%)
Seashore of Seto Inland Sea						
Kojima	46.0	5	230	248	477	100
Shibukawa	17.4	8	139	248	391	101
Tamano	15.3	8	122	248	381	103
Ushimado	12.8	8	102	248	351	100
Nishiwaki	20.9	8	167	248	419	101

The samples were adjusted to about pH 3 with dilute sulphuric acid after sampling, and filtered through a membrane filter (0.45µm). An appropriate amount of sample was transferred to a 25 ml test tube, 1 ml of 4.5 M sulphuric acid was added, and the mixture was heated at 95°C for 30 min. The seawater samples were diluted to 25 ml with distilled water. A 10 ml sample was used, except for Yoshii River sample D and the Kojima seawater for which 5 ml samples were used. In the recovery tests, the river water was diluted to 10 ml with distilled water after the addition of 248 ng of phosphorus (as phosphate), and the seawater was diluted to 25 ml with distilled water after the addition of phosphorus.
*The values are the means of three determinations.
Reproduced from Motomizi, S., et al. (1984) Talanta, **31**, 235, Pergamon Publications Ltd, by courtesy of authors and publishers.

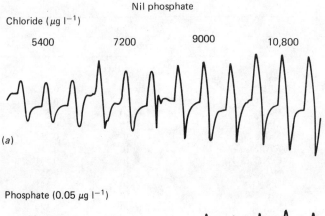

Figure 2.14 Interference by chloride in the autoanalyser determination of phosphate

Table 2.9 Dependence of chloride content of wash liquid on chloride content of sample

Chloride content of wash (mg l⁻¹)	Chloride content of sample (mg l⁻¹)
0	0–1800
4850	1800–7200
10000	7200–12600
15200	12600–18000

differs from flow injection analysis in the sense that whereas in the latter technique the sample plug is injected into a flowing stream of reagent, in the former technique plugs of reagent are injected into a continuous stream of the sample. Under these conditions the amount of sample in the zone of the reagent will increase as the dispersion increases. The sample will become well mixed with the reagent as its concentration increases so that well-formed peaks will result. Thus an analysis can be successfully performed by this technique with a sample concentration in the zone of the reagent that is typically in the range from 67% to 90% of the sample concentration in the carrier stream.

Johnson and Petty [111] adapted reverse flow injection analysis to the well-known Murphy and Riley [112] colorimetric phosphomolybdate reduction method for the determination of phosphate.

Figure 2.15 shows flow sheets for the determination of phosphate by flow injection analysis and reversed flow injection analysis. A block diagram of the apparatus used for the determination of phosphate in seawater is shown in Figure 2.15 (b).

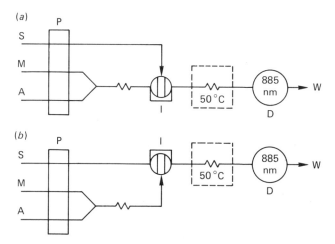

Figure 2.15 Analytical manifolds for the determination of phosphate by flow injection analysis (*a*) and reverse flow injection analysis (*b*). The symbols S, M and A are the seawater, mixed reagent and the ascorbic acid solutions. The pump, injection valve and detector are represented by P, I and D respectively. W = Waste (Reproduced from Johnson, K.S. and Pettrey, R.L. (1982) *Analytical Chemistry*, **54**, 1185, American Chemical Society, by courtesy of authors and publishers.)

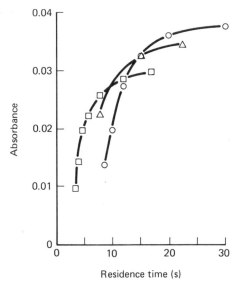

Figure 2.16 Effect of residence time on peak height. The squares (0.5 mm i.d. tubing) and circles (0.8 mm i.d.) were obtained by changing the flow rate through a 1 m reaction tube. The square at the extreme right corresponds to 0.7 ml/min^{-1}. The triangles were obtained by using a 0.8 mm i.d. tube and a flow rate of 2.0 ml/min^{-1}. Residence time was varied by changing the tube length (Reproduced from Johnson, K.S. and Pettry, R.L. (1982) *Analytical Chemistry*, **54**, 1185, American Chemical Society, by courtesy of authors and publishers.)

The effects of residence time on the peak height of a $2.5\,\mu$mol$\,$l^{-1} phosphate standard are shown in Figure 2.16. The residence time was defined as the time from injection of the reagent to appearance of the maximum signal at the detector. Residence times were varied by changing the flow rate (1.0–$3.5\,$ml$\,$min^{-1} in $0.5\,$ml$\,$min^{-1} increments) and the length of the reaction tube (0.5, 1.0 and $1.5\,$m). The peak height increased rapidly with residence time in both 0.5 and $0.8\,$mm i.d. tubes and then levelled off after $15\,$s. The change is about the same when the residence time is increased by using longer tubes. This suggests the increased peak height is due mainly to an increased extent of reaction.

Flynn and Meehan [113] have described a solvent extraction phosphomolybdate method using iso amyl alcohol for monitoring the concentration of ^{32}P in sea and coastal waters near nuclear generating stations.

Tyree and Bynum [100] have described an ion-chromatographic method for the determination of phosphate and nitrate in seawater.

2.10 Arsenate

Haywood and Riley [114] have described a spectrophotometric method for the determination of arsenic in seawater. Adsorption colloid flotation has been employed to separate phosphate and arsenate from seawater [115].

These two anions, in $500\,$ml filtered seawater, are brought to the surface in less than $5\,$min, by use of ferric hydroxide (added as $0.1\,$MFeCl$_2$ ($2\,$ml) as collector, at pH 4, in the presence of sodium dodecyl sulphate (added as 0.05% ethanolic solution ($4\,$ml)) and a stream of nitrogen ($15\,$ml$\,$min^{-1}). The foam is then removed and phosphate and arsenate are determined spectrophotometrically [116]. Recoveries of arsenate and arsenite exceeding 90% were obtained by this procedure.

2.11 Sulphate

Sulphate has been determined in seawater by photometric titration with hydrochloric acid in dimethyl sulphoxide [117]. The sample ($5\,$ml) is added slowly to dimethyl sulphoxide ($230\,$ml) and titrated with $0.02\,$M hydrochloric acid (standardized against sulphate) with bromocresol green as indicator. Since borate, carbonate and bicarbonate interfere, a separate determination of alkalinity is necessary.

Lebel [118] has described an automated chelometric method for the determination of sulphate in seawater. This method utilizes the potentiometric end-point method for back-titration of excess barium against EDTA following precipitation of sulphate as barium sulphate. An amalgamated silver electrode was used in conjunction with a calomel reference electrode in an automatic titration assembly consisting of a $2.5\,$ml electrical autoburette and a pH meter coupled to a recorder. Recovery of added sulphate was between 99 and 101% standard deviations of successive analyses less than 0.5% of the mean.

Polarography has also been used to determine sulphate [119]. In this method sulphate is precipitated as lead sulphate in a 20% alcoholic medium, followed by polarographic determination of the excess lead. It is claimed that there is little interference in this method; 25–30 samples can be analysed in a 5 hour period with a precision of 0.2% or less.

In this procedure up to 70 ml seawater, 10 ml 0.01 M lead nitrate, 20 ml 95% ethanol, 0.2 ml of 0.1% methyl red maximum current suppressor and two drops 3 M nitric acid are added to a 100 ml volumetric flask which is then diluted to 100 ml with distilled water.

The dropping mercury electrode had a drop time of 3–4 seconds under an open head of 50 cm Hg. A saturated calomel electrode was the reference electrode. Before recording, the solutions were well shaken. After recording, the electrodes were well rinsed with distilled water and wiped dry. A starting potential of 0.2 V was used and the solutions were degassed with dry nitrogen for 2 minutes prior to recording. A full-scale sensitivity of 10 μA was used.

As an indication of the accuracy of the technique the correlation coefficients for the calibration curves indicate a precision of 0.2% or less. The standard deviation of the measured samples is 0.70%.

The precision of the technique indicates its suitability for sulphate analysis. The technique appears not to be subject to any interferences normally present in seawater samples.

2.12 Alkalinity

Various workers have discussed the determination of total alkalinity and carbonate [120–122] and the carbonate–bicarbonate ratio [122] in seawater. A typical method utilizes an autoanalyser. Total alkalinity (T m-equiv. per litre) is found by adding a known (excess) amount of hydrochloric acid and titrating back with sodium hydroxide solution; a pH meter that records directly and after differentiation is used to indicate the end-point. Total carbon dioxide (C m-equiv. of HCO_3^- per litre) is determined by mixing the sample with dilute sulphuric acid and segmenting it with carbon dioxide free air, so that the carbon dioxide in the sample is expelled into the air segments; the air is then separated from the sample and passed into buffered phenolphthalein solution, thereby lowering the pH and diminishing the colour of the phenolphthalein. The reduction in colour is measured colorimetrically (540 nm). The concentration of carbonate is given by 2 (T − C) m-equiv. l^{-1}, and the concentration of bicarbonate is 20 − T m-equiv. l^{-1}.

A computer program has been used to calculate the magnitude of systematic errors incurred in the evaluation of equivalence points in hydrochloric acid titrations of total alkalinity and carbonate in seawater by means of Gran plots. Hansson [123] devised a modification of the Gran procedure that gives improved accuracy and precision. The procedure requires approximate knowledge of all stability constants in the titration.

Among the possible analytical methods for alkalinity determination, Gran-type potentiometric titration [124] combined with a curve-fitting algorithm is considered a suitable method in seawaters because it does not require a priori knowledge of thermodynamic parameters such as activity coefficients and dissociation constants which must be known when other analytical methods for alkalinity determination are applied [125–128].

Alkalinity determination in hypersaline solutions by the Gran-type titration is subject to a number of errors which can usually be neglected in lower ionic strength solutions. For example, the pH readings along the titration path may be inaccurate due to the marked difference between the ionic strength and composition of the sample and the standard buffers used to calibrate the glass and reference electrode pair [129].

Ben Yaakov and Lorch [130] identified the possible error sources encountered during an alkalinity determination in brines by a Gran-type titration and analysed the possible effects of these errors on the accuracy of the measured alkalinity. Special attention was paid to errors due to possible non-ideal behaviour of glass-reference electrode pair in brine. The conclusions of the theoretical error analysis were then used to develop a titration procedure and an associated algorithm which may simplify alkalinity determination in highly saline solutions by overcoming problems due to non-ideal behaviour and instability of commercial pH electrodes.

The titration procedure used was that described by Ben Yaakov et al. [131] in which 100 ml seawater or brine sample is titrated in 30–50 burette increments with 3–5 ml standard 0.1 M hydrochloric acid using a pH electrode.

The accuracy of the method was tested experimentally by running duplicate titrations on distilled water, artificial seawater with and without sulphate and artificial Dead Sea waters. For each run, alkalinity was calculated by two methods: (1) by the conventional Gran plot which presumes that the glass electrode is properly calibrated and (2) by the method that applies the titration data for *in situ* calibration of the glass electrode by the slope correction algorithm. The precision of the latter method when applied in the distilled water runs was found to be significantly better than the conventional method. This should be attributed to the non-stability of the glass electrode which is corrected for by the proposed algorithm.

2.13 Silicate

Spencer and Brewer [77] have reviewed methods for the determination of silicate in seawater. Various workers [132–135] have studied the application of molybdosilicate spectrophotometric methods to the determination of silicate in seawater. In general, these methods give anomalous results due, it is believed, to erratic blanks and uncertainty regarding the structure of the silicomolybdate formed.

A more promising approach to the analysis of silicate is that of Thomsen and Johnson [136], who used flow injection analysis. These workers achieved an analysis rate of 80 samples per hour with a detection limit of 0.5 M Si and a precision better than 1% at concentrations above $10 \,\mu mol \, l^{-1}$ Si.

Brzezinski and Nelson [140] have described a solvent extraction procedure for the spectrophotometric determination of nanomolar concentrations of silicic acid in seawater.

Beta silicomolybdic acid was formed by the reaction of silicic and molybdic acids at low pH. The combined acid was extracted into n-butanol and reduced with a mixture of p-methylaminophenol sulphate and sulphite. After phase separation, the samples were cleared by chilling in ice followed by centrifugating, and the colour was determined spectrophotometrically at 810 nm. The molar absorbance of the mixed acid was 229 000 in seawater and the precision was plus or minus $2.5 \, nmol \, l \, 2^{-1}$ silicon for concentrations less than or equal to standard aqueous analyses. Sensitivity in seawater was 70% of that in distilled water because of a significant salt effect. Natural concentrations of arsenate, arsenite and germanic acid caused negligible interference. Phosphate interference was equivalent to $10–12 \, nmol \, l^{-1}$ silicon over a broad range of phosphate concentrations.

Flow injection analysis has been used [141] to determine reactive silicate in seawater in amounts down to 0.5 μmol l^{-1} Si with a relative precision of better than 1% at concentrations above 10 μmol l^{-1} Si.

2.14 pH

Yaakov and Ruth [137,138] have described an improved *in situ* pH sensor for oceanographic and limnological applications. They report that an accuracy of 0.002 pH can be achieved.

Khoo *et al.* [139] have reported on the measurement of standard potentials, hence hydrogen concentrations in seawaters of salinities between 20 and 45o/oo at 5–40°C.

2.15 Suspended solids

Suspended solids determinations on non-saline samples are a routine procedure that does not present any problems. The solids, after filtration on a 0.45 nm glass fibre disc are dried at 105°C to obtain the moisture content and at 450–500°C to obtain the organic content. However, the application of this procedure to saline samples does present some problems, owing to errors caused by the occlusion of sea salts on the filter disc and the filtered solids. One innovation that has been adopted to correct for high solids contents on such samples has been to filter the sample through a double layer of glass fibre paper. The weight increase of the lower paper is due to the occluded salts whilst that on the top paper is due to salts plus sample solids. The corrected solids content is then obtained by subtracting the weight increase of the lower disc from that of the upper disc. Unfortunately, results obtained by this modified procedure are still unreliable becoming more so at higher salt concentrations in the sample.

A further simple innovation gives much more reliable solids results. In this the glass fibre disc mounted in its porcelain, glass or plastic holder is first wetted with distilled water and then the sample filtered through and washed with several small portions of distilled water without allowing the disc to become dry. It is believed that filling with distilled water the air spaces in the annulus of the disc trapped in the holder is the reason why better results are obtained by this procedure.

References

1. Jagner, D. and Aren, K. *Analytica Chimica Acta*, **52**, 491 (1970)
2. Meyer, R.E., Posey, F.A. and Lantz, P.M. *Journal of Electroanalytical Chemistry*, **19**, 99 (1968)
3. Noborn, O. *Japan Analyst*, **21**, 780 (1972)
4. Grasshoff, K. and Wenck, A. *Mercantile Journal Cons. Perm. International Exploration*, **34**, 522 (1972)
5. Kolthoff, J.M. and Yutzy, H. *Industrial and Chemical Engineering Chemistry, Analytical Editor*, **9**, 75 (1937)
6. Foti, S.C. *Report of the Atomic Energy Commission, USA*, AD 734383 (1971)
7. Rose, O. and Cuttita, A. *Advanced X-Ray Analysis*, **11**, 23 (1968)
8. Liebhafsky, H.A., Pfeiffer, H.G. and Zermany, P.D. *Applied Spectroscopy*, **26**, 311 (1972)
9. Walters, F.H. *Analytical Letters*, **17**, 1681 (1984)
10. Crompton, T.R. *Chemical Analysis of Organo-metallic Compounds*, Vol. 5, Academic Press, New York, p. 132

11. Sugawara, A. *Analytical Abstracts*, **3**, 1169 (1956)
12. Matthews, A.D. and Riley, J.P. *Analytica Chimica Acta*, **51**, 295 (1970)
13. Shizuo, T. *Analytica Chimica Acta*, **55**, 444 (1971)
14. Truesdale, V.W. *Marine Chemistry*, **6**, 253 (1978)
15. Truesdale, V.W. *Marine Chemistry*, **6**, (1978)
16. Truesdale, V.W. *Deep Sea Research*, **21**, 761 (1974)
17. Truesdale, V.W. and Smith, J.W. *Analyst (London)*, **100**, 111 (1975)
18. Truesdale, V.W. and Smith, J.S. *Marine Chemistry*, **7**, 133 (1979)
19. Herring, J.R. and Liss, P.S. *Deep Sea Research*, **21**, 777 (1974)
20. Schnepfe, M.M. *Analytical Chimica Acta*, **58**, 83 (1972)
21. Fletsch, R.A. and Richards, F.A. *Analytical Chemistry*, **42**, 1435 (1970)
22. Greenhalgh, R. and Riley, J.P. *Analytica Chimica Acta*, **25**, 679 (1961)
23. Wilkniss, P.K. and Linnenbon, W.J. *Limnology and Oceanography*, **13**, 530 (1968)
24. Brewer, P.G., Spencer, D.W. and Wilkniss, P.E. *Deep Sea Research*, **17**, 1 (1970)
25. Warner, T.B. *Science*, **165**, 178 (1969)
26. Windom, H.L. *Limnology and Oceanography*, **16**, 806 (1971)
27. *Orion, Research Analytical Methods Guide*. 6th edn, August (1973)
28. Anfalt, T. and Jagner, D. *Analytical Chimica Acta*, **53**, 13 (1971)
29. Rix, C.J., Bond, A.H. and Smith, J.D. *Analytical Chemistry*, **48**, 1236 (1976)
30. Jagner, T. *Analytica Chimica Acta*, **50**, 15 (1970)
31. Barendrecht, E. In *Electroanalytical Chemistry*, (ed. A.J. Bond), Vol. 2, Marcel Dekker, New York, p. 53 (1967)
32. Gran, G. *Analyst (London)*, **77**, 661 (1952)
33. Brand, H.J.D. and Rechnitz, G.A. *Analytical Chemistry*, **42**, 1172 (1970)
34. Baumann, E.W. *Analytica Chimica Acta*, **42**, 127 (1970)
35. Craggs, A., Moody, G.J. and Thomas, J.D.R. *Journal of Chemical Education*, **81**, 541 (1974)
36. Liberti, A. and Mascini, M. *Analytical Chemistry*, **41**, 676 (1969)
37. Baumann, E.W. *Analytica Chimica Acta*, **54**, 189 (1971)
38. Frani, M.S. and Ross, J.W. *Science*, **154**, 1553 (1966)
39. Riley, J.P. and Chester, R. In *Introduction to Marine Chemistry*, (eds J.P. Riley and R. Chester) Academic Press, London, pp. 70–80 (1971)
40. Bond, A.M. and Hefter, G. *Journal of Inorganic and Nuclear Chemistry*, **33**, 429 (1971)
41. Greenhalgh, R. and Riley, J.P. *Analytica Chimica Acta*, **25**, 179 (1961)
42. Warner, T.B. *Deep Sea Research*, **18**, 1255 (1971)
43. The Chemical Society (London)*Special Publication*, No. 17 (1964)
44. Prochazkova, L. *Analytical Chemistry*, **36**, 865 (1964)
45. Johnston, R. *International Conference on Exploration of the Sea*, c.m. 1966, N:11 (1966)
46. Strickland, J.D.H. and Austin, K.H. *Journal of Conservation of International Mercantile Exploration*, **24**, 446 (1959)
47. Newell, B.S. *Journal of the Marine Biological Association, UK*, **47**, 271 (1967)
48. Emmet, R.T. *Naval Ship Research and Development Centre, Report 2570* (1968)
49. Koroleff, F. Information on techniques and methods for sea water analysis. *Interlaboratory Report*, **3**, 19 (1970)
50. Solorzano, L. *Limnology and Oceanography*, **14**, 799 (1969)
51. Head, P.C. *Deep Sea Research*, **18**, 531 (1971)
52. Grasshoff, K. and Johansson, H. *Journal of Conservation of International Mercantile Exploration*, **34**, 516 (1972)
53. Liddicoat, M.I., Tibbits, S. and Butler, E.I. *Limnology and Oceanography*, **20**, 131 (1975)
54. Slawyk, G. and MacIsaac, J.J. *Deep Sea Research*, **19**, 521 (1972)
55. Benesch, R. and Mangelsdorf, P. *Helgolander wiss Meeresunters*, **23**, 365 (1972)
56. Harwood, J.E. and Huyser, D.J. *Water Research*, **4**, 695 (1970)
57. Matsunaga, K. and Nishimura, M. *Japan Analyst*, **29**, 993 (1971)
58. Richards, F.A. and Kletsch, R.A. In *Recent Researches in the fields of Hydrosphere, Atmosphere and Nuclear Geochemistry* (eds Y. Maysho and T. Kryama), Maruzen, Tokyo, pp. 65, 81 (1964)
59. Matsunaya, K. and Nishimura, M. *Analytica Chimica Acta*, **73**, 204 (1974)
60. Harwood, J.E. and Kuhn, A.L. *Water Research*, **4**, 805 (1970)
61. Degobbis, D. *Limnology and Oceanography*, **18**, 146 (1973)
62. Dal Pont, G., Hogan, M. and Newell, B. *Laboratory Techniques in Marine Chemistry. II. Determination of Ammonia in Sea water and the Preservation of Samples for Nitrate Analysis.* Commonwealth Scientific and Industrial Research Organization (Australia) Division of Fisheries and Oceanography Report No. 55, Marine Laboratory Cromella, Sydney (1974)

63. Berg, B.R. and Abdullah, M.I. *Water Research*, **11**, 637 (1977)
64. Hampson, B.L. *Water Research*, **11**, 305 (1977)
65. Grasshof, K. and Johanssoen, H. *Journal of Conservation and International Mercantile Exploration*, **36**, 90 (1974)
66. Mantoura, R.F.C. and Woodward, E.M.S. *Estuarine Coastal and Shelf Science*, **17**, 219 (1983)
67. Loder, T.C. and Gilbert, P.M. *Blank and Salinity Corrections for Automated Nutrient Analysis of Estuarine and Sea Waters*, University of New Hampshire Contribution, UNH-SG-JR-101 to Technicon International Congress, December 13–15 (1976)
68. Grasshoff, K. and Johanssen, H. *Journal du Conseil International pour L'exploration de la Mer*, **34**, 516 (1972)
69. Berg, B.R. and Abdullah, M.F. *Water Research*, **11**, 637 (1977)
70. Truesdale, V.W. *Analyst (London)*, **96**, 584 (1971)
71. Le Corre, P. and Treguer, P. *Journal du Conseil*, **38**, 147 (1978)
72. Matsumaga, K. and Nishimura, M. *Analytica Chimica Acta*, **73**, 204 (1974)
73. McLean, J.D., Stenger, V.A., Reim, R.E., Long, M.W. and Hiller, T.A. *Analytical Chemistry*, **50**, 1309 (1978)
74. Gilbert, T.R. and Clay, A.M. *Analytical Chemistry*, **45**, 1757 (1973)
75. Degobbis, D. *Limnology and Oceanography*, **18**, 146 (1973)
76. Willason, S.W. and Johnson, K.S. *Marine Biology*, **91**, 285 (1986)
77. Spencer, D.W. and Brewer, P.G. *Critical Review of Solid State Service*, **1**, 409 (1970)
78. Pett, S.K.W. *Water Research Centre Technical Report*, TR 120 (1979)
79. Matsunaga, K., Oyama, T. and Nishimura, M. *Analytica Chimica Acta*, **58**, 228 (1972)
80. Gianguzza, A. and Orecchio, S. *Inquinamento*, **24**, 94 (1982)
81. Wada, E. and Hattori, A. *Analytica Chimica Acta*, **56**, 233 (1971)
82. Matsunaga, K. and Nishimura, N. *Analytica Chimica Acta*, **45**, 350 (1969)
83. Malhotra, S.K. and Zanoui, A.D. *Journal of the American Water Works Association*, **62**, 568 (1970)
84. Lambert, R.S. and Du Bois, R.J. *Analytical Chemistry*, **43**, 955 (1971)
85. Mertens, J. and Massart, D.L. *Bulletin of the Belgium Chemical Society*, **80**, 151 (1971)
86. Hoather, R.C. Applications of spectrophotometry for the examination of waters. *Proceedings of the Society of Water Treatment and Examination*, **2**, 9 (1953)
87. Hoather, R.C. and Rackman, R.F. *Analyst (London)*, **84**, 548 (1959)
88. Goldman, E. and Jacobs, R.J. *Journal of the American Waterworks Association*, **53**, 187 (1961)
89. Morries, E. *Proceedings of the Society of Water Treatment and Examination*, **20**, 132 (1971)
90. Brown, L. and Bellinger, E.L. *Water Research*, **12**, 223 (1978)
91. Garsede, C. *Marine Chemistry*, **11**, 159 (1982)
92. Cox, R.D. *Development of analytical methodologies for parts per billion level determination of nitrate, nitrite and N-Nitroso group content*, PhD Thesis, University of Iowa (1980)
93. Anderson, L. *Analytica Chimica Acta*, **110**, 123 (1979)
94. Ranger, C.B. *Analytical Chemistry*, **53**, 20A (1981)
95. Ruzicka, J. and Hansen, E.H. *Analytical Chemistry*, **99**, 37 (1978)
96. Johnson, K.S. and Petty, R.L. *Limnology and Oceanography*, **28**, 1260 (1983)
97. Betteridge, D. *Analytical Chemistry*, **50**, 832A (1978)
98. Bajic, S.J. and Jaselskis, B. *Talanta*, **32**, 115 (1985)
99. Hydes, D.J. and Hill, N.C. *Estuarine, Coastal and Shelf Science*, **21**, 127 (1985)
100. Tyree, S.Y. and Bynum, M.A.O. *Limnology and Oceanography*, **29**, 1337 (1984)
101. Pakalus, P. and McAllister, B.R. *Journal of Marine Research*, **30**, 305 (1972)
102. Hosokawa, I. and Ohshima, F. *Water Research*, **7**, 283 (1973)
103. Isaeva, A.B. *Zhur Analit Khim*, **24**, 1854 (1969)
104. Ducret, L. and Drouillas, M. *Analytica Chimica Acta*, **21**, 86 (1959)
105. Sudakov, F.P., Klitina, V.I. and Ya. Dan'shova, T. *Zhur Analit Khim*, **21**, 1333 (1966)
106. Babko, A.K., Shkaravskii, Yu.F. and Kulik, V.I. *Zh. Analit. Khim.*, **21**, 196 (1966)
107. Matsuo, T., Shida, J. and Kurihara, W. *Analytica Chimica Acta*, **91**, 385 (1977)
108. Kirkbright, G.S., Narayanaswamy, R. and West, T.S. *Analytical Chemistry*, **43**, 1434 (1971)
109. Motomizu, S., Wakimoto, T. and Toei, K. *Analytical Chemistry*, **138**, 329 (1982)
110. Motomizu, S., Wakimoto, T. and Toel, K. *Talanta*, **31**, 235 (1984)
111. Johnson, K.S. and Petty, R.L. *Analytical Chemistry*, **54**, 1185 (1982)
112. Murphy, J. and Riley, J.P. *Analytica Chimica Acta*, **27**, 31 (1962)
113. Flynn, W.W. and Meehan, W. *Analytica Chimica Acta*, **63**, 483 (1973)
114. Haywood, M.G. and Riley, J.P. *Analytica Chimica Acta*, **85**, 219 (1976)
115. Chaime, F.E. and Zeitlin, H. *Separation Science*, **9**, 1 (1974)

116. Johnson, O. *Analytical Abstracts*, **22**, 2768 (1972)
117. Jagner, D. *Analytica Chimica Acta*, **52**, 483 (1970)
118. Lebel, J. *Marine Chemistry*, **9**, 237 (1980)
119. Luther, G.W. and Meyarson, A.L. *Analytical Chemistry*, **47**, 2058 (1975)
120. Keir, R.S. Kounaves, S.P. and Zirino, A.L. *Analytica Chimica Acta*, **91**, 181 (1977)
121. Johannson, A., Johansson, S. and Gran, G. *Analyst (London)*, **108**, 1086 (1972)
122. Stuart, W.A. and Lister, A.R. *Process Technol. Div. Atomic Energy Research Establishment AERE Harwell Berks, UK, Report UK Atomic Energy Authority, AERE-M2250-5PP*
123. Hansson, I. and Jagner, D. *Analytica Chimica Acta*, **65**, 363 (1973)
124. Gran, G. *Analyst (London)*, **77**, 661 (1952)
125. Culberson, C., Pytkowicz, R.M. and Hawley, J.E. *Journal of Marine Research*, **28**, 15 (1970)
126. Pearson, F. *Journal of Water Pollution Control*, **53**, 1243 (1981)
127. Dickson, A.G. *Deep Sea Research*, **28A**, 609 (1981)
128. Sass, E. and Ben-Yaakov, S. *Marine Chemistry*, **5**, 183 (1977)
129. Bates, R.G. *Determination of pH–Theory and Practice*, Wiley, New York, (1964)
130. Ben Yaakov, S. and Lorch, Y. *Marine Chemistry*, **13**, 293 (1983)
131. Ben-Yaakov, S., Raviv, R., Guterman, H., Dayan, A. and Lazar, B. *Talanta*, **29**, 267 (1982)
132. Liss, P.S. and Spencer, C.P. *Journal of the Marine Biological Association, UK*, **49**, 589 (1969)
133. Strickland, P. and Parsons, O. *Bulletin of the Fisheries Research Board, Canada*, **125**, 50–55 (1965)
134. Armstrong, A. *Journal of the Marine Biology Association of the UK*, **30**, 149 (1951)
135. Grasshoff, O. *Oceanographic Abstracts*, **11**, 597 (1964)
136. Thomsen, J. and Johnson, K.S. *Analytical Chemistry*, **55**, 2378 (1983)
137. Yaakov, S.B. and Ruth, E. *Limnology and Oceanography*, **19**, 144 (1974)
138. Yaakov, S.B. and Ruth, E. *Water Pollution Abstracts*, Abstract No. 1757, **42** (1969)
139. Khoo, K.H., Ramette, W., Culbeeson, C.H. and Bates, R.G. *Analytical Chemistry*, **49**, 29 (1977)
140. Brzezinski, M.A. and Nelson, D.M. *Marine Chemistry*, **19**, 139 (1986)
141. Thomson, J., Johnson, K.S. and Petty, R.L. *Analytical Chemistry*, **55**, 2378 (1983)

Chapter 3

Dissolved gases

Spencer and Brewer [1] have reviewed methods for the determination of dissolved gases in seawater (O_2, CO_2, N_2, H_2S, inert gases).

3.1 Free chlorine

Amperometric titration procedures

Although the amperometric titrimetric method has been accepted as a standard method for the determination of total residual chlorine in chlorinated effluents [2], recent reports [3,4] have suggested that in the case of chlorinated seawater, significant errors may occur if certain precautions are not taken. Furthermore, somewhat opposing views were presented in these reports on what the optimal procedure may be.

When molecular chlorine or hypochlorine is added to seawater, the following rapid reactions (reactions 3.1 and 3.2) ensue and hypobromite is formed [5]:

$$Cl_2 + H_2O \rightleftharpoons H^+ + Cl^- + HClO \tag{3.1}$$
$$HClO + Br^- \rightarrow HBrO + Cl^- \tag{3.2}$$

There is enough bromide in seawater at 35‰ salinity to convert $60\,mg\,l^{-1}$ of Cl_2 to hypobromite.

In the amperometric titration for the determination of total residual chlorine in seawater, tri-iodide ion are generated by the reaction between hypochlorite and/ or hypobromite with excess iodide at pH 4 (reactions 3.3 and 3.4). The pH is buffered by adding an acetic acid–sodium acetate buffer to the sample.

$$HXO + H^+ + 2I^- \rightarrow I_2 + H_2O + X^- \tag{3.3}$$
$$I_2 + I^- \rightarrow I_3^- \tag{3.4}$$

where X may be Cl or Br. The tri-iodide ion is then titrated with phenyl arsine oxide (reaction 3.5) and the end-point is determined amperometrically.

$$C_6H_5AsO + I_3^- + 2H_2O \rightarrow C_6H_5As(OH)_2O + 2H^+ + 3I^- \tag{3.5}$$

Carpenter, Moore and Macalady [3] suggest that this method may have underestimated the true value by a factor of three or more. The error was proposed to be due to the oxidation of iodide to iodate by molecular bromine and/or hypobromous acid (reaction 3.6).

$$3HBrO + I^- \rightarrow IO_3^- + 3H^+ + 3Br^- \tag{3.6}$$

and iodate does not revert directly with phenylarsine oxide and, at pH 4, the reaction between iodate and excess iodide to form tri-iodide ion (reaction 3.7) is sluggish:

$$IO_3^- + 8I^- + 6H^+ \rightarrow 3I_3^- + 3H_2O \tag{3.7}$$

Thus an apparent loss in total residual chlorine may be observed.

Carpenter, Moore and Macalady [3] proposed that, in order to overcome this difficulty, higher acidity and higher potassium iodide concentrations or a back titration procedure should be used.

Goldman, Quinby and Capuzzo [4], on the other hand, suggest that the order of the addition of the reagents for generating tri-iodide ions is crucial for obtaining accurate results. If the acidic buffer is added first, at pH 4, molecular bromine may be formed (reaction 3.8):

$$HBrO + 2H^+ + Br^- \rightarrow Br_2 + H_2O \tag{3.8}$$

and its loss by volatilization may cause an apparent loss in total residual chlorine. Thus they recommended that potassium iodide should be added before the acidic buffer solution. When potassium iodide is added to the chlorinated solution, hypoiodite will be formed (reaction 3.9).

$$HXO + I^- \rightarrow HIO + X^- \tag{3.9}$$

The disproportionation of hypoiodite to form iodate (reaction 3.10) is believed to be slow in slightly alkaline solutions such as seawater [6,7].

$$3IO^- \rightarrow IO_3^- + 2I^- \tag{3.10}$$

Upon subsequent acidification at pH 4, hypoiodite is converted to molecular iodine (reaction 3.11)

$$2HIO + 2H^+ \rightarrow 2H_2O + I_2 \tag{3.11}$$

which is, in turn, converted to tri-iodide ion (reaction 3.4). Since tri-iodide is a heavy ion, its loss by volatilization will be negligible if the titration follows the addition of the reagents promptly. The presently accepted standard procedure [2] calls for the addition of potassium iodide before the acidic buffer solution. However, this point is not stressed and its importance is never discussed. Goldman, Quinby and Capuzzo [4] further reported that the backward titration procedure does not significantly increase the total residual chlorine concentration.

Since reaction 3.6 is favoured in basic solutions, if Carpenter, Moore and Macalady [3] are correct, the longer molecular bromine and/or hypobromous acid are allowed to react with iodide at the natural pH 8 of seawater, the greater the amount of iodate will be formed. Consequently, there will be a greater apparent loss of total residual chlorine. This would imply that adding the acidic buffer prior to the addition of potassium iodide may minimize this source of error, because in an acidic solution reaction 3.6 will be forced to the left. On the other hand, if Goldman, Quinby and Capuzzo [4] imply that, if acid is added first, the apparent loss of total residual chlorine should increase with faster stirring rate and with longer time for stirring between the addition of the two reagents. The apparently contradictory implications for these two hypotheses were tested by Wong [8].

Wong [8] found that in the determination of residual chlorine in seawater by the amperometric titrimetric method, potassium iodide must be added to the sample before the addition of the pH 4 buffer and the addition of these two reagents should not be more than 1 minute apart. Serious analytical error may arise if the order of the addition of the reagents is reversed. There is no evidence suggesting the formation of iodate by the reaction between hypobromine and iodide. For

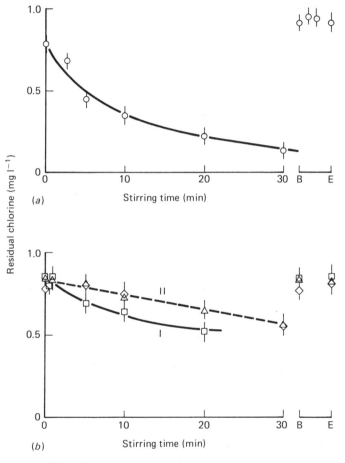

Figure 3.1 The effect of stirring time on the residual chlorine concentration in a borate solution with bromide determined by amperometric titrations. The concentration is that of a 20 ml aliquot after dilution to 200 ml and various treatments as described in the text. B and E represent titrations with the reference method at the beginning and end of the series of titrations. Data points between B and E represent titrations during the experiment. (a) The pH 4 buffer was added before the addition of potassium iodide; samples were stirred in titrator. (b) Potassium iodide was added before the addition of the acidic buffer. □ = Samples stirred in titrator; △ = samples stirred on stirrer; ◊ = samples stirred on stirrer and titrated at pH 2 (Reproduced from Wong, G.T.F. (1980) *Water Research*, **14**, 51, Pergamon Publications, by courtesy of author and publishers.)

concentrations of residual chlorine below $1\,mg\,l^{-1}$ iodate, which occurs naturally in seawater, may cause serious analytical uncertainties.

In the sodium borate solution containing bromide, when the pH 4 buffer is added before the potassium iodide solution, titrations give low total residual chlorine concentrations (Figure 3.1(a)). This loss increases with the amount of stirring time between the addition of the reagents. Even for a stirring time of 10 s, there is a loss of about 17% of the total residual chlorine. If the solution were stirred for 30 min, 85% of the total residual chlorine would have disappeared. The concentration of total residual chlorine determined by the reference method does not change throughout the experiment. This implies that this loss of chlorine does not occur in the reaction vessel, but in the titration cell as a result of the analytical procedure.

If potassium iodide is added first, and then the solution is stirred, acidified and titrated, the loss of residual chlorine is reduced although it is still significant (Figure 3.1(b)). The loss again increases monotonically with stirring time. About 38% of the residual chlorine is lost after the sample is stirred for 20 min. However, for a stirring time of 1 min or less, the loss is not detectable within the uncertainty of the analytical method. There is a loss of chlorine whether the sample is stirred in the titrator or on a stirrer (Figure 3.1(b)) although the loss seems smaller in the latter case. For a stirring time of 20 min, only 24% of the residual chlorine is lost. Moreover, titrations performed at pH 2 and pH 4 yield the same residual chlorine concentrations.

Wong [8] reported that the losses of chlorine are not related to the formation of iodate by the oxidation of iodide by hypobromite. The presence of iodate in seawater may cause significant uncertainty in the determination of small quantities of residual chlorine in water. Determinations of residual chlorine at the $0.01\,mg\,l^{-1}$ level are of questionable significance.

Carpenter and Smith [9] and Wong [8] pointed out that iodate is also a natural constituent of seawater. Thus it may cause a variable positive blank of up to $0.1\,mg\,l^{-1}$ $(1.4\,\mu equiv\,l^{-1})$ residual chlorine. This source of error may be safely neglected only for concentrations above $2\,mg\,l^{-1}$ $(28\,\mu equiv\,l^{-1})$. Carpenter and Smith proposed that iodate may be converted completely to triiodide by using a lower pH or a larger excess of iodide in the titration. However, other side reactions such as the oxidation of iodide by nitrite may interfere with the analysis under these modified conditions. As this presumably standard method is being applied to samples with lower and lower concentrations, interferences and blanks previously considered minor may become significant.

Wong [10,11] has studied this in further detail and has found that carrying out the titration at pH 2 yields a true concentration of total residual chlorine after correction for naturally occurring iodate.

The effectiveness of sulphamic acid in this method for the removal of the nitrite interference is shown in Figure 3.2. In this experiment, all the solutions contained $30\,\mu mol\,l^{-1}$ nitrite and about $0.5\,\mu mol\,l^{-1}$ of iodate. The absorbance of the solution at 353 nm decreased with increasing amounts of added sulphamic acid. A constant absorbance was recorded when 3 ml or more of 1% (w/v) sulphamic acid was added to the solution, and this absorbance was identical with that in a sample containing the same amount of iodate and no nitrite. A concentration of nitrite of $30\,\mu mol\,l^{-1}$ is unlikely to occur in estuarine water and seawater:

$$NH_2SO_3^- + NO_2^- \rightarrow N_2 + SO_4^{2-} + H_2O$$

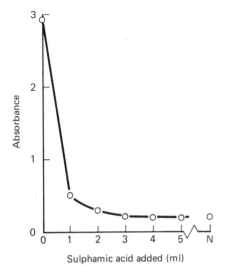

Figure 3.2 Absorbance of tri-iodide at 353 nm formed from a fixed concentration of iodate in the presence of 30 μmol^{-1} nitrite upon the addition of various volumes of 1% (w/v) sulphamic acid; N denotes the case where neither nitrite nor sulphamic acid was present in the solution (Reproduced from Wong, G.T.F. (1982) *Environmental Science and Technology*, **16**, 51, American Chemical Society, by courtesy of author and publishers.)

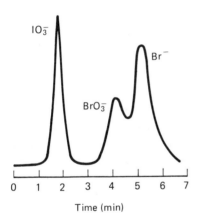

Figure 3.3 Determination of iodate, bromate and bromide (Reproduced from Carison, R.M. and Weberg, R.T. (1983) *Chemosphere*, **12**, 125, Pergamon Publications, by courtesy of authors and publishers.)

Carlson and Weberg [12] have also studied the interference from iodate during the iodometric determination of residual chlorine in seawater. These workers confirmed that due to the presence of naturally occurring iodate, results from the iodometric determinations carried out at pH 2 were up to 20% higher than those obtained in determinations carried out at pH 4. These workers seem to have been unaware of the results obtained by Wong as discussed above.

Carlson and Weberg [12] used mobile phase ion-chromatography to separate iodate, bromate, bromide and iodide (Figure 3.3). Ion-chromatography of a mixture of iodide and bromide which had been treated with hypochlorite clearly showed the presence of iodate and the absence of bromate. These results demonstrated that iodate generates iodine from iodide very quickly at pH 2 ($IO_3^- + 5I^- + 6H^+ \rightarrow 3I_2 + 3H_2O$) and relatively slowly at pH 4; while bromate fails to react with iodide at pH 4 and does so only slowly at pH 2 ($BrO_3^- + 6I^- + 6H^+ +$

$Br^- + 3I_2 + 3H_2O$). Moreover, chromatographic analysis (C-18 column) of an iodate solution $(10^{-4} M)$ after iodide addition (24 min pH 4) confirmed the continued presence of iodate. The concentration of iodate after this period (40% of original) agrees with the corresponding titration results.

The practical application of these observations is to minimize the effect of iodate by rapidly carrying out the iodometric titration of chlorine residual in seawater at pH 4. Moreover, if desired, a titration correction curve can be generated using iodate at the specific concentration of iodide in the sample in question as there appears to be a complete conversion of seawater iodide to iodate in the presence of excess chlorine.

Other methods for the determination of chlorine in seawater or saline waters are based on the use of barbituric acid [13] and on the use of residual chlorine electrodes [14] or amperometric membrane probes [15,16]. In the barbituric acid method [12], chlorine reacts rapidly in the presence of bromides and has completely disappeared after 1 minute. This result, verified in the range pH 7.5–9.4, proves the absence of free chlorine in seawater. A study of the colorimetric deterioration of free halogens by the diethylparaphenylene-diamine technique shows that the titration curve of the compound obtained is more like the bromine curve than that of chlorine. The author suggests drawing a calibration curve for different concentrations of chlorine in solution containing 60 mg bromide per litre.

The work carried out by Dimmock and Midgley [14–16] at the Central Electricity Board, UK is concerned with analysis of cooling waters, saline and non-saline. Although the analysis of seawater is not specifically discussed their work may be of some relevance.

3.2 Ozone

Sheeter [17] has discussed an ultraviolet method for the measurement of ozone in seawater. Crecelius [18] has discussed oxidation products obtained (bromine, hypobromous acid, bromate) when bromides in seawater are oxidized by ozone.

3.3 Nitric oxide

Cohen [19] used electron capture gas chromatography to determine traces of dissolved nitric oxide in seawater. Precision and accuracy are, respectively, 2% and 3%.

3.4 Hydrogen sulphide

Flow injection analysis has been used for the automated determination of hydrogen sulphide in seawater [20]. A low sensitivity flow injection analysis manifold for concentrations up to $200 \, \mu mol \, l^{-1}$ hydrogen sulphide had a detection limit of $0.12 \, \mu mol \, l^{-1}$. Sulphide standards were calibrated by colorimetric measurement of the excess tri-iodide ion remaining after reaction of sulphide with iodine. The coefficient of variation was less than 1% at concentrations greater than $10 \, \mu mol \, l^{-1}$. The method was fast, accurate, sensitive enough for most natural waters and could be used both for discrete and for continuous analysis.

References

1. Spencer, D.W. and Brewer, P.C. *Critical Review Solid State Science*, **1**, 409 (1970)
2. Rand, M.C., Greenberg, A.E. and Taras, M.J. (eds) *Standard Methods for the Examination of Water and Waste Water*, 14th edn, American Public Health Association, Washington, DC (1976)
3. Carpenter, J.H., Moore, C.A. and Macalady, D.L. *Environmental Science and Technology*, **11**, 992 (1977)
4. Goldman, J.C., Quinby, H.L. and Capuzzo, J.M. *Water Research*, **13**, 315 (1979)
5. Wong, G.T.F. and Davidson, J.A. *Water Research*, **11**, 971 (1977)
6. Li, C.H. and White, C.F. *Journal of the American Chemical Society*, **65**, 335 (1943)
7. Truesdale, V.W. *Deep Sea Research*, **21**, 761 (1974)
8. Wong, G.T.F. *Water Research*, **14**, 51 (1980)
9. Carpenter, J.H. and Smith, C.A. In *Water Chlorination–Environmental Impact and Health Effects* (eds R.L. Jolley, H. Gorchev and D.H. Hamilton), Vol. 2, Ann Arbor Science, Ann Arbor, MI, pp. 195–208 (1978)
10. Wong, G.T.F. *Environmental Science and Technology*, **16**, 785 (1982)
11. Wong, G.T.F. *Deep Sea Research*, **24**, 115 (1977)
12. Carlson, R.H. and Weberg, R.T. *Chemosphere*, **12**, 125 (1977)
13. Figuet, J.M. *Techniques et Sciences, Municipales*, **73**, 239 (1978)
14. Dimmock, N.A. and Ridgeley, D. *Talanta*, **29**, 557 (1982)
15. Dimmock, N.A. and Midgley, D. *Water Research*, **13**, 1101 (1979)
16. Dimmock, N.A. and Midgley, D. *Water Research*, **13**, 1317 (1979)
17. Sheeter, H. *Water Research*, **7**, 729 (1973)
18. Crecelius, E.A. *Journal of the Fisheries Research Board, Canada*, **36**, 1006 (1979)
19. Cohen, Y. *Analytical Chemistry*, **49**, 1238 (1977)
20. Sakamoto, O., Arnold, C.M., Johnson, V.S. and Beehler, C.L. *Limnology and Oceanography*, **31**, 894 (1986)

Chapter 4

Metals

Many of the published methods for the determination of metals in seawater are concerned with the determination of a single element. Single-element methods are discussed firstly in Section 4.1 by alphabetical order. However, much of the published work is concerned not with the determination of a single element but with the determination of groups of elements (Sections 4.2–4.9). This is particularly so in the case of new techniques such as graphite furnace atomic absorption spectrometry (Section 4.2), Zeeman background corrected atomic absorption spectrometroscopy (Section 4.3) and inductively coupled plasma spectrometry (Section 4.4). This also applies to other techniques, such as voltammetry, polarography (Section 4.5), neutron activation analysis (Section 4.6), X-ray fluorescence (Section 4.7), spectroscopy and isotope dilution techniques (Section 4.8).

The background concentrations at which metals occur in seawater is extremely low and much work has been done on pre-concentration procedures in attempts to lower detection limits for these metals. Various pre-concentration techniques, including hydride generation used before atomic absorption spectrometry, are discussed in Section 4.9.

Methods for determining metals in seawater have been published by the Standing Committee of Analysts (i.e., the blue book series; HMSO, London); they are not reproduced in this book as they are available elsewhere. These methods are based on chelation of the metals with an organic reagent followed by atomic absorption spectroscopy. Revisions of these methods are in the course of preparation, but were not published at the time of writing.

Two alternative methods are available for most elements: atomic absorption spectroscopy and scanning voltammetry.

4.1 Single items

Alkaline earths

Calcium
Blake, Bryant and Waters [1] have described a flame photometric method for the determination of calcium in solutions of high sodium content. The method was applied to simulated seawater. In the method Chelex-100 chelating resin (Na^+ form) (20 g) is stirred with 2 N hydrochloric acid (15 ml) for 5 min, the acid is

decanted and the resin is washed with water (2 × 25 ml), stirred with 2 N sodium hydroxide (15 ml) for 5 min and again washed with water (2 × 25 ml). The procedure is repeated five times then the resin is dried at 100°C. A neutral solution (100 ml) containing up to 50 p.p.m. of calcium and up to 4% of sodium is passed through a column of the resin, a specified amount of hydrochloric acid (pH 2.4) is passed through and the percolate containing the sodium is discarded. Elution is then effected with 2 N hydrochloric acid (5 ml) and the column is washed with water (25 ml), the combined eluate and washings are diluted to 100 ml and calcium is determined by flame photometry at 622 nm. There is no interference from magnesium, zinc, nickel, barium, mercury, manganese copper or iron present separately in concentrations of 25 p.p.m. or collectively in concentrations of 5 p.p.m. each. Aluminium depresses the amount of calcium found.

Jagner [2] used computerized photometric titration in a high precision determination of calcium in seawater.

Calcium is titrated with EGTA (1,2-bis-(2-aminoethoxyethane NN$N'N'$-tetra-acetic acid) in the presence of the zinc complex of zincon as indirect indicator for calcium. Theoretical titration curves are calculated by means of the computer program HALTAFALL in order to assess accuracy and precision. The method gives a relative precision of 0.00028 when applied to estuarine water of 0.05–0.35% salinity.

The complexometric titration is at present considered to be the best method for the determination of calcium but investigators have differed in the end-point detection technique used and in their evaluation of interference by other alkaline earth elements. Studies using different end-point techniques, some of which also considered magnesium to calcium ratios in seawater, do not agree on the effect of magnesium on the titration of calcium with EGTA (1,2-bis(2-ammoethoxy ethane) NN,N'-N'-tetra-acetic acid). Table 4.1 lists the bindings of some of these studies;

Table 4.1 Recent reported studies on determination of calcium using EGTA titration

Reference	Method	Conclusion
3	Hg-electrode	No Mg interference at seawater ratios
4	Zn-Zincon	Positive Mg interference from Mg:Ca = 1:5 and higher
5	Zn-Zincon	No Mg interference at seawater ratios
6	Theoretical, Zn-Zincon	Titration error if Mg > Ca
7	Various chemical visual indicators	No Mg interference at seawater ratios when end-points sharp
8	Zn-Zincon	Mg interference of +0.729% on Ca titre; Sr interference of +0.388% on Ca titre
9	GHA	Mg interference of −0.23% on Ca titre; Sr interference of +0.77% on Ca titre
10	Stability constants 'conditional constants'	Sr interference; increased titration error at seawater ratios – dependent on end-point sensitivity
11	Ca-Red	Sr interference of +0.37% on Ca titre
12	Computer simulated curves of Zn-Zincon	Mg interference at seawater ratios
13	Amalgamated-Ag electrode	No Mg interference; Sr interference of = 0.9% on Ca titre
14	Ca-ion selectode	No Mg interference
15	Ca-ion selectode	No Mg interference; no Sr interference

the references cited report that magnesium has no effect, causes a positive interference, and, in one case, has a negative interference.

In most cases where strontium interference was evaluated, a positive interference was found, but the degree of correction (of the calcium titre) varied from about -0.38% in several studies to -0.77% and -0.88% in other investigations which claim that all or nearly all strontium is co-titrated.

In the light of these observations Olson and Chen [16] decided to use a correction factor for use in their visual end-point calcium titration method involving titration with EGTA. They found that interferences by magnesium and strontium were insignificant at the molar ratios normally found in seawater, but is more serious in samples containing higher ratios of magnesium or strontium to calcium. An average value of 0.02103 was obtained for the ratio of calcium to chlorinity in samples of standard seawater.

They used the titration method of Tsunogai, Nishimura and Nakaya [9]. The titrant solutions were standardized against calcium carbonate of primary standard quality (99.9975% purity) rather than zinc, and the EGTA (Eastman Chemicals) was used without further purification.

Twenty-five millilitres of a titrant strong enough to complex about 98% of the dissolved calcium were added to samples of 25 ml. GHA-propanol reagent (4 ml) and the borate buffer (4 ml) were added to this solution. This was stirred rapidly for about 3 min and the amyl-alcohol (5 ml) added. The solution was then stirred vigorously and titrated with dilute EGTA under fluorescent lighting via a micrometer piston-buret (2.5 ml capacity) until a faint pink colour remained. At this point the titration became a series of small additions with vigorous stirring followed by periods in which the immiscible layers separated and the organic layer was checked for remaining red colour. This process was continued until all the red colour was gone. Reagent blanks, analysed with each batch of samples, had 50 ml of distilled-deionized water in place of sample and initial titrant. The blank volume was subtracted from the dilute titrant volume in calculating calcium concentration. Reagent blanks were typically less than $1 \, \mu\text{mol} \, \text{l}^{-1}$.

Table 4.2 shows the actual and measured concentrations for solutions of varied calcium content and 'salinity'. Table 4.3 shows the amount of calcium measured in solutions containing calcium and each individual alkaline earth in various ratios. Table 4.4 shows the amount of calcium measured in solutions of the 'salinity' matrix and of the individual salts. The solutions of the high purity calcium carbonate were accurate to about $\pm 4 \, \mu\text{mol} \, \text{l}^{-1}$ and those of 'salinity' matrix are probably accurate to $\pm 6 \, \mu\text{mol} \, \text{l}^{-1}$. The corrected values for calcium have had the appropriate amount of calcium subtracted as indicated in Table 4.4; calcium impurities were consistent with those listed for the reagents used.

Table 4.2 shows that the presence of normal concentrations of sodium, magnesium and strontium have no net effect on the determination of calcium above the approximate level of accuracy of about 0.1% so that no correction factor seems necessary. A sufficient amount of titrant must be added to complex at least 98% of dissolved calcium before the buffer is added; this apparently reduces the loss of calcium by co-precipitation with magnesium hydroxide.

Interference effects begin to appear at higher magnesium or strontium molar ratios. Tsunogai, Nishimura and Nakaya [9] found the interference of magnesium to be negative and for strontium, related to the extraction into the organic layer of the calcium GHA complex. They found a positive interference for strontium at twice the seawater molar ratios. Therefore the interferences of the individual

Table 4.2 Study of net interference on calcium determination in artificial seawater. Major groupings of calcium concentration approximate that found at 30‰, 35‰ and 40‰ salinity

cCa*	cNa:cCa†	cMg:cCa†	10^3 cSr:cCa†	cCa* measured	cCa‡ corrected	Δ§ (%)
8845	0	0	0	8841	8841	−0.05
8841	45.6	0	0	8838	8834	−0.08
8841	45.6	5.2	8.8	8838	8833	−0.09
8943	0	0	0	8946	8946	0.03
8935	45.1	0	0	8934	8930	−0.06
8935	45.1	5.1	8.7	8934	8929	−0.07
10321	0	0	0	10322	10322	0.01
10305	45.6	0	0	10314	10310	0.05
10305	45.6	5.2	8.8	10316	10311	0.06
10423	0	0	0	10419	10419	−0.04
10415	45.1	0	0	10426	10422	0.07
10415	45.1	5.1	8.7	10426	10421	0.06
11782	0	0	0	11785	11785	0.03
11782	45.6	0	0	11786	11782	0.00
11782	45.6	5.2	8.8	11784	11775	−0.06
11902	0	0	0	11906	11906	0.03
11902	45.1	0	0	11910	11906	0.03
11902	45.1	5.1	8.7	11906	11897	−0.04

*Calcium concentrations reported in μmol l⁻¹, actual concentration based on in vacuo mass.
†All ratios are molar: ratios approximate those of seawater.
‡Correction by subtraction of appropriate blank solution calcium concentration, 4 μmol l⁻¹ from solutions containing sodium, 5 μmol l⁻¹ from the lower two 'salinity' matrices and 9 μmol l⁻¹ from the high 'salinity' matrix.
§Δ = 100 (corrected − actual)/actual.
Reproduced from Olsen, F.J. and Chen, C.T.A. (1982) *Limnology and Oceanography*, **27**, 375, courtesy of authors and publishers.

Table 4.3 Study of separate alkaline earth elements interference

Solution composition (μmol l⁻¹)	cCa measured	cCa corrected	Δ* (%)
10399 Ca	10402	10402	+0.03
10399 Ca, 54000 Mg	10396(±5)†	10395	−0.04
10399 Ca, 54000 Mg, 470000 Na	10401	10396	−0.03
10399 Ca, 108000 Mg	10340(±10)‡	10338	−0.59
10399 Ca, 108000 Mg, 470000 Na	10396	10390	−0.09
10399 Ca, 216000 Mg	10180(±20)§	10176	−2.10
10399 Ca, 216000 Mg, 470000 Na	10300(±10)	10292	−1.00
10399 Ca	10402	10402	+0.03
10399 Ca, 91 Sr	10405	10404	+0.05
10399 Ca, 91 Sr, 470000 Na	10400	10395	−0.04
10399 Ca, 182 Sr	10430(±10)‖	10429	+0.28
10399 Ca, 182 Sr, 470000 Na	10417	10411	+0.12

*Δ = 100 (corrected − actual)/actual
†End-point colour change different – less sharp.
‡End-point much less sharp – organic layer clear, bulk solution orange-pink.
§End-point much less sharp – nearly no colour extracted into organic layer after bulk titrant addition.
‖End-point colour change slightly different – greenish.
Reproduced from Olsen, E.J. and Chen, C.T.A. (1982) *Limnology and Oceanography*, **27**, 375, by courtesy of authors and publishers.

Table 4.4 Study of calcium impurities in reagents used to prepare salt matrix

Solution composition (μmol l^{-1})*	cCa measured (μmol l^{-1})
403000 Na, 46000 Mg, 78 Sr	6
470000 Na, 53000 Mg, 91 Sr	4
537000 Na, 60800 Mg, 104 Sr	9
91 Sr	≤ 1
53000 Mg	1
470000 Na	4

*The three mixed salt solutions approximate 30‰, 35‰ and 40‰ salinity.
Reproduced from Olsen, E.J. and Chen, C.T.A. (1982) *Limnology and Oceanography*, **27**, 375, by courtesy of authors and publishers.

alkaline earth elements on the calcium titration found by Olson and Chan [16] are consistent in direction, though clearly not in magnitude with those that were reported by Tsunogai, Nishimura and Nakaya [9]. The presence of sodium (chloride) in the solutions also seems to diminish these interference effects in both cases. Although no explanation was found for the reduced interference effect when sodium is present, it does suggest the advantage of either standardizing the titrant against a seawater matrix calcium standard or of having some matrix available to evaluate individual interference effects with a procedure to be used for seawater.

Calcium and magnesium
Atomic absorption spectrophotometry [17,18] and probe photometric methods [19] have been used in the determination of calcium and magnesium in seawater.

Strontium
Carr [20] has studied the effects of salinity on the determination of strontium in seawater by atomic absorption spectrometry using an air–acetylene flame. Using solutions containing 7.5 mg l^{-1} strontium and between 15 and 14% sodium chloride he demonstrated a decrease in adsorption with increasing sodium chloride concentration. To overcome this effect a standard additions procedure is recommended.

Barium
Epstein and Zander [21] used graphite furnace atomic absorption spectrometry for the direct determination of barium in sea- and estuarine water. Roe and Froelich [22] achieved a detection of 30 pg barium for 50 μl injections of seawater using direct injection graphite furnace atomic absorption spectrometry.

Total alkaline earths

Jagner and Kerstein [23,24] used computer-controlled high-precision complex-iometric titration for the determination of the total alkaline earth metal concentration in seawater. Total alkaline earths were determined by photometric titration with EDTA with Eriochrome black as indicator. The method gave a value of 63.32 mmol kg^{-1} for the total alkaline earth metal concentration in standard seawater of 3.5% salinity. The precision was about 0.01%.

Alkali metals

Sodium
Polarimetry [25] and amperometric [26] methods have been used to determine sodium in seawater. In the indirect polarimetry method [25] sodium is precipitated as the zinc uranyl acetate salt and the uranium present in the precipitate is determined polarimetrically after reaction with (+)-tartaric acid. The sample is diluted to contain 0.1–1% (w/v) of sodium. A portion (1–2 ml) is treated with saturated aqueous zinc uranyl acetate (10–20 ml) and the mixture evaporated to the volume of reagent solution added. It is cooled then the precipitate is filtered off and washed with reagent solution (5 × 2 ml) and with saturated ethanolic zinc uranyl acetate (5× 2 ml). The precipitate is dissolved in water, 1 M tartaric acid (15 ml) added, the pH adjusted to 5 with aqueous sodium hydroxide and diluted to 50 ml. The optical rotation is measured in a 20 cm tube and the sodium content of the sample determined by reference to a calibration graph which is rectilinear over the range 1.62–16.2 mg sodium per 50 ml. The maximum error was 2.2%.

In the indirect amperometric method [26] saturated uranyl zinc acetate solution is added to the sample containing 0.1–10 mg sodium. The solution is heated for 30 minutes at 100°C to complete precipitation. The solution is filtered and the precipitate washed several times with 2 ml of the reagent and then five times with 99% ethanol saturated with sodium uranyl zinc acetate. The precipitate is dissolved and diluted to a known volume. To an aliquot containing up to 1.7 mg zinc 1M tartaric acid 2–3 ml and 3M ammonium acetate 8–10 ml are added and pH adjusted to 7.5–8.0 with 2M aqueous ammonia. The solution is diluted to 25 ml and an equal volume of ethanol added. It is titrated amperometrically with $0.01\,MK_4Fe(CN)_6$ using a platinum electrode. Uranium does not interfere with the determination of sodium.

Potassium
Polarography has also been applied to the determination of potassium in seawater [27]. The sample (1 ml) is heated to 70°C and treated with 1 ml 0.1 M sodium tetraphenylborate. The precipitated potassium tetraphenylborate is filtered off, washed with 1% acetic acid and dissolved in 5 ml acetone. This solution is treated with 3 ml 0.1 M thallium nitrate and 1.25 ml 2M sodium hydroxide, and the precipitate of thallium tetraphenylborate is filtered off. The filtrate is made up to 25 ml and, after de-aeration with nitrogen, unconsumed thallium is determined polarographically. There is no interference from 60 mg sodium, 0.2 mg calcium or magnesium, 20 μg barium or 2.5 μg strontium. Standard deviations at concentrations of 375, 750 and 1125 μg potassium per ml were 26.4, 26.9 and 30.5 respectively. Results agreed with those obtained by flame photometry.

Potentiometric titration has been applied to the determination of potassium in seawater [28–30]. Torbjoern and Jaguer [28,29] used a potassium selective valinomycin electrode and a computerized semiautomatic titrator. Samples were titrated with standard additions of aqueous potassium so that the potassium to sodium ion ratio increased on addition of the titrant and the contribution from sodium ions to the membrane potential could be neglected. The initial concentration of potassium ions was then derived by the extrapolation procedure of Gran.

Marquis and Lebel [30] precipitated potassium from seawater or marine sediment pore water using sodium tetraphenylborate, after first removing halogen ions with silver nitrate. Excess tetraphenylborate was then determined by silver nitrate titration using a silver electrode for end-point detection.

The content of potassium in the sample is obtained from the difference between the amount of tetraphenyl boron measured and the amount initially added.

To test the reproducibility of the method, Marquis and Level [30] carried out a series of replicate measurements of potassium on a sample of standard seawater of 35% salinity. Table 4.5 shows that the results obtained give an acceptable K : Cl ratio of 0.0206 [31]. The standard deviation for the ten replicates is ±1.0%.

To ascertain that the precipitation of potassium is complete, i.e. that the amount of tetraphenylborate ions titrated does indeed constitute the excess and that there is no interference by other ions, different known amounts of potassium sulphate were added to 1 ml standard seawater of known potassium content. The results in Table 4.6 show that the amount of potassium recovered varies from 98 to 102%. This confirms that the recovery is quantitative and that there is no systematic variation related to the amount of potassium added.

Ward [32] evaluated various types of potassium ion-selective electrodes for the analysis of seawater. Three types of potassium ion-selective electrodes were evaluated for their suitability for continuous monitoring and *in situ* measurement applications in water of varying salinities and at temperatures of 10°C and 25°C. The three types comprised a glass membrane single electrode, a glass-membrane combination electrode and a liquid-ion exchange electrode. Although all three

Table 4.5 Replicate determinations of potassium in a seawater sample

Sample weight (g)	NaTPB added (g)	Volume of dilute $AgNO_3$ (cm^3)	$[K^+]$ (mg kg^{-1})	K/Cl (mg/‰)
1.0032	1.9892	1.753	403	0.0208
1.0016	2.0125	1.783	399	0.0206
1.0102	2.0033	1.771	398	0.0205
1.0074	1.9941	1.758	402	0.0208
1.0039	1.9924	1.764	396	0.0204
1.0065	2.0119	1.787	392	0.0202
1.0072	1.9894	1.750	405	0.0209
1.0157	1.9826	1.747	397	0.0205
1.0117	1.9886	1.753	399	0.0206
1.0027	1.9912	1.755	403	0.0208

[Strong $AgNO_3$] = 46.086 g kg^{-1}; [dilute $AgNO_3$] = 95.1928 g of strong $AgNO_3$ l^{-1}; [NaTPB] = 9.568 g kg^{-1}
Mean potassium concentration = 399 mg kg^{-1}; standard deviation = 3.8 mg kg^{-1} (1.0%)
Reproduced from Marquis. G. and Lebel. J. (1981) *Analytical Letters*, **14**, 913. Marcel Dekker, by courtesy of authors and publishers.

Table 4.6 Recovery of potassium added to 1 ml seawater samples

K^+ sample (ml)	K^+ added (mg)	K^+ total (mg)	K^+ titrated (mg)	K^+ recovery (%)
0.405	0.217	0.622	0.614	99
0.406	0.217	0.622	0.612	98
0.400	0.177	0.577	0.573	99
0.406	0.177	0.583	0.593	102
0.404	0.130	0.534	0.538	101
0.405	0.128	0.533	0.525	98

$[K^T]$ added = 0.9979 g kg^{-1}.
Reproduced from Marquis. G. and Lebel. J. (1981) *Analytical Letters*, **14**, 913. Marcel Dekker, by courtesy of authors and publishers.

electrode systems performed well in fresh water, the results obtained with the liquid-ion exchange electrode in seawater were significantly better than those with glass membranes. An accuracy of 5% could be achieved under certain conditions but response times generally exceeded 10 minutes and glass-membranes electrodes were sensitive to external motion and flow variations.

Lithium
Benzwi [33] determined lithium in Dead Sea water using atomic absorption spectrometry. The sample was passed through a 0.45 μm filter and lithium was then determined by the method of standard additions. Solutions of lithium in hexanol and 2-ethylhexanol gave greatly enhanced sensitivity.

Rona and Schmuckler [34] used gel-permeation chromatograph to separate lithium from Dead Sea brine. The elements emerged from the column in the order potassium, sodium, lithium, magnesium and calcium and it was possible to separate a lithium-rich fraction also containing some potassium and sodium but no calcium and magnesium.

Wiernik and Amiel [35] used neutron activation analysis to measure lithium and its isotopic composition in Dead Sea brines.

Rubidium
Schoenfeld and Held [36] used a spectrochemical method to determine rubidium in seawater. They determined concentrations of rubidium in the range 0.008–0.04 $\mu g\,ml^{-1}$ in the presence of varying proportions and concentrations of other salts as internal standard. The coefficient of variation ranged from 7 to 25% for simulated-seawater standards.

Caesium
Nuclear activities such as electricity production by nuclear power plants or accidents such as occurred at Chernobyl release radionuclides, including caesium, into the environment. The caesium concentration in these matrices is very low, so that in addition to a sensitive analytical method, it is necessary to make use of an enrichment technique to bring the caesium concentration within the scope of the analytical method.

Atomic-absorption spectrometry is suitable as a method of analysis of the concentrate and is applicable to radioactive and non-radioactive forms of the element. Methods of analysis of radiocaesium are discussed in Chapter 5.

For the enrichment of caesium from seawater and other types of sample Frigieri *et al.* [37] used ammonium hexacyanocobalt ferrate. This was chosen because it can be employed in strongly acidic solutions, with the exception of concentrated sulphuric acid.

Atomic absorption spectrometry has been used to determine caesium in seawater [38]. The method uses preliminary chromatographic separation on a strong cation-exchange resin, ammonium hexacyanocobalt ferrate, followed by electrothermal atomic absorption spectrometry. The procedure is convenient, versatile and reliable, although decomposition products from the exchanger, namely iron and cobalt, can cause interference.

Caesium is fully retained by a chromatographic column of ammonium hexacyanocobalt ferrate and can then be recovered by dissolution of the ammonium hexacyanocobalt ferrate in hot 12 M sulphuric acid.

As iron and cobalt both interfere with the determination of caesium, using the

852.1 nm caesium line, these elements were removed in a preliminary separation and then caesium determined.

Aluminium

Aluminium has been determined by spectrophotometric methods using aluminium [39,40] oxine [41,42] Eriochrome Cyanine R [43] and Chrome Azurol S [44] fluorimetric methods using Pontachrome Blue Black R [45,46] Lumogallion [47–49] and salicylaldehyde semicarbazone [52–54], gas chromatographic methods [55,56] emission spectroscopy [57] and neutron activation analysis [58,59]. Most of these methods necessitate pre-treatment steps and special and expensive instruments, require large volumes of sample solution and are time consuming. For instance, although the fluorimetric method using Lumogallion reported by Hydes and Liss is sensitive and rapid, the fluorescence spectrophotometer used is not as popular an instrument as the spectrophotometer. Dougan and Wilson [60] have also reported the spectrophotometric determination of aluminium (at concentrations of 0.05 and $0.3 \, mg \, l^{-1}$) in water with pyrocatechol violet, and Henriksen and co-workers [61,62] have improved the method to some extent, but these procedures are not satisfactory for the concentrations of aluminium normally found in seawater (about $2 \, \mu g \, l^{-1}$).

An example of a gas chromatographic method is that of Lee and Burrell [55]. In this method the aluminium is extracted by shaking a 30 ml sample (previously subjected to ultraviolet radiation to destroy organic matter) with $0.1 \, M$ trifluoroacetylacetone in toluene for 1 h. Free reagent is removed from the separated toluene phase by washing it with $0.01 \, M$ aqueous ammonia. The toluene phase is injected directly on to a glass column (15 cm × 6 mm) packed with 4.6% of DC 710 and 0.2% of Carbowax 20 M on Gas-Chrom Z. The column is operated at 118°C with nitrogen as carrier gas (285 ml per min) and electron-capture detection. Excellent results were obtained on 2 μl of extract containing 6 pg of aluminium.

Spencer and Sachs [63] determined particulate aluminium in seawater by atomic absorption spectrometry. The suspended matter was collected from seawater (at least 2 litres) on a 0.45 μm membrane filter, the filter was ashed, and the residue was heated to fumes with 2 ml concentrated hydrofluoric acid and one drop of concentrated sulphuric acid. This residue was dissolved in 2 ml 2 M hydrochloric acid and the solution was diluted to give an aluminium concentration in the range $5–50 \, \mu g \, l^{-1}$; potassium chloride was added to give 1 mg aluminium per ml. Atomic-absorption determination of potassium was carried out with a nitrous oxide acetylene flame. The effects of calcium, iron, sodium and sulphate alone and in combination on the aluminium absorption were studied.

Weisel, Duce and Fasching [64] determined aluminium, lead, vanadium in North Atlantic seawater after co-precipitation with ferric hydroxide.

Korenaga, Motomizum and Toei [65] have described an extraction procedure for the spectrophotometric determination of trace amounts of aluminium in seawater with pyrocatechol violet. The extraction of ion-associate between the aluminium/pyrocatechol violet complex and the quaternary ammonium salt, zephiramine (tetradecyldimethylbenzyl ammonium chloride), is carried out with 100 ml seawater and 10 ml chloroform. The excess of reagent extracted is removed by back-washing with $0.25 \, M$ sodium bromide solution at pH 9.5.

The calibration graph at 590 nm obeyed Beer's law over the range 0.13–1.34 μg aluminium. The apparent molar absorptivity in chloroform was $9.8 \times 10^4 \, l \, mol^{-1} \, cm^{-1}$.

Several ions–such as manganese, iron (II), iron (III), cobalt, nickel, copper, zinc, cadmium, lead and uranyl–react with pyrocatechol violet and to some extent are extracted together with aluminium. The interferences from these ions and other metal ions generally present in seawater could be eliminated by extraction with diethyldithiocarbamate as masking agent. With this agent most of the metal ions except aluminium were extracted into chloroform and other metal ions did not react in the amounts commonly found in seawater.

The apparent aluminium content of seawater stored in ordinary containers such as glass and polyethylene bottles decreases gradually, as shown in Figure 4.1, but

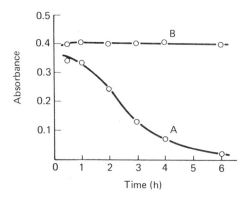

Figure 4.1 Stability of aluminium in seawater. (*a*) No sulphuric acid added; (*b*) $0.5 \, ml \, l^{-1}$ concentrated sulphuric acid added. Absorbances of both solutions measured against reagent blank. Sample: shore of Seto Inland Sea at Sanban, Okayama Prefecture, Japan, sampled on 17 September 1973. This sample contained $96 \, \mu g \, l^{-1}$ iron

if the samples are acidified with $0.5 \, ml \, l^{-1}$ concentrated sulphuric acid the aluminium content remains constant for at least 1 month. Accordingly, samples should be acidified immediately after collection. However, the aluminium could be recovered by acidifying the stored samples and leaving them for at least 5 h.

Some total aluminium results obtained for various seawater samples are given in Table 4.7.

Van der Berg, Murphy and Riley [66] determined aluminium in seawater by anodic stripping voltammetry. They give details of a procedure for the determination of dissolved aluminium in natural waters, including seawater, by complexation with 1,2-dihydroxyanthraquinone-3-sulphonic acid, collection of the complex on a hanging mercury drop electrode, and determination by cathodic stripping voltammetry. The advantages of this method over other techniques are indicated and optimal conditions are described. The total time required was 10–15 minutes per sample and the limit of detection was $1 \, nmol \, l^{-1}$ aluminium for an adsorption time of 45 seconds. No serious interferences were found, but ultraviolet irradiation was recommended for samples with a high organic content.

Table 4.7 Determination of total aluminium in seawater with pyrocatechol violet

Sample source		Date of sampling	Test volume (ml)	Aluminium content* (μg l^{-1})	Recovery† (%)
Sea	Location				
Seashore of Seto Inland Sea (Okayama Prefecture)	Shibukawa	July 16, 1977	25	21 ± 0	97.9
		May 4, 1978	50	12.4 ± 0.5	102.7
	Uno	August 24, 1978	25	30 ± 1	–
	Mizushima	August 24, 1978	25	28 ± 1	95.8
	Kojiina	August 24, 1978	25	15.9 ± 0.4	–
	Sanban	September 17, 1978	10	63 ± 2	99.1
	Kukui	September 17, 1978	10	44 ± 1	98.5
Offshore of Seto Inland Sea (Kagawa Prefecture)	Teshima	April 23, 1978	100	6.4 ± 0.2	100.6
	Naoshima	April 23, 1978	100	7.7 ± 0.3	–
Seashore of Pacific Ocean (Ehime Prefecture)	Tanohama	August 23, 1977	50	11.0 ± 0.3	–
	Funakoshi	August 29, 1977	50	13.2 ± 0.5	–
Seashore of Japan Sea (Tottori Prefecture)	Aoya	August 27, 1977	100	9.1 ± 0.4	–

*Mean ± SD (n = 3).
†0.335 μg aluminium was added to the samples.
Reproduced from Korenga, T., et al. (1980) Analyst (London), **105**, 328, Royal Society of Chemistry, by courtesy of authors and publishers.

Antimony

Sturgeon, Willie and Berman [67] preconcentrated antimony (III) and antimony (V) from coastal and seawaters by adsorption of their ammonium pyrrolidine diethyldithiocarbamate chelates onto ^{18}C bonded silica prior to determination by graphite furnace atomic absorption spectrometry. A detection limit of 0.05 μg l^{-1} was achieved.

Sturgeon, Willie and Berman [68] have described a hydride generation atomic absorption spectrometric method for the determination of antimony in seawater. The method uses formation of stibene using sodium borohydride. Stibine gas was trapped on the surface of a pyrolytic graphite coated tube at 250°C and antimony determined by atomic absorption spectrometry. An absolute detection limit of 0.2 ng was obtained and a concentration detection limit of 0.04 μg l^{-1} obtained for 5 ml sample volumes.

The determination of antimony is discussed further in the later section on hydride methods.

Arsenic

Afansev, Ryabinin and Azhipa [69] have described an extraction photometric method for the determination of arsenic at the μg l^{-1} range in seawater. This method uses diantipyrilmethane as the chromogene reagent. The coefficient of variation is 2.5% for antimony concentrations in the 1.5–5 μg l^{-1} range. Good agreement was obtained with results obtained by neutron activation analysis.

The neutron activation method for the determination of arsenic and antimony in seawater has been described by Ryabinin et al. [70]. After co-precipitation of arsenic acid and antimony in a 100 ml sample of water by addition of a solution of ferric iron (10 mg iron per litre) followed by aqueous ammonia to give a pH of 8.4. The precipitate is filtered off and, together with the filter-paper, is wrapped in polyethylene and aluminium foil. It is then irradiated in a silica ampoule in a neutron flux of 1.8×10^{13} neutrons cm^{-2} s^{-1} for 1–2 h. Two days after irradiation, the γ-ray activity at 0.56 MeV is measured with use of a NaI(Tl) spectrometer coupled with a multi-channel pulse-height analyser, and compared with that of standards.

Burton and co-workers [71] have studied the distribution of arsenic in the Atlantic Ocean. Samples from 1000 m and above were filtered through acid-washed 0.45 μm Sartorius membrane filters. Analyses on samples from depths below 1000 m were made on unfiltered water.

Aliquots of 50 ml were placed in a round-bottomed flask, fitted with a modified Dreschel head and an injection syringe in a side arm. Concentrated hydrochloric acid 20 ml, 1 ml of 1 M ascorbic acid solution and 1 ml 1 M potassium iodide solution were added. The solution was stood for 30 min to allow reduction of As^V to As^{III} which was necessary to ensure quantitative recovery of inorganic arsenic as arsine under the conditions used in the subsequent step. With nitrogen passing through the flask at a flow rate of 150 ml min^{-1}, 0.5 ml 8% w/v sodium borohydride solution was added from the syringe. The arsine evolved was trapped in 2 ml of a solution containing 0.7% w/v potassium iodide, and excess iodine, over a period of 3 min.

The concentrates were subsequently analysed for arsenic using a Varian-Techtron AA5 atomic absorption spectrophotometer fitted with a Perkin-Elmer

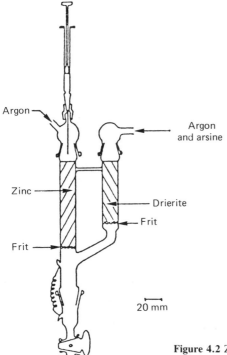

Figure 4.2 Zinc reductor column for generation of arsine by electrothermal atomic absorption spectrophotometry

HGA 72 carbon furnace, linked to a zinc reductor column for the generation of arsine (Figure 4.2). A continuous stream of argon was allowed to flow with the column connected into the inert gas line between the HGA 72 control unit and the inlet to the furnace. Calcium sulphate (10–20 mesh) was used as an adsorbent to prevent water vapour entering the carbon furnace. The carbon tube used was of 10 mm i.d. and had a single centrally located inlet hole.

A wide range of elements was tested for interfering effects; the only significant interferences found were at concentrations much higher than those encountered in seawater. No significant difference in the results was found when a sample of seawater was analysed in the way described and also by the same procedure but using the method of standard additions.

The determination of arsenic is discussed further in the later section on hydride methods.

Bismuth

Gilbert and Hume [72] Florence [73] and Eskilsson and Jaguer [74] have applied anodic stripping voltammetry to the determination of bismuth in seawater. Gilbert and Hume [72] and Florence [73] investigated the electroanalytical chemistry of bismuth (III) in the marine environment using linear-sweep anodic stripping voltammetry and a film of mercury on a glassy carbon [73] or a graphite [72] substrate as working electrode. Gillain et al. [75] used differential-pulse anodic stripping voltammetry with a hanging mercury drop electrode for the simultaneous determination of antimony (III) and bismuth (III) in seawater.

In the method of Gilbert and Hume [72], the sample contained in a silica cell, was purged and stirred by passage of purified nitrogen. A platinum counter-electrode was used. The reference-electrode consisted of a silver wire, previously anodized in seawater, held in a borosilicate-glass tube containing a small untreated portion of the sample that was separated from the sample being analysed by a plug of unfused Vycor. To diminish the effect of the steeply rising background current $(0.1 \mu A\,s^{-1})$ on the stripping peaks, a compensating circuit was devised. Bismuth was deposited at $-0.4\,V$ from seawater made 1 M in hydrochloric acid and gave a stripping peak of $-0.2\,V$, the height of which was proportional to concentration without interference from antimony or metals normally present. Antimony was deposited at $-0.5\,V$ from seawater made 4 M in hydrochloric acid and gave a stripping peak at $-0.3\,V$, the area of which was proportional to the sum of antimony and bismuth. By use of the standard-addition technique, satisfactory results were obtained for the concentration ranges $0.2–0.09\,\mu g\,kg^{-1}$ for bismuth and $0.2–0.5\,\mu g\,kg^{-1}$ for antimony.

Florence [73] carried out anodic stripping voltammetry of bismuth in a weakly acidic medium, with a polished vitreous-carbon electrode mercury-plated in situ. The limit of detection is 5 ng bismuth per litre. Seawater was found to contain $0.02–0.11\,\mu g$ bismuth per litre in surface samples.

Since computerized potentiometric stripping analysis [76–78] is, in many respects, a simpler analytical technique than linear-sweep or differential-pulse anodic stripping voltammetry, the optimum experimental conditions for the potentiometric stripping determination of bismuth (III). Although only data for the determination of bismuth (III) in seawater are reported in this paper, the optimum experimental conditions with respect to sample matrix, interferences, limits of detection and other experimental parameters can be applied to samples other than saline waters. During the course of his work it became apparent that the surface Kattegatt samples analysed during this investigation contained approximately one order of magnitude less bismuth (III) than the results obtained hitherto by electroanalytical [72,73,75] and ion-exchange [79] techniques. Because the direct determination of such low concentrations of bismuth in seawater by means of potentiometric stripping analysis would be somewhat time consuming, a simple pre-concentration technique was used. This technique was based on the co-precipitation of bismuth (III) with magnesium hydroxide, thus taking advantage of the naturally high magnesium concentration in seawater [80,81].

Figure 4.3 shows the potentiometric stripping curve obtained in the direct determination after 4 min of pre-electrolysis at −0.90 V versus SCE in an acidified Kattegatt surface seawater sample (curve a). Curves (b) and (c) of Figure 4.3

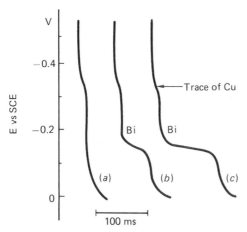

Figure 4.3 Potentiometric stripping curves registered after 4 min of pre-electrolysis at −0.90 V versus SCE before (curve *a*) and after standard additions corresponding to 12 and 24 nmol l⁻¹ bismuth (III) (curves *b* and *c*, respectively)

represent the potentiometric stripping curves recorded under the same experimental conditions after standard additions corresponding to 12 and 24 nmol l⁻¹ bismuth (III) respectively. Obviously pre-electrolysis for 4 min is not sufficient for the direct determination of bismuth (III) in an unpolluted seawater sample.

The detection limit in potentiometric stripping analysis decreases linearly with increasing time of pre-electrolysis. In practice, pre-electrolysis periods of more than 1 hour are seldom exploited. In order to determine the detection limit after 1 hour of pre-electrolysis an acidified Kattegatt seawater sample was analysed before and after a standard addition corresponding to 0.5 nmol l⁻¹ bismuth (III). Prior to the addition of bismuth (III) no measurable stripping signal was obtained. Twenty consecutive pre-electrolysis/stripping cycles after the standard addition of 0.5 nmol l⁻¹ bismuth (III) yielded an average bismuth stripping signal equal to 36 ms (Figure 4.3) with a standard deviation of 2.5 ms. Thus the detection limit after 1 hour of pre-electrolysis is 0.07 nmol l⁻¹ at the 2 SD level.

Table 4.8 Equilibrium data for bismuth (III) and antimony (III) in aqueous chloride medium

Equlibrium	Log of stability constant
$Bi^{3+} + OH^- \rightleftharpoons BiOH^{2+}$	12.4
$Bi^{3+} + 3OH^- \rightleftharpoons Bi(OH)_3$	32
$SbOH^{2+} + OH^- \rightleftharpoons Sb(OH)_2^+$	15.4
$Sb(OH)_2^+ + OH^- \rightleftharpoons Sb(OH)_3$	12.8
$Bi^{3+} + Cl^- \rightleftharpoons BiCl^{2+}$	2.2
$Bi^{3+} + 2Cl^- \rightleftharpoons BiCl_2^+$	3.5
$Bi^{3+} + 3Cl^- \rightleftharpoons BiCl_3$	5.8
$Bi^{3+} + 4Cl^- \rightleftharpoons BiCl_4^-$	6.8
$Bi^{3+} + 5Cl^- \rightleftharpoons BiCl_5^{2-}$	7.3
$Bi^{3+} + Cl^- + H_2O \rightleftharpoons BiOCl(s) + 2H^+$	−6.5
$Sb(III)^* + Cl^- \rightleftharpoons Sb(III)Cl$	2.3
$Sb(III)^* + 2Cl^- \rightleftharpoons Sb(III)Cl_2$	3.5
$Sb(III)^* + 3Cl^- \rightleftharpoons Sb(III)Cl_3$	4.2
$Sb(III)^* + 4Cl^- \rightleftharpoons Sb(III)Cl_4$	4.7

*Chemical form not stated.
Reproduced from Eskilsson, H. and Jaquer, D. (1982) *Analytica Chimica Acta*, **138**, 27, Elsevier Science Publishers, Amsterdam, by courtesy of authors and publishers.

The two most important parameters influencing the optimum conditions for the potentiometric stripping determination of bismuth (III) are the irreversible behaviour of the bismuth (III)-hydroxy complexes and possible interference from reversible antimony (III)-chloro complexes. The stability constants in the Bi^{III} OH^- I^- and Sb^{III}-OH^-I^- systems are summarized in Table 4.8. Polynuclear complexes have not been included in Table 4.8 because these are of minor importance at the low total bismuth (III) and antimony (III) concentrations in seawater. As can be seen from Table 4.8 soluble bismuth (III)-chloro complexes will be the predominant bismuth III species in seawater at pH values below 2, the chloride concentration in seawater being approximately $0.5 \, mol \, l^{-1}$. It can also be concluded from Table 4.8 that reversible antimony (III)-chloro complexes will be predominantly antimony (III) species only in very acidic solution and in the presence of high chloride concentration, i.e. in concentrated hydrochloric acid media. This is in agreement with previous electroanalytical results [72,73,75] and has also been confirmed experimentally by potentiometric stripping experiments. These experiments showed that a 100-fold amount of antimony (III) did not interfere with the bismuth stripping signal in seawater acidified to pH 1 with hydrochloric acid.

As indicated in Figure 4.3 a pre-electrolysis potential of $-0.24 \, V$ versus SCE would be sufficient for the determination of bismuth (III) in acidified seawater. At this potential copper would not be co-deposited (cf. Figure 4.3). However, because the potentiometric stripping sensitivity for bismuth increases on decreasing pre-electrolysis potential down to $-0.7 \, V$ versus SCE, and because copper (II) does not interfere with the bismuth determination, the optimum pre-electrolysis potential was found to be -0.7 to $-1.2 \, V$ versus SCE. In this potential region the potentiometric stripping sensitivity for bismuth (III) was constant.

The efficiency of the bismuth (III) co-precipitation procedure was investigated by adding 1, 2, 3, 4, 6 and 10 pmol bismuth (III) to 200 ml subsamples of a Kattegatt surface water prior to magnesium hydroxide precipitation. The recovery was in the range 80–110% for all samples.

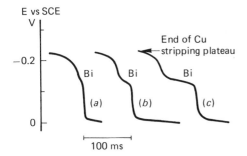

E vs SCE
V

−0.2

End of Cu
←stripping plateau

Bi   Bi   Bi

(a)   (b)   (c)

0

100 ms

Figure 4.4 Potentiometric stripping curves obtained after 8 min of pre-electrolysis at −0.90 V versus SCE in dissolved magnesium hydroxide precipitates. In curve (*a*) no bismuth (III) had been added to the seawater subsample prior to precipitation. Curves (*b*) and (*c*) correspond to standard additions of 10 and 30 pmol l^{-1} bismuth (III)

Table 4.9 Total bismuth (III) concentrations in three different Kattegatt surface seawater samples

Station location	*Mean direct determination*[*]		*Determination after co-precipitation*		
	(mol l^{-1})	(μg l^{-1})	*No. of determinations*	*Mean* (pmol l^{-1})	*Mean ± SD* (ng l^{-1})
N 56° 33.3′ E 12° 53.6′	<0.5	<0.1	8	13 ± 7	2.8 ± 1.5
N 57° 36.6′ E 11° 53.4′	<0.5	<0.1	4	7 ± 3	1.5 ± 0.5
N 57° 38.9′ E 11° 52.2′	<0.5	<0.1	4	7 ± 1	1.5 ± 0.2

[*]n = 2.
Reproduced from Eskiltson, H., Jaquer, D. (1982) *Analytica Chimica Acta*, **138**, 27, Elsevier Science Publishers, Amsterdam, by courtesy of authors and publishers.

Figure 4.4 shows the potentiometric stripping curves obtained after pre-electrolysis for 8 min at −0.90 V versus SCE in dissolved precipitates obtained from three seawater subsamples to which 0, 10 and 30 pmol bismuth (III) had been added prior to precipitation (curves a–c respectively). As can be seen by comparison with Figure 4.3 the potentiometric stripping sensitivity for bismuth (III) in the dissolved precipitates is approximately twice that in acidified seawater.

The results obtained from three different Kattegatt locations are summarized in Table 4.9 from which it is also possible to estimate the precision of the co-precipitation method. Table 4.9 shows that the total bismuth (III) concentration in the samples examined is less than 20 pmol l^{-1}. This is more than one order of magnitude less than the concentration levels found with electroanalytical and ion-exchange techniques. Since all samples examined were coastal surface waters, the bismuth (III) concentrations (5–12 pmol l^{-1}) indicated in Table 4.9 might well be due to local contamination.

Boron

Various chromogenic reagents have been used for the spectrophotometric determination of boron in seawater. These include curcumin [82,83] nile blue [84] and more recently 3,5 di-tert butylcatechol and ethyl violet [85]. Uppstroem [82] added anhydrous acetic acid (1 ml) and propionic anhydride (3 ml) to the aqueous sample (0.5 ml) containing up to 5 mg of boron per litre as H_3BO_3 in a polyethylene beaker. After mixing and the dropwise addition of oxalyl chloride (0.25 ml) to catalyse the removal of water, the mixture is set aside for 15–30 min and cooled to room temperature. Subsequently concentrated sulphuric acid - anhydrous acetic acid (1:1) (3 ml) and curcumin reagent (125 mg curcumin in 100 ml anhydrous acetic acid) (3 ml) are added and the mixed solution is set aside for at least 30 min. Finally 20 ml standard buffer solution (90 ml of 96% ethanol, 180 g ammonium acetate (to destroy excess of protonated curcumin) and 135 ml anhydrous acetic acid diluted to 1 litre with water) is added, the mixture is cooled to room temperature and the extinction is measured at 545 nm. For less than 0.01 mg boron per litre, the coloured complex must be concentrated: a portion of sample (2–10 ml) in which the colour reaction has taken place is diluted with water (100 ml) and the complex is extracted into 5 or 10 ml of extractant (100 ml isobutyl methyl ketone, 150 ml of chloroform and 1 g phenol). The extinction of the organic phase is measured at 545 nm. The colour of the complex is stable for about 2 h. Interference is caused by germanium and fluoride. Small amounts of water are tolerated but they reduce the efficiency of the method.

A curcumin based automated version of the above procedure [82] has been described [83]. Determinations can be made in the range 0.1–6 mg boron per litre. At a level of $3 \, mg \, l^{-1}$ the coefficient of variation was 1.5% and the detection limit was $0.01 \, mg \, l^{-1}$. Up to 240 samples per hour can be processed by this procedure.

In the Nile blue spectrophotometric method 10 ml 2% aqueous hydrofluoric acid is added to 10 ml sample contained in a polyethylene bottle. The mixture is shaken for about 2 h. Aqueous ferrous sulphate 10% 10 ml and 1 ml 0.1% aqueous Nile blue A are added, then extracted with o-dichlorobenzene (10 ml and 3 × 5 ml). The combined organic extracts are diluted to 50 ml with the solvent and the extinction measured at 647 nm. Interference from chloride ions up to $100 \, mg \, l^{-1}$ can be eliminated by precipitation as silver chloride.

Marcantoncetos, Gamba and Marnier [86] have described a phosphorimetric method for the determination of traces of boron in seawater. This method is based on the observation that in the 'glass' formed by ethyl ether containing 8% of sulphuric acid at 77K, boric acid gives luminescent complexes with dibenzoylmethane. A 0.5 ml sample is diluted with 10 ml 96% sulphuric acid and to 0.05–0.3 ml of this solution 0.1 ml 0.04 M-dibenzoylmethane in 96% sulphuric acid is added. The solution is diluted to 0.4 ml with 96% sulphuric acid, heated at 70°C for 1 hour, cooled, ethyl ether added in small portions to give a total volume of 5 ml and the emission measured at 77K at 508 nm, with excitation at 402 nm. At the level of 22 ng boron per ml 100-fold excesses of 33 ionic species give errors of less than 10%. However, tungsten and molybdenum both interfere.

Atomic absorption spectrometry has been used for the rapid determination of boron in seawater [87,88].

Tsaikov [89] has described a coulometric method for the determination of boron in coastal seawaters. This method is based on the potentiometric titration of boron with electrogenerated hydroxyl ions, after removal of the cation component by ion-exchange. The method has good reproducibility and is more accurate than other methods; it is fairly rapid (25–30 min per determination).

Cadmium

Various workers have discussed the application of graphite furnace atomic absorption spectrometry to the determination of cadmium in seawater [90–110].

In the determination of cadmium in seawater for both operational reasons and ease of interpretation of the results, it is necessary to separate particulate material from the sample immediately after collection. The 'dissolved' trace metal remaining will usually exist in a variety of states of complexation and possibly also of oxidation. These may respond differently in the method, except where direct analysis is possible with a technique using high-energy excitation, such that there is no discrimination between different states of the metal. The only technique of this type with sufficiently low detection limits is carbon furnace atomic absorption spectrometry, which is subject to interference effects from the large and varying content of dissolved salts.

Batley and Farrah [95] and Gardner and Yates [93] used ozone to decompose organic matter in samples and thus break down metal complexes prior to atomic absorption spectrometry. By this treatment, metal complexes of humic acid and EDTA were broken down in less than 2 minutes. These observations lead Gardner and Yates [93] to propose the following method for the determination of cadmium in seawater.

The sample is filtered immediately after collection, acidified to about pH 2, and transferred to a 1-litre Pyrex storage bottle. Prior to extraction the sample is ozonized in the sample bottle for 30 min. Nitrogen is passed through the sample for 5 min to remove excess ozone, then the pH is carefully raised to about 5 by addition of ammonia solution and about 5 ml Chelex 100 resin in the ammonia form is added. After stirring for at least 1 h, the resin is collected in a Pyrex chromatography column and washed with the calculated quantity of an appropriate buffer to elute calcium and magnesium. After further washing with 50 ml deionized water, the resin is eluted with 2 M nitric acid, to a volume of 25 ml. The eluate is analysed by graphite furnace atomic absorption spectrometry.

Danielson, Magnusson and Westerlund [96] have described a method for the determination of cadmium in seawater. The samples were analysed by graphite furnace atomic absorption spectroscopy after a two-stage extraction. Replacing the acetate buffer and performing the extraction in a clean room with Teflon utensils significantly improved blank levels. Extractions were performed on board ship immediately after sampling and the extracts brought home for analysis. An aliquot of the sample was also transferred into carefully cleaned Teflon FEP bottles and acidified with 1 ml nitric acid per litre. The nitric acid had been purified by sub-boiling distillation. These samples were extracted about 2 months after sampling at the shore laboratory. The same method was used with the exception that extra ammonia was added to the buffer to compensate for the acidification. The method was applied to arctic seawaters and showed a profile of cadmium with sampling depth range from $0.133 \, \text{nmol} \, l^{-1}$ cadmium at the surface to $0.205 \, \text{nmol} \, l^{-1}$ cadmium at 2000 m.

As cadmium is one of the most sensitive graphite furnace atomic absorption determinations it is not surprising that this is the method of choice for the determination of cadmium in seawater. Earlier workers separated cadmium from the seawater salt matrix prior to analysis. Chelation and extraction [97–104], ion-exchange [98,100,101,105] and electrodeposition [106,107] have all been studied.

The direct determination of cadmium in seawater is particularly difficult because the alkali and alkaline earth salts cannot be fully charred away at temperatures that will not also volatilize cadmium. Most workers in the past [101,108–111] who have attempted a direct method have volatilized the cadmium at temperatures which would leave sea salts in the furnace. This required careful setting of temperatures and was disturbed by situations that caused temperature settings to change with the life of the furnace tubes.

Lundgren, Lundmark and Johansson [108] showed that the cadmium signal could be separated from a 2% sodium chloride signal by atomizing at 820°C, below the temperature where the sodium chloride was vaporized. This technique has been called selective volatilization. They detected $0.03 \, \mu g \, l^{-1}$ cadmium in the 2% sodium chloride solution. They used an infra-red optical temperature monitor to set the atomization temperature accurately.

Campbell and Ottaway [112] also used selective volatilization of the cadmium analyte to determine cadmium in seawater. They could detect $0.04 \, \mu g \, l^{-1}$ cadmium (2 pg in 50 μl) in seawater. They dried at 100°C and atomized at 1500°C with no char step. Cadmium was lost above 350°C. They could not use ammonium nitrate because the char temperature required to remove the ammonium nitrate volatilized cadmium also. Sodium nitrate and sodium and magnesium chloride salts provided reduced signals for cadmium at much lower concentrations than their concentration in seawater if the atomization temperature was in excess of 1800°C. The determination required lower atomization temperatures to avoid atomizing the salts. Even this left the magnesium interference which required the method of additions.

Guevremont, Sturgeon and Berman [92] used a direct, selective volatilization determination of cadmium in seawater. They used 20 μl seawater samples, $1 \, g \, l^{-1}$ of EDTA an atomization ramp from 250°C to 2500°C in 5 s, and the method of additions. Their detection limit was $0.01 \, \mu g \, l^{-1}$ (0.2 pg in 20 μl) the characteristic amount was 0.7 pg/0.0044 A. The EDTA promoted the early atomization of cadmium below 600°C. Their test seawater sample ($0.053 \mu g \, l^{-1}$) was confirmed by other methods. These authors were unable to separate reliably the cadmium and background signals by using the method of Campbell and Ottaway; the EDTA made this possible.

Guevremont, Sturgeon and Berman [92] studied the use of different matrix modifiers in the graphite furnace gas method of determination of cadmium in seawater. These included citric acid, lactic acid, aspartic acid, histidine and EDTA. The addition of less than 1 mg of any of the compounds to 1 ml seawater significantly decreased matrix interference. Citric acid achieved the highest sensitivity and reduction of interference, with a detection limit of 0.01 μg cadmium per litre.

In similar work, Sturgeon et al. [101] compared direct furnace methods with extraction methods for cadmium on two coastal seawater samples. They found 0.2 and $0.05 \% . 112 g \, l^{-1}$ cadmium and could have measured cadmium down to $0.01 \, \mu g \, l^{-1}$. They used $10 \, \mu g \, l^{-1}$ ascorbic acid as a matrix modifier. Various organic matrix modifiers were studied by Guevremont [91] for this analysis. He found citric acid to be somewhat preferable to EDTA, aspartic acid, lactic acid, and histidine. The method of standard additions was required. The standard deviation was better than $0.01 \, \mu g \, l^{-1}$ in a seawater sample containing $0.07 \, \mu g \, l^{-1}$. Generally, he charred at 300°C and atomized at 1500°C. The method required compromise between char and atomization temperatures, sensitivity, heating rates, etc. but the analytical results seemed precise and accurate. Nitrate added as sodium nitrate delayed the cadmium peak and suppressed the cadmium signal.

Sperling [109] has reported extensively on the determination of cadmium in seawater as well as in other biological samples and materials. He added ammonium persulphate which permitted charring seawater at 430°C without loss of cadmium. For work below $2 \mu g l^{-1}$ cadmium in seawater he recommended extraction of the cadmium to separate it from the matrix [102,110,111]. He found no change in the measured levels over many months when the seawater was stored in high density polyethylene or polypropylene.

Table 4.10 Zeeman graphite furnace conditions

	Dry	Char	Atomize	Clean-out	Cool
Temp, °C	160	550	1600	2600	20
Ramp, s	1	1	0	1	1
Hold, s	60	45	5	6	20
Int gas flow, ml/min	300	300	0	300	300
Recorder, s			−5		

Reproduced from Pruzkowska, F., et al. (1983) Analytical Chemistry, **55**, 182, American Chemical Society, by courtesy of authors and publishers.

Pruszkowska, Carnrick and Slavin [111] described a simple and direct method for the determination of cadmium in coastal water utilizing a platform graphite furnace and Zeeman background correction. The furnace conditions are summarized in Table 4.10. These workers obtained a detection limit of $0.013 \mu g l^{-1}$ in $12 \mu l$ samples or about 0.16 pg cadmium in the coastal seawater sample. The characteristic integrated amount was 0.35 pg cadmium per 0.0044 A s. A matrix modifier containing di-ammonium hydrogen phosphate and nitric acid was used. Concentrations of cadmium in coastal seawater were calculated directly from a calibration curve. Standards contained sodium chloride and the same matrix modifier as the samples. No interference from the matrix was observed.

Seawater samples usually contain a total of 2–3% of several alkali and alkaline earth salts with sodium chloride as a main constituent. A $2 \mu l$ sample of seawater charred at 700°C has a background signal so high over 2 A that even the Zeeman correction system cannot handle it (Figure 4.5). The large amounts of sodium chloride present in seawater are reportedly volatilized below 950°C [113] but even with ammonium phosphate, the matrix modifier recommended for cadmium, it is not possible to char at so high a temperature. Figure 4.5 shows that $200 \mu g$ of diammonium hydrogen phosphate reduced the SB signal of $2 \mu l$ seawater to 0.5 A but $500 \mu g$ ammonium nitrate reduced the background more effectively to 0.16 A. No reduction of the cadmium signal occurred in the presence of ammonium nitrate if the char temperature was below 600°C, and phosphate was used as a matrix modifier. If ammonium nitrate was used without phosphate the cadmium was lost at temperatures below 500°C. The addition of the phosphate stabilized the cadmium while the ammonium nitrate promoted the release or conversion of the

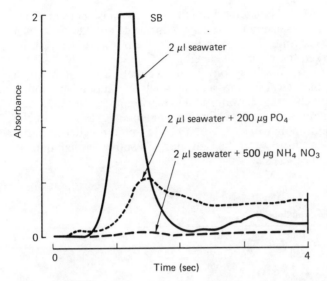

Figure 4.5 Background (SB) profiles for $2\,\mu l$ seawater alone, with $200\,\mu g$ $(NH_4)_2HPO_4$ and with $500\,\mu g$ NH_4NO_3. The char temperature was 700°C and the atomization temperature was 1700°C. The signals from the Data System 10 reported here are called ZAA signals for the analytical result and SB signals (single beam) for the backgrounds. The SB signals are expressed in absorbance units (A) and the ZAA signals are usually in absorbance units – seconds (A-s). The SB signal is signal plus background, but for the small analyte signals of this study, the SB signal is effectively background. The actual integrated absorbance signals that were used were calculated by software on the Data Station 10 from signals transmitted by the Zeeman/5000. The plots shown in later figures show typical signals but were not used for quantitative evaluation

Table 4.11 Quantitative results and recoveries

	Mean ($\pm SD$) amount found ($\mu g\ l^{-1}$)		Percentage recovery
	Prieszouska et al.	*NRC*	
Sandy Cove No. 8	0.040 ± 0.006	0.05 ± 0.01	103
Sandy Cove No. 9	0.058 ± 0.007		95
Bermuda	0.029 ± 0.004	0.029 ± 0.004	103
NRC	0.049 ± 0.005		94

Reproduced from Pruzkowska, F., *et al.* (1983) *Analytical Chemistry*, **55**, 182, American Chemical Society, by courtesy of authors and publishers.

bulk of the background producing material. The addition of the phosphate produced a background signal that appeared much later than the cadmium peak.

It was shown that 1.25 mg ammonium nitrate is enough to keep the background signal below 1.5 A and there are no large differences in background absorbances for amounts from 1.25 to 7.5 mg ammonium nitrate.

It was also shown that 2% nitric acid reduced the background to a level that can be handled by the Zeeman correction system. From 4% to 8% nitric acid the changes in background signal shapes were not very large.

The results of the determination of cadmium in four seawater samples are shown in Table 4.11. The determination was done directly from the calibration curve and $12\,\mu l$ seawater was used for each run.

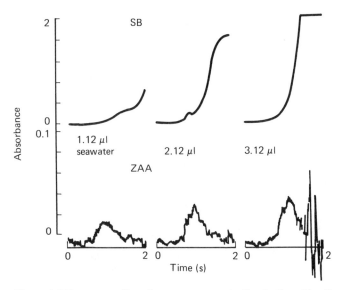

Figure 4.6 Zeeman profiles of a seawater sample (Sandy Cove No. 9) and SB profiles. The first pair of profiles represent a single 12 µl aliquot, the second pair, two aliquots, and the third pair, three aliquots. The modifier was 200 µg $(NH_4)_2HPO_4$, 8% HNO_3 and 5 µg $Mg(NO_3)_2$. The char temperature was 550°C, and the atomization temperature 1600°C

The detection limit for cadmium in seawater, calculated as 2 SD for low concentrations, was $0.013\,\mu g\,l^{-1}$. With 12 µl of seawater, this corresponded to 0.16 pg of cadmium in the seawater.

Typical sample (ZAA) signals and background (SB) signals for a seawater sample are shown in Figure 4.6. Brewer [114] has used electrically vaporized thin gold film atomic emission spectrometry to determine cadmium at the 10 ppb level in highly acidic saline solutions following pre-concentration with a microload of strong-base anion exchange resin.

Kounaves and Zirino [115] studied cadmium-EDTA complex formation in seawater using computer assisted stripping polarography. They showed that the method is capable of determining the chemical speciation of cadmium in seawater at concentrations down to 10^{-8} M.

Turner, Robinson and Whitfield [116] studied the automated electrochemical stripping of cadmium in seawater.

Further work on the co-determination of cadmium with other elements in seawater is discussed later.

Cerium

This is discussed under rare earths.

Chromium

Total chromium
Reported concentrations of chromium in open ocean waters range from 0.07 to $0.96\,\mu g\,l^{-1}$ with a preponderance of values near the lower limit. Methods used for

the determination of chromium at this concentration have generally used some form of matrix separation and analyte concentration prior to determination [117–120], electroreduction [121,122] and ion-exchange techniques [123,124].

Whereas it is desirous to utilize analytical schemes that permit elucidation of the various chromium species particularly since Cr^{VI} is acknowledged to be a toxic form of this element, it is useful to have the capability of rapid, total chromium measurement where speciation is a matter of secondary importance.

Determination of chromium by many of the methods cited above is problematic. Variable and non-quantitative recovery with chelation-solvent extraction techniques necessitates use of the method of additions [120]. Co-precipitation techniques require lengthy processing times and extensive sample manipulation. Ion-exchange suffers from slow uptake and release kinetics, necessitating total destruction and solubilization of the resin [124] or complex apparatus and multicomponent eluting solutions.

Diphenylcarbazone and diphenylcarbazide have been widely used for the spectrophotometric determination of chromium [125]. Only relatively recently, however, has the nature of the complexation reactions been elucidated. Cr^{III} reacts with diphenylcarbazone whereas Cr^{VI} reacts (probably via a redox reaction combined with complexation) with diphenylcarbazide [126]. Although speciation would seem a likely prospect with such reactions, commercial diphenylcarbazone is a complex mixture of several components, including diphenylcarbazide, diphenylcarbazone, phenylsemicarbazide, and diphenylcarbadiazone with no stoichiometric relationship between the diphenylcarbazone and diphenylcarbazide [127]. As a consequence, use of diphenylcarbazone to chelate Cr^{III} selectively also results in the sequestration of some Cr^{VI}. Total chromium can be determined with diphenylcarbazone following reduction of all chromium to Cr^{III}.

Use of immobilized chelating agents for sequestering trace metals from aqueous and saline media presents several significant advantages over chelation-solvent extraction approaches to this problem [128,129]. With little sample manipulation, large preconcentration factors can generally be realized in relatively short times with low analytical blanks.

As a consequence of these considerations Willie, Sturgeon and Berman [130] developed a new approach to the determination of total chromium. This involves preliminary concentration of dissolved chromium from seawater by means of an immobilized diphenylcarbazone chelating agent, prior to determination by atomic absorption spectrometry. A Perkin-Elmer Model 500 atomic absorption spectrometer fitted with a HGA-500 furnace with Zeeman background correction capability was used for chromium determinations. Chromium was first reduced to Cr^{III} by addition of 0.5 ml aqueous sulphur dioxide and allowing them to stand for several minutes. Aliquots of seawater were then adjusted to pH 9.0 ± 0.2 by using high-purity ammonium hydroxide and gravity fed through a column of silica at a nominal flow rate of 10 ml min^{-1}.

The sequestered chromium was then eluted from the column with 10.0 ml 0.2 M nitric acid. More than 93% of chromium was recovered in the first 5 ml of eluate by this method. Extraction of 80 ng spikes of Cr^{III} from 200 ml aliquots of seawater was quantitative. Neither Cr^{III} nor Cr^{VI} could be quantitatively extracted.

Results for the analysis of a near-shore sample of seawater and open ocean trace metal reference seawater, NASS-1 are given in Table 4.12.

Isotope dilution gas chromatography–mass spectrometry has also been used for the determination of ppb of total chromium in seawater [131–133]. The samples

Table 4.12 Concentration analysis (μg l^{-1}) of seawater for total chromium

Trial	Coastal water (salinity = 29.5‰)	Open-ocean, NASS-1 (salinity = 35.0‰)
1	0.100	0.19
2	0.096	0.15
3	0.095	0.18
4		0.19
Mean ($\pm$SD)	0.097 $\pm$ 0.003	0.18 $\pm$ 0.02
Accepted value	0.10 $\pm$ 0.01	0.184 $\pm$ 0.016

Reproduced from Willie, S.N., *et al.* (1983) *Analytical Chemistry*, **55**, 981, American Chemical Society, by courtesy of authors and publishers.

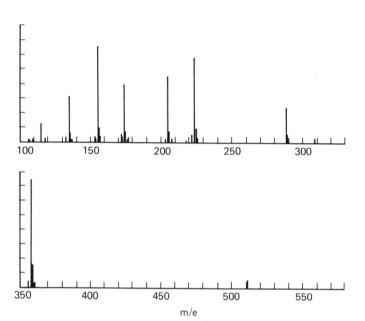

Figure 4.7 Mass spectrum of Cr(tra)$_3$.

were reduced to ensure Cr^{III} and then extracted and concentrated as tris (1,1,1-trifluoro-2,4-pentanediono) chromium (III) (Cr(tfa)$_3$) into hexane. The Cr(tfa)$_2$ + mass fragments were monitored into a selected ion monitoring (SIM) mode.

Isotope dilution techniques are attractive because they do not require quantitative recovery of the analyte. One must, however, be able to monitor specific isotopes which is possible by using mass spectrometry.

In this method, chromium is extracted and pre-concentrated from seawater with trifluoroacetylacetone H(tfa) which complexes with trivalent but not hexavalent chromium. Chromium reacts with trifluoroacetylacetone in a 1 : 3 ratio to form an octahedral complex, Cr(tfa)$_3$. The isotopic abundance of its most abundant mass fragment, Cr(tfa)$_2$ + was monitored by a quadrupole mass spectrometer.

A mass spectrum of Cr(tfa)$_3$ is shown in Figure 4.7. The isotopic distribution of the Cr(tfa)$_2$ + fragment (m/e 358 and 359 here) is evident. This is readily calculable

Table 4.13 Natural abundance of Cr(tfa)

m/e	% calculated	% measured
356	3.8	3.8
357	0.5	0.6
358	75.2	74.9
359	17.0	16.6
360	3.5	4.0

Reproduced from Siu, K.W.M., et al. (1983) Analytical Chemistry, 55, 473, American Chemical Society, by courtesy of authors and publishers.

Table 4.14 Abundance of Cr(tfa)$_2^+$ for the chromium-53 spike

m/e	% calculated	% measured
358	3.1	3.3
359	86.5	86.7
360	9.9	9.1
361	0.5	0.9

Reproduced from Siu, K.W.M., et al. (1983) Analytical Chemistry, 55, 473, American Chemical Society, by courtesy of authors and publishers.

if the individual elemental abundances are known. Assuming the isotopic abundance of ^{12}C and ^{13}C to be 98.89% and 1.11 and ^{50}Cr, ^{52}Cr, ^{53}Cr and ^{54}Cr to be 4.31, 83.76, 9.55 and 2.38% respectively, and neglecting any isotopic abundances less than 1%, one can obtain a set of calculated abundances for the Cr(tfa)$_2$ + ion. These and the measured isotopic abundances (by SIM) are listed in Table 4.13. The agreement between the two sets is excellent. The same calculation can be made for the chromium 53 spike solution by using isotopic abundances given by the supplier: ^{52}Cr, 3.44%, ^{53}Cr, 96.4% and ^{54}Cr, 0.18%. Table 4.14 lists the calculated and the measured isotopic abundances for the spike solution.

A series of typical chromatograms of a spiked seawater sample is shown in Figure 4.8. The geometric isomers of chromium trifluoracetylacetone are not fully resolved.

Table 4.15 shows results of two seawater sample analyses. Agreement with data obtained by isotope dilution spark source mass spectrometry [134] and graphite furnace [135] was excellent.

Trivalent chromium
The chemiluminescence technique has been used to determine trivalent chromium in seawater. Chang et al. [136] showed luminal techniques for determination of chromium (III) were hampered by a salt interference–mainly from magnesium ions (Figure 4.9). Elimination of this interference is achieved by seawater dilution and utilizing bromide ion chemiluminescence signal enhancement. The chemiluminescence results were comparable with those obtained by a graphite furnace flameless atomic absorption analysis for the total chromium present in samples. The detection limit is 3.3×10^{-9} mol l^{-1} (0.2 ppb) for seawater with a salinity of 35% with 0.5 M bromide enhancement.

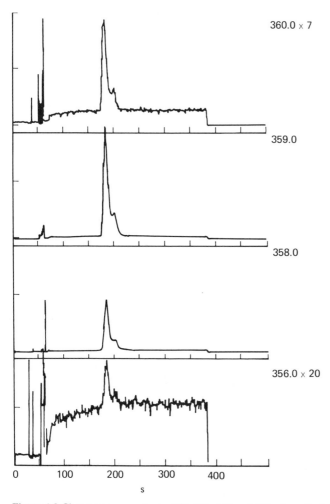

Figure 4.8 Chromatograms of m/m 356, 358, 359 and 360 of a spiked seawater sample. Multipliclation factor

Table 4.15 Mean (±SD) chromium concentration in seawater (μg l⁻¹) (n ⩾ 3)

ID-GC/MS	ID-SSMS	GFAAS
0.177 ± 0.009	0.17 ± 0.03	0.19 ± 0.03
0.19 ± 0.01*	0.18 ± 0.01	ND

*Seawater reference material NASS-1.
ND = not determined.
Reproduced from Siu, K.W.M., et al. (1983) Analytical Chemistry, **55**, 473, American Chemical Society, by courtesy of authors and publishers.

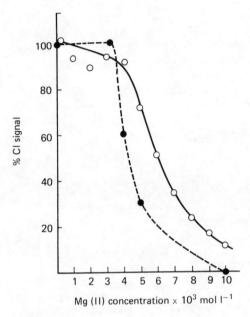

Figure 4.9 Mg^{II} interference of Cl analysis for Cr. —○—○— = in the presence of $0.3\,M\,Br^-$; –-●–-●–-● = in the absence of Br^-. $Cr^{III} = 6 \times 10^{-8}$ M. EDTA $= 2.5 \times 10^{-3}$ M

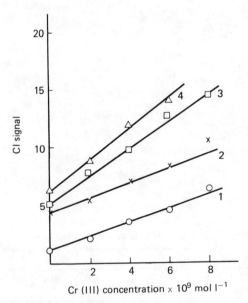

Figure 4.10 Cl analysis for Cr in natural seawater. Curve 1: calibration curve of Cr (no seawater). Curves 2–4: standard addition curves with 2,3 and 4 ml seawater added, respectively. $Br^- = 0.30$ M. EDTA $= 2.5 \times 10^{-3}$ M

The effect of calcium interference is somewhat different. At its concentration in seawater, 0.010 M, calcium ion had no effect upon chemiluminescence analysis of a 6×10^{-8} M Cr^{III} solution in the absence of bromide ion. The chemiluminescence signal dropped to zero, however, if the calcium ion concentration was increased to 0.013 M. In the presence of 0.3 M bromide ion, no interference was observed for analysis of 6×10^{-8} M Cr^{III} when the calcium concentration was less than or equal to 0.002 M. The chemiluminescence signal increased linearly with increasing calcium ion concentration when the calcium concentration exceeded 0.002 M.

The combined effect of cation interference for both Mg^{II} and Ca^{II} is almost identical with the solid curve in Figure 4.9 indicating that the magnesium ion interference is the dominant one. Figure 4.10 shows calibration curves obtained upon spiking a seawater sample with Cr^{III}.

Tri- and hexa-valent chromium
Various workers have discussed the separate determination of Cr^{III} and Cr^{VI} in seawater [137–141].

Cranston and Murray [138,139] took the samples in polyethylene bottles that had been precleaned at 20°C for 4 days with 1% distilled hydrochloric acid. Total chromium (Cr^{VI}) + Cr^{III} + Cr_p (particulate chromium), was co-precipitated with iron (II) hydroxide and reduced chromium (Cr^{III} + Cr_p) was co-precipitated with iron (III) hydroxide. These co-precipitation steps were completed within minutes of sample collection to minimize storage problems. The iron hydroxide precipitates were filtered through 0.4 μm Nuclepore filters and stored in polyethylene vials for later analyses in the laboratory. Particulate chromium was also obtained by filtering unaltered samples through 0.4 μm filters. In the laboratory the iron hydroxide co-precipitates were dissolved in 6 M distilled hydrochloric acid and analysed by flameless atomic absorption. The limit of detection of this method is about 0.1–0.2 nmol l^{-1}. Precision is about 5%.

Organic forms of chromium
In the determination of the two oxidation states of chromium the calculation of one oxidation state by difference presupposes that the two oxidation states in question were statistically the only contributors to the total concentration. Because of this, contributions from other possible species such as organic complexes were generally not considered. It has recently been suggested [142] however, that this presumption may not be warranted and that contributions from organically bound chromium should be considered. This arises from the reported presence of dissolved organic species in natural waters which form stable soluble complexes with chromium and which may not readily be amenable to determination by procedures commonly in use. The results of research into the valency of chromium present in seawater has not always been consistent. For instance, Grimaud and Michard [143] reported that chromium (III) predominates in the equatorial region of the Pacific Ocean, whereas Cranston and Murray [144] found that practically all chromium is in the hexavalent state in the north-east Pacific. Organic Cr^{III} complexes may be formed under the conditions prevailing in seawater as well as inorganic Cr^{III} and Cr^{VI} forms. Inconsistencies in earlier research may therefore be at least partly due to the fact that the possibility of organic chromium species was ignored [145,146].

Nakayama *et al.* [147] have described a method for the determination of chromium (III), chromium (VI) and organically bound chromium in seawater.

They found that seawater in the sea of Japan contained about 9×10^{-9} M dissolved chromium. This is shown to be divided as about 15% inorganic Cr^{III}, about 25% inorganic Cr^{VI} and about 60% organically bound chromium.

These workers studied the co-precipitation behaviours of chromium species with hydrated iron III and bismuth oxides.

The collection behaviour of chromium species was examined as follows. Seawater (400 ml) spiked with 10^{-8} M Cr^{III}, Cr^{VI} and Cr^{III} organic complexes labelled with ^{51}Cr was adjusted to the desired pH by hydrochloric acid or sodium hydroxide. An appropriate amount of hydrated iron (III) or bismuth oxide was added; the oxide precipitates were prepared separately and washed thoroughly with distilled water before use [148]. After about 24 h, the samples were filtered on 0.4 μm Nuclepore filters. The separated precipitates were dissolved with hydrochloric acid and the solutions thus obtained were used for γ-activity measurements. In the examination of solvent extraction, chromium was measured by using ^{51}Cr, while iron and bismuth were measured by electrothermal atomic absorption spectrometry. The decomposition of organic complexes and other procedures were also examined by electrothermal atomic absorption spectrometry.

Collection of Cr^{III} and Cr^{VI} with hydrated iron^{III} or bismuth oxide
Only Cr^{III} co-precipitates quantitatively with hydrated iron (III) oxide at the pH of seawater, around 8. To collect Cr^{VI} directly without pretreatment, e.g. reduction to Cr^{III}, hydrated bismuth oxide, which forms an insoluble compound with Cr^{VI} was used. Cr^{III} is collected with hydrated bismuth oxide (50 mg 400 ml^{-1} seawater) to collect Cr^{VI} in seawater about pH 4 and Cr^{VI} is collected below pH 10. Both Cr^{III} and Cr^{VI} are thus collected quantitatively at the pH of seawater around 8.

Collection of Cr^{III} organic complexes with hydrated iron (III) or bismuth oxide
The percentage collection of Cr^{III} with hydrated iron (III) oxide may decrease considerably in the neutral pH range when organic materials capable of combining with Cr^{III} such as citric acid and certain amino acids, are added to the seawater[149]. Moreover, synthesized organic Cr^{III} complexes are scarcely collected with hydrated iron (III) oxide over a wide pH range [149].

As it was not known what kind of organic matter acts as the major ligand for chromium in seawater Nakayama *et al.* [147] used EDTA and 8-quinolinol-5-sulphonic acid to examine the collection and decomposition of organic chromium species, because these ligands form quite stable water-soluble complexes with Cr^{III} although they are not actually present in seawater. Both these Cr^{III} chelates are stable in seawater at pH 8.1 and are hardly collected with either of the hydrated oxides. The organic chromium species were then decomposed to inorganic chromium (III) and chromium (VI) species by boiling with 1 g ammonium persulphate per 400 ml^{-1} seawater acidified to 0.1 M with hydrochloric acid. Iron and bismuth, which would interfere in atomic absorption spectrometry, were 99.9% removed by extraction from 2 M hydrochloric acid solution with a p-xylene solution of 5% tri-iso-octylamine. Cr^{III} remained almost quantitatively in the aqueous phase in the concentration range 10^{-9}–10^{-6} M, whether or not iron or bismuth was present. However, as about 95% of Cr^{VI} was extracted by the same method, samples which may contain Cr^{VI} should be treated with ascorbic acid before extraction so as to reduce Cr^{VI} to Cr^{III}.

When the residue obtained by the evaporation of the aqueous phase after the extraction was dissolved in 0.1 M nitric acid and the resulting solution was used for electrothermal atomic absorption spectroscopy, a negative interference, which was seemingly due to residual organic matter, was observed. This interference was successfully removed by digesting the residue on a hot plate with 1 ml of concentrated hydrochloric acid and 3 ml of concentrated nitric acid. This process had the advantage that the interference of chloride in the atomic absorption spectroscopy was eliminated during the heating with nitric acid.

Table 4.16 shows examples of the vertical distribution of each chromium species in the Japan Sea and the Pacific Ocean; samples were collected during the summer of 1979.

Table 4.16 Vertical distribution of chromium species in the Sea of Japan and in the Pacific Ocean

Depth (m)	$Cr(III)^*$ $(\times 10^{-9} M)$	$Cr(VI)$ $(\times 10^{-9} M)$	$Organic\ Cr$ $(\times 10^{-9} M)$	$Total\ Cr$ $(\times 10^{-9} M)$
Japan Sea (44°11.9′N 138°56.4′E; depth 3447 m)				
0	1.3	2.1	4.9	8.3
10	1.4	1.7	5.9	9.0
51	1.6	1.8	4.3	7.7
102	1.2	1.7	5.3	8.2
152	1.3	1.8	4.2	7.3
203	1.2	1.8	4.6	7.6
403	1.4	2.9	5.0	9.3
602	1.1	2.3	3.7	7.1
1000	1.5	2.4	3.5	7.4
1427	1.1	3.0	4.2	8.3
1920	1.8	1.7	6.2	9.7
2417	1.1	2.1	4.6	7.8
2916	1.2	–	–	9.1
3165	1.4	1.7	5.0	8.1
Mean	1.3	2.1	4.8	8.2
Max.	1.8	3.0	6.2	9.7
Min.	1.1	1.7	3.5	7.1
Pacific Ocean (32°19.3′N 137°33.5′E; depth 4079 m)				
0	1.4	2.0	5.0	8.4
10	1.4	–	–	8.1
49	1.3	2.5	4.7	8.5
98	0.96	2.4	–	–
143	1.0	2.8	4.1	7.9
197	1.2	2.1	5.5	8.8
394	1.2	2.6	4.9	8.7
591	1.5	2.4	5.0	8.9
985	1.1	3.6	5.9	10.6
1477	1.1	3.3	6.2	10.6
1804	1.2	4.5	6.0	11.7
2299	1.4	4.0	5.4	10.8
2803	1.2	3.7	5.2	10.1
3303	1.5	3.4	–	–
3801	1.0	3.3	5.3	9.6
4050	1.1	3.7	5.8	10.6
Mean	1.2	3.1	5.2	9.5
Max.	1.5	4.5	6.2	11.7
Min.	0.96	2.0	4.1	7.9

*Inorganic.

Reproduced from Nakayama, E., *et al.* (1981) *Analytica Chimica Acta*, **131**, 247, Elsevier Science Publishers, Amsterdam, by courtesy of authors and publishers.

Table 4.17 Literature data on the chromium contents of seawater

Reference	Date	Locations	Methods	Concentration ($\mu g\ l^{-1}$)
150	1950	Japanese coastal	$Al(OH)_3$ co-precipitation	Cr^{III} 0.04–0.06
151	1966	British coastal	$Fe(OH)_3$ co-precipitation Acidic reduction	Cr^{III} 0.46 Cr^{VI} 0.6
152,153	1967	Mediterranean	$Fe(OH)$ co-precipitation Acidic reduction	Cr^{III} 0.02–0.25 Cr^{VI} 0.05–0.38
154	1970	Pacific Ocean	$BiOH(NO_3)_2$ co-precipitation Acidic reduction	Cr^{III}? $Cr^{III}+$ Cr^{VI} 0.13–0.96 Cr^{VI}? Organic Cr 0.07–0.27
155	1974	Pacific Ocean	$Fe(OH)_3$ co-precipitation $Fe(OH)_3$ reduction	Cr^{III} 0.24–0.52 Cr^{VI} 0.03–0.11
157	1978	Pacific Ocean	$Al(OH)_3$ co-precipitation Acidic reduction	Total 0.06–0.96 Cr^{III} 17–99%
154	1980	Pacific Ocean and Japan Sea	$Fe(OH)_3$ co-precipitation $Fe(OH)_3$ reduction	Cr^{III} 0.005 Cr^{VI} 0.15 Cr^{III} 0.06 Cr^{VI} 0.14 Organic Cr 0.26 Total 0.46

For comparison, some of the results reported by other workers for chromium concentrations in seawater are listed in Table 4.17. In most of these methods, co-precipitation with hydrated iron (III) oxide was used to separate Cr^{III} from Cr^{VI} and the Cr^{VI} concentration was subsequently determined by suitable reduction of Cr^{VI} to Cr^{III} before a further co-precipitation. In others, hydrated iron (II) oxide served as both reductant and carrier. Ishibashi and Shigematsu [150] used co-precipitation with aluminium hydroxide and did not employ reduction, so that the value reported most likely corresponds to inorganic Cr^{III} alone; in fact, the present value for inorganic Cr^{III} is in remarkable agreement.

In Chuecas and Riley's study [151] the samples were stored for a long time under acidic conditions before analysis, so that Cr^{VI} could have been reduced to Cr^{III} and any organic chromium dissociated with the result that all chromium species would have been determined as Cr^{III}. When a sample is reduced under acidic conditions, organic chromium is likely to dissociate partly, initially increasing the apparent concentration of Cr^{VI}. When the analytical procedure described earlier [154] was re-examined the value for Cr^{III} was found actually to be the sum of Cr^{III} and Cr^{VI} while the value for Cr^{VI} was partly organic chromium. For the same reason, the Cr^{VI} values determined by Fukai [152], Fukai and Vas [153] and Yamamoto, Kadowski and Carpenter [156] probably include organic chromium species. When an iron (II) precipitate is used, there seems to be little chance of determining the organic chromium species as Cr^{VI}. The value for Cr^{VI} reported by Cranston and Murray [157] agrees quite well with the value for Cr^{VI} reported by them although the value for Cr^{III} is lower. The results obtained by Grimaud and Michard [143] for Cr^{III} differ considerably but the discrepancies cannot be discussed because details of the analytical procedure were not given. It seems reasonable to conclude that the inconsistency of past results concerning the dominant chromium species and the total chromium concentration in seawater can be attributed, at least partly, to the fact that the presence of organic chromium species was not considered properly.

Mullins [158] has described a procedure for determining the concentrations of dissolved chromium species in seawater. Chromium (III) and chromium (VI) separated by co-precipitation with hydrated iron (III) oxide and total dissolved chromium are determined separately by conversion to chromium (VI), extraction with ammonium pyrrolidine diethyl dithiocarbamate into methyl isobutyl ketone and determination by atomic absorption spectroscopy. The detection limit is $40 \, \text{ng} \, \text{l}^{-1}$ Cr. The dissolved chromium not amenable to separation and direct extraction is calculated by difference. In the waters investigated, total concentrations were relatively high $(1-5 \, \mu\text{g} \, \text{l}^{-1})$ with chromium (VI) the predominant species in all areas sampled with one exception, where organically bound chromium was the major species.

A standard contact time of $4 \, \text{h}$ was found to be necessary for the quantitative co-precipitation of chromium on ferric oxide. The results of triplicate determinations of samples taken from six locations in the Sydney area are listed in Table 4.18. The r.s.d. values for the determinations of chromium (III), chromium (VI)

Table 4.18 Determination of dissolved chromium species in some seawaters

Location	Chromium found ($\mu\text{g} \, \text{l}^{-1}$)*			
	Cr(III)	Cr(VI)	Cr bound	Cr total
Port Hacking	0.27 ± 0.02	0.49 ± 0.03	0.56 ± 0.07	1.32 ± 0.05
Georges River	0.42 ± 0.04	0.89 ± 0.04	0.42 ± 0.08	1.72 ± 0.06
Drummoyne Bay	0.32 ± 0.03	0.95 ± 0.04	0.69 ± 0.10	1.96 ± 0.07
Botany Bay	0.45 ± 0.04	1.26 ± 0.06	0.71 ± 0.03	2.41 ± 0.09
Cooks River	0.51 ± 0.04	2.98 ± 0.11	0.88 ± 0.10	4.37 ± 0.06
Parramatta River	0.88 ± 0.02	3.17 ± 0.06	0.82 ± 0.09	4.87 ± 0.11

*All results are the mean (±SD) of three measurements.
Reproduced from Mullins, T.L. (1984) *Analytica Chimica Acta*, **165**, 97, Elsevier Science Publishers, Amsterdam, by courtesy of authors and publishers.

and total dissolved chromium were generally 10.0%, 5.0% and 5.0% respectively. From these results, the r.s.d. for the calculated concentration of the bound species was 20%.

Ahern *et al.* [159] have discussed the speciation of chromium in seawater. The method used co-precipitation of trivalent and hexavalent chromium, separately, from samples of surface seawater and determination of the chromium in the precipitates and particulate matter by thin-film X-ray fluorescence spectrometry. An ultraviolet irradiation procedure was used to release bound metal. The ratios of labile trivalent chromium to total chromium were in the range 0.4–0.5 and the totals of labile tri- and hexavalent chromium were in the range $0.3-0.6 \, \mu\text{g} \, \text{l}^{-1}$. Bound chromium ranged from 0 to $3 \, \mu\text{g} \, \text{l}^{-1}$, and represented 0–90% of total dissolved chromium. Acidification of the samples in the usual manner for the determination of trace metals altered the proportion of trivalent to hexavalent chromium.

Cobalt

Various methods have been proposed for the determination of traces of cobalt in seawater and brines, most necessitating pre-concentration. Solvent extraction followed by spectrophotometric measurements [160–167] is the most popular method but has many sources of errors; the big difference in the volumes of the two phases results in mixing difficulties, and the solubility of the organic solvent in the aqueous phase changes the volume of organic phase resulting in decreased reproducibility of the measurements. In many cases, excess of reagent and various metal complexes are co-extracted with cobalt and cause errors in determining the absorbance of the cobalt complex.

The procedure of Kentner and Zeitlin [160] is as follows: to a filtered 750 ml sample of seawater add 20% aqueous sodium citrate (25 ml) 30% aqueous hydrogen peroxide (1 ml) and 1% ethanolic 1-nitroso-2-naphthol (treated with activated charcoal and filtered) (1 ml) and set aside for 10 min. Extract the cobalt complex into chloroform and back-extract the excess of reagent into basic wash solution (mix 1 M sodium hydroxide (50 ml), 20% aqueous sodium citrate (10 ml) and 30% aqueous hydrogen peroxide (1 ml) with water to produce 100 ml (3 × 5 ml), shaking for 60 s for each extraction. Extract copper and nickel from the organic phase into 2 M hydrochloric acid (5 ml), back-extract any released reagent into basic wash solution (5 ml) and wash the chloroform phase again with 2 M hydrochloric acid (5 ml). Dilute the organic phase to 50 ml (for 30 ml cells) or 30 ml (for 20 ml cells) and measure the extinction at 410 nm against a blank in 10 cm cells.

In another spectrophotometric procedure Motomizu [161] adds to the sample (2 litres) 40% (w/v) sodium citrate dihydrate solution (10 ml) and a 0.2% solution of 2-ethylamino-5-nitrosophenol in 0.01 M hydrochloric acid (20 ml). After 30 min, add 10% aqueous EDTA (10 ml) and 1,2-dichloroethane (20 ml), mechanically shake the mixture for 10 min, separate the organic phase and wash it successively with hydrochloric acid (1:2) (3 × 5 ml), potassium hydroxide (5 ml) and hydrochloric acid (1:2) (5 ml); filter and measure the extinction at 462 nm in a 50 mm cell. Determine the reagent blank by adding EDTA solution before the citrate solution. The sample is either set aside for about 1 day before analysis (the organic extract should then be centrifuged) or preferably, it is passed through a 0.45 μm membrane-filter. The optimum pH range for samples is 5.5–7.5. From 0.07 to 0.16 μg l^{-1} of cobalt was determined; there is no interference from species commonly present in seawater.

Harvey and Dutton [168] determined nanogram amounts of cobalt in seawater after concentration on manganese dioxide formed by photochemical oxidation of divalent manganese in a photochemical reactor. The sample (1 litre) containing 100 μg manganese was irradiated in a Hanovia reactor fitted with a 2 W low-pressure Hg discharge lamp radiating mainly at 254 and 185 nm. The manganese dioxide deposit that adhered to the silica jacket of the reactor was dissolved in 0.15 M hydrochloric acid containing a trace of sulphur dioxide, the solution was evaporated to dryness and the residue was dissolved in 4 ml or 0.625 M hydrochloric acid; 1 ml 5 M aqueous ammonia and 0.1 ml of 0.1% dimethylglioxime' in ethanol were added, and cobalt was determined by pulse polarography. The polarograph was operated in the derivative mode, starting at −1.0 V and a 50 mV pulse height and 1 s mercury drop life were used.

Kouimtzis, Apostolopoulou and Staphilakis [169] described a spectrophotometric method for down to 1 μg l^{-1} cobalt in seawater in which the cobalt is extracted with 2,2' dipyridyl-2-pyridylhydrazone (DPPH) [170,172–175] and the cobalt complex is back-extracted into 20% perchloric acid and this solution is evaluated spectrophotometrically at 500 nm. This avoids many of the sources of error that occur in earlier procedures.

Isshiki and Nakayama [171] have discussed the selective concentration of cobalt in seawater by complexation with various ligands or sorption on macroporous XAD resins. Complexed cobalt is collected after passage through a small XAD resin packed column.

Copper

The importance of complexing agents in the mineral nutrition of phytoplankton and other marine organisms has been recognized for more than 20 years. Complexing agents have been held responsible for the solubilization of iron and therefore, its greater biological availability [176]. In contrast, complexing agents are assumed to reduce the biological availability of copper and minimize its toxic effect [177–190]. Experiments with pure cultures of phytoplankton in chemically defined media have demonstrated that copper toxicity is directly correlated with cupric ion activity and independent of the total copper concentration. In these experiments, cupric ion (Cu^{2+}) concentrations was varied in media containing a wide range of total concentrations through the use of artificial complexing agents. When Cu^{2+} concentration was calculated for earlier experiments with phytoplankton in defined media, it appeared that Cu^{2+} was toxic to a number of phytoplankton species in concentrations as low as $10^{-6}\,\mu mol\,l^{-1}$ [186]. Since copper concentrations in the world ocean typically range from 10^{-4} to $10^{-1}\,\mu mol\,l^{-1}$ complexing agents and other materials affecting the solution chemistry of copper must maintain the Cu^{2+} activity at sublethal levels.

Copper may exist in particulate, colloidal and dissolved forms in seawater. In the absence of organic ligands, or particulate and colloidal species, carbonate and hydroxide complexes account for more than 98% of the inorganic copper in seawater [191,192]. The Cu^{2+} concentration can be calculated if pH, ionic strength and the necessary stability constants are known [191–194]. In most natural systems, the presence of organic materials and sorptive surfaces significantly alters speciation and decreases the utility of equilibrium calculations. Analytical difficulties in the measurement of Cu^{2+} and copper associated with naturally occurring ligands has encouraged numerous workers to introduce the 'complexation capacity' concept [178,195–197]. Functionally, the copper complexing capacity of a water sample is the ability of the sample to remove added copper from the free ion pool [198]. Analytically, complexation capacity measurements depend on quantitation of the complexing ability of an operationally defined group of ligands. The assumption is made that unknown ligands may be classed into meaningful groups on the basis of the physical properties of their metallo-complexes (e.g. lability to anodic stripping voltammetry (ASV), chelating resins, or ultraviolet radiation). Schemes to determine the concentration of copper associated with different classes have been proposed as useful ways to address complexing capacity questions in natural systems [199,200] as have various titrametric techniques [201,202]. Different analytical procedures measure the copper chelating capacity of slightly different classes of ligands and there is some overlap in the complexes included in classes defined by different techniques. For example, while there is a small fraction of organic material in seawater which forms ASV-labile complexes not dissociated by Chelex resin [203] most ASV-labile complexes are also labile to chelating resins [204].

Work on the determination of copper in seawater is predominantly concerned with speciation. Before discussing this the limited amount of work on the determination of total copper, i.e. work not concerned with speciation, is discussed below.

Determination of total copper

Abraham, Winpe and Ryan [205] determined total copper in seawater spectrophotometrically using quinaldehyde 2-quinolyl-hydrazone as chromogenic reagent. This method is capable of determining copper at the ppb level.

Atomic absorption spectrometry has been used to determine copper [206,207].

Muzzaralli and Rocchetti [206] showed that Chitosan is superior to Dowex A700 ion-exchange resin and modified celluloses for the collection of copper from unoxidized seawater. In this procedure the sample is passed through a column (30 × 3 mm) packed with chitosan (100 mg: 100–200 mesh) and the chelated copper eluted with a 1% solution of 1.10-phenanthroline (20 ml) at 50°C or with 1 M sulphuric acid (20 ml). Place an aliquot (20 μl) in a hot graphite analyser programmed to dry for 20 s, char for 20 s and atomize for 10 s. Determine the amount of copper present by comparison with standards. Average recoveries from the column were 100% and the coefficient of variation was = 7.5% for 3.4 μg of copper per litre.

A hanging mercury drop electrodeposition technique has been used [207] for a carbon filament flameless atomic absorption spectrometric method for the determination of copper in seawater. In this method, copper is transferred to the mercury drop in a simple three-electrode cell (including a counter-electrode) by electrolysis for 30 min at −0.35 V versus the SCE. After electrolysis, the drop is rinsed and transferred directly to a prepositioned water-cooled carbon-filament atomizer and the mercury is volatilized by heating the filament to 425°C; copper is then atomized, and determined by atomic absorption. The detection limit is 0.2 μg copper per litre simulated seawater.

Ion-selection electrodes have been used for the potentiometric determination of the total cupric ion content of seawater [208]. Down to $2 \mu g \, l^{-1}$ cupric copper could be determined by this procedure.

Background copper levels in seawater have been measured by electron spin resonance techniques [209]. The copper was extracted from the seawater into a solution of 8-hydroxyquinoline in ethyl propionate (3 ml extractant per 100 ml seawater) and the organic phase (1 ml) was introduced into the electron spin resonance tube for analysis. Signal-to-noise ratio was very good for the four-line spectrum of the sample and of the sample spiked with 4 and 8 ng Cu^{2+}, the graph of signal intensity versus concentration of copper was rectilinear over the range 2–10 $\mu g \, l^{-1}$ of seawater, and the coefficient of variation was 3%.

Traces of copper and lead are separated [210] from macro amounts of calcium, magnesium, sodium and potassium by adsorption from the sample on to active carbon modified with hydroxyquinoline, dithizone or diethyldithiocarbamate.

Marvin, Proctor and Neal [211] have discussed the effects of sample filtration on the determination of copper in seawater and concluded that glass filters could seriously affect the reliability of subsequent analysis.

Copper speciation
Wood, Evans and Alberts [212] have described an ion-exchange technique for the measurement of the copper complexing capacity of seawater samples taken in the Sargasso Sea and continental shelf samples. This technique measures the copper complexing capacity of relatively strong dissolved and colloidal organic complexing agents in natural seawater. The technique was used to compare the copper complexing capacity of strong organic dissolved and colloidal complexing agents in those samples. They also analysed the relationship between the copper complexing capacity of a specific group of complexing agents and the concentration of two large heterogeneous pools of potential complexing agents; dissolved organic carbon and total particular material. The copper complexing capacity of these samples ranged from 0.014 to 1.681 μmol Cu per litre on the inner shelf, from 0.043 to 0.095 mol Cu per litre in mid- and outer shelf waters and from 0.010 to 0.036 μmol Cu per litre at the Sargasso Sea stations.

The ion-exchange procedure used by Wood, Evans and Alberts [212] to estimate copper complexation capacity was a modification of that used by Stolzberg and Rosin [213] and Giesy [214]. Excess Cu^{2+} is added to the filtered samples and allowed to equilibrate with available ligands; the sample is then passed through a column packed with Nahelex resin (Biorad 100–200 mesh) and Cu^{2+} and Cu associated with weak or rapidly dissociating complexes are removed by the resin and Cu remaining in the sample after chromatography provides a quantitative measure of the Cu-chelating capacity of strong ligands remaining in the sample. The procedure has the advantage that complex formation proceeds at seawater pH in a relatively undisturbed sample. However, the procedure also depends on the assumption that essentially all the Cu ligands in the sample are associated with Cu. This involves the reaction:

$$mCu^{2+} + L^{n-} = Cu_mL^{(n-2m)-}$$

All chromatography was conducted at flow rates greater than $20\,ml\,cm^{-2}\,s^{-1}$ since slower flow rates resulted in complex dissociation (Figure 4.11).

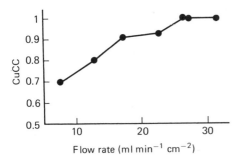

Figure 4.11 Copper complexing capacity (CuCC) as a function of sample flow rate through the ion-exchange column. A Gulfstream water sample was filtered through an acid-washed precombusted Reeve Angel glass-fibre filter (984H) and spiked with $0.78\,\mu mol\,l^{-1}$ Cu ($1.57\,\mu mol\,l^{-1}\,dm^{-3}$ final concentration) and $0.2\,\mu mol\,l^{-1}$ EDTA ($0.4\,\mu mol\,l^{-1}\,dm^{-3}$ final concentration). Samples were equilibrated overnight and chromatographed as described in the text but with variable flow rates. CuCC is expressed as a proportion of the maximum observed CuCC ($0.265\,\mu mol\,l^{-1}$ Cu dm^{-3}). Assuming 1:1 stoichiometry of the EDTA: Cu complex, these results indicate that under the analytical conditions used, approximately half the Cu complexed with EDTA passes through the column and contributes to CuCC estimates. Experiments conducted with glycine showed that none of the weak complexes formed with this ligand passed through the column. From these data we conclude that most of the ligands contributing to the CuCC measured in natural waters formed relatively strong (K_3 18) complexes

The concentration of copper in the column eluent was determined by flame atomic absorption spectroscopy of samples which were preconcentrated with ammonium pyrrolidine dithiocarbamate (APDC) and methyl isobutyl ketone. The pH of the acidified sample was adjusted to pH 2.5–3.5 using 400 µl 8 M ammonium acetate (Chelex cleaned).

Zorkin, Grill and Lewis [215] developed a procedure to estimate the amount of biologically active copper in seawater based on the assumption that the divalent copper ion was the most toxic species and its concentration could be related to the amount of metal sorbed on a sulphuric-acid cation-exchange resin. The method was tested by application to artificial seawater containing copper and the organic ligands EDTA, NTA, histidine, and glutamic acid. Other experiments showed there was a correlation between the inorganic copper fraction determined by the ion-exchange procedure, and the toxic fraction of copper quantified by a bioassay using the marine diatom.

In recent years much of the work concerned with the speciation of copper and other trace metals in natural waters has been done using anodic stripping voltammetry. This work has primarily progressed in two directions: studies of the shift in trace metal peak potentials with changing concentrations of ligands [216–220] and studies of changes in metal peak height or peak area under differing experimental conditions. Variants of the second approach include pH titrations [220–222] and compleximetric titrations [223] in which natural or added ligands are quantitatively titrated with metal ions or, alternatively, metal ions are titrated with ligands [223–225]. In this technique, the electrolysis potention is set at a value which presumably discriminates between the 'free' (i.e. rapidly reducible) metal and the complexed metal which is reduced at a much slower rate. This approach has been used extensively to estimate the 'complexation capacity' of natural waters.

Techniques based on the shift of the peak potential depend on the degree of reactivity of the oxidized metal with the ligand of interest in the reaction layer. They can describe the species undergoing reduction, i.e. the speciation in the natural medium, only indirectly and by assuming reversibility. Thus they are more suitable for model studies [216,226] and for the determination of stability constants in known media [217,218,227] than for direct determination of natural speciation. On the other hand, methods dependent on peak height or peak area can give direct information on the natural species as long as a direct proportionality exists between the quantity of metal reduced during electrolysis and the metal oxidized from the amalgam. One relatively novel form of anodic stripping voltammetry which gives information about the species undergoing reduction is stripping polarography, sometimes called pseudopolarography [228–230].

In this technique, peak heights or peak areas obtained by anodic stripping voltammetry are plotted against the applied electrolysis potential. These plots have the sigmoidal shape of ordinary d.c. polarograms but without the residual current component and present the possibility of extending existing polarographic methodology to trace metals at the part per billion (μg l^{-1}) level. The shapes of the plots indicate the degree of reversibility of the species undergoing reduction and may be useful for their identification.

Zirino and Kounaves [231] applied this technique to a study of the reduction of copper in seawater. Figure 4.12 shows three plots of 6 μg l^{-1} copper added to unfiltered seawater from San Diego Bay and analysed under varying conditions. Each of the experimental points represents the copper peak current obtained by anodic stripping voltammetry after a 5-min electrolysis at a hanging mercury drop

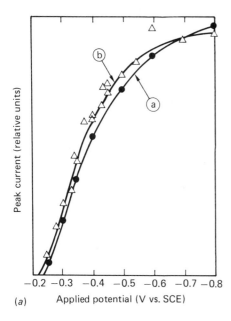
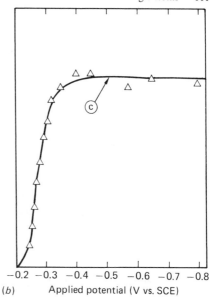

Figure 4.12 Stripping polarographic plots of 6 ppb Cu in seawater (not to scale). (*a*) Raw seawater, pH 8. (*b*) Ultraviolet-oxidized seawater, pH 8. (*c*) Raw seawater, pH 3

electrode. The plots obtained for copper at pH 8 (Figure 4.12, curves a and b) feature broadly curving slopes and no distinct limiting plateau is reached, even at the highest applied potential. The shapes of these 'waves' indicate an irreversible reduction. On the other hand, the reduction of copper at pH 3 is quasi-reversible with $E3/4 - E1/4 = 42\,mV$.

Potentiometric stripping analysis has been applied by Sheffrin and Williams [232] to the measurement of copper in seawater at environmental pH. The advantage of this technique is that it can be used to specifically measure the biologically active labile copper species in seawater samples at desired pH values. The method was applied to seawater samples that had passed a $0.45\,\mu m$ Millipore filter. Samples were studied both at high and at low pH values.

These workers used a radiometer ISS 820 ion-scanning system [233–236] equipped with a glassy carbon electrode to determine copper at the 2–$200\,\mu g\,l^{-1}$ level in nitrogen purged $0.45\,\mu m$ Millipore filtered seawater to which had been added 5 ppm mercury.

The speciation of copper is different at high and low pH. At pH 1.0 most of the copper will be labile and a total copper concentration will be measured. At pH 7.7 there should be a smaller proportion of labile copper, as much will be complexed in various forms, depending on the constituents of the seawater.

Because of this complexation capacity, any standard addition performed at high pH will not return 100% of the spike, so a true value for the copper concentration cannot be calculated. Therefore, after an initial measurement at high pH the sample was acidified to pH 1.0 with 0.5 ml acid and another trace obtained. This compared the amount of copper released at low pH with the labile fraction at high pH. Standard additions were performed on the sample at low pH so almost all of

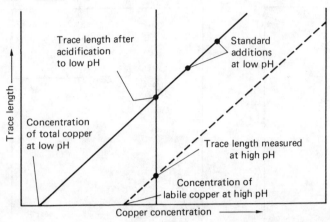

Figure 4.13 Measurements of labile copper at high pH

Table 4.19 Analysis of 'poor quality' seawater by potentiometric stripping analysis and atomic absorption spectroscopy and 'good quality' seawater by potentiometric stripping analysis

Analysis	Measured copper concentration (ppb)	pH	Sample treatment	Electrolysis voltage	Plating time (min)
Potentiometric stripping analysis of 'poor quality' seawater	8.3	1.7	Acidified	−0.6	8
	8.6	1.1	Acidified	−1.1	8
	6.8	1.7	Acidified	−0.6	8
	9.6	1.7	Acidified	−0.6	16
	8.6	1.3	Acidified and boiled (15 min)	−0.6	32
	14.0	1.6	Acidified and boiled (15 min)	−0.6	8
	15.8	1.6	Acidified and boiled (15 min)	−0.6	8
	Mean (±SD) = 10.24 ± 3.3				
AAS/ATDC -IBMK* analysis of 'poor quality' seawater	35 samples: Mean (±SD) = 12.4 ± 3.1				
Potentiometric stripping analysis of 'good quality' seawater	1.4	1.3	Acidified and boiled (15 min)	−0.6	32
	2.2	0.9	Acidified	−0.6	32
	8.6	1.3	Acidified	−0.6	32
	0.36	7.0	Ultraviolet irradiation (67 min)	−0.6	32

*Ammonium tetramethylenedithiocarbamate - isobutyl methyl ketone.
Reproduced from Sheffrin, V. and Williams, E.E. (1982) *Analytical Proceedings*, **19**, 483, Society for Analytical Chemistry, by courtesy of authors and publishers.

the spike was returned. This allowed an estimate to be made of the percentage of total copper that was labile at high pH and the quantification of this fraction in $\mu g\,l^{-1}$. This is illustrated graphically in Figure 4.13.

The analysis of total copper by potentiometric stripping analysis depends on the way any bound copper is released; that is, whether the sample is acidified and to what pH, whether it is acidified and boiled, or treated with ultraviolet radiation and then acidified. Results obtained using these different methods are given in Table 4.19 which compares the analysis of a 'poor quality' i.e. low pH, aquarium seawater by potentiometric stripping analysis and by atomic-absorption spectrometry after extraction with ammonium tetramethylene-dithiocarbamate-isobutyl methyl ketone [237,238] and also a potentiometric stripping analysis of a 'good quality' i.e. higher pH seawater. The difference between the results for the 'poor quality' seawater analysed by the two techniques was not significant.

Shuman and Michael [239,240] introduced a technique that has sufficient sensitivity for kinetic measurement at very dilute solutions. It combines anodic scanning voltammetry with the rotating disk electrode and provides a method for measuring kinetic dissociation rates *in situ* and a method for distinguishing labile and non-labile complexes kinetically, consistent with the way they are defined.

Odier and Plichon [241] used a.c. polarography to determine the chemical form and concentration of copper in seawater. The shift of the $E_{1/2}$ of reduction of Cu^{II} determined by a.c. polarography serves to identify the inorganic complexes of copper and to determine their formation constants. They showed that copper is present in seawater mainly as Cu^{2+}, $CuCl^+$ and $(Cu(HCO_3)_2 (OH))^-$. Copper down to $3 \mu g\, l^{-1}$ was determined in seawater by a.c. polarography after acidifying to pH 5 by passage of carbon dioxide.

Turner, Robinson and Whitfield [116] studied the application of the automated electrochemical stripper to the determination of copper in seawater.

Neutron activation analysis has been used [242] to determine total copper in seawater.

Cathodic stripping voltammetry has been used to determine copper species in seawater [487,488]. Van der Berg [243] determined copper in seawater by cathodic stripping voltammetry of complexes with catechol.

A reduction current occurred when a solution of catechol and copper was subjected to cathodic stripping voltammetry at a hanging mercury drop electrode. The composition of the adsorbed film and the optimal conditions for its formation were investigated. The phenomenon was used to determine copper in seawater using a.c. polarography to measure complex adsorption. Currents were detected at the very low copper concentrations that occurred in uncontaminated seawater. Competition for copper ions by natural organic complexing ligands was evident at low concentrations of catechol. The method was more sensitive and had a shorter collection period than the rotating disc electrode DPASV technique, with comparable accuracy.

Nelson [244] studied voltammetric measurement of copper II–organic interactions in estuarine waters. Based on results of previous studies on the effects of organic matter on adsorption of copper at mercury surfaces, Nelson developed a method to evaluate the interactions between divalent copper and organic ligands, based on ligand exchange. The copper/organic species competed with glycine, which formed copper glycinate and these two complexes could be distinguished voltammetrically, since copper glycinate gave a higher surface excess of copper at a gelatin-coated hanging-drop mercury electrode. The method was applied successfully to both chloride media and estuarine waters. It was demonstrated that estuarine waters contained two types of ligand capable of binding divalent copper; humic material with polyelectrolyte type binding, and discrete ligands with stability constants of about 1000 million. The extent of binding by humic material decreased down the estuary as a result of dilution and increased salinity.

Germanium

This is discussed later in this chapter under hydride generation techniques.

Gold

Pilipenko and Pavlova [245] determined traces of gold in seawater using a 'spot' photometric method. This method is based on the catalysis (by $AuCl_2SO_4^-$) of the oxidation of Fe^{II} by Ag^I with production of metallic silver. The sample is filtered through paper, and the paper is dried and decomposed with sulphuric acid, nitric acid, hydrofluoric acid, water solution (1 : 1 : 1 :1). The residue is dissolved in a few drops of aqua regia, this solution is evaporated and the residue is dissolved in one drop of 0.05 M sulphuric acid. This solution is applied to filter paper and to the resulting spot are applied drops of phosphate–citrate buffer solution (pH 2.4) of 0.72 M ferrous sulphate, in 0.05 M sulphuric acid and of 0.1 M silver nitrate; 15 s after the silver nitrate solution has been applied, the reflectance of the spot (due to metallic silver) is measured with a suitable instrument. The reflectance is proportional to the amount of gold on the paper from 3 to 60 pg. As little as $0.0025 \, \mu g \, l^{-1}$ of gold can be determined in seawater.

Indium

Matthews and Riley [246] determined indium in seawater at concentrations down to $1 \, ng \, l^{-1}$. Pre-concentration of metals on a cation exchanger was followed by separation of alkali metals and alkaline-earth metals by retention of indium as a chloro-complex on an anion exchanger. Samples of indium containing eluate were then concentrated and irradiated in a thermal-neutron flux of 5×10^{12} neutrons cm^{-2} s^{-1} for several weeks and the resulting 1.98 MeV β-radiation of the long-lived ^{114m}In daughter nuclide was counted. Minor elements were removed by a series of post-irradiation solvent-extraction stages.

Iron

Atomic absorption spectrometry coupled with solvent extraction of iron complexes has been used to determine down to $0.5 \, \mu g \, l^{-1}$ iron in seawater [247,248]. Hiire [247] extracted iron as its 8-hydroxyquinoline complex. The sample is buffered to pH 3–6 and extracted with a 0.1% methyl isobutyl ketone solution of 8-hydroxyquinoline. The extraction is aspirated into an air–acetylene flame and evaluated at 248.3 nm.

Moore [248] used the solvent extraction procedure of Danielson, Magnusson and Westerlund [249] to determine iron in frozen seawater. To a 200 ml aliquot of sample was added 1 ml of a solution containing sodium diethyldithiocarbamate (1% w/v) and ammonium pyrrolidine dithiocarbamate (1% w/v) in 1% ammonia solution and 0.65 ml 1 M hydrochloric acid to bring the pH to 4. The solution was extracted three times with 5 ml volumes of 1,1,2 trichloro-1,2,2 trifluorethane and the organic phase evaporated dryness in a silica vial and treated with 0.1 ml Ultrex hydrogen peroxide (30%) to initiate the decomposition of organic matter present. After an hour or more, 0.5 ml 0.1 M hydrochloric acid was added and the solution irradiated with a 1000 W Hanovia medium pressure mercury vapour discharge tube at a distance of 4 cm for 18 min. The iron in the concentrate was then compared with standards in 0.1 M hydrochloric acid using a Perkin-Elmer Model 403 Spectrophotometer fitted with a Perkin-Elmer graphite furnace (HGA 2200).

The coefficient of variation of analyses was 21% for seven subsamples containing 1.6 nmol Fe l^{-1} and 30% for eight sub-samples at 0.6 nmol Fe l^{-1} per litre. The detection limit was estimated to be 0.2 nmol Fe l^{-1} per litre.

The efficiency of the extraction procedure was tested using seawater spiked with iron-59, which indicated a recovery of 97% and with stable iron of 86%.

Radioisotope dilution using the chelating agent bathophenanthroline has been used to determine down to $5\,\mu g\,l^{-1}$ iron in seawater [250].

Shriadah and Ohzeki [251] determined iron in seawater by densitometry after enrichment as a bathophenanthroline disulphonate complex on a thin-layer of anion exchange resin. Seawater samples (50 ml) containing iron (II) and iron (III) were diluted to 150 ml with water followed by sequential addition of 20% hydrochloric acid (100 μl), 10% hydroxylammonium chloride (2 ml), 5 M ammonium solution (to pH 3.0 for iron (III) reduction) bathophenanthroline disulphonate solution (1.0 ml) and 10% sodium acetate solution (2.0 ml) to give a mixture with a final pH of 4.5. A macroreticular anion exchange resin, Amberlyst A27, in the chloride form was added, the resultant coloured thin layer was scanned by a sensitometer and the absorbance measured at 550 nm.

Lead

Various workers [252–256] have applied graphite furnace atomic absorption spectrometry to the determination of lead in seawater.

The large amount of sodium chloride in seawater samples causes non-specific absorption [257–261] which can be only partially compensated by background correction. In addition the seawater matrix may give rise to chemical as well as physical interferences related to the complex physico-chemical phenomena [262–264] associated with vaporization of metals and of the matrix itself.

Several matrix modifiers, which alter the drying or charring properties of the sample matrix, have been tested [265–269] to reduce non-specific absorption. However, the matrix modification methods do not permit determinations of the indigenous lead in seawater because of the relatively high detection limit and poor precision. Yet gross chemical manipulations of the samples should be avoided to prevent contaminations which can be dramatic when the analyte is present at $\mu g\,l^{-1}$ or sub-$\mu g\,l^{-1}$ level.

With the temperature-controlled heating method described by Lundgren, Lundmark and Johansson [270] the heating rate can be chosen independently of the final temperature, thus permitting a selective volatilization. However, this method cannot be used successfully for the determination of lead in strong sodium chloride solutions like seawater, because the temperature at which atomization of lead is rapid coincides with the volatilization temperature of sodium chloride. Ashing of seawater samples by a hydrogen diffusion flame [271] was successful in the direct determination of Fe, Ni and Cu but cannot be applied for lead because hydrogen is not sufficiently effective as a suppressor for the Pb-NaCl system.

In order to overcome the problem of the high non-specific absorption, alternative procedures have been tested, which involve prior separation of the trace metals from the salt matrix. Examples of extraction of trace metals from seawater as chelates with subsequent determination by electrothermal atomic absorption spectrometric procedures have been described [272,273] but these, and similar methods, are seldom effective and satisfactory when the matrix is very complex and the analyte concentration very low.

In contrast, the coupling of electrochemical and spectroscopic techniques, i.e. electrodeposition of a metal followed by detection by atomic absorption spectrometry, has received limited attention. Wire filaments, graphite rods, pyrolytic graphite tubes and hanging drop mercury electrodes have been tested [274–285] for electrochemical preconcentration of the analyte to be determined by atomic absorption spectroscopy. However, these *ex situ* pre-concentration methods are often characterized by unavoidable irreproducibility, contaminations arising from handling of the support and detection limits unsuitable for lead detection at sub-ppb levels.

These drawbacks could be certainly avoided by performing *in situ* deposition. The sole attempt in this direction was made by Torsi [286] who set up an apparatus which permitted both *in situ* electrochemical pre-concentration of the analyte from a flowing solution and almost complete suppression of matrix effects because the matrix could be removed by suitable washing. The feasibility of this approach was successfully tested with respect to lead determinations in different salt solutions (mainly ammonium acetate) and some preliminary results were reported for real samples, such as seawater and urine [286].

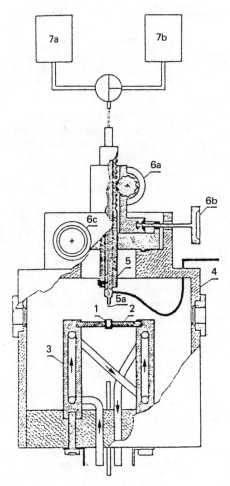

Figure 4.14 Overall view of the apparatus. 1 = Vitreous carbon crucible; 2 = graphite rod; 3 = water-cooled, steel column electrical leads; 4 = plexiglas cover; 5 = feeder; 5a = feeder tip; 6(a–c) = slide knobs; 7(a,b) = washing and sample solution reservoirs. The glassy carbon crucible (1) was 8 mm high, 5 mm o.d., 3 mm i.d., 6 mm deep (Le Carbon Lorraine, Paris) graphite rods (2), which keep the crucible firmly in position. Water-cooled stainless steel columns (3) press the graphite rods against the crucible by two screws hidden inside and act also as electrical leads. The plexiglas box (4) allowed the use of the controlled inert atmosphere necessary to avoid drastic reduction of the absorption signal caused by oxygen. The solution feeder (5) can be moved up and down by means of a knob (6a) into a metal block attached to the upper part of the plexiglas box. A three-way stopcock at the cylinder top connects, by Teflon tubing (1.5 mm o.d., o.8 mm i.d.), the feeder tip (5a) to the washing and sample solution reservoirs (see below). Other knobs (6b and c) enable the feeder to be moved in the horizontal plane. The three knobs permit a micrometric spatial adjustment of the feeder tip

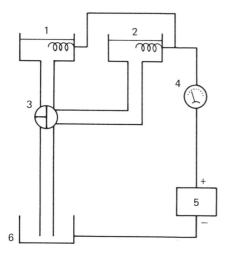

Figure 4.15 Electrolysis circuit layout. 1, 2 = Sample and washing solution compartments; 3 = three-way stopcock; 4 = ammeter; 5 = 500 V d.c. variable power supply; 6 = crucible

Torsi *et al.* [286,287] have carried out a systematic investigation to establish the potentialities of such an apparatus. The apparatus is basically an electrothermal device in which the furnace (or the rod) is replaced by a small crucible made of glassy carbon. Figure 4.14 represents an overall view of the apparatus. Figure 4.15 shows a block diagram of the electrolysis circuit; the crucible (6) acts as a cathode while the anode is a platinum foil dipped into either the sample solution reservoir (1) or the washing solution reservoir (2). The pre-electrolysis was performed at constant current by a 500 V d.c. variable power supply (5). Under these conditions, the cathode potential is not controlled so that other metals can be co-deposited with lead. There is no great need to control the deposition potential, because the spectral selectivity is sufficiently good to prevent interferences by other metals during the atomic absorption step.

A typical measurement was performed as follows. The feeder was lowered into the crucible and the sample solution (seawater) was allowed to flow under an inert atmosphere with the suction on. A constant current was applied for a predetermined time. When the pre-electrolysis was over, the flow was changed from the sample to the ammonium acetate washing solution, while the deposited metals were maintained under cathodic protection. Ammonium acetate was selected for its low decomposition temperature, and a $0.2 \, ml \, l^{-1}$ concentration was used to ensure sufficient conductivity. At this point the feeder tip was raised to the highest position and the usual steps for an electrothermal atomic absorption spectrometry measurement were followed: drying for 30 s at 90°C, ashing for 30 s at 700°C and atomization for 8 s at 1700°C with measurement at 283.3 nm. A typical atomization signal obtained in this way is shown in Figure 4.16. As can be seen, the baseline increases smoothly with time as a consequence of an upward lift of the crucible caused by thermal expansion of the material.

A calibration curve for lead in seawater obtained by the standard addition method is shown in Figure 4.17. A deviation from linearity is observed at higher lead concentrations. The estimated value for the original sample was found to be $0.51 \, \mu g \, l^{-1}$ with confidence limits at the 95% confidence level of $\pm \, 0.036 \, \mu g \, l^{-1}$, compared with a value of $0.65 \, \pm \, 0.08 \, \mu g \, l^{-1}$ obtained by anodic scanning

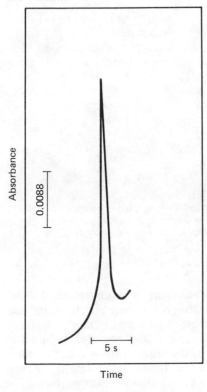

Absorbance

0.0088

5 s

Time

Figure 4.16 Typical recorder trace of seawater containing 2.8 ngPb^{2+}per ml after a 2 min electrolysis time

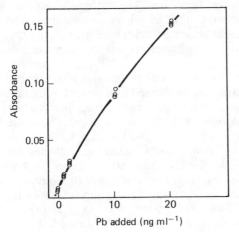

Absorbance

0.15

0.10

0.05

0   10   20

Pb added (ng ml^{-1})

Figure 4.17 Calibration curve for lead in seawater, pH 1.9

voltammetry. This value is well within the normal range reported in the literature for the natural lead content of unpolluted seawater. A detection limit of 0.03 ng ml^{-1} was obtained.

Halliday, Houghton and Ottaway [288] have described a simple rapid graphite furnace method for the determination of lead in amounts down to 1 µg l^{-1} in polluted seawater. The filtered seawater is diluted with an equal volume of deionized water, ammonium nitrate added as a matrix modifier and aliquots of the solution injected into a tantalum-coated graphite tube in a HGA-2200 furnace atomizer. The method eliminates the interference normally attributable to the ions commonly present in seawater. The results obtained on samples from the Firth of Forth were in good agreement with values determined by anodic stripping voltammetry.

Hirao *et al.* [289] concentrated lead in seawater using a chloroform solution of dithizone and determined it in amounts down to $40\,\mu g\,l^{-1}$ by graphite furnace atomic absorption spectrometry. Lead in 1 kg acidified seawater was equilibrated with ^{212}Pb of a known radioactivity, extracted with dithizone in chloroform, back-extracted with 0.1 M hydrochloric acid and subjected to graphite furnace atomic absorption spectrometry by a two-channel spectrometer. Recovery yield of lead was found to be 60–90% from the radioactivity of ^{212}Pb in the back-extract. Lead concentrations were thus determined with about 10% precision.

Early work [290] on the application of cyclic and anodic stripping voltammetry to the determination of lead showed a poor correlation between peak current values and Pb^{II} concentration at high pH values. This is due to the low electrochemical activity of PbOH.

Acebal, DeLuca and Rebello [291] discussed the quantitative behaviour of lead (and copper) when voltammetric determinations are done at mercury film electrodes and hanging mercury drop electrodes. The samples were collected in polyethylene bottles and, generally, were not acidified immediately after collection. This might place some doubt on the results reported.

Voltammetric measurements were done with a PAR Electrochemical System (Model 174-A) and a saturated calomel reference electrode, a platinum wire auxiliary electrode and a glassy carbon rod (PAR 0333) coated with a mercury film as the working electrode. A Pyrex glass cell was used for measurements with the hanging drop mercury electrode; either this cell or a Teflon cell was used with the mercury film electrode. No advantage concerning adsorption or contamination was found when a Teflon cell was used for seawater at pH 2. Stirring was done magnetically. Nitrogen with a maximum oxygen content of 10 ppm was used as purging gas. Mercury (II) nitrate used to form *in situ* films on the glassy carbon rod was prepared from tridistilled mercury and nitric acid.

Experimental parameters used in all the determinations are given in Table 4.20. Unless otherwise mentioned, all potentials specified are referred to the saturated calomel electrode.

Table 4.20 Experimental conditions used for the voltammetry of lead

Electrode	HMDE	MFE	
Mode	D.p.a.s.v.	L.s.a.s.v.	D.p.a.s.v.
Peak potential (V/SCE)	−0.43	−0.44 to −0.47	−0.52 to −0.56
Stirring speed (rpm)	360	430	430
Mercury drop size (cm^2)	0.022	–	–
Pulse height (mV)	50	–	25–50
Pulse repetition (s)	0.5	–	0.5
Scan rate ($mV\,s^{-1}$)	5	50	5
Electrodeposition time (min)	5–15	5–30	30
Resting time (s)	30	30	30
pH	1.4–1.7	1.4–1.7	1.4–1.7
Mercury concentration	–	$(2.2–4.4) \times 10^{-5}$ M	$(2.2–4.4) \times 10^{-5}$ M
Electrodeposition potential (V/SCE)	−0.7, −0.9	−0.8	−0.8

Reproduced from Acebal, S.A. and De Luca Rebello, A. (1983) *Analytica Chimica Acta*, **148**, 71, Elsevier Science Publishers, Amsterdam, by courtesy of authors and publishers.

This work showed that application of the linear sweep mode at pH 1.5 is a fast and reliable way of dealing with interferences caused by organic materials in polluted water. The lower sensitivity of this mode limits its use to lead contents exceeding $0.1 \, \mu g \, kg^{-1}$ but such levels are commonly reached in polluted waters.

The determination of lead in seawater collected from 14 stations in Guanabara Bay gave values between 0.07 and $5.5 \, g \, kg^{-1}$.

Automated chemical stripping has been used for the determination of lead in seawater [116].

Quentel, Madoc and Courtotoupez [295] studied the influence of dissolved organic matter in the determination of lead in seawater by anodic stripping voltammetry.

Ultraviolet spectroscopy has been applied to the determination of lead and lead speciation studies [293]. Scaule and Patterson [294] used isotope dilution–mass spectrometry to determine the lead profile in the open North Pacific Ocean.

Further work on the determination of lead is discussed in the section on multielement analyses.

Manganese

The natural water chemistry of manganese, which is an important trace element both biologically and geologically is complicated by non-equilibrium behaviour. From thermodynamic considerations manganese dioxide (manganese (IV)) is expected to be the stable form of manganese in seawater [296]. However, seawater contains a significant quantity of dissolved manganese (II) which is only slowly oxidized (Murray and Brewer) [297]. Investigations of the oxidation of manganese (II) have shown that the process is autocatalytic, the product being a solid manganese oxide phase whose composition depends on the reaction conditions [297–300]. The heterogeneous nature of the oxidation process can thus account for the extremely slow oxidation of manganese (II) in seawater where concentrations of particulate matter can be relatively small [297].

Estuaries, in contrast, appear to be important sites for manganese redox reactions. Manganese maxima have been observed in several estuaries [301–303] and it has been suggested that these maxima result from a recycling of precipitated manganese [303]. The proposed mechanism is essentially a redox cycle in which dissolved manganese (II) is oxidized in the water column and precipitated. Reduction in anoxic sediments results in the subsequent release of manganese (II) to the water column.

The details of estuarine manganese chemistry are far from clear, however; Sholkovitz [304] notes that while adsorption onto colloidal humic acids or hydrous iron oxides is a major factor controlling the removal of many trace metals from estuarine waters manganese does not conform well to this pattern as it is known to behave independently of iron and associates only weakly with organic matter. Detailed investigation of estuarine manganese reactions requires analytical methods specific to the species involved; a requirement met only by electrochemical methods at natural concentration levels. Davison [305,306] has described the use of direct polarographic methods in the analysis of manganese in lake waters in the concentration range $0.1–5 \, mg \, l^{-1}$.

Manganese is of particular interest because of its central role in many marine geochemical processes and involvement in biological systems. Manganese and many other trace metals are present in open ocean waters at concentrations in the order of nmol l^{-1} or less, and it has only been in the past 5–10 years, when adequate contamination control measures have been applied during sampling and measurements, that accurate data have been obtained.

Graphite-furnace atomic absorption spectrometry, although element-selective and highly sensitive, is currently unable to directly determine manganese at the lower end of their reported concentration ranges in open ocean waters. Techniques that have been successfully employed in recent environmental investigations have thus used a preliminary step to concentrate the analyte and separate it from the salt matrix prior to determination by atomic absorption spectrometry.

The determination of manganese in seawater using graphite furnace atomic absorption spectrometry has been investigated by many workers [307–315]. If the seawater matrix is atomized along with the analyte, the result is a large background signal which is often beyond the correcting capabilities of current instrumentation. The presence of large amounts of chlorides has also been shown to provide interferences [316,317] usually making direct analysis difficult.

To reduce these problems, most workers either have used matrix modification [307,310,311,315] or have extracted the metal from the seawater matrix [309,313]. Few workers have been successful with the direct determination in seawater after volatilization of the matrix during the char program step [307,308,312,314,318]. Slavin and Manning [319–321] have shown that by using a furnace at steady-state temperature (the L'vov platform), the interference of chloride on manganese determination was greatly reduced as long as the background signal was within the limits that the deuterium arc background corrector could handle.

Segar and Gonzalez [308] attributed the reduced sensitivity for manganese in a seawater matrix to covolatilization of some manganese with the salt matrix. More recent work suggests that this reduced sensitivity is a vapour-phase binding of a portion of the manganese by chloride.

Ediger, Peterson and Kerber [322] showed that it was necessary to char away as much as possible of the seawater matrix to get maximum sensitivity for manganese and to be free of interference.

Segar and Cantillo [314] developed a direct method for manganese in seawater with a detection limit for manganese of about 0.3 µg l^{-1}. Only ordinary graphite tubes were available and they found that, as the tube aged, the analytical signal fell linearly at a rate of 50% per 100 firings. Since variations in salinity produced relatively large changes in signal, the method of standard additions was required.

McArthur [310] preferred to use ammonium nitrate matrix modification to determine manganese in seawater. Most of his paper discussed the charring process. He found considerable salinity dependence if charring was too rapid. There was considerable change in the salinity dependence with the age of the tubes.

Kingston et al. [313] resorted to extraction on Chelex 100 followed by stripping into nitric acid. The ammonium nitrate matrix modification technique was used by Montgomery and Peterson [315] for the determination of manganese in seawater. They showed that the pyrolytically coated tubes they used deteriorated very rapidly using the combination of ammonium nitrate and seawater. Manganese was determined in seawater (with copper and cobalt) by Hydes [311] after adding 1% ascorbic acid to the sample. He used the Perkin Elmer HGA-2100 furnace and found significant loss of manganese from seawater between 600°C and 900°C.

The direct furnace method of Sturgeon et al. [307] for manganese was very similar to the method of Segar and Cantillo [312,314]. The Sturgeon detection limit was 0.22 µg l^{-1} for manganese in seawater, using 20 µl samples in the HGA-2200 furnace and pyrolytically coated graphite tubes. They found a loss in sensitivity during the life of the tubes. They had to use the method of additions to accommodate small residual interference.

Procedures using chelation followed by extraction have been described for manganese using the 8-hydroxy-quinoline-chloroform system [323,324]. Dithiocarbamate systems can simultaneously extract manganese as well as other trace metals under suitable conditions [325–327].

Statham [328] has optimized a procedure based on chelation with ammonium pyrrolidine dithiocarbamate and diethylammonium dithyldithiocarbamate for the preconcentration and separation of dissolved manganese from seawater prior to determination by graphite furnace atomic absorption spectrometry. Freon-TF was chosen as solvent because it appears to be much less toxic than other commonly used chlorinated solvents, it is virtually odourless, has a very low solubility in seawater, gives a rapid and complete phase separation and is readily purified. The concentrations of analyte in the back-extracts are determined by graphite furnace atomic absorption spectrometry (Table 4.21). This procedure concentrates the trace metals in the seawater 67.3 fold.

When a 350 ml seawater sample was spiked with ^{54}Mn and taken through the chelation, extraction and back-extraction procedures, the observed recovery of the radio-tracer was 100.6%. Estimates of blanks and detection limits for manganese based on sets of both ship-bound and shore laboratory separations are given in Table 4.22.

Table 4.21 Spectrometer and furnace settings

	Cd	Mn
Wavelength (nm)	228.8	279.5
Bandpass (nm)	0.5	0.5
Lamp current (ma)	3	5
Furnace programs:		
(temperature in °C)		
dry	135°/35 s LR7*	135°/35 s LR7
ash	400°/10 s LR4	1000°/10 s LR4
atomize	1800°/4 s TC†	2600°/4 s TC
clean	2000°/3 s TC	2700°/3 s TC
Injection volume (μl)	5	15

*Linear ramp rate.
†Optical feedback temperature control.
Reproduced from Statham, P.J. (1985) *Analytica Chimica Acta*, **169**, 149, Elsevier Science Publishers, by courtesy of author and publishers.

Table 4.22 Blanks and detection limits for manganese (nmol l^{-1}) (350 ml sample)

Mean blank	n*	DL†
0.15	5	0.11
0.13	5	0.10
0.07	10	0.10

*Number of observations.
†Detection limit defined as three times the standard deviation of the blank.
Reproduced from Statham, P.J. (1985) *Analytica Chimica Acta*, **169**, 149, Elsevier Science Publishers, Amsterdam, by courtesy of author and publishers.

Table 4.23 Precision of the developed technique for manganese in seawater

Concentration (nmol l⁻¹)	n^*	RSD (%)†
0.96	5	10
3.24	4	5

*Number of observations.
†Relative standard deviation.
Reproduced from Statham, P.J. (1985) *Analytica Chimica Acta*, **169**, 149, Elsevier Science Publishers, Amsterdam, by courtesy of author and publishers.

Estimates of the precision of the developed technique for dissolved manganese are given in Table 4.23; all samples were natural seawaters.

The accuracy of the technique is demonstrated by data from the ICES fifth round intercalibration exercise for trace metals in seawater [329].

Klinkhammer [330] has described a method for determining manganese in a seawater matrix for concentrations ranging from about 30 to 5500 ng l⁻¹. The samples are extracted with 4 nmol l⁻¹ 8-hydroxyquinoline in chloroform and the manganese in the organic phase is then back-extracted into 3 M nitric acid. The manganese concentrations are determined by graphite furnace atomic absorption spectrophotometry. The blank of the method is about 3.0 ng l⁻¹ and the precision from duplicate analyses is ± 9% (1 SD).

The theoretical yield of the method is less than 100% since only 80–90% of the aqueous phase is removed after the back-extraction. The actual yield obtained by ^{54}Mn counting was 69.5 ± 7.8% and this can be allowed for in the calculation of results. Environmental Protection Agency standard seawater samples of known manganese content (4370 ng l⁻¹) gave good manganese recoveries (4260 ng l⁻¹).

Burton [331] has also described an atomic absorption method for the determination of down to 0.3 nmol l⁻¹ manganese in seawater.

Samples for the analysis of manganese were pressure filtered through 0.4 μm nucleopore filters. To 350 ml filtrate, 20 ml an aqueous solution of the complexing agents (2% w/v in both ammonium and diethyl ammonium diethyl dithiocarbamate) were added, and the solution extracted first with 35 ml and then with 20 ml Freon for 6 min. The combined extracts were shaken with 100 μl of concentrated nitric acid for 30 s. After standing for 5 min 5 ml distilled water was added and the solution shaken for 30 s. The aqueous phase was separated and combined with that from a further back-extraction using the same procedure. The combined aqueous solutions were returned to the shore laboratory and manganese determined by electrothermal atomic absorption spectrophotometry.

Bender and co-workers [332,333] determined total and soluble manganese in seawater. The samples were collected into 500 ml polyethylene bottles. All samples were brought to pH 2 with nitric acid free of trace metals and stored in individual zip-lock plastic bags to minimize contamination.

When the samples were returned to the laboratory the pH was adjusted to approximately pH 8 using concentrated ammonia (ultrapure, G. Frederick Smith). Twenty ml of chelating cation exchange resin in the ammonia form (Chelex-100, 100–200 mesh, Bio-Rad) was added to the samples and they were batch extracted on a shaker table for 36 hours. The resin was decanted into columns and the manganese eluted using 2 N nitric acid [334]. The eluant was then analysed by graphite furnace atomic absorption spectrophotometry. Replicate analyses of samples indicate a precision of about 5%.

Graphite furnace atomic absorption spectrometry with the L'vov platform and Zeeman background correction has been applied to the determination of down to $0.02\,\mu g\,l^{-1}$ manganese in seawater [335].

Knox and Turner [336] have described a polarographic method for manganese (II) in estuarine waters which covers the lower concentration range $10{-}300\,\mu g\,l^{-1}$. The method, which is specific to manganese (II) and its labile complexes, is used in conjunction with a colorimetric technique to compare the levels of manganese (II) and total dissolved manganese in an estuarine system. They showed that polarographically determined manganese (II) can vary widely from 100% to less than 10% of the total dissolved manganese, determined spectrophotometrically at 45 nm by the formaldoxine method [337] calibrated in saline medium to overcome any salt effects. It is suggested that the manganese not measured by the polarographic method is in colloidal form.

Polarography of Mn^{II} was carried out using a PAR 174 A Polarographic Analyser (Princeton Applied Research Corporation) in the differential pulse mode in conjunction with the cell and electrode system shown in Figure 4.18. The Luggin

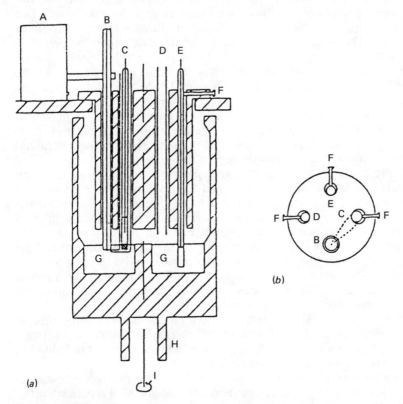

Figure 4.18 Diagrammatic representation of the cell and electrode system used for the analysis of manganese (II): (a) side view, (b) plan view. The cylindrical Electrocell (McKee-Pederson Instruments) has been modified to accommodate a dropping mercury electrode. The cell is approximately 5 cm in diameter and 7 cm high. Hatching = Perspex cell and electrode holder; cross-hatching = porous Vycor plug; A = PAR 174/70 drop timer; B = dropping mercury electrode; C = Ag/AgCl/0.1 M NaCl reference electrode with Luggin capillary; D = nitrogen inlet; E = platinum foil counter-electrode; F = fixing screw; G = 15 ml sample volume; H = attachment to Electrocell motor; I = axis of rotation

capillary arrangement ensures accurate potential control in low salinity waters where the supporting electrolyte concentration is low. A standard dropping mercury electrode mounting (PAR 174/70 drop timer head) is combined with a modified version of the rotatable Electrocell (McKee-Pederson Instruments [338]).

Figure 4.19 presents total manganese and manganese (II) obtained in estuary surveys. On several occasions the polarographically active manganese (manganese II) is significantly less than the total dissolved manganese. Generally the difference between manganese (II) and total manganese is greatest at low salinities.

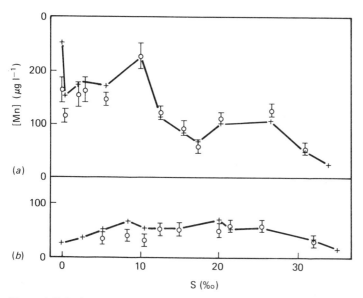

Figure 4.19 Surface manganese concentrations in the Tamar Estuary (SW England) (a) 12 July 1978 (b) 11 September 1978. + —— + = Total dissolved manganese (colorimetric); $\bigcirc$ = Polarographically determined manganese (II) with 95% confidence limits; $\dagger$ = Manganese (II) below polarographic detection limit ($10\,\mu g\,l^{-1}$)

Biddle and Wehry [339] carried out fluorimetric determination of manganese II in seawater via catalysed enzymatic oxidation of 2,3-diketogulonate. The detection limit was $8\,\mu mol\,l^{-1}$ Mn (II).

Olaffson [340] has described a semi-automated determination of manganese in seawater using leucomalachite green. The autoanalyser had a 620 nm interference filter and 50 min flow cells. Findings indicated initial poor precision was related to pH, temperature and time variations. With strict controls on sample acidity and reaction conditions, the semi-automated method had high precision, at least as good as that achieved by preconcentration and atomic absorption procedures and provided precise, rapid, ship-board information on the continental distribution of manganese and on anomalies associated with geothermal sea-floor activity. The method was not suitable for estuarine samples nor quite sensitive enough for study of the open-ocean manganese distribution.

Neutron activation analysis has been used to determine total manganese in seawater [242,341].

Wiggins, Duffey and El Kady [341] used neutrons from the thermal column of a 10 kW pool-type research reactor and from a 120 μg [252] Cf source to study the prompt-photon emission resulting from neutron capture in manganese nodules (terromanganese oxides) from the ocean floor. Spectra were recorded with a Ce(Li) detector and a 1024-channel analyser. Complex spectra were obtained by irradiation of seawater, but it was possible to detect and estimate manganese in nodules in a simulated marine environment by means of the peaks at 7.00, 6.55, 6.22 and 6.04 μV.

Mercury

Atomic absorption and atomic fluorescence techniques using closed system reduction–aeration have been applied widely to determine mercury concentrations in natural samples [342–356].

Typical of these methods is that of Topping and Pirie [352] in which the mercury is concentrated by drawing air for 5 h (600 ml min^{-1}) through a mixture of the sample (4 litres) and 20% stannous chloride solution, in 5 M hydrochloric acid (45 ml) and absorbing mercury vapour from the air stream in 20 ml 2% $KMnO_4$ solution: 50% (v/v) sulphuric acid (1:1). To the absorption solution was added 15 ml of the stannous chloride solution and this mixture was aerated at 2 l min^{-1}. The air and mercury vapour are passed through a 15 cm gas cell (with silica windows) in an atomic-absorption spectrophotometer for measurement at 253.65 nm. Samples containing down to 2 ng Hg l^{-1} could be analysed by this procedure.

Olaffson [356] described a similar procedure in which the sample (450 ml) is acidified with nitric acid, aqueous stannous chloride is added and the mercury is entrained in a stream of argon into a silica tube wound externally with resistance wire and containing pieces of gold foil, on which the mercury is retained. The tube and its contents are then heated electrically to about 320°C and the vaporized mercury is swept by argon into a 10 cm silica absorption-cell in an atomic-absorption spectrophotometer equipped with a recorder. The absorption (measured at 253.7 nm) is directly proportional to the amount of mercury in the range 0–24 ng per sample.

In many applications such as the analysis of mercury in open ocean seawaters where the mercury concentrations can be as small as 10 ng l^{-1} [352,356–360], a preconcentration stage is generally necessary. A preliminary concentration step may separate mercury from interfering substances, and the lowered detection limits attained are most desirable when sample quantity is limited. Concentration of mercury prior to measurement has been commonly achieved either by amalgamation on a noble metal metal [344,351,353,356] or by dithizone extraction [346,356,359] or extraction with sodium diethyldithiocarbamate [359]. Preconcentration and separation of mercury has also been accomplished using a cold-trap at the temperature of liquid nitrogen.

Voyce and Zeitlin [361] have used adsorption colloid flotation to determine mercury in seawater. The sample 500 ml is treated with concentrated hydrochloric acid, an aqueous solution of cadmium sulphate and a fresh aqueous solution of sodium sulphate are added. The pH is adjusted to pH 1.0 and the poured into a flotation cell with a nitrogen flow of 10 ml min^{-1}. Ethanolic octadecyltrimethylammonium chloride is injected and the froth dissolved in aqua regia in a flameless atomic absorption cell.

Following reduction of mercury with stannous chloride the mercury vapour is flushed from the system.

To determine organically bound mercury, the sample is treated (500 ml) with 0.5 M sulphuric acid aqueous potassium permanganate and set aside for 24 h. Aqueous hydroxylammonium chloride is added and the determination completed as above. Calculate the amounts of mercury in the samples by reference to the standard absorptions. Average recoveries of 0.05 µg mercury were 88%.

Fitzgerald, Lyons and Hunt [362] have described a method based on cold trap preconcentration prior to gas phase atomic absorption spectrometry for the determination of down to $2\,ng\,l^{-1}$ mercury in seawater.

The cold-trap is created by the immersion in liquid nitrogen of a glass U-tube packed with glass beads (80/100 mesh). After reduction, purging, and trapping, the mercury is removed from the glass column by controlled heating, and the gas phase absorption of eluting mercury is measured. This procedure has been employed for both shipboard and laboratory analyses of mercury in seawater.

The mercury analyses were conducted using a Coleman Instruments mercury analyser (MAS-50) equipped with a recorder. The aqueous sample solution was contained in a 250 ml Pyrex glass bubbler placed at one end of a sampling train employing nitrogen as the purging and carrier gas. A schematic diagram of the entire system is shown in Figure 4.20.

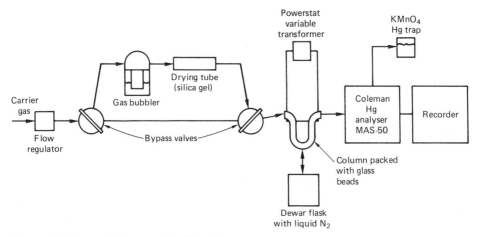

Figure 4.20 Schematic diagram of the Hg cold-trap pre-concentration system for the determination of mercury by gas phase absorption

Fitzgerald, Lyons and Hunt [362] showed that the most significant quantities of mercury occurred in the waters of the Atlantic Oceans continental shelf and slope $21-78\,ng\,l^{-1}$ compared with open ocean samples ($2-11\,ng\,l^{-1}$).

These workers distinguished between inorganic mercury obtained by direct analysis on the sample as received and organic mercury (the difference between total mercury, obtained upon ultraviolet irradiation of the sample) and inorganic mercury.

To preserve natural water samples for mercury analysis, much care must be exercised to prevent loss of mercury during storage [363–365]. Coyne and Collins

Table 4.24 Summary of results (concentrations in ngl⁻¹)

Lab. no. (ref.)*	Pretreatment	Pre-concentration	Method	Blank	Det. lim.	Seawater x	Seawater s	SIGN/NS	Seawater + spike I (15.4 µg l⁻¹) x	s	% rec.	Seawater + spike II (143 µg l⁻¹) x	s	% rec.
1(1)	None	None	CVAA NaBH₄	?	5	5.3	0.8	NS	21.0	0	102	108	4	72
2	Br and HNO₃, 45°C, overnight	None	CVAA SnCl₃	?	20	ND	ND		ND	ND		173	29	
3(2)	None	None	NA	0–3	1–5	10.6	5.8	SIGN	24.5	2.4	90	132	13	85
4	None	None	CVAA SnCl₂	0	2.5	7.2	1.5	NS	20.7	0.6	88	133	1.5	88
5	None	Org. extr.	AA furnace	10	5	27.2	2.5	SIGN	41.6	3.5	94	145	13	82
6	None?	None	CVAA SnCl₂	9	18	31.7	11.2	SIGN	45.0	5.0	86	195	5	114
7	None	Au amalg.	CVAA SnCl₂	?	30	63.6	9.9	NS	83.5	10.0	129	148	12	59
8	None	Au amalg.	CVAA SnCl₂	1–2	2?	14.9	5.6	SIGN	36.4	1.6	140	204	13	132
9	None	None	CVAA SnCl₂	2.5	5.0	11.1	3.6	NS	23.3	3.1	79	152	7.2	99
10	100 ml sample + 2 ml KMnO₄ (5%) + 2 ml K₂S₂O₈ (5%) + 2 ml H₂SO₄ (conc.) 80°C, 2 h	None	CVAA HO·NH₃Cl SnCl₂	20	20	17.5	5.0		62.5	19	292	136	11	83
11	None	Au amalg.	CVAA SnCl₂	0.5	1.7	2.9	0.7	SIGN	17.9	3.3	97	146	7	100
12(3,4)	None	Ag amalg.	CVAA SnCl₂	5.4	0.2	7.1	1.9	SIGN	26.5	2.5	126	110	5	72
13	Br and HNO₃ added, 45°C, 16 h	None	CVAA SnCl₂	5.0	10.0	<10			10.0	0		70	0	
14	Oxidation KMnO₄, H₂SO₄	None	CVAA SnCl₂	<20	20				33.3	5.8		130	10	
16(5)	Oxidation H₂SO₄, HNO₃, H₂O₂	None	CVAA SnCl₂	225	50				658	30		385	46	
17	None	None	DPASV	<6	6	29.2	7.1	SIGN	153	23	806	343	53	219
18	None	Au amalg.	CVAA SnCl₂	1.1	0.7	2.2	0.4	NS	17.5	0.6	100	130	2	89

Ref	Preparation	Collection	Method											
20(6,7)	None	SnCl₂ Coll Hg in brominating sol.	CVAF	0.4–1.0	0.6	2.4	0.4	NS	17.2	2.9	96	115	5	100
22	500 ml sample 10 ml KMnO₄ (2%) + 10 ml H₂SO₄ (50%)	SnCl₂ Coll. Hg in KMnO₄/H₂SO₄ sol.	CVAA SnCl₂	20	16	44	4.6		37	4.2		166		85
24(4)	None	Au amalg.	CVAA SnCl₂	0.3–1.9	0.4	2.5	0.4	NS	18.1	0.8	101	113	4	98
25	None	Au amalg.	CVAA SnCl₂	1.4	1.0	0.7	0.3	NS	21.0	1.3	132	76	9.9	53
26	200 ml sample + 9 ml H₂SO₄ + KMnO₄ (5%) + K₂S₂O₈ (5%) + 12% NaCl/HO·NH₃Cl	None	CVAA SnCl₂	5	2	3	1	NS	17.5	1.9	91	153	11	105
29(8)	None	Au amalg.	CVAA SnCl₂	0	1.0	2.9	1.1	NS	16.9	1.6	91	111	7	96
30(9)	(a) None (b) 300 ml sample + 3 ml 18 N H₂SO₄, heat	Au amalg.	CVAA SnCl₂	0.1	0.5	2.4	0.3		19.6	1.0	112	113	3	98
31	None	Ionic Sr:3 chel. res.	CVAA NaOH, HO·NH₃Cl SnCl₂	2–3	2–3	14.9			45.4	3.3	198	246	16	162
32	None	None	CVAA SnCl₂	0	1	3.8	0.5	NS	110	55	105	200	0	74
33	100 ml sample, H₂SO₄/HNO₃/KMnO₄/K₂S₂O₈ oxidation, HO·NH₃Cl + N₂ flush	None	CVAA SnCl₂	50	50	93.8	50	NS	110	55	105	200	0	74
34	None	None	CVAA SnCl₂	0	0.7	2.1	0.2	NS	19.4	0.9	112	150	3	104
35	14 ml HNO₃ (conc.) containing 0.5% K₂Cr₂O₇ per litre sample	None	CVAA SnCl₂	1.6	1.6	3.9	0.4	NS	24.5	0.9	134	176	5	121
36(4)	None	Au amalg.	CVAA SnCl₂	0	2	8.2	1.3	SIGN	26.2	1.2	117	164	8	109

*AA, atomic absorption; CVAA, cold vapour atomic absorption; CVAF, cold vapour atomic fluorescence; DPASV, differential pulse anodic stripping voltametry (gold disk electrode); NA, neutron activation (electrolytic deposition on gold); ND, not detectable; NS, no significant difference between seawater duplicates; SIGN, significant difference between seawater duplicates.
*1, reference 370; 2, 371; 3, 372; 4, 373; 5, 374; 6, 375; 7, 376; 8, 377; 9, 378.
Reproduced from Olafsen, J. (1982) Marine Chemistry, 11, 129, Elsevier Science Publishers, Amsterdam, by courtesy of authors and publishers.

[363] recommended preacidification of the sample bottle with concentrated nitric acid to yield a final pH of 1 in the sample solution. However, when this procedure was used for storage of seawater, fresh water, and distilled deionized water in low density polyethylene storage containers, abnormally high absorption was observed. This absorption at the mercury wavelength (253.7 nm) is due to the presence of volatile organic plasticizer material and any polyethylene residue leached by the concentrated nitric acid and by the acid solution at pH 1. Those procedures employing acidified sample storage in polyethylene bottles and gas phase mercury detection may be subject to artificial mercury absorption due to the presence of organic material.

Reported mercury values in the oceans determined since 1971 have spanned three orders of magnitude caused in part at least due to errors due to incorrect sampling [366]. Olafsson [367] has attempted to establish reliable data on mercury concentrations obtained in cruises in North Atlantic water.

The sampling, storage and analytical methods used by Olafsson [367] in this study have been evaluated. The Hydros-Bios water bottles used were modified by replacing internal rubber rings with silicone rubber equivalents. At the commencement of a cruise the water bottles were cleaned by filling with a solution of the detergent Deacon 90. Samples for the analysis of mercury were drawn into 500 ml Pyrex vessels and acidified to pH 1 with nitric acid (Merck 457), containing less than 0.05 nmol l^{-1} mercury impurities. The Pyrex bottles were precleaned with both nitric acid and a solution of nitric and hydrofluoric acids (10 : 1) and subsequently stored up to the time of sampling holding a small volume of nitric acid. Ashore, reactive mercury was determined by cold vapour atomic absorption after preconcentration by amalgamation on gold [368]. A 20 cm long optical cell and a Varian AA6 spectrophotometer were employed. The total mercury concentration was similarly determined following 1 hour ultraviolet irradiation of a duplicate sample using a 500 W low-pressure mercury lamp (Hanovia) and immersion irradiation equipment. The precision of the mercury determination assessed by analysing 19 replicates over a period of 107 days had been found to be ± 2.0 pmol l^{-1} for a concentration of 12.5 pmol l^{-1}.

Olafsson [369] has reported on the results obtained in an international intercalibration for mercury in seawater. Sixteen countries participated in this exercise, which involved analysis of a seawater and seawater spiked with 15.4 and 143 ng l^{-1} mercury. The results show for the majority of calibrations, good accuracy and precision in the recovery of spikes but serious errors in the low level determinations on the seawater (Table 4.24).

Since the intercalibration samples had been ultraviolet irradiated, the majority of participants in this exercise preferred to analyse the samples without any pretreatment. Ten did, however, employ oxidizing pretreatment and with three exceptions the results suggest that this approach should be taken with great caution.

Half of the participants have pre-concentrated mercury from the seawater prior to determination, eleven by amalgamation on gold, one by amalgamation on silver, two by collection into oxidizing solutions, one by organic extraction and one by ion-exchange chromatography.

Reduction to metallic mercury was used by an overwhelming proportion of the participants and with stannous chloride as reductant in all but one case, where sodium borohydride was used. In all cases but four, the participants used cold-vapour atomic absorption for final determination. This makes comparison of detection techniques difficult, but the good results obtained by cold-vapour atomic fluorescence are of interest and the spurious results obtained by differential pulse anodic stripping voltammetry may be indicative of the risk of mercury contamination in polarographic laboratories.

Filippelli [379] determined mercury at the subnanogram level in seawater using graphite furnace atomic absorption spectrometry. Mercury (II) was concentrated using the ammonium tetramethylenedithiocarbamate (ammonium pyrrolidine-dithiocarbamate, APDC)–chloroform system, and the chloroform extract was introduced into the graphite tube. A linear calibration graph was obtained for 5–1500 ng of mercury in 2.5 ml chloroform extract. Because of the high stability of the Hg^{II}–APDC complexes, the extract may be evaporated to obtain a crystalline powder to be dissolved with a few microlitres of chloroform.

About 84% of mercury was recovered in a single extract (97% in two extractions). The calibration graph was prepared by plotting the peak height against amount of mercury added to 500 ml distilled water. The optimized experimental conditions are as follows: lamp current, 6 mA; wavelength, 253.63 nm; drying, 100°C for 10 s; ashing, 200°C for 10 s; atomization, 2000°C for 3 s; and purge gas, nitrogen 'stopped flow'.

The coefficient variation of this method was about 2.6% at the $1 \mu g \, l^{-1}$ mercury level. The calibration graph is linear over the range 5–1500 μg mercury.

Other techniques that have been used include subtractive differential pulse voltammetry at twin gold electrodes [380], anodic stripping voltammetry using glassy-carbon electrodes [385], X-ray fluorescence analysis [381] and neutron activation analysis [382].

Fujita and Iwashima [383] pre-concentrated mercury compounds in seawater by first forming the diethyldithiocarbamate and then concentrating this on XAD-2 resin. The resin was eluted with methanol/3 M hydrochloric acid; the organic mercury was extracted with benzene and then back-extracted with cysteine

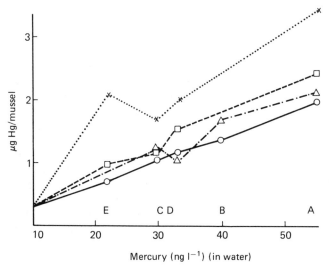

Figure 4.21 Relationship between mean total mercury concentration in water at the cage positions (A–E) and the mercury loadings per mussel after different exposure times. ○ = 20 days; △ = 55 days; □ = 106 days; × = 153 days

solutions. The organic mercury in the cysteine solution and the total mercury adsorbed on the resin were determined by flameless atomic absorption spectrometry. The method was applied to determinations of mercury levels in seawater in and around the Japanese archipelago. The lower limit of detection in seawater is $0.1\,\mathrm{ng}\,l^{-1}$ for organic mercury, using $80\,l$ samples.

Agemian and Da Silva [384] have described an automated method for total mercury in saline waters.

Bio-assay methods have been used to obtain estimates of low mercury concentrations $(5-20\,\mu g\,l^{-1})$ in seawater [386]. This method is useful for detecting comparatively small enhancements over background mercury concentrations in estuarine and seawater.

This method consists of suspending 70 Mussels Mytilus edulis each of a standard weight for a standard time, in a plastic coated wire cage 2 m below the surface. Mercury in the mussels was determined by cold vapour atomic absorption spectrometry [387,388]. The procedure is calibrated by plotting determined mercury content of mussels against the mercury content of the seawater in the same area (Figure 4.21).

Molybdenum

A limited amount of work has been carried out on the determination of molybdenum in seawater by atomic absorption spectrometry [389,390,391] and graphite furnace atomic absorption spectrometry [392]. In a recommended procedure [392] a 50 ml sample of seawater at pH 2.5 is passed through a column of 0.5 g p-aminobenzylcellulose, then the column is left in contact with 1 M ammonium carbonate for 3 h, after which three 5 ml fractions are collected. Finally, molybdenum is determined by atomic absorption at 313.2 nm with use of the hot-graphite-rod technique. At the $10\,\mathrm{mg}\,l^{-1}$ level, the standard deviation was $0.13\,\mu g$.

Emerick [393] showed that sulphate interferes with the graphite furnace atomic absorption determination of molybdenum in aqueous solutions with concentrations of only 0.1% (w/v) sodium sulphate causing complete elimination of the molybdenum absorbance peak in solutions free of other salts. Matrix modification with 0.5% (w/v) $CaCl_2.2H_2O$, in a volume equal to the sample, facilitates the determination of molybdenum in the presence of solutions containing as high as 0.4% (w/v) sodium sulphate. The need for matrix modification for molybdenum determination in natural waters appears to exist when sulphate greatly exceeds the equivalent calcium content. The routine use of a volume of 0.5% $CaCl_2.2H_2O$ equal to sample volume is recommended in the determination of molybdenum in seawater.

Nakahara and Chakrabarti [389] showed that the seawater salt matrix can be removed from the sample by selective volatilization at 1700–1850°C but the original presence of sodium chloride, sodium sulphate and potassium chloride causes a considerable decrease in molybdenum absorbance and magnesium chloride and calcium chloride a pronounced enhancement. The presence of magnesium chloride prevents the depressive effects. Samples of less than 50 μl can be analysed directly without using a background corrector with a precision of 10%.

These workers conclude that the selective volatilization technique is highly suitable for the determination of traces of molybdenum in synthetic (and most probably real) seawater samples. It has the advantages of freedom from contamination and loss during sample preparation and is faster, and cheaper, than procedures using separations.

The sensitivity achieved should allow seawater samples to be analysed for molybdenum, because the concentration of molybdenum in seawater is usually 2.1–18.8 µg l^{-1}. The selected temperature of 1700–1850°C during the charring stage permits separation of the seawater matrix from the analyte prior to atomization with the Perkin Elmer Model 603 atomic absorption spectrometer equipped with a heated graphite atomizer (HGA-2100).

Thorium hydroxide has been used as a collector of molybdenum from seawater [394]. To a 500 ml sample of seawater add 9 M sulphuric acid (1 ml) and 0.1 M thorium nitrate (3 ml) and adjust the pH to 6.0 with dilute aqueous ammonia. After 30 min collect the precipitate on an 8 µm Millipore filter, dissolve it in concentrated hydrochloric acid (2–3 ml), evaporate the solution to dryness and dissolve the residue in 6 M hydrochloric acid (5 ml). Dilute the solution to 40 ml with water, add 0.5 M ferric-chloride (one drop) and determine the molybdenum by the thiocyanate method spectrophotometrically.

Anion exchange resins have been used to preconcentrate molybdenum in seawater prior to its spectrophotometric determination as the Tiron complex [395–397].

Kuroda and Kawabuchi have concentrated molybdenum by anion-exchange from seawater containing acid and thiocyanate [398] or hydrogen peroxide [398,399] and determined it spectrophotometrically. Korkisch, Godl and Gross [400] have concentrated molybdenum from natural waters on Dowex 1-X8 in the presence of thiocyanate and ascorbic acid. A sodium citrate and ascorbic acid system has also been worked out for the concentration of molybdenum on Dowex 1-X8 (citrate form) as a citrate complex from tap and mineral waters.

Co-precipitation [401–405] and solvent extraction [406–410] have often been used to pre-concentrate molybdenum from seawater. The other pre-concentration methods available for molybdenum include co-crystallization [410,404,405,411], sorption on chitosan and modified cellulose [412,396,413,414], cation-exchange sorption on ZeoKarb 225 [417] and Chelex 100 [415,416] concentration on Sephadex G-25 [418] and (as the pyrrolidine dithiocarbamate complex) on charcoal [418].

Chou and Lum-Shui-Chan [419] investigated the use of atomic absorption in conjunction with solvent extraction using 1% oxime in methyl isobutyl ketone for preconcentration. The detection limit is 3 µg l^{-1}, in which a pre-concentration factor of 20 is employed. The disadvantages of the above system are that it requires a 100 ml sample and there are interferences, although some of these interferences can be eliminated [420].

In a method described by Kiriyama and Kuroda [421] molybdenum is sorbed strongly on Amberlite CG 400 (Cl form) at pH 3 from seawater containing ascorbic acid and is easily eluted with 6 M nitric acid. Molybdenum in the effluent can be determined spectrophotometrically with potassium thiocyanate and stannous chloride. The combined method allows selective and sensitive determination of traces of molybdenum in seawater. The precision of the method is 2% at a molybdenum level of 10 µg l^{-1}.

To evaluate the feasibility of this method Kiriyama and Kuroda [421] spiked a known amount of molybdenum and analysed it by the procedure and the results are given in Table 4.25. As can be seen, the recoveries for the 8 12g molybdenum added to 5000 or 1000 ml samples are satisfactory.

Table 4.25 Determination of molybdenum in saline water (0.5M NaCl) and seawater

Sample*	Sample volume (litres)	Mo added µg	Mo found µg	Original content (µg l^{-1})
Saline water	1.0	0	0.12, 0.05	av. 0.09
	1.0	8.48	8.42, 8.25	
			8.18, 8.56	av. 8.37 ± 0.15†
			8.42	
Seawater A	0.5	0	4.27, 4.51	8.54, 9.02
			4.39	8.78
	1.0	0	8.90	8.90
	0.5	4.24	8.73	8.98
	0.5	8.48	12.80	8.64
				av. 8.81 ± 0.19
Seawater B	0.5	0	4.85, 4.70	9.70, 9.40
			4.85	9.70
	1.0	0	9.48	9.48
	0.5	4.24	9.02	9.56
	0.5	8.48	13.20	9.44
				av. 9.55 ± 0.13
Seawater C	0.5	0	4.11, 4.08	8.22, 8.16
			4.29	8.58
	1.0	0	8.54	8.54
	0.5	4.24	8.44	8.40
	0.5	8.48	12.70	8.44
				av. 8.39 ± 0.17

*Seawater A: collected at Kamoike Harbour, Kagoshima Bay, Japan, on 23 June 1983, Salinity 33.48%
Seawater B: collected on the shore at Kushikino, East China Sea, on 30 June 1983, Salinity 34.03%
Seawater C: collected at Yamagawa Harbour, Kagoshima Bay, on 6 July 1983, Salinity 31.55%
†Average recovery of total present after addition of Mo.
Reproduced from Kiriyama, T. and Kuroda, R. (1984) *Talanta*, **31**, 472, by courtesy of authors and publishers.

Shriadah, Katoaka and Ohzeki [422] determined molybdenum VI in seawater by densitometry after enrichment as the Tiron complex on a thin layer of anion exchange resin. There were no interferences from trace elements or major constituents of seawater, except for chromium and vanadium. These were reduced by the addition of ascorbic acid. The concentration of dissolved molybdenum (VI) determined in Japanese seawater was 11.5 µg l^{-1} with a relative standard deviation of 1.1%.

An adsorbing colloid formation method has been used to separate molybdenum from seawater prior to its spectrophotometric determination by the thiocyanate procedure [423].

Kuroda and Tarui [424] developed a spectrophotometric method for molybdenum based on the fact that Mo^{VI} catalyses the reduction of ferric iron by divalent tin ions. The plot of initial reaction rate constant versus molybdenum concentration is rectilinear in the range 0.01–0.3 mg l^{-1} molybdenum. Serveral elements interfere, vis titanium, rhenium, palladium, platinum, gold, arsenic, selenium and tellurium.

Various other techniques have been used to determine molybdenum including adsorption voltammetry [425], electron-paramagnetic resonance spectrometry [426] and neutron activation analysis [427,428]. EPR spectrometry is carried out on the isoamyl alcohol soluble Mo(SCN)$_5$ complex and is capable of detecting 0.46 mg l^{-1} molybdenum in seawater. Neutron activation is carried out on the β naphthoin oxime [428] complex and the pyrrolidone dithiocarbamate and diethyldithiocarbamate complex [427]. The neutron activation analysis method [428] was capable of determining down to 0.32 µg Mo per litre seawater.

Monien *et al.* [429] have compared results obtained in the determination of molybdenum in seawater by three methods based on inverse voltammetry, atomic absorption spectrometry and X-ray fluorescence spectroscopy. Only the inverse voltammetric method can be applied without prior concentration of molybdenum in the sample and a sample volume of only 10 ml is adequate. Results of determinations by all three methods on water samples from the Baltic sea are reported, indicating their relative advantages with respect to reliability.

Van der Bery [430] carried out direct determinations of molybdenum in seawater by adsorption voltammetry. The method is based on complex formation of molybdenum (VI) with 8-hydroxyquinoline (oxine) on a hanging mercury drop electrode. The reduction current of adsorbed complex ions was measured by differential pulse adsorption voltammetry. The effects of variation of pH and oxine concentration and of the adsorption potential were examined. The method was accurate up to $300 \, nmol \, l^{-1}$. The detection limit was $0.1 \, nmol \, l^{-1}$.

Nickel

The concentration of nickel in natural waters is so low that one or two enrichment steps are necessary before instrumental analysis. The most common method is graphite furnace atomic absorption after pre-concentration by solvent extraction [431] or co-precipitation [432]. Even though this technique has been used successfully for the nickel analyses of seawater [433,434], it is vulnerable to contamination as a consequence of the several manipulation steps and of the many reagents used during preconcentration.

This element has been determined spectrophotometrically in seawater in amounts down to $0.5 \, \mu g \, l^{-1}$ as the dimethylglyoxime complex [435,436]. In one procedure [435] dimethylglyoxime is added to a 750 ml sample and the pH adjusted to 9–10. The nickel complex is extracted into chloroform. After extraction into 1 M hydrochloric acid, it is oxidized with aqueous bromine, adjusted to pH 10.4 and dimethylglyoxime reagent added. It is made up to 50 ml and the extinction of the nickel complex measured at 442 nm. There is no serious interference from iron, cobalt, copper, or zinc but manganese may cause low results.   In another procedure [436] the sample of seawater (0.5–3 litres) is filtered through a membrane-filter (pore size 0.7 µm) which is then wet-ashed. The nickel is separated from the resulting solution by extraction as the dimethylglyoxime complex and is then determined by its catalysis of the reaction of tiron and diphenylcarbazone with hydrogen peroxide with spectrophotometric measurement at 413 nm. Cobalt is first separated as the 2-nitroso-1-naphthol complex and is determined by its catalysis of the oxidation of alizarin by hydrogen peroxide at pH 12.4. Sensitivities are $0.8 \, \mu g l^{-1}$ (nickel) and $0.04 \, \mu g l^{-1}$ (cobalt).

Rampon and Cavelier [437] used atomic absorption spectrometry to determine down to $0.5 \, \mu g \, l^{-1}$ nickel in seawater. Nickel is extracted into chloroform from seawater (500 ml) at pH 9–10, as its dimethylglyoxime complex. Several extractions and a final washing of the aqueous phase with carbon tetrachloride are required for 100% recovery. The combined organic phases are evaporated to dryness and the residue is dissolved in 5 ml of acid for atomic-absorption analysis.

Lee [438] described a method for the determination of nanogram or subnanogram amounts of nickel in seawater. Dissolved nickel is reduced by sodium borohydride to its elemental form which combines with carbon monoxide to form nickel carbonyl. The nickel carbonyl is stripped from solution by a helium–carbon monoxide mixed gas stream, collected in a liquid nitrogen trap, and atomized in a quartz tube burner of an atomic absorption spectrophotometer. The sensitivity of the method is 0.05 ng of nickel. The precision for 3 ng nickel is about 4%. No interference by other elements is encountered in this technique.

Between 0.3 and 0.6 μg l^{-1} nickel was found by this method, in a vertical profile of water samples taken down to 1200 m in the Santa Catalina Basin.

Rare earths

Shigematsu *et al*. [439] determined cerium fluorimetrically at the 1 μg l^{-1} level in seawater. Quadrivalent cerium is co-precipitated with ferric hydroxide and the precipitate is dissolved in hydrochloric acid and interfering ions are removed by extraction with isobutyl methyl ketone. The aqueous phase is evaporated almost to dryness with 70% perchloric acid, then diluted with water and passed through a column of bis-(2-ethylhexyl) phosphate on poly (vinyl chloride), from which Ce^{IV} is eluted with 0.3 M perchloric acid. The eluate is evaporated, then made 7 M in perchloric acid and treated with Ti^{III} and the resulting Ce^{III} is determined spectrofluorimetrically at 350 nm (excitation at 255 nm).

Elderfield and Greaves [440] have described a method for the mass spectrometric isotope dilution analysis of rare earth elements in seawater. In this method, the rare earth elements are concentrated from seawater by co-precipitation with ferric hydroxide and separated from other elements and into groups for analysis by anion exchange [441–446] using mixed solvents. Results for synthetic mixtures and standards show that the method is accurate and precise to $\pm$1%; and blanks are low (e.g. 10^{-12} moles La and 10^{-14} moles Eu). The method has been applied to the determination of nine rare earth elements in a variety of oceanographic samples. Results for North Atlantic ocean water below the mixed layer are (in 10^{-12}mol kg^{-1}) 13.0 La, 16.8 Ce, 12.8 Nd, 2.67 Sm, 0.644 Eu, 3.41 Gd, 4.78 Dy, 407 Er and 3.55 Yb, with an enrichment of rare earth elements in deep ocean water, by two times for the light rare earth elements and by 1.3 times for the heavy rare earth elements.

Elution–volume calibrations were performed using radioactive tracers of the rare earth elements and ^{133}Ba, with atomic-absorption or flame-emission analysis of iron, sodium, potassium, calcium and magnesium. As shown in Figure 4.22 any barium added to the second columns is eluted at the start of the 'light rare earth element fraction' (see below). To ensure barium removal the sample can be put through the first column again.

Rhenium

Matthews and Riley [447] have described the following procedure for determining down to 0.06 μg l^{-1} rhenium in seawater. From 6 to 8 μg l^{-1} rhenium was found in Atlantic seawater. The rhenium in a 15 litre sample of seawater, acidified with hydrochloric acid, is concentrated by adsorption on a column of De-Acidite FF anion-exchange resin (Cl- form), followed by elution with 4 M nitric acid and evaporation of the eluate. The residue (0.2 ml) together with standards and blanks,

137

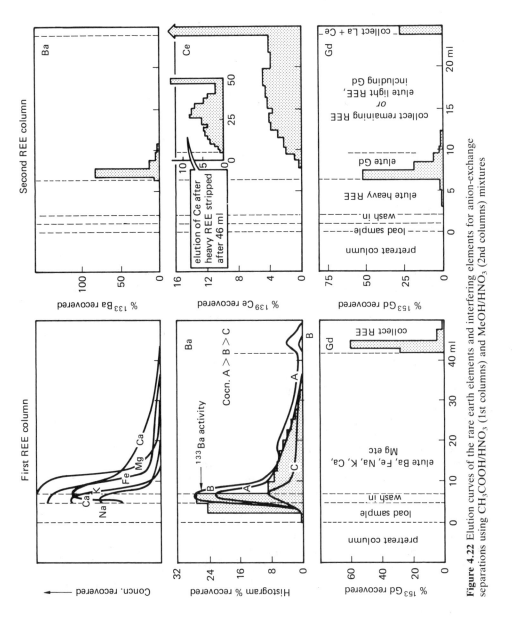

Figure 4.22 Elution curves of the rare earth elements and interfering elements for anion-exchange separations using CH_3COOH/HNO_3 (1st columns) and $MeOH/HNO_3$ (2nd columns) mixtures

is irradiated in a thermal neutron flux of at least 3×10^{12} neutrons $cm^{-2} s^{-1}$ for at least 50 h. After a decay period of 2 days, the sample solution and blank are treated with potassium perrhenate as carrier and evaporated to dryness with a slight excess of sodium hydroxide. Each residue is dissolved in 5 M sodium hydroxide. Hydroxylammonium chloride is added (to reduce Tc^{VII}) which arises are ^{99m}Tc from activation of molybdenum present in the samples, and the Re^{VII} is extracted selectively with ethyl methyl ketone. The extracts are evaporated, the residue is dissolved in formic acid–hydrochloric acid (19:1), the rhenium is adsorbed on a column of Dowex 1 and the column is washed with the same acid mixture followed by water and 0.5 M hydrochloric acid; the rhenium is eluted at 0°C with acetone–hydrochloric acid (19:1) and is finally isolated by precipitation as tetraphenylarsonium perrhenate. The precipitate is weighed to determine the chemical yield and the ^{186}Re activity is counted with an end-window Geiger–Muller tube. The irradiated standards are dissolved in water together with potassium perrhenate. At a level of 0.057 $\mu g l^{-1}$ rhenium the coefficient of variation was ±7%.

Selenium

In recent years, the physiological role of selenium as a trace element has created considerable speculation and some controversy. Selenium has been reported as having carcinogenic as well as toxic properties; other authorities have presented evidence that selenium is highly beneficial as an essential nutrient [448,449]. Its significance and involvement in the marine biosphere is not known. A review of the marine literature indicates that selenium occurs in seawater as selenite ions (SeO_3^{2-}) with a reported average of 0.2 $\mu g l^{-1}$ [450].

Various techniques have been applied to the determination of selenium in seawater including flameless atomic absorption spectrometry [451–453], gas chromatography [454–456] and adsorption collard flotation [457].

Neve, Hanocq and Molle [453] digested the sample with nitric acid. After digestion the sample is reacted selectively with an aromatic o-diamine and the reaction product is detected by flameless atomic absorption spectrometry, after the addition of nickel (III) ions. The detection limit is 20 mg l^{-1} and both selenium (IV) and total selenium can be determined. There was no significant interference in a saline environment with three times the salinity of seawater.

Sturgeon, Willie and Berman [452] preconcentrated selenium (IV) by adsorption of their ammonium pyrollidine diethyl dithiocarbamate chelates on to C_{18} bonded silica prior to desorption and determination by graphite furnace atomic adsorption spectrometry. The detection limit was 7 ng l^{-1} selenium (IV), based on a 300 ml water sample.

Shimoishi [454] determined selenium by gas chromatography with electron capture detection. To 50–100 ml seawater was added 5 ml concentrated hydrochloric acid and 2 ml 1% 4-nitro-o-phenylenediamine and, after 2 h, the product formed was extracted into 1 ml of toluene. Wash the extract with 2 ml 7.5 M hydrochloric acid, then inject 5 μl into a glass gas–liquid chromatography column (1 × 4 mm) packed with 15% of SE-30 on Chromosorb W (60–80 mesh) and operated at 200°C with nitrogen (53 min^{-1}) as carrier gas. There is no interference from other substances present in seawater.

Measures and Burton [455] used gas chromatography to determine selenite and total selenium in seawater. Siu and Berman [456] determined selenium (IV) in seawater by gas chromatography after co-precipitation with hydrous ferric oxide.

After co-precipitation, selenium is derivatized to 5-nitropiaz-selenol, extracted into toluene, and quantified by electron capture detection. The detection limit is $5\,ng\,l^{-1}$ with 200 ml samples and the precision at the 0.025 µg Se per litre level is 6%.

Ferric hydroxide co-precipitation techniques are lengthy, 2 days being needed for a complete precipitation. To speed up this analysis, Tzeng and Zeitlin [457] studied the applicability of an intrinsically rapid technique, namely adsorption colloid flotation. This separation procedure uses a surfactant–collector–inert gas system, in which a charged surface-inactive species is adsorbed on a hydrophobic colloid collector of opposite charge; the colloid with the adsorbed species is floated to the surface with a suitable surfactant and inert gas, and the foam layer is removed manually for analysis by a methylene blue spectrometric procedure. The advantages of the method include a rapid separation, simple equipment and excellent recoveries.

These workers used the flotation unit that was devised by Kim and Zeitlin [458].

The efficiency of the flotation procedure was studied by preparing two sets of seawater samples. To one set (A) was added 5 ml of standard selenium solution, the flotation procedure was carried out, and the concentration of selenium determined. Set B was treated identically except that the standard selenium solution was added to the foams after flotation of the unspiked seawater samples; the spiked foams were then analysed for selenium as described. The results of these tests are summarized in Table 4.26. The recovery of the selenium was found to be $100 \pm 10\%$ (at the 95% confidence level).

Table 4.26 Flotation recovery

	Set A			Set B	
Sample	Min⁻¹	pH	Sample	Min⁻¹	pH
1	0.130	5.3	1	0.130	5.3
2	0.130	5.3	2	0.132	5.3
3	0.129	4.0	3	0.124	4.0
4	0.135	4.0	4	0.126	4.0

Reproduced from Tseng, J.H. and Zeitlin, H. (1978) *Analytica Chimica Acta*, **101**, 71, Elsevier Science Publishers, Amsterdam, by courtesy of authors and publishers.

Using this method Tzeng and Zeitlin [457] found $0.40 \pm 0.12\,\mu g\,l^{-1}$ selenium in seawater.

Other methods that have been used to determine selenium in seawater include spectrophotometry [459–462] and neutron activation analysis [463–464].

The determination of selenium is discussed further in the later section on pre-concentration (hydride generation methods).

Silver

Kawabuchi and Riley [465] used neutron activation analysis to determine silver in seawater. Silver in a 4 litre sample of seawater was concentrated by ion exchange on a column (6 cm × 0.8 cm) containing 2 g of De-acidite FF-IP resin, previously treated with 50 ml 0.1 M hydrochloric acid. The silver was eluted with 20 ml 0.4 M aqueous thiourea and the eluate was evaporated to dryness, transferred to a silica irradiation capsule, heated at 200°C and ashed at 500°C. After sealing, the capsule was irradiated for 24 h in a thermal-neutron flux of 3.5×10^{12} neutrons $cm^{-2} s^{-1}$, and after a decay period of 2–3 days, the ^{110m}Ag arising from the reaction $^{199m}Ag(n,\gamma)$ ^{110m}Ag was separated by a conventional radiochemical procedure. The activity of the ^{110m}Ag was counted with an end-window Geiger–Muller tube, and the purity of the final precipitate was checked with a Ge(Li) detector coupled to a 400-channel analyser. The method gave a coefficient of variation of $\pm 10\%$ at a level of 40 ng silver per litre.

Tellurium

Petit [466] has described a method for the determination of tellurium in seawater at picomolar concentrations.

Tellurium (VI) was reduced to tellurium (IV) by boiling in 3 M hydrochloric acid. After pre-concentration by co-precipitation with magnesium hydroxide, tellurium was reduced to the hydride by sodium borohydrate at 300°C for 120 s then 257°C for 12 s. The hydride was then analysed by atomic absorption spectroscopy. Recovery was 90–95% and the detection limit $0.5 \, pmol \, l^{-1}$.

Tin

In an early method Kodama and Tsubota [467] determined tin in seawater by anion-exchange chromatography and spectrophotometry with catechol violet.

After adjusting to $2 \, mol \, l^{-1}$ in hydrochloric acid 500 ml of the sample is adsorbed on a column of Dowex 1-X s resin (Cl- form) and elution is then effected with 2 M nitric acid. The solution is evaporated to dryness after adding 1 M hydrochloric acid and the tin is again adsorbed on the same column. Tin is eluted with 2 M nitric acid. Tin is determined in the eluate by the spectrophotometric catechol violet method. There is no interference from 0.1 mg of each of aluminium, manganese, nickel, copper, zinc, arsenic, cadmium, bismuth and uranium, any titanium, zirconium and antimony are removed by the ion exchange. Filtration of the sample through a Millipore filter does not affect the results, which are in agreement with those obtained by neutron-activation analysis.

Brinckmann and co-workers [468] used a gas chromographic method with or without hydride derivatization for determining volatile organotin compounds (e.g. tetramethyltin), in seawater. For non-volatile organotin compounds a direct liquid chromatographic method was used. This system employs a 'Tenax-GC' polymeric sorbent in an automatic purge and trap (P/T) sampler coupled to a conventional glass column gas chromatograph equipped with a flame photometric detector (FPD). Figure 4.23 is a schematic of the P/T-GC-FPD assembly with typical operating conditions. Flame conditions in the FPD were tuned to permit maximum response to SnH emission in a H-rich plasma, as detected through narrow band-pass interference filters (610 ± 5 nm) [469]. Two modes of analysis were used: (1) volatile stannanes were trapped directly from sparged 10–50 ml water samples with no pretreatment; and (2) volatilized tin species were trapped from the same or replicate water samples following rapid injection of aqueous excess sodium borohydride solution directly into the P/T sparging vessel immediately prior to beginning the P/T cycle [470].

For either ion-exchange resolution of aqueous cations, $R_n Sn^{(4-n)+}$ aq [471] or their separation as ion-pairs, $[R_n Sn^{(4-n)+} X^- 4-n]^0$, on reverse bonded-phase columns [472], the method is restricted to 'free' tin analytes. Unlike the vigorous

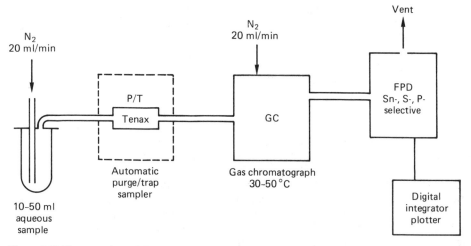

Figure 4.23 The purge/trap GC-FPD system and operating conditions

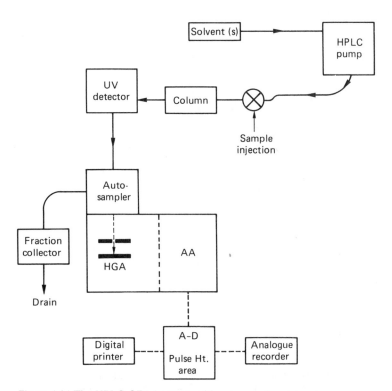

Figure 4.24 The HPLC-GFAA system with automated peripherals

hydride derivatization used in the gas chromatography–flame photometric detector method, common high performance liquid chromatography solvent combinations or their ionic addends usually will not provide sufficient co-ordination strength to labilize organotin ions strongly bound to solids in environmental samples. Moreover, the high performance liquid chromatography separations require that injected samples be free of particulates that may clog the column or pumping system.

On the other hand, high performance liquid chromatography if coupled with a sensitive element-specific detection system such as atomic absorption spectrometry offers a valuable tool for organotin speciation in complex fluids (especially for high organic loadings) not readily amenable to gas phase derivatization methods. Figure 4.24 shows a schematic of the basic high performance liquid chromatography set-up described by Brinckmann [468] coupled to a graphite furnace atomic absorption spectrometry in a manner giving automatic periodic (typically 45 s intervals) sampling of the resolved eluents for tin-specific determination [472]. Injected sample volumes may vary from 10 to 500 μl. Consequently, system sensitivity is broad and samples can be very representative.

Mixtures of R_3Sn^+ compounds (R = n-butyl, phenyl, cyclo-hexyl) were separated by ion-exchange–high performance liquid chromatography–graphite furnace atomic absorption spectrometry [448]. The small spread in calibration slopes in Figure 4.25 signifies similar efficiencies for their separation and column

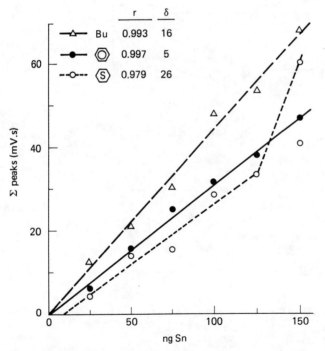

Figure 4.25 Calibration curves for R_3Sn^+ (R=butyl, phenyl, c-hexyl) separated by HPLC-GFAA with strong cation exchange (SCX) columns using MeOH/H$_2$0/NH$_4$OAc eluents [40] are shown with respective correlation coefficients (r) and system detection limits (δ) (95% confidence level) [51]

recovery, as well as graphite furnace sensitivities. Figure 4.25 shows that considerably more sensitivity is possible with P/T-gas chromatography–flame photometric detector speciation of related organotin species known [473,477,479] to occur in environmental media. Much greater divergence in the P/T gas chromatography–flame photometric detector system calibration slopes (ratios > 25) is obtained, probably a result of different rates of hydride derivatization during the fixed P/T purge time (10 min), different partition coefficients affecting the rates at which end species is sparged from the solution [475] or different retentivities on the Tenax-GC sorbent [476]. On the basis of the values obtained for the gas chromatographic method (Figure 4.26) with 10 ml sample volumes, nominal working ranges of 10–40 ng l^{-1} organotin are feasible.

Both systems are capable of at least a ten-fold increase in sensitivity with only minor changes in procedure and equipment. For high performance liquid chromatography–graphite furnace atomic absorption spectrometry, this can be achieved by both increasing injected sample size and optimizing flow rates with a

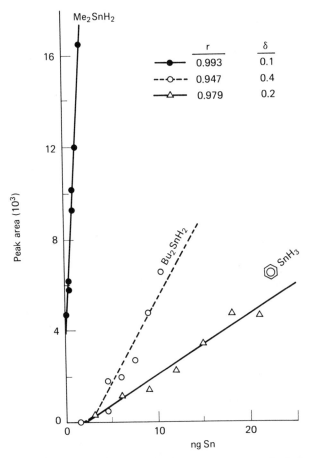

Figure 4.26 Calibration curves for three aqueous organotins indicated separated by P/T-GC-FPD using the hydride generation mode are shown with respective r and δ estimates.

graphite furnace–atomic absorption spectrometry thermal programme designed to give maximum atomization efficiency for a specific organotin analyte [473,477]. For high performance liquid chromatography–graphite furnace atomic absorption spectrometry, improvements are realized by adjusting purge flow rate and time while altering sodium borohydride additions to optimize evolution of a given organotin analyte [475]. Also, both increasing sample volumes [477,479] and operating the Tenax-GC trap at subambient temperatures [480,481] will yield lower working ranges.

Brickmann [468] generated calibration curves by the P/T gas chromatography–PFD method for borohydride reductions of Sn^{IV}, Sn^{II} and Me_2Sn^{2+} species to SnH_4, SnH_8 or Me_2H_2, respectively in distilled water, 0.2 M sodium chloride and bay water. All three analytes showed a substantial increase in their calibration slopes in going from distilled water to 0.2 M sodium chloride solution, the latter approximating the salinity and ionic strength common to estuarine waters. Presumably these effects could arise from formation of chlorohydroxy tin species favouring more rapid hydridization (see equations 4.1 and 4.2) [473,474] as well as the more propitious partition coefficients for dynamic gas stripping of the volatile tin hydrides from saline solutions [475,476]. In the typical laboratory distilled water calibration solutions only 16% of Sn^{II} was recovered as SnH_4, compared with Sn^{IV} though this sensitivity ratio can probably be altered somewhat with pH changes [477,478]. However, in spiking anaerobic pre-purged Chesapeake Bay water with these three tin species, a striking reversal occurred in overall relative sensitivities, i.e. calibration slopes. Brickmann [468] found that not only was Me_2SnH_2 generation repressed by 50% but that, very significantly, SnH_4 formation from Sn^{IV} was reduced by 15-fold as compared with the sodium chloride medium.

$$RnSn_4 \cdot _{-n}^{+}(aq) + excess\ BH_4^- \rightarrow RnSnH_{4-n} \qquad (4.1)$$

$$4HSnO_2^- + 3BH_4^- + 7H^+ + H_2O \rightarrow 3H_3BO_3 + 4SnH_4 \qquad (4.2)$$

The overall effect of estuarine water on the hydridization process is thus one of reducing yields of the three tin species tested. It is expected that not only the dissolved and particulate organics and chloride influence formation of Sn-H bonds, but that other aquated metal ions play an important role, too. Several workers have reported that, for example, As^{III}, As^V, Cu^{II}, Co^{II}, Ni^{II}, Hg^{II}, Pb^{II} and Ag^I interfere by unknown means at low concentrations [477,482].

In summary, the hydride generation method cannot adequately differentiate between aquated Sn^{IV} and Sn^{II} which may coexist in certain, especially anaerobic, environments found in marine waters. Previous reports [477,479] or inorganic tin, speciated as 'tin (IV)', should probably be regarded as 'total reducible inorganic tin' until more discriminatory techniques become available.

Dogan and Haerdi [483] and Bergerioux and Haerdi [484] determined total tin in seawater by graphite furnace atomic absorption spectrometry. These workers added 0.1–1.0 ml 0.25 M 1, 10-phenanthroline and 0.1–1.0 ml 0.2 M tetraphenyl boron (both the reagents were freshly prepared) to 50–1000 ml of water sample which had been previously filtered through 0.45 μm millipore filter. The pH of this solution was adjusted to 5.0 before addition of co-precipitating reagents. The precipitate thus obtained was either filtered or centrifuged and dissolved in a 1–5 ml aliquot of ammoniacal alcohol (methanol, ethanol or iso-propanol) solution, pH = 8–9 or in Lumatom®. For large volumes of water, the dissolution of the co-precipitate must be carried out with Lumatom® since a precipitate is formed due to other ions present with ammoniacal alcohol solution.

Tungsten

Van der Sloot and Das [485] have described a method for the determination of tungsten in seawater.

Titanium

Yong, Shih and Yeh [486] have described a spectrophotometric method for the determination of dissolved titanium in seawater after pre-concentration using sodium diethyldithiocarbamate.

Uranium

Quinoline–chloroform extraction has been used to separate uranium from seawater but with limited success [487].

Van der Berg and Huang [488] determined uranium (VI) in seawater by cathodic stripping voltammetry at pH 6.8 of uranium (VI)-catechol ions. A hanging mercury drop electrode was used. The detection limit was $0.3 \, nmol \, l^{-1}$ after a collection period of 2.5 min. Interference by high concentrations of iron (III) was overcome by selective adsorption of the uranium ions at a collection potential near the reduction potential of iron (III). Organic surfactants reduced the peak heights for uranium by up to 75% at high concentrations. EDTA was used to eliminate competition by high concentrations of copper (II) for space on the surface of the drop.

Vanadium

Nishimura et al. [489] described a spectrophotometric method using 2-pyridyl azoresorcinol for the determination of down to $0.025 \, \mu g \, l^{-1}$ vanadium in seawater. The vanadium was determined as its complex with 4-(2-pyridylazo) resorcinol formed in the presence of 1,2-diaminocyclohexane-NNN'N'-tetra acetic acid. The complex was extracted into chloroform by coupling with zephiramine. Difficulties due to turbidity in the chloroform layer and incomplete masking of some cations by 2-pyridylazoresorcinol were overcome by addition of potassium cyanide and washing the chloroform layer with sodium chloride solution. The extinction of the chloroform layer was measured at 560 nm against water as was that of a blank prepared with vanadium-free artificial seawater. Sixteen foreign ions were investigated and no interferences were found at 5–100 times their usual concentration in seawater.

Kiriyama and Kuroda [490] combined ion-exchange pre-concentration with spectrophotometry using 2-pyridyl azoresorcinol in the determination of vanadium in seawater.

The sample (2 litres) made 0.1 M in hydrochloric acid filtered, and made 0.1 M in ammonium thiocyanate, is passed through a column of Dowex 1-X8 resin (SCN form). The vanadium is retained and is eluted with concentrated hydrochloric acid. Thiocyanate in the eluate are decomposed by heating with nitric acid and the solution is evaporated to fuming with sulphuric acid. A solution of the residue is neutralized with aqueous ammonia and evaporated nearly to dryness. The residue is treated with water and aqueous sodium hypobromite and after 30 min with phenol, phosphate buffer solution of pH 6.5 and aqueous 1.2-diaminocyclohexane-NN$N'N'$-tetra-acetic acid, and the vanadium is determined spectrophotometrically at 545 nm with 4-(2-pyridylazo) resorcinol. Vanadium was determined in seawater at levels of 1.65 μg l^{-1}. After boiling such samples under reflux with potassium permanganate and sulphuric acid (to establish the concentration of organically bound vanadium), values for vanadium were 30–60% higher than corresponding values obtained without oxidation.

Monien and Stangel [491] studied the performance of a number of alternative chelating agents for vanadium and their effect on vanadium analysis by atomic absorption spectrometry with volatilization in a graphite furnace. Two promising compounds were evaluated in detail, namely 4-(2-pyridylazo) resorcinol in conjunction with tetraphenylarsonium chloride and tetramethylenedithiocarbamate. These substances, dissolved in chloroform, were used for extraction of vanadium from seawater, and after concentrating the organic layer 5 μl were injected into a pyrolytic graphite furnace coated with lanathanum carbide. For both reagents a linear concentration dependence was obtained between 0.5 and 7 μg l^{-1} after extraction of a 100 ml sample.

Using the 2-pyridyl azoresorcinol-tetraphenyl-arsonium chloride system a concentration of 1 μg l^{-1} could be determined with a relative standard deviation of 7%.

Van der Berg and Huang [492] carried out direct electro-chemical stripping of dissolved vanadium in seawater using cathodic stripping. Voltammetry was performed with a hanging mercury drop electrode. The detection limit was 0.3 nmol l^{-1} after a collection period of 2 min.

Two methods for the determination of vanadium in seawater have been studied which use neutron activation analysis and atomic absorption spectrometry [493].

In the atomic absorption spectrometric procedure [493], potassium thiocyanate (10 g) and ascorbic acid (5 g to reduce to V^{VI}) were dissolved in 1 litre of seawater

Table 4.27 Results of vanadium determinations in seawater samples

Sample	Volume (litres)	Vanadium concentration (μg l^{-1})	
		NAA	AAS
Pacific Ocean (Scripps Pier)	1		1.80
	1		2.00* (2.0)
	2		1.90*
	3		1.73
	0.098	1.99†	
	0.098	2.00§ (0.20)	
Adriatic Sea			
(Shore near Lignano Sabbiadoro, Italy)	3		1.71
	0.041	1.69	
(Shore near Ancona, Italy)	3		1.73
	0.043	1.64	

*Results after subtraction of the quantity of vanadium added to the sample before the ion-exchange or co-precipitation step. The amount added (in μg) is shown in parentheses.
†Average of 12 determinations, standard deviation 0.10 μg l^{-1}.
§Average of two samples, average deviation 0.01 μg l^{-1}.
Reproduced from Weiss, H.V., et al. (1977) Talanta, 24, 509, by courtesy of authors and publishers.

and the solutions were left to stand for 2–3 h. These samples (1–3 litres) were passed through a Daroex 1-X8 ion-exchange column at a flow rate of 1.7 ml min^{-1}. The resin was then washed with 20 ml distilled water and vanadium eluted with 150 ml eluent solution.

The vanadium eluate was slowly evaporated under an infra-red lamp, the residue dissolved in 10 ml 6 M hydrochloric acid containing 1 ml the aluminium chloride solution [494] and vanadium determined by atomic absorption spectrophotometry. For calibration, suitable standard solutions were aspirated before and after each batch of samples.

The analysis of the seawater samples by both methods is shown in Table 4.27. The average concentration and standard deviation of the Pacific Ocean waters (μg l^{-1}) were 2.00 ± 0.09 by neutron activation analysis and 1.86 ± 0.12 by atomic absorption spectrometry. For the Adriatic water the corresponding values were about 1.7 μg l^{-1}. The difference between the values for the same seawater is within the range to be expected from the standard deviations observed.

Though the neutron activation analysis is inherently more sensitive than the atomic absorption spectrometry, both procedures give a reliable measurement of vanadium in seawater at the natural levels of concentration.

Zinc

Muzzarrelli and Sipos [495] showed that a column of critosan (15 × 10 mm) can be used to concentrate zinc from 3 litres seawater before determination by anodic-stripping voltammetry with a composite mercury–graphite electrode. Zinc (and lead) are eluted from the column by 50 ml 2 M ammonium acetate, copper by 10 ml of 0.01 M EDTA and cadmium by 3 ml 0.1 M potassium cyanide.

Anodic stripping voltammetry has been used [496] using a tubular mercury–graphite electrode to determine zinc in seawater. Zinc concentrations of 1×10^{-9} M can be detected within 5 min using this system.

Graphite furnace atomic adsorption spectrometry has also been used to determine zinc [497–499] in seawater with a detection limit of 2 μg [497]. Guevremont [498] has discussed the use of organic matrix modifiers from the direct determination of zinc.

Adsorption colloid flotation using dodecylamine as surfactant has been used [499] to separate zinc with 95% efficiency from seawater.

Van der Berg [500] determined zinc complexing capacity in seawater by cathodic stripping voltammetry of zinc-ammonium pyrrolidine dithiocarbamate complex ions.

The successful application of cathodic stripping voltammetry, preceded by adsorptive collection of complexes with ammonium pyrrolidine dithiocarbamate for the determination of zinc complexing capability in seawater is described. The reduction peak of zinc was depressed as a result of ligand competition by natural organic material in the sample. Sufficient time was allowed for equilibrium to occur between the natural organic matter and added ammonium pyrrolidine dithiocarbamate. Investigations of electrochemically reversible and irreversible complexes in seawater of several salinities are detailed, together with experimental measurements of ligand concentrations and conditional stability constants for complexing ligands. Results obtained were comparable with those obtained by other equilibrium techniques but the above method had a greater sensitivity.

Van der Berg [501] also reported a direct determination of sub-nanomolar levels of zinc in seawater by cathodic stripping voltammetry. The ability of ammonium pyrrolidine dithiocarbamate to produce a significant reduction peak in the presence of low concentrations of zinc was used to develop at levels two orders of magnitude below those achieved with anodic stripping voltammetry. Interference from nickel and cobalt ions could be overcome by using a collection potential of 1.3 V and that of organic complexing material by ultraviolet irradiation. Zinc could be determined in seawater and fresh water. Zinc and nickel could be determined simultaneously by using dimethylgloxime at a collection potential of -0.7 V, followed by ammonium pyrrolidine dithiocarbamate at -1.3 V. The sensitivity for this determination was $3 \, \text{pmol} \, \text{l}^{-1}$.

Zirconium

Zirconium has been separated from seawater with 60–70% efficiency by co-precipitation with ferric hydroxide prior to determination by Alizarin Red S spectrophotometric method [502].

4.2 Graphite furnace atomic absorption spectrometry

The majority of the work on multi-element analysis in seawaters has been carried out using the graphite furnace technique as this has the additional sensitivity over the direct technique that is required in seawater analysis.

Theoretical treatments of graphite furnace atomic absorption spectrometry include a study of background signals due to sea salts [503], pyrometric measurement of furnace temperature [504] and methods of introducing the sample into the furnace [505].

Cadmium and lead

Bengtsson, Danielsson and Magnusson [506] found that the high background adsorption of solutions of trace metals containing up to $400 \, \text{mg} \, \text{l}^{-1}$ can be easily minimized by addition of 2% v/v nitric acid. Of the several agents added in an attempt to eliminate the decrease in sensitivity caused by the salt and the variability in sensitivity between graphite tubes, only lanthanum added at $1 \, \text{g} \, \text{l}^{-1}$ was effective for both lead and cadmium.

Cadmium and zinc

Campbell and Ottaway [507] have described a simple and rapid method for the determination of cadmium and zinc in seawater, using atomic absorption spectrometry with carbon furnace atomization. Samples, diluted $1 + 1$ with de-ionized water, are injected into the carbon furnace and atomized in a HGA-72 furnace atomizer under gas-stop conditions. A low atomization temperature of 1492°C is used to separate the atomic absorption signals from background absorption. Detection limits (2SD) of $0.04 \, \mu\text{g} \, \text{l}^{-1}$ for cadmium and $1.7 \, \mu\text{g} \, \text{l}^{-1}$ for zinc are reported. These limits appear to be adequate for all but the cleanest seawater samples. The use of standard addition is essential because of the interference from magnesium chloride and also when samples of varying salinity have to be analysed.

Lead, cobalt and nickel

These elements have been determined in seawater by atomic absorption spectrometry after electrodeposition on pyrolytic graphite coated tubes [508]. The tubular, pyrolytic graphite-coated furnace has been incorporated in a flow-through cell for the electrodeposition with mercury of heavy metals from seawater. After plating, the furnace is transferred to an atomic absorption spectrometer for atomization of the deposited metals. The flow assembly was tested for the analysis of lead in seawater, comparing results with those obtained by anodic stripping voltammetry. The technique is applied to the determination in seawater of both labile and total cobalt and nickel. These metals are irreversibly deposited on graphite and have poor sensitivity using anodic stripping voltammetry but are readily measured by atomic absorption spectrometry. Measurements are reproducible with a relative standard deviation of 15%. For 15- and 10-minute depositions, respectively, copper and nickel characteristic concentrations are $0.02\,\mu g\,l^{-1}$.

Cadmium, lead and chromium

Stein, Canelli and Richards [509] have described a simplified, sensitive and rapid method for determining low concentrations of cadmium, lead and chromium in estuarine waters. To minimize matrix interferences, nitric acid and ammonium nitrate are added for cadmium and lead; nitric acid only is added for chromium. Then 10, 20 or 50 μl of the sample or standard (the amount depending on the sensitivity required) is injected into a heated graphite atomizer, and specific atomic absorbance is measured. Analyte concentrations are calculated from calibration curves for standard solutions in demineralized water for chromium or an artificial seawater medium for lead and cadmium.

Detection limits ($\mu g\,l^{-1}$) were 0.1 for cadmium, 4 for lead and 0.2 for chromium. For cadmium (0.5 and 5 $\mu g\,l^{-1}$) lead (4 and 50 $\mu g\,l^{-1}$) and chromium (1 and 10 $\mu g\,l^{-1}$) in half-strength artificial seawater, the relative standard deviations (n = 10) were 20 and 9.5, 18 and 10.4 and 25 and 8.0% respectively.

Iron, chromium and manganese

A graphite furnace procedure has been described [510] for the direct determination of iron, chromium and manganese in seawater and estuarine waters in which the interference normally associated with the presence of sodium chloride is eliminated. The technique requires only very small sample volumes (10–20 μl) for the atomization stage. The reproducibility of the method was very good.

Iron, manganese and zinc

Sturgeon *et al.* [511] and Segar and Cantillo [512] described a direct determination of these elements by graphite furnace atomic absorption spectrometry. A combination of furnace tube redesign, selective volatilization and matrix modification techniques allows all three elements to be determined by the method of standard additions. Lower limits of determination of 0.2, 0.2 and 0.4 $\mu g\,l^{-1}$;

sensitivities of 0.4, 0.2 and 0.07 $\mu g\,l^{-1}$; and precisions of determination of 4.5, 3 and 11% (at $2\,\mu g\,l^{-1}$ level) are obtained for iron, manganese and zinc.

Copper, iron and manganese

Montgomery and Peterson [513] showed that ammonium nitrate used as a matrix modifier in seawater analysis to eliminate the interference of sodium chloride, degrades the pyrolytic coating on graphite-furnace tubes. The initially increased sensitivities for copper, manganese and iron are maintained for up to 15 atomizations. There is then a rapid decline to a constant lower sensitivity. The characteristics depend strongly on the particular lot of furnace tubes. To decrease the sodium chloride interference without using matrix modifier, estuarine samples must be diluted (1 + 1) with pure water. Blanks and standards are prepared and diluted with sample water containing low amounts of trace metals to match the sample matrix.

Iron and manganese have been determined in saline pore water [514] by the following technique.

100 μl of the pre-acidified pore water (diluted with Q water if necessary) was pipetted by an Eppendort pipette into a 10 ml volumetric Pyrex flask. To this flask 50 μl of nitric acid was added and the solution was then brought to volume with Q-water. Standards were made up by adding various amounts to $1\,mg\,l^{-1}$ stock metal solutions, 50 μl of nitric acid and 100 μl of a seawater solution of approximately the same salinity as the samples to be analysed. This final addition ensures that the standards are of approximately the same ionic strength and contain the same salts as the samples.

The samples were analysed by injecting 25 μl aliquots into a HGA 2000 Perkin-Elmer graphite furnace attached to a Jarrell-Ash 82-800 Double Beam Atomic Absorption Spectrophotometer. Graphite tubes in the furnace were replaced after 75–100 analyses. Metal concentrations were determined by comparing the peak heights of the samples to the standard curve established by the determination of at least five known standards. The detection limits of this technique for 1% absorption were: Fe, $0.9\,\mu mol\,l^{-1}$ and Mn, $0.2\,\mu mol\,l^{-1}$. The coefficient of variation was: Fe, ±11% at $6.5\,\mu mol\,l^{-1}$ and Mn, +12% at $11.8\,\mu mol\,l^{-1}$.

Cadmium, copper and silver

These elements have been determined by an ammonium pyrrolidine dithiocarbamate chelation followed by a methyl isobutyl ketone extraction of the metal chelate from the aqueous phase [515,516] followed by graphite furnace atomic absorption spectrometry.

The detection limits of this technique for 1% absorption were determined to be: Cu, $0.03\,\mu mol\,l^{-1}$; Cd, $2\,nmol\,l^{-1}$; Ag, $2\,nmol\,l^{-1}$.

Lead, manganese, vanadium and molybdenum

Tominaga, Bansho and Umezaki [517,518] studied the effect of ascorbic acid on the response of these metals in seawater obtained by graphite-furnace, atomic absorption spectrometry from the point of view of variation of peak times and the sensitivity. Matrix interferences from seawater in the determination of lead, manganese, vanadium and molybdenum were suppressed by addition of 10% (w/

v) ascorbic acid solution to the sample in the furnace. Matrix effects on the determination of cobalt and copper could not be removed in this way. These workers propose a direct method for the determination of lead, manganese, vanadium and molybdenum in seawater.

Copper, iron, manganese, cobalt, nickel and vanadium

Segar and Gonzalez [519] carried out a direct determination of these elements in seawater using a graphite atomizer and a deuterium background connector. Sea salts are volatilized at a lower temperature than is required for the volatilization of the above elements.

Nickel, copper, molybdenum and manganese

Hayase, Shitashima and Tsuibota [520] first extracted the seawater sample with chloroform to remove dissolved organic matter prior to analysis of the aqueous phase by graphite furnace atomic absorption spectrometry. Seawater samples at pH 3 and at pH 8 were extracted with chloroform, evaporated to dryness and the residue treated with nitric acid. Acid solutions were subjected to metal analyses by graphite furnace atomic absorption spectrometry.

4.3 Zeeman graphite furnace atomic absorption spectrometry

The widespread use of graphite furnace systems has greatly expanded the requirements for accurate background correction in atomic absorption measurements. Correction for background absorption is most commonly achieved using continuum sources. While adequate in many cases, the continuum source technique has several inherent limitations. The intensity of the continuum sources is not always adequate; inaccurate correction is possible if the background is structured; plus it is necessary to maintain correct optical alignment between the source and continuum lamps. The combined effect of these limitations means that it is not always possible to obtain accurate correction for applications with high background levels.

The need for improved background correction performance has generated considerable interest in applying the Zeeman effect, where the atomic spectral line is split into several polarized components by the application of a magnetic field. With a Zeeman effect instrument background correction is performed at, or very close to, the analyte wavelength without the need for auxiliary light sources. An additional benefit is that double-beam operation is achieved with a very simple optical system.

In 1971 Hedeishi and McLaughlin [521] first reported the application of the Zeeman effect for the determination of mercury. Numerous workers have since investigated the technique utilizing systems in which the magnetic field is applied directly to the light source [522–530] or to the atom source [531–534]. There are several possible design approaches for a Zeeman effect atomic absorption spectrophotometer. The magnetic field may be fixed or modulated, the field may be aligned in a direction transverse (perpendicular) or longitudinal (parallel) to the optical path and the field may be applied to the light source or the atom source.

Fernandez *et al.* [535] of Perkin Elmer Ltd. reported results obtained in investigating several of these possible design approaches. Background correction performance will probably not be significantly influenced by the position or type of magnet used. However, this is not the case regarding sensitivity and analytical range, where the design employed will have a significant impact on performance. These workers developed a Zeeman effect instrument capable of providing improved background correction performance with minimal sacrifice in analytical sensitivity and working range and this was incorporated into the Hadel 5000 instrument. This design utilizes a modulated, transverse field applied to the graphite tube. Comparisons of analytical sensitivity and working range versus standard atomic absorption performance with the system is reported for the determination of manganese in seawater [536].

Further improvements in the technique [537] include the use of a L'vov platform to achieve a temperature that is constant in time and improved pyrolytically coated graphite tubes. To achieve improved performance requires a fast spectrophotometer, rapid heating of the furnace, integration of the absorbance signals and usually an appropriate matrix modifier. All of these aspects of the analytical system must be carefully integrated with an understanding of the role played by the system. These workers give guidelines for optimizing each part of the system.

In addition to manganese, discussed below [517,536] the Zeeman technique has been used for the determination of other elements [538] in seawater. De Kersabiec, Blanc and Pinta [539] have described a Zeeman method for the determination of copper, lead, cadmium, cobalt, nickel and strontium in brines in the soil water adjacent to the Red Sea.

Manganese
To determine manganese [517,536] several factors had to be controlled carefully to obtain reliable results against simple standards that were independent of salinity and variations in matrix composition. Use of the L'vov platform and integration of the absorbance signal reduced the sensitivity to matrix composition. Pyrolytically coated graphite reduced variations that depend upon the life of the tubes. The tubes appeared to fall by intercalation of the Na or NaCl matrix. The char temperature must not vary outside the range of 1100–1300°C (see below). Zeeman background correction permitted use of larger seawater samples. The detection limit of the procedure using 20 μl samples was 0.1 μg l^{-1} (2 pg) manganese. By use of the Zeeman background corrector, less than 0.02 μg l^{-1} manganese was detected in seawater using a 75 μl sample.

4.4 Inductively coupled plasma atomic emission spectrometry (ICP AES)

The d.c. plasma was introduced as an excitation source for atomic emission spectrometry by Margoshes and Scribner [540] and Korolev and Vainshtein [541]. Modified designs have been characterized by a number of other authors [542–548]. Commercial equipment is now available from several manufacturers. The principle of the plasma torch arrangement used in these instruments is illustrated in Figure 4.27 [549].

Winge *et al.* [549] have investigated the determination of twenty or more trace elements in saline waters by the inductively coupled plasma technique. They give

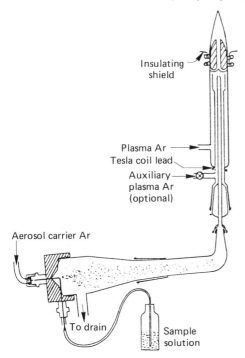

Insulating shield

Plasma Ar
Tesla coil lead
Auxiliary
plasma Ar
(optional)

Aerosol carrier Ar

To drain

Sample solution

Figure 4.27 Torch and sample aerosol generation system (QVAC 127 system)

details of experimental procedures, detection limits and precision and accuracy data. The technique when applied directly to the sample is not sufficiently sensitive for the determination of many of the elements at the low concentrations at which they occur in seawater and for these samples pre-concentration techniques are required. However, it has the advantages of being amenable to automation and the capability of analysis for several elements simultaneously.

Work on this technique published in 1979 [550] on the application of the Spectroscan d.c. plasma emission spectrometer confirmed that for the determination of cadmium, chromium, copper, lead, nickel and zinc in seawater the method was not sufficiently sensitive as its detection limits are just approaching the levels found in seawater. High concentrations of calcium and magnesium increased both the background and elemental line emission intensities. The extension of inductively coupled plasma atomic emission spectrometry to seawater analysis has been slow for two major reasons. The first is that the concentrations of almost all the trace metals of interest are $1 \, \mu g \, l^{-1}$ or less, below detection limits attainable with conventional pneumatic nebulization. The second is that the seawater matrix, with some 3.5% dissolved solids, is not compatible with most of the sample introduction systems used with ICPs. Thus direct multi-elemental trace analysis of seawater by ICP AES is impractical, at least with pneumatic nebulization. In view of this, a number of alternative strategies can be considered:

1. Pre-concentration and removal of the metals of interest from the seawater matrix prior to ICP analysis.

2. The use of ultrasonic nebulization with aerosol desolvation.
3. A combination of the above two strategies.

Owing to inadequate detection limits by direct analysis various workers examined pre-concentration procedures including dithiocarbamate pre-concentration [551–554], ion-exchange pre-concentration [555–557] chelation solvent extraction [556], co-precipitation [558] and pre-concentration in silica immobilized 8-hydroxyquinoline [559].

Iron, manganese, zinc, copper and nickel

Berman, McLaren and Willie [555] have shown that if a seawater sample is given a 20-fold pre-concentration by one of the above techniques, then reliable analysis can be performed by ICP AES for the following elements (i.e. concentration of element in seawater is more than five times the detection limit of the method): iron, manganese, zinc, copper and nickel. Lead, cobalt, cadmium, chromium and arsenic are below the detection limit and cannot be determined reliably by ICP AES. These latter elements would need an at least 100-fold pre-concentration before they could be reliably determined.

Ion-exchange preconcentration ICP AES

Berman, McLaren and Willie [555] attempted to determine the nine above elements in seawater by a combination of ion-exchange pre-concentration on Chelex 100 [556,560] and ICP AES using ultrasonic nebulization. Pre-concentration factors of between 25 and 100 were obtained by this technique.

Table 4.28 compares the ICP AES results with data generated for the same sample by two other independent methods–isotope dilution spark source mass spectrometry (IDSSMS) and graphite furnace atomic absorption spectrometry (GFAAS). The IDSSMS method also uses a pre-concentration of the metals and matrix separation using the ion-exchange procedure, following isotope addition. The atomic absorption determinations were preceded by an MIBK extraction [556]. In general the agreement is very good, but one discrepancy merits comments. The spark source mass spectrometry result for manganese is not as

Table 4.28 Analysis of Sandy Cove seawater

Element	ICP-AES	GFAAS	IDSSMS
Mn	1.5 ± 0.1	1.4 ± 0.2	1.8 ± 0.2*
Fe	1.5 ± 0.6	1.5 ± 0.1	1.4 ± 0.1
Zn	1.5 ± 0.4	1.9 ± 0.2	1.6 ± 0.1
Cu	0.7 ± 0.2	0.6 ± 0.2	0.7 ± 0.1
Ni	0.4 ± 0.1	0.33 ± 0.08	0.37 ± 0.02
Pb	–	0.22 ± 0.04	0.35 ± 0.03
Cd	–	0.24 ± 0.04	0.28 ± 0.02
Cr	–	–	0.34 ± 0.06
Co	–	–	0.02*

Precision expressed as 95% confidence intervals.
*Spark source mass spectrometry, internal standard method.
Reproduced from Berman, S.S., et al. (1980) Analytical Chemistry, **52**, 488, American Chemical Society, by courtesy of authors and publishers.

reliable as the other data by this method. Since manganese is monoisotopic, a less accurate internal standardization method of calibration has to be used. The ICP AES result for manganese is in close agreement with the GFAAS result.

Pre-concentration factor 25, i.e. 250 ml seawater concentrated to 10 ml nitric acid extract.

Manganese, iron, zinc, copper and nickel can be determined by pre-concentration ICP AES giving good agreement with the other methods. No consistently reliable results were obtained for cadmium, chromium, cobalt and lead when samples up to 1 litres were processed (i.e. no pre-concentration). Chromium is only weakly retained by the Chelex 100 resin under the conditions used so that the seawater concentration (about $0.3 \mu g l^{-1}$) is not sufficiently enhanced.

Cadmium, zinc, lead, iron, manganese, copper, nickel and cobalt

Sturgeon et al. [556] compared five different analytical methods in a study of trace metal contents of coastal seawater. Analysis for cadmium, zinc, lead, iron, manganese, copper, nickel, cobalt and chromium was carried out using isotope dilution spark source mass spectrometry (EDSSMS), graphite furnace atomic absorption spectrometry (GFAAS) and inductively coupled plasma emission spectrometry (ICPES) following trace metal separation pre-concentration (using ion-exchange and chelation solvent extraction) and direct analysis (by GFAAS).

Table 4.29 gives results obtained on a sample of seawater. Overall, there is good agreement in elemental analysis obtained by the various methods.

Although ICP AES is a multi-element technique, its inferior detection limits, relative to graphite furnace atomic absorption spectrometry, would necessitate the processing of large volumes of seawater, improvements in pre-concentration procedures in use up to this time or new alternate pre-concentration procedures such as carrier precipitation (see below).

Table 4.29 Analysis of seawater sample

Element	Concentration ($\mu g l^{-1}$)			
	GFAAS		ICPES ion exchange	IDSSMS ion exchange
	Direct	Chelation–extraction		
Fe	1.6 ± 0.2*	1.5 ± 0.1	1.5 ± 0.6	1.4 ± 0.1
Mn	1.6 ± 0.1	1.4 ± 0.2	1.5 ± 0.1	ND
Cd	0.20 ± 0.04	0.24 ± 0.04	ND	0.28 ± 0.02
Zn	1.7 ± 0.2	1.9 ± 0.2	1.5 ± 0.4	1.6 ± 0.1
Cu	ND	0.6 ± 0.2	0.7 ± 0.2	0.7 ± 0.1
Ni	ND	0.33 ± 0.08	0.4 ± 0.1	0.37 ± 0.02
Pb	ND	0.22 ± 0.04	ND	0.35 ± 0.03
Co	ND	0.018 ± 0.008†	ND	0.020 ± 0.003‡

*Precision expressed as standard deviation.
†Pre-concentrated 100-fold by Chelex 100 ion exchange.
‡Spark source mass spectrometry, internal standard method.
§Weekly retained by Chelex 100 resin not repeated.
Reproduced from Sturgeon, R.E., et al. (1980) Analytical Chemistry, 52. 1585, American Chemical Society, by courtesy of authors and publishers.

Carrier precipitation ICP AES
Chromium, manganese, cobalt, nickel, copper, cadmium and lead

Hiraide *et al.* [558] developed a multi-element pre-concentration technique for chromium (III), manganese (II), cobalt, nickel, copper (II), cadmium and lead in artificial seawater using co-precipitation and flotation with indium hydroxide followed by ICP AES. The metals are simultaneously co-precipitated with indium hydroxide adjusted to pH 9.5 with sodium hydroxide ethanolic solutions of sodium oleate and dodecyl sulphate added, and then floated to the solution surface by a steam of nitrogen bubbles. Cadmium may be completely recovered without the co-precipitation of magnesium. Concentrations of heavy metals (chromium (III), manganese (II), cobalt, nickel, copper (II), cadmium and lead) in 1200 ml of artificial seawater were increased 240-fold, while those of sodium and potassium were reduced to 2–5% and those of magnesium, calcium and strontium to 50%.

Down to $1\,\mu g\,l^{-1}$ of the above-mentioned heavy metals can be determined by this procedure. However, it is emphasized that real seawater samples were not included in this study.

Lead, zinc, cadmium, nickel, manganese, iron, vanadium and copper

Diethyldithiocarbamate and pyrrolidine carbodithioate chelation ICP AES.

Sugimae [552] developed a method for these elements in which they chelated with diethyldithiocarbamic acid and the chelates extracted with chloroform and the chelate decomposed prior to determination. When 1 litre water samples are used, the lowest determinable concentrations are: Mn, $0.063\,g\,l^{-1}$; Zn, $0.13\,\mu g\,l^{-1}$; Cd, $0.25\,\mu g\,l^{-1}$; Fe, $0.25\,\mu g\,l^{-1}$; V, $0.38\,\mu g\,l^{-1}$; Ni, $0.5\,\mu g\,l^{-1}$; Cu, $0.5\,\mu g\,l^{-1}$; Pb, $2.5\,\mu g\,l^{-1}$. Above these levels, the relative standard deviations are better than 12% for the complete procedure.

Cadmium, lead, zinc, iron, copper, nickel, molybdenum and vanadium

Muyazaki *et al.* [553] found that di-isobutyl ketone is an excellent solvent for the extraction of the 2,4-pyrrolidone dithiocarbamate chelates of these elements from seawater.

Unlike halogenated solvents, it does not produce noxious substances in the inductively coupled plasma, has a very low aqueous solubility and gives 100-fold concentration in one step. Detection limits are reported in Table 4.30. The results indicate that the proposed procedure should be useful for the precise determination of metals in oceanic water, although a higher sensitivity would be necessary for lead and cadmium.

A comparison of the results obtained for eight elements in coastal Pacific Ocean water by ICP AES and by atomic absorption spectrometry is shown in Table 4.31. The results for cadmium, lead, copper, iron, zinc and nickel are in good agreement. For iron, the data obtained by the solvent extraction-ICP method are also in good agreement with those determined directly by ICP AES. In most of the results shown in Table 4.31 the relative standard deviations were 4% for all elements except cadmium and lead, which had relative standard deviations of about 20% owing to the low concentrations determined.

Table 4.30 Detection limits

Line measured	Wavelength (nm)	Detection limit (μg l^{-1})	Concentration in seawater (μg l^{-1})
Cd^{II}	226.50	0.022	0.05
Pb^{II}	220.35	0.60	0.03
Cu^{I}	324.75	0.051	3
Zn^{II}	202.55	0.084	4
Fe^{II}	259.94	0.075	3
V^{II}	311.07	0.035	1.5
Mo^{II}	202.03	0.21	10
Ni^{II}	221.65	0.11	2

*Concentration giving three times the standard deviation of the background signal. Concentration factor = 100.
Reproduced from Mujazaki, A., et al. (1982) Analytica Chimica Acta, **144**, 213, Elsevier Science Publishers, Amsterdam, by courtesy of authors and publishers.

Table 4.31 Results for river and seawaters

Sample	Concentration found (μg l^{-1})*							
	Cd		Pb		Cu		Zn	
	ICP	AAS	ICP	AAS	ICP	AAS	ICP	AAS
Sea								
D	0.02 ± 0.005	0.02	0.7 ± 0.17	0.6	2.0 ± 0.1	2.8	2.0 ± 0.1	2.2
E	0.02 ± 0.005	0.02	ND	ND	1.3 ± 0.1	0.9	2.1 ± 0.1	3.0
F	0.04 ± 0.01	0.03	0.7 ± 0.10	0.8	0.3 ± 0.05	0.3	1.9 ± 0.08	2.4
G	–	–	–	–	–	–	–	–

Sample	Fe			Ni		Mo	V
	ICP	ICP†	AAS	ICP	AAS	ICP	ICP
Sea							
D	140 ± 5	140 ± 6	140 ± 6	0.15 ± 0.02	0.16	12.0 ± 0.4	1.29 ± 0.04
E	300 ± 10	300 ± 9	300 ± 10	–	–	11.9 ± 0.4	1.40 ± 0.04
F	320 ± 9	310 ± 9	310 ± 9	–	–	11.2 ± 0.4	1.87 ± 0.05
G	–	–	–	0.90 ± 0.04	0.93	–	–

*Mean and standard deviation of 5 measurements.
†Determined directly by ICPAES.
ND = not detected.
Reproduced from Mujazaki, A., et al. (1982) Analytica Chimica Acta, **144**, 213, Elsevier Science Publishers, Amsterdam, by courtesy of authors and publishers.

Application of ICP AES to non-oceanic highly saline samples

Bloekaert, Leis and Lagua [554] applied ICP AES with ammonium pyrrolidine thiocarbamate pre-concentration to the determination of cadmium, copper, iron, manganese and zinc in highly saline waste waters. The application of ICP AES to the analysis of brines containing up to 37.5 mmol l^{-1} sodium chloride has been discussed [562,563]. Detection limits as low as 4 μg l^{-1} have been claimed.

4.5 Voltammetric methods

The relative advantages and disadvantages of voltammetric and atomic absorption methodologies are shown in Tables 4.32 and 4.33. It is concluded that for laboratories concerned with aquatic chemistry of metals, and this includes seawater analysis, instrumentation for both atomic absorption spectrometry (including potentialities for graphite furnace atomic absorption spectrometry as well as hydride and cold vapour techniques) and voltammetry should be available. This offers a much better basis for a problem-orientated application of both methods, and provides the important potentiality to compare the data obtained by one method with that obtained in an independent manner by the other; an approach that is now common for the establishment of accuracy in high-quality trace analysis.

Table 4.32 Voltammetric methods

Advantages	Disadvantages
Simultaneous analysis for several elements per run	Prior photolytic decompositing of dissolved organic matter required for many types of sample including seawater
Substance specific	Suspended particulates need prior digestion
Suitable for speciation studies	Applicable to limited range of metals e.g. Cu, Pb, Cd, Zn, Ni, Co
3–4 elements per hour	

Table 4.33 Graphite furnace atomic absorption methods

Advantages	Disadvantages
High analysis rate	Non-specific absorption
3–4 elements per hour	Spectral interferences
Applicable to many more metals than voltammetric methods	Element losses by molecular distillation before atomization
Superior to voltammetry for mercury and arsenic particularly in ultra-trace range	Limited dynamic range
	Contamination sensitivity
	Element specific (or one element per run)
	Not suitable for speciation studies in seawater
	Prior separation of sea salts from metals required
	Suspended particulates need prior digestion
	About three times more expensive than voltammetric equipment
	Inferior to voltammetry for cobalt and nickel

Anodic stripping voltammetry

Earlier work on the application of anodic stripping voltammetry to the determination of methods in seawater is reviewed in Table 4.34.

A variety of electrodes have been used in this technique including rotating mercury coated vitreous carbon [667,670], wax impregnated graphite cylinders [669] and hanging mercury drops [664–666].

It was the development of the rotating glassy carbon electrode with a preplated or coplated mercury film that gave this technique the sensitivity and resolution required for use in seawater.

Table 4.34 Metals in seawater – anodic stripping voltammetry

Zn, Cd, Pb, Cu		1–10 nm mol l^{-1}	664
Zn, Cd, Pb, Cu	Zn ⎫		
	Pb ⎬	0.01–0.1 mg l^{-1}	665
	Cu ⎭		
	Cd	0.001–0.1 mg l^{-1}	
Cd, Pb	Cd	0.18 μg l^{-1}	666
	Pb	0.21 μg l^{-1}	
Bi, Cu, Pb, Cd, Zn		–	667
Cu, Pb, Cd, Zn		–	668
Tl		0.2–1 μmol l^{-1}	669
Pb, Cd		1 μg l^{-1}	670
Pb		4 μg l^{-1}	671

Anodic stripping voltammetry using a rotating glassy carbon electrode has been extensively used to study metal organic interactions. The instrumentation is adaptable to use at sea and does not generally require any chemical pretreatment of samples prior to analysis. This permits rapid analysis in *in situ* pH, thus minimizing any changes in speciation due to storage [667–676].

A typical electrode tip consists of a 6 mm glassy carbon disc sealed with Teflon, that had been polished with diamond polishing compound. The reference electrode is a Ag/AgCl type inserted into an acid-cleaned Vycor tip Teflon bridge tube containing clean seawater. High purity argon which is passed through a high-temperature catalytic scrubber to remove oxygen and then is rehumidified using an in-line natural seawater bubbler prior to use is commonly used as a purge gas. Electronic interfaces between the polarograph and the electrode permits the polarograph to control all steps of the analysis automatically including purging. This sample is placed in an acid-cleaned Teflon polarography cell that has been copiously rinsed with the sample to be analysed. Samples are analysed as quickly as possible after collection at natural pH for free and other easily reducible species using a pre-plated mercury-film technique.

Because of differing sensitivities and natural levels of free or anodic scanning voltammetric labile metal, cadmium and copper in seawater are analysed using a 10-minute plating time, a -1.0 V plating potential and scanning in 6.67 mV s^{-1} increments. Zinc determinations can be made on a fresh aliquot of sample to eliminate any possible effects due to Cu-Zn intermetallic complex formation. Zinc is analysed by plating at -1.25 V for 5 minutes. The remaining operating conditions are the same as described above for copper and lead. The detection limits of this system were, approximately: Zn, 0.02 nmol kg^{-1}; Cd, 0.02 nmol kg^{-1}; and Cu, 0.3 nmol kg^{-1}.

Figure 4.28 is a standard addition curve at natural pH for zinc in a seawater sample for two seawater samples. There are two striking points in this plot. The first is that it required additions of metal equivalent to 1.5 nmol kg^{-1} to give detectable stripping current signals at natural pH. The second is that the plots for natural pH appear to be straight lines with a positive *x*-intercept. The plots indicate that no anodic scanning voltammetric labile metal is initially present in the sample and that interaction is occurring such that the free metal added is not initially labile.

Although anodic stripping voltammetry is one of the few techniques suitable for the direct determination of heavy metals in natural waters [677–693], it is not readily adaptable to *in situ* measurements. Lieberman and Zirino [694] examined

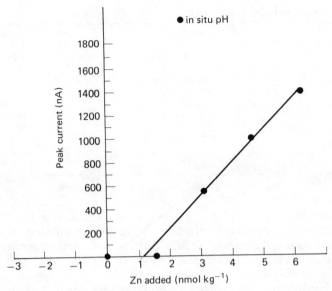

Figure 4.28 ASV standard addition curve for zinc in a sample collected in March 1980 at 130 m depth, 0600 hours local time, in the Bay of Campeche

a continuous flow system for the anodic stripping voltammetry determination of zinc in seawater, using a tubular graphite electrode predeposited with mercury. A limitation of the approach was the need to pump seawater to the measurement cell, while the method required the removal of oxygen with nitrogen, before measurements.

Recently, Batley and Matousek [695,696] examined the electrodeposition of the irreversibly reduced metals cobalt, nickel and chromium on graphite tubes for measurement by electrothermal atomization. This method offered considerable potential for contamination-free preconcentration of heavy metals from seawater. Although only labile metal species will electrodeposit, it is likely that, at the natural pH, this fraction of the total metal could yet prove to be the most biologically important [697].

Batley [698] examined the techniques available for the *in situ* electrodeposition of lead and cadmium in seawater. These included anodic scanning voltammetry at a glass carbon thin film electrode and the hanging drop mercury electrode in the presence of oxygen and, *in situ* electroposition on mercury coated graphite tubes.

Batley [698] found that *in situ* deposition of lead and cadmium on a mercury coated tube was the more versatile technique. The mercury film, deposited in the laboratory, is stable on the dried tubes which are used later for field electrodeposition. The deposited metals were then determined by electrothermal atomic absorption spectrometry.

Studies with spiked seawater showed that low concentrations of cadmium and lead could be measured in the presence of oxygen by using differential pulse anodic scanning voltammetry at the hanging mercury drop electrode. The presence of oxygen resulted in a highly sloping baseline (Figure 4.29) giving rise to greater analytical errors. In samples buffered to pH 4.8, peak heights and peak potentials

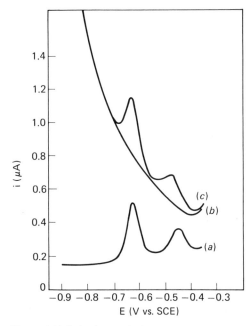

Figure 4.29 Stripping peaks for cadmium and lead in seawater, pH 7.8, at HMDE. Differential pulse mode, 35 mV modulation, 3 min deposition at −0.9 V versus SCE. (*a*) Deoxygenated solution containing 2.2 µg Cd per litre and 4.1 µg Pb per litre. (*b*) Blank containing dissolved oxygen. (*c*) Blank containing dissolved oxygen 2.2 µg Cd per litre and 4.1 µg Pb per litre

Table 4.35 Effect of oxygen and pH on anodic stripping voltammetry calibration plots for Cd and Pb in seawater*

Metal	pH	Electrode	Slope of calibration plot (µA l^{-1})	
			O_2 present	O_2 absent
Cd	4.8	HMDE	0.133 ± 0.015	0.139 ± 0.009
Cd	7.8	HMDE	0.140 ± 0.011	0.147 ± 0.005
Pb	4.8	HMDE	0.072 ± 0.007	0.068 ± 0.005
Pb	7.8	HMDE	0.040 ± 0.004	0.033 ± 0.002
Cd	4.8	GCE†	1.50 ± 0.08	1.50 ± 0.07
Cd	7.8	GCE†	1.24 ± 0.06	1.36 ± 0.08
Pb	4.8	GCE†	0.710 ± 0.021	0.762 ± 0.034
Pb	7.8	GCE†	0.382 ± 0.013	0.234 ± 0.032

*Differential pulse mode, 25 mV modulation amplitude, 3 min deposition at −0.9 V versus SCE.
†Preformed film by 3-min deposition at −0.3 V versus SCE, from 7.2 × 10^{-4} M Hg^{2+} in 0.016 M acetate, pH 4.8.
Reproduced from Batley, G.E. (1981) *Analytica Chimica Acta*, **124**, 121, Elsevier Science Publishers, Amsterdam, by courtesy of authors and publishers.

did not differ significantly before and after oxygen removal (Table 4.35). For samples at the natural pH 7.8, although the cadmium wave was unchanged, the lead wave in the presence of oxygen was greater in height by 21% and shifted by 15 mV to a more negative potential (Figure 4.29 and Table 4.35).

At the glassy carbon electrode, using both *in situ* and preformed mercury films, similar results were obtained, but the sloping baseline interference observed at the hanging mercury drop electrode was less evident because of the higher stripping

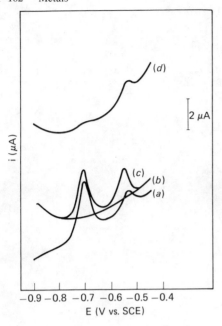

Figure 4.30 Stripping peaks for cadmium and lead in seawater, pH 7.8 at a performed mercury film on GCE. Differential pulse mode, 25 mV modulation, Deposition at -0.9 V versus SCE. (*a*) Deoxygenated solution, 2.2 µg Cd per litre, 4.1 µg Pb per litre, 3 min deposition. (*b*) In presence of dissolved oxygen, blank, 3-min deposition. (*c*) As for (*b*) plus 2.2 µg Cd per litre, 4.1 µg Pb per litre. (*d*) As for (*b*), 30 min deposition.

currents. Preformed film data are shown in Table 4.35 and illustrated in Figure 4.30. At the natural pH, the lead wave was increased in height in the presence of oxygen by more than 60% (Table 4.35) with a negative potential shift of 17 mV. The peak heights showed good linear relationships with solution lead concentration up to 10 µg l^{-1} and with deposition times up to 24 min.

For the analysis of seawater at the natural pH a standard deviation of $\pm$ 14% was obtained which was within the precision of the technique.

Calibration graphs obtained for lead in seawater at pH 7.8 showed a 48% increase in slope in the presence of oxygen. This increase, though less than that for the glassy carbon electrode, is greater than that observed at the hanging mercury drop electrode.

The *in situ* electrodeposition technique was applied to the determination of lead in saline waters of the Port Hacking Estuary near Sydney. Graphite tubes precoated with mercury were used in the immersible Perspex electrode probe. For natural lead concentrations, depositions in excess of 15 min were required to give absorbance values greater than 0.1 during atomization (Table 4.36). Blank values for the coated-tubes were low, but increased for tubes immersed in the sampled water at a controlled potential below that required for lead deposition (-0.3 V versus Ag/AgCl), for deposition times similar to those used for lead determinations. Results for lead showed good agreement with those for labile lead determined independently by stripping voltammetry.

The limits of detection for metals in seawater using *in situ* graphite tube electrodeposition will be governed by the deposition time. Unlike laboratory analyses, there will be no depletion of metals from the solution when lengthy deposition times are used, since fresh sample is being continuously pumped through the electrode. It should therefore be possible to detect the extremely low

Table 4.36 Determination of lead in seawater (Port Hacking Estuary) by *in situ* graphite electrodeposition tube

Deposition time (min)	Deposition potential (V *vs. Ag/AgCl*)	Absorbance	Deposition time (min)	Deposition potential (V *vs. Ag/AgCl*)	Absorbance
15	−0.85	0.162	15	−0.30	0.115
15	−0.85	0.178	15	−0.30	0.060
15	−0.85	0.150	30	−0.30	0.095
30	−0.85	0.273	30	−0.30	0.090
30	−0.85	0.354	0	−	0.025
30	−0.85	0.305	0	−	0.054
			0	−	0.018

Calculated labile Pb concentration = 0.15 μg l^{-1}.
Measured anodic stripping voltammetry labile Pb concentration = 0.13 μg l^{-1}.
Reproduced from Batley, G.E. (1981) *Analytica Chimica Acta*, **124**, 121, Elsevier Science Publishers, Amsterdam, by courtesy of author and publishers.

metal concentrations in open ocean water. For lead it was found, for example, that a 2 h deposition in the presence of oxygen gave a measured lead atomization absorbance equivalent to twice the blank value, for seawater containing 10 ng Pb per litre.

Scarponi *et al.* [699] studied the influence of an unwashed membrane filter (Millipore type HA, 47 mm diameter) on the cadmium, lead and copper concentrations of filtered seawater. Direct simultaneous determination of the metals was achieved at natural pH by linear-sweep anodic stripping voltammetry at a mercury film electrode. These workers recommended that at least 1 litre of seawater be passed through uncleaned filters before aliquots for analysis are taken; the same filter can be re-used several times and only the first 50–100 ml of filtrate need be discarded. Samples could be stored in polyethylene containers at 4°C for 3 months without contamination, but losses of lead and copper occurred after 5 months storage.

Brugman, Magnusson and Westerlund [700] compared results obtained by anodic scanning voltammetry and atomic absorption spectrometry in the determination of cadmium, copper, lead, nickel and zinc in seawater. Three methods were compared. Two consisted of atomic absorption spectroscopy but with pre-concentration using either freon or methyl isobutyl ketone and anodic stripping voltammetry was used for cadmium, copper, and lead only. Inexplicable discrepancies were found in almost all cases. The exceptions were the cadmium results by the two methods and the lead results from the Freon with atomic absorption spectrometric methods and the anodic scanning voltammetric methods.

Clem and Hodgson [701] discuss the temporal release of traces of cadmium and lead in bay water from EDTA, ammonium pyrrolidine diethyldithiocarbamate, humic acid and tannic acid after treatment of the sample with ozone–anodic scanning voltammetry was used to determine these elements.

Differential pulse anodic stripping voltammetry

For most tasks in the trace chemistry of natural waters, voltammetric determination requires pre-concentration, because in a group of simultaneously determined ecotoxic heavy metals one usually has levels below 0.1 μg l^{-1}.

Electrochemical pre-concentration can be attained in the following two different ways, depending whether differential pulse stripping voltammetry (differential pulse anodic scanning voltammetry) or adsorption differential pulse voltammetry has been applied.

Heavy metals capable of forming amalgams–i.e. Cu, Pb, Cd, Zn, etc.–are plated at a stationary mercury electrode consisting of a hanging mercury drop electrode with adjustment of a rather negative potential of -1.2 V versus the Ag/AgCl reference electrode for several minutes. To speed up mass transfer, the solution is stirred with a magnetic bar at 900 rpm. Their pre-concentration is achieved by the accumulation of the heavy plated metals in the mercury drop. Subsequently the stirring is terminated, and after a quiescent period of 30 s the potential is scanned into the anodic direction in the differential pulse mode. At the respective redox potential the plated heavy metal is reoxidized and the corresponding current is recorded (Figure 4.31). The voltammetric peak heights obtained are proportional to the bulk concentrations of the respective trace metals in the analyte.

The hanging drop mercury electrode can usually be applied down to trace levels of 0.1–0.05 μg l^{-1}.

At lower ultra-trace levels the less voluminous mercury film electrode has to be used. It consists of a mercury film of only several hundred Å thickness on a glassy carbon electrode as support. The fabrication of this glassy carbon electrode is

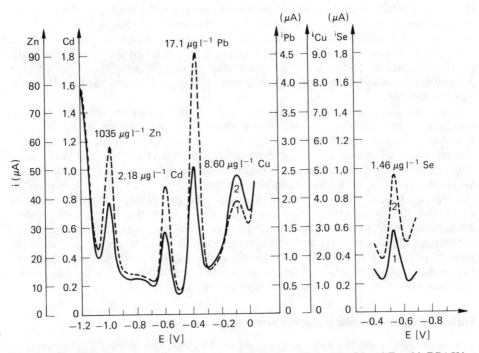

Figure 4.31 Voltammogram of the simultaneous determination of Cu, Pb, Cd and Zn with DPASV at the HMDE and subsequent determination of Se^{IV} by DPCSV in the same run in rain water at an adjusted pH of 2. Pre-concentration time for DPASV 3 min at -1.2 V, for DPCSV 5 min at -0.2 V. 1 = Original analyte. 2 = After first standard addition. Total analysis time with two standard additions, 30–40 min

critical for obtaining an optimal mercury film electrode suitable to perform determinations down to $1 \, ng \, l^{-1}$ or below.

Certain trace substances such as Se^{IV} can be determined in a similar manner by differential cathodic stripping voltammetry (DPCSV). For selenium a rather positive pre-concentration potential of $-0.2 \, V$ is adjusted. Se^{IV} is reduced to Se^{2-} and Hg from the electrode is oxidized to Hg^{2+} at this potential. It forms, with Se^{2-} on the electrode, a layer of insoluble HgSe and in this manner the pre-concentration is achieved. Subsequently the potential is altered in the cathodic direction in the differential pulse mode. The Hg^{II} resulting peak produced by the Hg^{II} reduction is proportional to the bulk concentration of Se^{IV} in the analyte. Cathodic stripping voltammetry has been discussed by Van der Berg [702]. This procedure has recently made rapid progress and could now be used to determine lead, cadmium, copper, zinc, uranium, vanadium, molybdenum, nickel and cobalt in water, with great sensitivity and specificity, allowing study of metal speciation directly in the unaltered sample. The technique used pre-concentration of the metal at a higher oxidation state by adsorption of certain surface-active complexes, after which its concentration was determined by reduction. The reaction mechanisms, effect of variation of the adsorption potential, maximal adsorption capacity of the hanging mercury drop electrode, and possible interferences are discussed.

Adsorption differential pulse voltammetry

A number of heavy metals are not capable of forming stable amalgams. Adsorption of suitable species has to be utilized for their electrochemical pre-concentration at the electrode interface. Thus, for example, in the determination of copper and nickel, the sample is adjusted to pH 9.2 with ammonia buffer and dimethylglyoxime added to form the copper and nickel chelates. Dimethylglyoxime transforms a certain amount into the chelates $Ni(DMG)_2$ while another fraction of the overall concentrations of Ni and Co exists as ammonia complexes in the analyte. Then a potential of $-0.7 \, V$ is adjusted for several minutes at the hanging mercury drop electrode. This potential in the range of the zero charge potential of the mercury electrode is most favourable for the adsorption of the dimethylglyoxime chelates of both heavy metals. To speed up mass transfer, the solution is stirred at 900 rpm during the adsorption time. The adsorbed amount of the chelates is proportional to their bulk concentration. As all the complex equilibria in the analyte between the various complexes of nickel and cobalt with ammonia and dimethyl glyoxime as ligands are adjusted, the adsorbed amount of their dimethylglyoxime chelates is also proportional to the total bulk concentrations of both heavy metals in the analyte. The proviso is that the adsorbed amounts of the dimethylglyoxime chelates should correspond to the rising part of the adsorption isotherms, and full coverage of the electrode surface is avoided. By this adsorption, substantial pre-concentration is attained at the electrode surface within a few minutes' adsorption time. Then the electrode potential is made more negative in the differential pulse mode until the reduction potentials of nickel and cobalt are reached; and the peaks produced by the reduction of the adsorbed chelate species $Ni(DMG)_2$ and $Co(DMG)_2$ are recorded (Figure 4.32).

The ultimate determination limits obtainable by voltammetric methods are given in Table 4.37. It is emphasized that these are practical limits, due to the presently possible minimization of the blanks; suitable differential pulse voltammetric methods could in principle reach still lower determination limits.

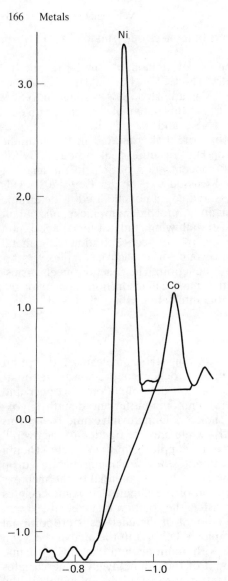

Figure 4.32 Simultaneous determination of Ni and Co in seawater from the Baltic sea by ADPV at the HMDE; two standard additions; pre-concentration time 2 min; total analysis time with two standard additions 20 min; $0.554\,\mu g\,ml^{-1}$ Ni and $0.094\,\mu g\,l^{-1}$ Co; polarograph EG and G, PAR 384 B with electronic subtraction of stored blank ammonia buffer

Nurnberg [703] has studied in great detail various aspects which are important to obtaining reliable results by voltammetric methods. These include sampling, sample pre-treatment steps, optimum pH adjustment, sample storage, decomposition of dissolved organic matter, the voltammetric determination, digestion of filtered-off suspended matter and automation. These details are extremely interesting for anyone who is considering setting up voltammetric methods in their laboratory. Nurnberg [703] also discusses results obtained by these techniques on seawater samples from the Mediterranean and Belgian, Dutch and German coastal zones of the North Sea, the Norwegian Sea and North Atlantic, the Pacific, the Arctic Oceans and the Weddell Sea.

Table 4.37 Determinations limits (μg l⁻¹ ppb) of differential pulse methods for an RSD of +20% (in parentheses for an RSD of ⩽±10%)

Method	Cu	Pb	Cd	Zn	Ni	Co	Hg	Se^{IV}	As^{III}
DPSV/ HMDE	0.05 (0.10)	0.02 (0.30)	0.02 (0.10)	0.02 (0.50)	–	–	–	0.10 (1.00)	–
DPSV/ MFE	0.007 (0.05)	0.001 (0.0015)	0.0003 (0.0015)	–	–	–	–	–	–
DPSV/ AuE	0.02 (0.10)	–	–	–	–	–	0.04 (0.20)	–	0.1 (2.0)
ADPV HMDE	–	–	–	–	0.001 (0.02)	0.001 (0.02)	–	–	–
SDPSV/ twin Au-E	–	–	–	–	–	–	0.001	–	–

DPSV = Differential pulse stripping voltammetry; HMDE = hanging mercury drop electrode; MFE = mercury film electrode; ADPV = adsorption differential pulkse voltammetry; Au-E = gold electrodes; SDPSV = differential pulse anodic stripping voltammetry. Reproduced from Batley, G.E. (1981) *Analytica Chimica Acta*, **124**, 121, Elsevier Science Publishers, Amsterdam, by courtesy of author and publishers.

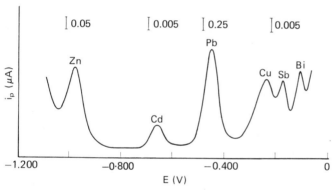

Figure 4.33 Voltammograms of Zn, Cd, Pb, Cu, Sb and Bi dissolved in seawater at pH 1 containing 2 *M* chloride. Deposition times 20 min (Zn, Cd, Pb) and 40 min (Cu, Sb, Bi)

Differential pulse anodic stripping voltammetry is a convenient technique for the simultaneous determination of heavy metals at the microtrace level. Gillain, Daychaerts, and Disteche [704] have described the possibility of determining directly and simultaneously six elements (Zn, Cd, Pb, Cu, Sb and Bi) in seawater by differential pulse anodic stripping voltammetry with the hanging mercury drop electrode. This electrode seems to be the most reliable one for routine analysis. It can be applied over a large potential range with high reproducibility and sensitivity.

Figure 4.33 shows a voltammogram for these six elements obtained on a sample of North Sea water. The sample was filtered through a 0.8 μm Millipore filter and stored in acid-cleaned polyethylene bottles at −20°C until the analysis is started. For a 60 min plating time, the detection limits were of the order of 0.1 μg l⁻¹ for zinc and copper, 0.01 μg l⁻¹ for cadmium and lead and 0.05 μg l⁻¹ for antimony and bismuth.

Mant, Nurnberg and Dryssen [705] and Valenta, Mart and Rutzel [706] have described two differential pulse anodic stripping voltammetric methods for the determination of cadmium, lead and copper in arctic seawater. After a previous plating of the trace metals into a mercury film on a rotating electrode with highly polished glassy carbon as substrate, they were stripped in the differential pulse mode. *In situ* plating was used.

The film formed during the blank test was left on the glassy carbon surface. Under addition of further mercury nitrate, 20–50 l of a $5\,g\,l^{-1}$ solution are added to 50 ml, analysis was performed. For samples where copper was expected to be close to the determination limit of $10\,ng\,kg^{-1}$, only the film plated during the blank test run was used, thus avoiding an increasing slope by the growth of the mercury film.

Pihlar, Valenta and Nurnberg [707] have described a sensitive differential pulse voltammetric method for the determination of nickel and cobalt in seawater. The pH of the sample is adjusted to 9.2–9.3 by adding an ammonia ammonium chloride buffer. Optimal ammonia buffer concentration is $0.1\,M$ for nickel concentrations below $10\,\mu g\,kg^{-1}$, and 20 μg of a $0.1\,M$ dimethylglyoxime solution in ethanol is added to a 50 ml sample. The analyte is de-aerated for 10 min. At the working electrode, a hanging mercury drop electrode, a potential of $-0.7\,V$ is adjusted and the nickel–dimethylglyoxime complex is adsorbed at the mercury surface. To speed up mass transfer the solution is stirred with a magnetic bar. Depending on the concentration of nickel (and cobalt) 5–10 min of adsorption time are needed. After a rest period of 30 s the voltammogram is recorded by scanning the potential into negative direction. Concentrations were evaluated by standard addition.

Other groups of elements that have been determined in seawater by differential pulse anodic stripping voltammetry include cadmium, copper and zinc [708]; copper, lead and cadmium [709]; and zinc, cadmium, lead and copper [710,711].

Krznaric [712] studied the influence of surfactants (EDTA, NTA) on measurements of copper and cadmium in seawater by differential pulse anodic stripping voltammetry. Adsorption of surfactants on to the electrode surface were shown to change the kinetics of the overall electrode charge and mass transfer, resulting in altered detection. Possible implications for studies on metal speciation in polluted seawater with high surfactant contents are outlined by this worker.

Bond, Heritage and Thormann [713] studied strategies for trace metal determination in seawater by anodic stripping voltammetry using a computerized multi-time domain measurement method. A microcomputer based system allowed the reliability of the determination of trace amounts of metals to be estimated. Peak height, width and potential were measured as a function of time and concentration to construct the database. Measurements were made with a potentiostat polarographic analyser connected to the microcomputer and a hanging drop mercury electrode. The presence of surfactants, which presented a matrix problem, was detected via time domain dependent results and non-linearity of the calibration. A decision to pre-treat the samples could then be made. In the presence of surfactants, neither a direct calibration mode or a linear standard addition method yielded precise data. Alternative ways of eliminating the interferences based either on theoretical considerations or destruction of the matrix needed to be considered.

Turner, Robinson and Whitfield [714] showed that rapid staircase stripping at 1–2 ms step widths provides a fast sensitive alternative to differential pulse stripping, for field use, particularly in the automated determination of copper.

Potentiometric stripping analysis

A new electrochemical method for the determination of lead, copper, zinc and cadmium has been introduced; potentiometric stripping analysis [715,716]. In some ways this technique resembles anodic stripping voltammetry. The analytical device is based on a 3-electrode system: (1) a glassy carbon electrode (serves as a cathode), (2) a saturated calomel electrode (which is the reference electrode) and (3) the counter-electrode during electrolysis made of platinum.

Analysis of metal ions in a sample solution is started by electrochemical formation of a mercury film on the glassy carbon electrode. Subsequently, the metal ions are reduced and amalgamated in the mercury film during the electrolysis step (plating). When the plating is terminated, the metals are stripped from the mercury film back into the solution by chemical oxidation. During this step the potential of the carbon electrode (against SCE) versus time is recorded. The metals are identified by their stripping potentials and the quantitative determination is obtained by measuring the stripping time for each metal.

Jagner, Jodrgdon snf Erdyrtlunf [717] used this technique to determine zinc, cadmium, lead and copper in seawater. Their method includes computerization of the potentiometric stripping technique. They compared results obtained with those obtained by solvent extraction–atomic absorption spectrometry and showed that the computerized potentiometric stripping technique is more sensitive and has advantages over anodic stripping voltammetry. Computerization makes deoxygenation of the sample unnecessary.

Water samples from the Arctic Sea were analysed by the potentiometric stripping technique. Lead (II) and cadmium (II) were determined after pre-electrolysis for 32 min at -1.1 V versus Ag/AgCl; the detection limits were 0.06 and 0.04 nmol l^{-1} respectively. Zinc (II) was determined after the addition of gallium (III) by pre-electrolysis for 16 min at -1.4 V versus Ag/AgCl; the detection limit was 0.25 nmol l^{-1}. Problems in the determination of copper (II) at the very low concentrations found in oceanic waters are outlined. The average zinc (II) and cadmium (II) and lead (II) concentrations in eight different samples were 2.5, 0.16 and 0.10 nmol l^{-1} as determined by potentiometric stripping analysis and 1.9, 0.16 and 0.09 nmol l^{-1} as determined by solvent extraction–atomic absorption spectrometry. The advantages of this computerized technique for the analysis of seawater are discussed.

Drabek et al. [718] applied potentiometric stripping analysis to the determination of lead, cadmium and zinc in seawater. The precision was evaluated by several duplicate determinations and was found to be in the range of 5–16% relative, depending on the concentration level. The accuracy of the method was evaluated by comparison with other conventional methods, i.e. atomic absorption spectroscopy and anodic stripping voltammetry and good agreement was found.

Procedure for determination of copper and lead in seawater

Acidified seawater (3 ml l^{-1} concentrated nitric acid) 25 ml is placed in the analyser. A Hg^{II} solution, usually 100 μl of a solution containing 1 mg l^{-1} mercury but depending on the metal concentration, is added together with an internal standard (copper).

To increase precision, especially at the lower concentration levels encountered in relatively non-polluted seawater, preparation, regeneration and precoating of the working electrode are carried out prior to each analysis. Likewise, adequate cleaning of all the equipment is a prerequisite. With respect to plating, a normal procedure using 10 precoating–stripping cycles is employed [716]. Simultaneously the sample is de-aerated by purging with helium.

After precoating and de-aeration of the sample, analysis is performed at −0.95 V versus SCE with an appropriate plating time (1–2 min). The method of standard addition is used. Analysis is performed in a cyclic mode and standards are added immediately after recording the stripping curve.

Procedure for determination of zinc in seawater

Prior to plating, the same analytical steps as for the above analysis are applied, the only exception being the use of lead instead of copper as the internal standard. Furthermore, gallium and sodium acetate are added, the former to prevent the formation of the inter-metallic zinc–copper compound and the latter to adjust pH to 4.7 which is favourable for zinc determination. Plating is carried out at −1.25 V versus SCE. Again the method of standard addition is applied.

Figure 4.34 shows typical stripping curves for cadmium, lead and zinc obtained from a 25 ml seawater sample. The sample was analysed as previously described. The concentrations found were: Pb^{II}, 7.1 $\mu g \, l^{-1}$; Cd^{II} 0.2 $\mu g \, l^{-1}$; and Zn^{II}, 4.1 $\mu g \, l^{-1}$. An internal standard was used to correct for variations in oxidation rate. Precision data are given in Table 4.38.

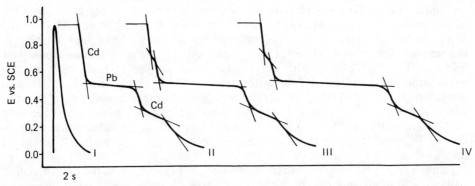

Figure 4.34 Potentiometric stripping curve for a 25 ml seawater sample. I = background, II = curve before standard addition, III and IV = curve after each addition of 5 ng Cd^{II} and 100 μg Pb^{II}. 25 ng Cu^{II} was used as internal standard. Plating time 32 min at − 95 V versus SCE

Table 4.38 Relative precision (in %) of the potentiometric stripping analysis

Concentration level ($\mu g \, l^{-1}$)	Pb	Cd	Zn
>15	5.2 (19)	–	4.7 (20)
1–15	16.5 (13)	6.3 (11)	9.4 (15)
0.1–1	–	10.3 (5)	–
0.03–0.1	–	16.0 (7)	–

Relative precision estimated from several duplicate determinations. Numbers in parentheses denote numbers of pairs used for the estimation.
Reproduced from Drabek, I., et al. (1983) International Journal of Environmental Analytical Chemistry, 15, 153, American Chemical Society, by courtesy of authors and publishers.

4.6 Neutron activation analysis

The application of neutron activation analysis to water samples has several advantages. Like inductively coupled plasma atomic absorption spectrometry it is a multi-element analysis technique and, as such, is useful for performing element scans when looking for unknown elements. It enables several elements to be determined in the same run. It is not limited to metallic elements. Its disadvantage is high cost and the probability that most laboratories cannot house the reactor equipment so that, in fact, samples have to be sent away for analysis to an establishment with such facilities. This delays results and demands careful consideration of sample preservation during the waiting period. The technique is intrinsically sensitive, and can be made more so by pre-concentrating the metals onto a resin such as Chelex 100 which can be directly analysed by the neutron activation technique. With these facilities analysis for many metals at the ultra-low background levels at which they occur in seawater becomes a possibility.

Table 4.39 Application of neutron activation analysis to the determination of metals in seawater

Elements	Sample pre-concentration	Reference
Co, Cr, Ca, Fe, Pb, Se, Sr	Sea salts	721
	Freeze dried	
As, Cu, Sb	Thionalide precipitation	722
Misc. elements	Frozen sample	723
Al, V, Cu, Mo, Zn, U	Organic co-precipitants (8-hydroxyquinoline)	724
Ag, Au, Cd, Ce, Co, Cr, Eu, Fe, Hg, La, Mo, Sc, Se, U, Zn, As, Sb	Adsorption and charcoal	725
Ba, Ca, Cd, Ce, Co, Cr, Cu, Fe, La, Mg, Mn, Sc, U, V, Zn	Adsorption of Chelex-100 glass powder	726
Hg, Au, Cu	Co-precipitation with lead diethyldithiocarbamate	727
12 elements	Chelating resin	728
Transition metals	Chelating resin	729
As, Mo, U, V	Colloid flotation	730
Th	Ion exchange chromatography	731
Co, Cu, Hg	Co-precipitation with lead pyrrolidine dithiocarbamate	732

Some applications of neutron activation analysis to seawater analysis are summarized in Table 4.39.

The application of neutron activation techniques to the measurement of trace metals in the marine environment has been reviewed by Roberton and Carpenter [719,720].

Stiller, Mantel and Rappaport [732] have described the determination of cobalt, copper and mercury in the Dead Sea by neutron activation analysis followed by X-ray spectrometry and magnetic deflection of β-ray interference.

The metals were co-precipitated with lead-ammonium pyrrolidine dithiocarbamate and detected by X-ray spectrometry following neutron activation. Magnetic fields deflect the β rays while the X-rays reach the silicon (lithium) detector unaltered. The detectors have low sensitivity to γ rays. The concentration of cobalt found by this method was $1.3\,\mu g\,l^{-1}$ about one fifth of that measured previously, while that of copper, $2.0\,\mu g\,l^{-1}$ agreed with results of some previous workers. The concentration of mercury was $1.2\,\mu g\,l^{-1}$.

Huh and Bacon [731] used neutron activation analysis to determine thorium-232 in seawater. Seawater samples were subjected to pre- and post-irradiation procedures. Separation and purification of the isotopes, using ion-exchange chromatography and solvent extraction, were performed during pre-irradiation. After irradiation protactinium-233 was extracted and counted. Yields were monitored with thorium-230 and protactinium-231 tracers. Thorium-232 concentrations were 27×10^{-7} dpm kg^{-1} for deep-water samples from below 400 m.

In a procedure [726] employing pre-concentration of the metals on a column containing a mixture of Chelex-100 and Pyrex glass powder the problems associated with swelling of the Chelex 100 were overcome and constant flow rates of sample down the column achieved.

The water samples were passed through the resin column and eluted with 100 ml 0.01 M nitric acid and the eluate was discarded. Trace elements were collected from the column by eluting with 50 ml 4 M nitric acid. The eluate was heated to reduce its volume, transferred to a volumetric flask and diluted to 10 ml.

A 3.5 ml portion in a 4 ml polyethylene vial was irradiated for 5 min (Figure 4.35). Another portion, 3.0 ml in a 3.5 ml silica vial, was irradiated for 3 days. After the short irradiation, 3 ml of the irradiated solution were transferred into an activity-free vial and submitted to γ-ray spectrometry with a Ge(Li) detector coupled with the 4000 channel analyser. After the long irradiation the sample was allowed to cool for 3 days, then the surface of the silica ampoule was cleaned with dilute nitric acid, and the sealed ampoule was placed in the counter (the background activity of the ampoule was negligible). γ-ray energy and peak-areas were calculated by computer. To determine the half-lives of the nuclides produced, the counting was repeated at appropriate intervals.

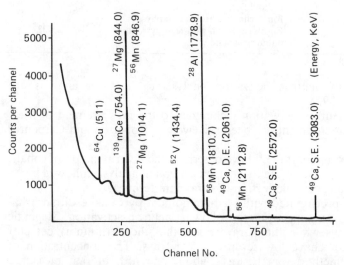

Figure 4.35 Gamma-ray spectrum of pre-concentrated river water after short irradiation. Irradiation time 3 min; thermal-neutron flux 1×10^{13} n cm^{-3} s^{-1}; decay time 3 min; counting time 100 s

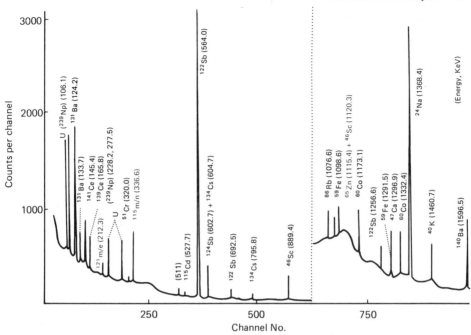

Figure 4.36 Gamma-ray spectrum of pre-concentrated seawater after long irradiation. Irradiation time 3 days; thermal-neutron flux $1 \times 10^{13}\,n\,cm^{-3}\,s^{-1}$; decay time 3 days; counting time 1000 s

The major interfering elements such as sodium, potassium and bromine and chlorine (Figures 4.35 and 4.36) were completely removed from the column with 100 ml 0.01 M nitric acid, whereas many trace elements were quantitatively retained. These elements were eluted with the succeeding 50 ml 4 M nitric acid.

The recovery ratios shown in Tables 4.40 and 4.41 indicate that the added traces of cadmium, cerium, copper, lanthanum, manganese, scandium and zinc are quantitatively recovered. The recoveries of barium, cobalt, bromium, iron, uranium and vanadium were also satisfactory.

Table 4.41 shows the recoveries of calcium and magnesium were very poor for seawater. The reasons for this may be connected with matrix effects.

In order to evaluate the precision of this method, replicate analyses were carried out by Lee *et al.* [726] by the proposed procedure, for trace elements in a seawater sample taken from the Kwangyang Bay. The results in Table 4.42 show satisfactory precision.

Lieser *et al.* [725] studied the application of neutron activation analysis to the determination of trace elements in seawater with particular reference to the limits of detection and reproducibility obtained for different elements when comparing different preliminary concentration techniques such as adsorption on charcoal, cellulose and quartz and complexing agents such as dithizone and sodium diethyldithiocarbamate.

Table 4.40 Separation of interfering elements from trace elements, % recovery

Isotopes	From effluent (2 litres)	From successive 20 ml portions of 0.01 M HNO₃					From successive 10 ml portions of 4 M HNO₃					Total
		1	2	3	4	5	1	2	3	4	5	
^{24}Na	95.1	3.9	0.5	0.3	0.1	0	0	0	0	0	0	99.9
^{82}Br	98.2	1.3	0.4	0.1	0	0	0	0	0	0	0	100.0
^{42}K	96.3	3.0	0.5	0.4	0.1	0	0	0	0	0	0	100.3
^{38}Cl	97.0	2.4	0.6	0.2	0	0	0	0	0	0	0	100.2
^{115m}Cd	0	0	0	0	0	0	18.1	37.0	38.4	6.2	0.3	100.0
^{64}Cu	0	0	0	0	0	1.4	31.2	43.1	21.7	2.5	0	99.9
^{56}Mn	0	0	0	0	0	0	21.8	33.0	32.7	10.3	1.9	99.7
^{64}Zn	0	0	0	0	0	0	19.9	38.1	39.9	2.2	0.1	100.2

Reproduced from Lee, C., et al. (1977) *Talanta*, **24**, 112, by courtesy of authors and publishers.

Table 4.41 Recovery of trace elements from spiked samples of stripped waters

Element	Seawater		
	Added (µg)	Found (µg)	Yield (%)
Ba	1.44	1.48	103
Ca	–	–	–
Cd	30.0	27.6	90.0
Ce	1.48	1.45	98.0
Co	1.50	1.49	99.3
Cr	230	230	100
Cu	189	187	98.9
Fe	615	615	100
La	0.830	0.832	100
Mg	–	–	–
Sb	–	–	–
Sc	0.140	0.142	101
U	0.160	0.157	98.1
V	27.0	27.0	100
Zn	401	397	99.0

Reproduced from Lee, C., et al. (1977) *Talanta*, **24**, 112, by courtesy of authors and publishers.

Table 4.42 Results of replicate analysis for trace elements in waters taken from the Kwangyang Bay and the Nakdong River, South Korea

Element	Seawater (µg l^{-1})
Ba	4.8 ± 0.8
Ca	–
Cd	0.20 ± 0.05
Ce	16.7 ± 0.7
Co	–
Cr	2.33 ± 0.20
Cu	1.1 ± 0.07
Fe	250 ± 10
La	0.72 ± 0.04
Mg	–
Mn	1.50 ± 0.04
Sc	0.098 ± 0.004
U	1.36 ± 0.1
V	2.14 ± 0.05
Zn	45.9 ± 1.4

Reproduced from Lee, C., et al. (1977) *Talanta*, **24**, 112, by courtesy of authors and publishers.

Table 4.43 Mean values and standard deviations found by adsorption on charcoal (KF = correction factor)

Element	Concentration in the standardized water samples (g l⁻¹)	Found by adsorption on activated charcoal, pH 8.5; three determinations			Found by adsorption on activated charcoal in presence of dithizone pH 8.5; seven determinations			Found by adsorption on activated charcoal in presence of NaDDTC, pH 5.5; seven determinations		
		g l⁻¹	%	KF	g l⁻¹	%	KF	g l⁻¹	%	KF
Ag	$1 \cdot 10^{-7}$	$(8.5 \pm 1.4) \cdot 10^{-8}$	85	1.18	$(8.5 \pm 0.3) \cdot 10^{-8}$	85	1.18	$(5.5 \pm 1.5) \cdot 10^{-8}$	55	1.82
As	$1 \cdot 10^{-6}$	$(6.6 \pm 1.2) \cdot 10^{-7}$	66	1.52	$(6.7 \pm 1.5) \cdot 10^{-7}$	67	1.49	$(3.2 \pm 1.9) \cdot 10^{-7}$	32	3.13
Au	$1 \cdot 10^{-9}$	$(1.1 \pm 0.3) \cdot 10^{-9}$	100	1.00	$(1.1 \pm 0.1) \cdot 10^{-9}$	100	1.00	$(1.0 \pm 1.7) \cdot 10^{-9}$	100	1.00
Br	$5 \cdot 10^{-2}$	$(5.3 \pm 1.0) \cdot 10^{-7}$	10^{-3}	–	$(2.1 \pm 0.5) \cdot 10^{-7}$	10^{-3}	–	$(3.6 \pm 1.7) \cdot 10^{-7}$	10^{-3}	–
Ca	$1 \cdot 10^{-1}$	$(1.2 \pm 2.3) \cdot 10^{-3}$	1	–	$(1.7 \pm 0.2) \cdot 10^{-3}$	2	–	$(1.6 \pm 0.8) \cdot 10^{-3}$	2	–
Cd	$1 \cdot 10^{-4}$	$(4.9 \pm 1.3) \cdot 10^{-7}$	49	2.04	$(9.5 \pm 2.7) \cdot 10^{-7}$	95	1.05	$(7.7 \pm 2.7) \cdot 10^{-7}$	77	1.30
Ce	$1 \cdot 10^{-6}$	$(8.7 \pm 2.1) \cdot 10^{-7}$	87	1.15	$(8.2 \pm 1.1) \cdot 10^{-7}$	82	1.22	$(4.3 \pm 1.6) \cdot 10^{-7}$	43	2.33
Co	$1 \cdot 10^{-6}$	$(4.0 \pm 0.6) \cdot 10^{-7}$	40	2.50	$(8.1 \pm 0.8) \cdot 10^{-7}$	81	1.23	$(7.6 \pm 1.0) \cdot 10^{-7}$	76	1.32
Cr	$1 \cdot 10^{-6}$	$(9.1 \pm 0.3) \cdot 10^{-7}$	91	1.10	$(9.6 \pm 0.3) \cdot 10^{-7}$	96	1.04	$(3.6 \pm 0.2) \cdot 10^{-7}$	36	2.78
Eu	$5 \cdot 10^{-7}$	$(5.2 \pm 0.4) \cdot 10^{-7}$	100	1.00	$(4.7 \pm 0.3) \cdot 10^{-7}$	95	1.05	$(3.8 \pm 1.7) \cdot 10^{-7}$	76	1.32
Fe	$1 \cdot 10^{-4}$	$(7.4 \pm 1.5) \cdot 10^{-5}$	74	1.35	$(7.7 \pm 0.8) \cdot 10^{-5}$	77	1.30	$(7.0 \pm 0.6) \cdot 10^{-5}$	70	1.43
Hg	$1 \cdot 10^{-7}$	$(9.7 \pm 0.2) \cdot 10^{-8}$	97	1.03	$(1.0 \pm 0.3) \cdot 10^{-7}$	100	1.00	$(1.0 \pm 0.1) \cdot 10^{-7}$	100	1.00
K	$4 \cdot 10^{-1}$	$(3.2 \pm 0.4) \cdot 10^{-5}$	10^{-2}	–	$(2.1 \pm 0.3) \cdot 10^{-5}$	10^{-2}	–	$(4.3 \pm 0.9) \cdot 10^{-5}$	10^{-2}	–
La	$1 \cdot 10^{-6}$	$(1.0 \pm 0.1) \cdot 10^{-6}$	100	1.00	$(1.0 \pm 0.1) \cdot 10^{-6}$	100	1.00	$(9.1 \pm 0.3) \cdot 10^{-7}$	91	1.10
Mo	$1 \cdot 10^{-6}$	$(5.0 \pm 3.3) \cdot 10^{-7}$	50	2.00	$(2.1 \pm 0.9) \cdot 10^{-7}$	21	4.76	$(1.0 \pm 0.1) \cdot 10^{-6}$	100	1.00
Na	5	$(1.4 \pm 0.3) \cdot 10^{-5}$	10^{-4}	–	$(1.6 \pm 0.2) \cdot 10^{-5}$	10^{-4}	–	$(3.2 \pm 1.5) \cdot 10^{-5}$	10^{-4}	–
Sb	$1 \cdot 10^{-6}$	$(1.8 \pm 0.5) \cdot 10^{-7}$	18	5.56	$(4.0 \pm 0.9) \cdot 10^{-7}$	40	2.50	$(5.6 \pm 2.0) \cdot 10^{-7}$	56	1.79
Sc	$2 \cdot 10^{-7}$	$(1.9 \pm 0.2) \cdot 10^{-7}$	95	1.05	$(2.0 \pm 0.1) \cdot 10^{-7}$	100	1.00	$(1.4 \pm 0.1) \cdot 10^{-7}$	70	1.43
Se	$1 \cdot 10^{-4}$	$(7.7 \pm 1.8) \cdot 10^{-7}$	77	1.30	$(6.1 \pm 1.7) \cdot 10^{-7}$	61	1.64	$(4.0 \pm 0.9) \cdot 10^{-7}$	40	2.50
U	$1 \cdot 10^{-7}$	$(1.1 \pm 0.1) \cdot 10^{-7}$	100	1.00	$(1.2 \pm 0.2) \cdot 10^{-7}$	100	1.00	$(7.8 \pm 0.2) \cdot 10^{-8}$	78	1.28
Zn	$1 \cdot 10^{-6}$	$(9.6 \pm 0.4) \cdot 10^{-7}$	96	1.04	$(1.0 \pm 0.1) \cdot 10^{-6}$	100	1.00	$(1.0 \pm 0.1) \cdot 10^{-6}$	100	1.00

Reproduced from Lieser, K.H., et al. (1977) Journal of Radioanalytical Chemistry, **37**, 717, by courtesy of authors and publishers.

Table 4.44 Determination of trace elements in seawater (mean ± SD). Samples taken at 54° 3′ north, latitude, 6° 30′ east, longitude, in the North Sea

Element	Water without suspended material; 5 determinations ($g\ l^{-1}$)	Suspended material; 5 determinations ($g\ l^{-1}$)	Water with suspended material; 3 determinations ($g\ l^{-1}$)
Ag	$(8.8 \pm 0.4) \cdot 10^{-9}$	$(3.6 \pm 0.3) \cdot 10^{-9}$	$(8.7 \pm 0.4) \cdot 10^{-9}$
As	$(3.5 \pm 0.3) \cdot 10^{-7}$	$(3.4 \pm 1.2) \cdot 10^{-8}$	$(4.7 \pm 0.7) \cdot 10^{-7}$
Au	$(3.5 \pm 0.3) \cdot 10^{-10}$	$(4.5 \pm 2.0) \cdot 10^{-10}$	$(3.9 \pm 0.2) \cdot 10^{-10}$
Ba	$(5.7 \pm 0.1) \cdot 10^{-7}$	$(4.1 \pm 0.6) \cdot 10^{-7}$	$(1.3 \pm 0.5) \cdot 10^{-6}$
Br	$(5.5 \pm 0.1) \cdot 10^{-7}$	$(4.0 \pm 3.8) \cdot 10^{-6}$	$(3.7 \pm 0.2) \cdot 10^{-6}$
Ca	$(3.6 \pm 0.2) \cdot 10^{-5}$	$(1.5 \pm 0.3) \cdot 10^{-4}$	$(2.0 \pm 0.5) \cdot 10^{-4}$
Cd	$<10^{-6}$	$<10^{-6}$	$<10^{-6}$
Ce	$(3.4 \pm 0.3) \cdot 10^{-4}$	$(2.4 \pm 1.5) \cdot 10^{-6}$	$(1.0 \pm 0.2) \cdot 10^{-6}$
Co	$(4.5 \pm 0.3) \cdot 10^{-4}$	$(2.3 \pm 0.8) \cdot 10^{-4}$	$(6.7 \pm 0.2) \cdot 10^{-8}$
Cr	$(1.4 \pm 0.1) \cdot 10^{-7}$	$(1.3 \pm 0.3) \cdot 10^{-7}$	$(2.9 \pm 0.2) \cdot 10^{-7}$
Eu	$(8.2 \pm 1.4) \cdot 10^{-10}$	$(4.4 \pm 1.9) \cdot 10^{-9}$	$(5.6 \pm 1.7) \cdot 10^{-9}$
Fe	$(1.5 \pm 0.2) \cdot 10^{-5}$	$(3.5 \pm 1.2) \cdot 10^{-5}$	$(6.8 \pm 1.5) \cdot 10^{-5}$
Hg	$(2.2 \pm 0.2) \cdot 10^{-6}$	$<5 \cdot 10^{-9}$	$<5 \cdot 10^{-9}$
K	$(3.6 \pm 1.9) \cdot 10^{-5}$	$(2.9 \pm 0.7) \cdot 10^{-5}$	$(4.7 \pm 1.4) \cdot 10^{-5}$
La	$(3.2 \pm 0.1) \cdot 10^{-8}$	$(2.8 \pm 2.1) \cdot 10^{-8}$	$(7.4 \pm 1.4) \cdot 10^{-8}$
Mo	$(4.4 \pm 1.4) \cdot 10^{-8}$	$(2.1 \pm 0.1) \cdot 10^{-8}$	$(6.1 \pm 0.5) \cdot 10^{-8}$
Na	$(3.6 \pm 0.3) \cdot 10^{-4}$	$(4.7 \pm 6.9) \cdot 10^{-4}$	$(1.7 \pm 0.7) \cdot 10^{-4}$
Sb	$(5.7 \pm 0.4) \cdot 10^{-9}$	$(2.3 \pm 0.8) \cdot 10^{-9}$	$(1.3 \pm 0.2) \cdot 10^{-8}$
Sc	$(4.5 \pm 0.3) \cdot 10^{-8}$	$(1.9 \pm 2.0) \cdot 10^{-8}$	$(2.3 \pm 0.4) \cdot 10^{-8}$
Se	$(4.5 \pm 0.3) \cdot 10^{-8}$	$(3.5 \pm 0.9) \cdot 10^{-8}$	$(6.3 \pm 0.7) \cdot 10^{-8}$
U	$(3.3 \pm 1.0) \cdot 10^{-8}$	$(1.5 \pm 0.8) \cdot 10^{-8}$	$(5.6 \pm 0.9) \cdot 10^{-8}$
Zn	$(2.3 \pm 0.1) \cdot 10^{-6}$	$(3.3 \pm 1.5) \cdot 10^{-7}$	$(3.9 \pm 0.2) \cdot 10^{-6}$

Reproduced from Lieser, K.H., et al. (1977) Journal of Radioanalytical Chemistry, **37**, 717, by courtesy of authors and publishers.

In these procedures 1 litre of seawater was shaken with 60 mg charcoal for 15 minutes. Complexing agents were added in amounts of 1 mg, dissolved in 1 ml acetone. The pH was 5.5 or it was adjusted to 8.5 by addition of 0.1 M ammonia. The charcoal was filtered off and irradiated. Results of three sets of experiments with charcoal alone, charcoal in the presence of dithizone and charcoal in the presence of sodium diethyldithiocarbamate are presented in Table 4.43. The following elements are adsorbed to an extent from 75 to 100%: silver, gold, cerium, cadmium, cobalt, chromium, europium, iron, mercury, lanthanum, scandium, uranium and zinc. The amount of sodium is reduced to about 10^{-6}, bromine to about 10^{-5} and calcium to about 10^{-2}.

Analyses of North Sea water and suspended solids obtained by this procedure are tabulated in Table 4.44.

Murthey and Ryan [730] used colloid flotation as a means of pre-concentration prior to neutron activation analysis for arsenic, molybdenum, uranium and vanadium. Hydrous iron (III) oxide is floated in the presence of sodium decyl sulphate with small nitrogen bubbles from 1 litre of seawater at pH 5.7. Recoveries of arsenic, molybdenum and vanadium were better than 95% whilst that of uranium was about 75%.

4.7 X-ray fluorescence spectrometry

This technique has received very limited application in seawater analysis. Pre-concentration by the ring-over technique [733] and by solvent extraction have both been used in order to improve the sensitivity of X-ray fluorescence spectrometry. Armitage and Zeitlin [733] converted uranium, copper, nickel, cobalt, iron and manganese to the 8-hydroxyquinolates and extracted these with chloroform. The extract was applied to a filter paper disc in a ring oven at 160°C and the metals separated prior to final determination by the X-ray technique.

Morris [734] separated microgram amounts of vanadium, chromium, manganese, iron, cobalt, nickel, copper and zinc from 800 ml seawater by precipitation with ammonium tetramethylenedithiocarbamate and extraction of the chelates at pH 2.5 with methylisobutyl ketone. Solvent was removed from the extract and the residue dissolved in 25% nitric acid and the inorganic residue dispersed in powdered cellulose. The mixture was pressed into a pellet for X-ray fluorescence measurements. The detection limit was 0.14 µg or better, when a 10 min counting period is used.

Knockel and Prange [735] converted metals in seawater into their diethyldithiocarbamates prior to X-ray fluorescence analysis of the separated solids. Membrane filtration of the precipitates resulted in carbamate-loaded filters, which could be directly measured by using radioisotope-excited X-ray fluorescence analysis. Furthermore, elution of Chromosorb columns loaded with the dithiocarbamate complex, by the passage of chloroform gave chloroform solutions in which the trace metals could be determined by X-ray fluorescence analysis using totally reflecting sample supports. Similarly the precipitate on the membrane filter could be dissolved in chloroform and determined in solution. The sensitivity of this method and the pH dependence of the reaction was also investigated.

Murata, Omatsu and Muskimoto [736] give details of equipment and a procedure for determination of traces of heavy metals by solvent extraction using di-isobutyl

Table 4.45 Results of analyses (µg l^{-1}) of liquid samples by the proposed method, with reference values obtained by atomic absorption spectrometry (AAS)

Ion analysed	Wastewater						Seawater	
	Sample A		Sample B		Sample C			
	This work*	AAS	This work†	AAS	This work†	AAS	This work†	AAS
Mn	120	130	–	–	240	240		
Fe	170	200	130	150	100	90	60	60
Co	220	240	–	–	–	–	–	–
Ni	130	140	20	24	70	80	–	–
Cu	–	–	40	40	30	30	20	20
Zn	–	–	140	140	60	50	–	–
Pb	–	–	70	70	50	40	–	–

Sample: concentrated 10-fold times.
*DDTC-IBMK extraction.
†DDTC-DIBK extraction.
Reproduced from Murata, M., et al. (1984) X-ray Spectrometry, 13, 83, by courtesy of authors and publishers.

ketone and isobutyl methyl ketone, combined with microdroplet analysis by X-ray fluorescence spectrometry using a specially designed filter paper, sodium diethyldithiocarbamate is used as chelating agent. The limits of detection for manganese, iron, cobalt, nickel, copper, zinc and lead were 15, 16, 8, 8, 13, 13 and $40 \, \mu g \, l^{-1}$ respectively for a $100 \, \mu l$ sample volume. The method was applied to analyses of seawater from Chirihama, Japan.

Table 4.45 shows that the results are in fair agreement with the reference values determined by atomic-absorption spectrometry.

Prange, Knockel and Michaelis [737] carried out multi-element determinations of dissolved heavy metals in Baltic seawater by total reflection X-ray fluorescence spectrometry. The metals were separated by chelation adsorption of the metal complexes on lipophilized silica-gel carrier and subsequent elution of the chelates by a chloroform/methanol mixture. Trace element loss or contamination could be controlled because of the relatively simple sample preparation. Aliquots of the eluate were then dispensed in highly polished quartz sample carriers and evaporated to thin films for spectrometric measurement. Recoveries, detection limits, and reproducibilities of the method for several metals were satisfactory.

4.8 Isotope dilution mass spectrometry

The determination of a number of elements in seawater has been separated by this technique including lithium [738], barium [739], lead [740], rubidium [741], uranium [742] and copper [743,744]. The technique has not been extensively applied.

The earlier stable isotope dilution mass spectrographic work was accomplished with a thermal ion mass spectrometer which had been specifically designed for isotope abundance measurements. However, Leipziger [745] demonstrated that the spark source mass spectrometer could also be used satisfactorily for this purpose. Although it did not possess the excellent precision of the thermal unit, Paulsen and co-workers [746] pointed out that it did have a number of important advantages.

In the analysis of seawater, isotope dilution mass spectrometry offers a more accurate and precise determination than is potentially possible with other conventional techniques, such as flameless atomic absorption spectrophotometry or anodic stripping voltimetry. Instead of using external standards measured in separate experiments, an internal standard, which is an isotopically enriched form of the same element, is added to the sample. Hence, only a ratio of the spike to the common element need be measured. The quantitative recovery necessary for the flameless atomic absorption and anodic scanning voltammetric techniques is not critical to the isotope dilution approach. This factor can become quite variable in the extraction of trace metals from the salt-laden matrix of seawater. Yield may be isotopically determined by the same experiment; however, by the addition of a second isotopic spike after the extraction has been completed.

An outline of the elements in seawater that may be analysed by isotope dilution techniques has been presented by Chow [747]. Most of the subsequent work has pertained to the analysis of lead in seawater [747–751]. The extension of the technique to the analysis of copper, cadmium, thallium and lead has been made by Murozumi [751] using a thermal source mass spectrometer; whereas, Mykytiuk, Russell and Sturgeon [752] employed a spark source mass spectrometer in the determination of iron, cadmium, zinc, copper, nickel, lead and uranium in seawater. It may be noted that thermal source mass spectrometry has two prime advantages over spark source mass spectrometry that may be of particular value in seawater analysis. One is the much higher precision (about 0.1%) capable of the instrument, a factor important in assessing the stability of the trace metals in seawater with time or the extraction technique itself. Secondly, the ratios to be determined may be measured at various filament currents (temperature) for an internal corroboration. This is of particular importance in detecting isobaric interferences that may lead to spurious ratios and hence misleading results.

Stuckas and Wong [753] have investigated the feasibility of using a thermal source mass spectrometer in the isotope dilution analysis of copper, cadmium, lead, zinc, nickel and iron in seawater. The approach basically follows that which had been successfully employed by the authors in the analysis of lead in seawater [750]. Herein, great importance was attached to the definition of the blank and the initial clean-up schedule. Once the operating parameters had been established by the analysis of pure isotopic spikes, the various components that constitute a blank were identified and minimized where possible using ultra-clean room techniques. The subsequent extractions of reagent water and seawater by dithizone and by ammonium pyrrolidine-diethyl-dithiocarbamate ion-exchange resins were evaluated for suitability for the isotope dilution mass spectrometric approach, yield and contamination levels. Corroboration of the results obtained on seawater were performed by two independent techniques, anodic scanning voltammetry and flameless atomic absorption spectrometry.

Satisfactory agreement was obtained in most cases (Table 4.46).

Table 4.46 Comparison of IDMS data with FAA and ASV approached on seawater (mean ($\pm$SD) concentrations (nmol kg^{-1}))

Method	Cu	Cd	Pb	Zn	Ni	Fe
'Soluble' IDMS	5.26 ± 0.09	0.708 ± 0.007	0.0423 ± 0.0019	11.18 ± 0.13	9.43 ± 1.06*	58 ± 6
'Total' IDMS	5.29 ± 0.13	0.691 ± 0.009	0.0892 ± 0.0046	12.73 ± 0.43	7.70 ± 0.42	62 ± 1
APDC/DDDC with FAA	5.35 ± 0.09	0.498 ± 0.089†	0.0381 ± 0.0100	11.01 ± 0.31	5.79 ± 0.17	50 ± 5
Chelex or APDC/DDDC with FAA	2.52		0.721 ± 0.027	0.087	11.62	10.39 ± 0.51
ASV	3.93 ± 0.60	0.649 ± 0.044	0.290 ± 0.034	12.47 ± 1.07		

*Dual values resulting from different extraction techniques.
†Yields were low and variable (85 + 14%); error quoted reflects average of FAA data alone.
APDC = Ammonium pyrollidine dithiocarbamate. ASV = Anodic scanning voltammetry. DDDC = Diethyldithiocarbamate. FAA = Flameless atomic absorption spectrometry. IDMS = Isotope dilution mass spectrometry.
Reproduced from Stukas, V.J. and Wong, C.S. (1981) *Trace Metals in Seawater*, Plenum Press, New York, by courtesy of authors and publishers.

Mykytiuk, Russell and Sturgeon [754] have described a stable isotope dilution spark source mass spectrometric method for the determination of cadmium, zinc, copper, nickel, lead, uranium and iron in seawater and compared results with those obtained by graphite furnace atomic absorption spectrometry and inductively coupled plasma emission spectrometry. These workers found that to achieve the required sensitivity it was necessary to pre-concentrate elements in the seawater using Chelex 100 [755] followed by evaporation of the desorbed metal concentrate onto a graphite or silver electrode for isotope dilution mass spectrometry.

Table 4.47 Analysis of seawater sample B (concentrations (ng ml^{-1}) expressed as means ± SD)

	IDSSMS ion exchange	GFAAS		ICPES ion exchange
		Solvent extraction	Ion exchange	
Fe	3.4 ± 0.3	3.2 ± 0.2	3.4 ± 0.4	3.2 ± 0.2
Cd	0.07 ± 0.01	0.06 ± 0.01	0.053 ± 0.007	ND
Zn	1.9 ± 0.1	1.8 ± 0.1	2.0 ± 0.1	1.6 ± 0.2
Cu	0.61 ± 0.04	0.5 ± 0.1	0.51 ± 0.03	0.73 ± 0.06
Ni	0.43 ± 0.03	0.46 ± 0.03	0.45 ± 0.05	0.38 ± 0.02
Pb	0.11 ± 0.02	0.06 ± 0.02	0.10 ± 0.01	ND
Co	0.028 ± 0.001*	0.015 ± 0.003	0.018 ± 0.008	ND
U	2.6 ± 0.2	ND	ND	ND

*By internal standard.
ND = not determined.
Reproduced from Mykytink, A.P., et al. (1980) *Analytical Chemistry*, **52**, 1281, American Chemical Society, by courtesy of authors and publishers.

Results obtained on a seawater sample by three procedures are given in Table 4.47. Isotope dilution results (IDSSMS) agree well with those obtained by graphite furnace atomic absorption spectrometry (GFAAS) and inductively coupled plasma emission spectrometry (ICPES).

The pre-concentration for all three techniques was achieved by Chelex 100 ion-exchange. However, since solvent extraction with ammonium pyrollidine dithiocarbamate is the most commonly accepted method, the values obtained using it with graphite furnace atomic absorption spectrometry are also included for comparison.

One of the advantages of the isotope dilution technique is that the quantitative recovery of the analytes is not required. Since it is only their isotope ratios that are being measured, it is necessary only to recover sufficient analyte to make an adequate measurement. Therefore, when this technique is used in conjunction with graphite furnace atomic absorption spectrometry, it is possible to determine the efficiency of the pre-concentration step. This is particularly important in the analysis of seawater where the recovery is very difficult to determine by other techniques since the concentration of the unrecovered analyte is so low. In using this technique, one must assume that isotopic equilibrium has been achieved with the analyte regardless of the species in which it may exist.

The thermal ion mass spectrometer was specifically developed for the measurement of isotope abundances and is capable of excellent precision. Although the spark source mass spectrometer used in this work lacks some of this precision, it proved itself to be very useful in stable isotope dilution work. It has a number of advantages including greater versatility, relatively uniform sensitivity and better applicability to a wide range of elements.

4.9 Metal pre-concentration methods

The considerable difficulty of trace element analysis in a high salt matrix such as seawater, estuarine water or brine is clearly reflected in the literature. The extremely high concentrations of the alkali metals, the alkaline earth metals and the halogens, combined with the extremely low levels of the transition metals and other elements of interest, make direct analysis by most analytical techniques difficult or impossible.

In the previous sections in this chapter, brief mention has been made particularly in connection with the inductively coupled plasma atomic absorption spectrometric technique, of the need to pre-concentrate seawater samples prior to the determination of metals. This is so that adequate detection limits can be achieved.

A variety of pre-concentration procedures has been used, including solvent extraction of metal chelates, co-precipitation, chelating ion-exchange, adsorption on to other solids such as silica-bonded organic complexing agents and liquid–liquid extraction.

An ideal method for the pre-concentration of trace metals from natural waters should have the following characteristics: it should simultaneously allow isolation of the analyte from the matrix and yield an appropriate enrichment factor; it should be a simple process, requiring the introduction of few reagents in order to minimize contamination, hence producing a low sample blank and a correspondingly lower detection limit; and it should produce a final solution that is readily matrix matched with solutions of the analytical calibration method.

There is much published work on pre-concentration techniques and this is discussed below. Pre-concentration is also discussed, incidentally, in some of the single element sections at the start of this chapter.

Concentration by chelation-solvent extraction-atomic-absorption spectrometry and chelation-solvent extraction graphite furnace atomic absorption spectrometry

The dithiocarbamate extraction method has been one of the most widely used techniques of pre-concentration for trace metal analysis by atomic absorption spectrometry [564–573]. This extraction method can be generally classified into two major categories. The first one comprises conversion of the metals to metal–dithiocarbamate chelates, then the extraction of the metal–dithiocarbamate complexes from a large volume of the aqueous phase into a smaller volume of oxygenated organic solvents such as methyl isobutyl ketone (thereby achieving concentration of metals) and then analysing the solvents directly [564–568]. The other one is to extract the metal complexes into oxygenated or chlorinated organic solvents such as chloroform, methylisobutyl ketone, etc. followed by a nitric acid back-extraction, and then analysing the trace elements in the acid solution. The latter category has been the subject of a number of reports [569–573]. There are several drawbacks associated with the acid back-extraction of metal dithiocarbamates. the kinetics is generally slow and the efficiency of acid extraction is poor for certain metals such as cobalt, copper and iron [571].

Lee and Burrell [574] have used a toluene solution of trifluoroacetyl acetone to extract cobalt, iron, indium and zinc from seawater.

A selection of chelation-solvent extraction methods is summarized in Table 4.48. It is seen that the majority of these use as the chelating agent, diethyldithiocarbamate, ammonium pyrrolidine dithiocarbamate or a mixture of both. Other chelating agents discussed include dithizone, 8-hydroxyquinoline and hexahydroazepine-1-carbodithioate, Freon, methylisobutyl ketone, chloroform, butyl acetate, xylene and carbon tetrachloride feature as extraction solvents. Detection limits (defined as 2 or 2.5 times the standard deviation of the blank) are in the ranges shown in Table 4.49 and as such are often suitable for the analysis of background levels in seawater.

Table 4.48 Pre-concentration of metals in seawater chelation solvent extraction techniques followed by direct atomic absorption spectrometry and graphite furnace atomic absorption spectrometry

Metals	Chelating agent	Solvent	Detection limit ($\mu g\ l^{-1}$)		Reference
DIRECT ATOMIC ABSORPTION SPECTROMETRY					
Mu, Fe, Co, Ni, Zn, Pb, Cu	hexahydro-azepine-l-carbodithioate	butyl-acetate	Mn Fe Co Ni Zn Pb Cu	0.2 1.5 0.6 0.6 0.4 2.6 0.5	575
Fe, Pb, Cd, Co, Ni, Cr, Mn, Zn, Cu	diethyldithio-carbamate	MIBK or xylene			576
Fe, Cu	ammonium pyrrolidine dithiocarbamate	MIBK	Cu Fe	<1 <1	577
Cd, Zn, Pb, Ca, Ni, Cu, Ag	dithizone	Chloroform	Ag Cd Zn Pb Cu Ni Co	0.05 0.05 0.6 0.04 0.06 0.3 0.04	573
Cd, Cu, Pb, Ni, Zn	(a) Ammonium dipyrrolidine dithiocarbamate (b) Ammonium dipyrrolidine dithiocarbamate plus diethyl dithiocarbamate	MIBK	Cu Cd Pb Ni Zn	10 2 4 16 30	579
GRAPHITE FURNACE ATOMIC ABSORPTION SPECTROMETRY					
Cu, Ni, Cd	Ammonium pyrrolidine dithiocarbamate				580
Ag, Cd, Cr, Cu, Fe, Ni, Pb, Zn	Ammonium dipyrrolidine dithiocarbamate	MIBK	Ag Cd Cr Cu Fe Ni Pb Zn	0.02 0.03 0.05 0.05 0.20 0.10 0.03 0.03	581 (2 × SD used)
Cu, Cd, Zn, Ni	Diethyl dithiocarbamate plus ammonium pyrrolidine dithiocarbamate	Chloroform	Cu Cd Zn Ni	1.0 0.2 2 10	582 (2.5 × SD used)
Cd, Pb, Ni, Cu, Zn	Ammonium pyrrolidine dithiocarbamate plus diethyl dithiocarbamate	Freon	Not stated		583
Cu	Ammonium pyrrolidine dithiocarbamate	MIBK	< 0.5		584

Table 4.48 (contd)

Metals	Chelating agent	Solvent	Detection limit (μg l^{-1})		Reference
GRAPHITE FURNACE ATOMIC ABSORPTION SPECTROMETRY contd					
Cd, Cu, Ni, Zn	Dithizone	Chloroform	Cu	0.006	585 (2
			Cd	0.0004	× SD
			Ni	0.032	used)
			Zn	0.016	
Cd, Cu, Fe	Ammonium pyrrolidine dithiocarbamate plus diethyl dithiocarbamate	Freon		Not stated	586
Cd, Zn, Pb, Cu, Fe, Mn, Co, Cr, Ni	Ammonium pyrrolidine N carbodithioate plus 8-hydroxy-quinoline	MIBK	Fe	0.08	587
			Cu	0.10	
			Pb	0.06	
			Cd	0.02	
			Zn	0.34	
Cd	Ammonium pyrrolidine dithiocarbamate	Carbon tetrachloride	Cd	0.006	588 590
Cd, Zn, Pb, Fe, Mn, Cu, Ni, Co, Cr	Dithiocarbamate	MIBK		Not stated	589
Cd, Co, Cu, Fe, Mn, Ni, Pb, Zn	Ammonium pyrrolidine dithiocarbamate	Chloroform	Cd	<0.0001	591
			Cu	<0.012	(2 ×
			Fe	<0.02	SD
			Mn	<0.004	used)
			Ni	<0.012	
			Pb	<0.016	
			Zn	<0.08	
Cd, Co, Cu, Fe, Mn, Ni, Pb, Zn	Ammonium pyrrolidine dithiocarbamate	Chloroform	Cd	0.02	592
			Cu	0.24	
			Fe	0.24	
			Mn	0.02	
			Ni	0.08	
			Pb	0.04	
			Zn	1.0	
Mn, Cd	Ammonium pyrrolidine dithiocarbamate and diethyl-ammonium diethyldithio-carbamate	Freon	Mn	0.07	593
			Cd	0.027	(2 × SD used)
Cd, Cu, Fe, Pb, Ni, Zn	Ammonium pyrrolidine dithiocarbamate and diethyl-ammonium diethylammonium diethyldithio-carbamate	Freon		Not quoted	594

Table 4.49 Detection limits for metals in seawater

	Lowest detection limit reported ($\mu g\ l^{-1}$)	Reference	Highest detection limit reported ($\mu g\ l^{-1}$)	Reference
Manganese	0.004	591	0.2	575
Iron	0.02	591	1.5	575
Cobalt	0.04	578	0.6	575
Nickel	0.012	591	16	579
Lead	0.016	591	4	579
Copper	0.006	585	10	579, 582
Silver	0.02	581	0.05	578
Cadmium	0.0001	591	2	579
Zinc	0.016	585	30	579
Chromium	0.05	581		

The most sensitive methods for all ten elements are covered by four references [578,581,585,591], two of which use a chloroform solution of dithizone [578,585] and two of which use a methylisobutyl ketone solution or chloroform solution of ammonium pyrrolidine dithiocarbamate [581,591].

Lo et al. [592] have recently developed a new method of back-extracting metals from their solution as a metal chelate in an organic solvent.

This procedure uses dilute mercury (II) solution instead of nitric acid. This back-extraction method is based on the fact that the extraction constant of the mercury (II) ammonium pyrollidine dithiocarbamate complex is much greater than most of the common trace metals of environmental importance. The substitution of mercury (II) for other metals in the form of dithiocarbamate complex is extremely fast and the efficiency of recovery is nearly 100% for a number of metals including cobalt, copper and iron. In addition, the back-extracted solution contains a low concentration of mercury (II) which is virtually interference free in graphite furnace atomic absorption spectrometry due to its high volatility. This two-step pre-concentration method pre-concentrates a number of trace metals such as cadmium, cobalt, copper, iron, manganese, nickel, lead and zinc in seawater by graphite furnace.

Concentration on ion-exchange resins

Studies of the use of ion-exchange resins for the pre-concentration of metals from seawater have been mainly concerned with the use of Chelex 100 resin [595–600].

A limited amount of work has been carried out using polyacrylamidoxine resin [601] and Amberlite XAD-7 [602] and Amberlite XAD-1 [603] resins.

Colella, Siggia and Barnes [601] agitated seawater samples with polyarylamidox-ime resin preparatory to the determination of iron (III), copper (II), cadmium (II), lead (II) and zinc (II). Metals were removed from the filtered OH resin by equilibrating with 1:1 hydrochloric acid/water mixture and their concentrations determined by atomic absorption spectrometry. Metal concentrations as determined by the resin method were in good agreement with the values determined directly on samples by either differential pulse polarography or differential pulse anodic stripping voltammetry.

Wan, Chaing and Corsini [602] determined conditions for the direct pre-concentration of cadmium, manganese, chromium, copper, nickel, iron, cobalt and lead from seawater samples using a two column Amberlite XAD-7 resin system. Low breakthrough volumes in the presence of humic materials necessitated their prior removal at a pH of 1–2 prior to pre-concentration of the trace metals on a second column of XAD-7 at pH 8. Metals were subsequently desorbed from the second column with 1% nitric acid by means of a precolumn of XAD-7. The final effluent for measurement by graphite-furnace atomic absorption spectrometry is readily matrix matched and permits use of the standard calibration curve procedure. Pre-concentration factors of 40 were obtained by this procedure permitting the analysis of coastal seawaters for the eight elements mentioned earlier.

Mackay [603] investigated the suitability of Amberlite XAD-1 resin for studying trace metal speciation in seawater. At low flow rates and at loading capacities far below theoretical values, the adsorption of these metals is not reproducible and the results are reminiscent of the behaviour observed when the adsorption capacity is being exceeded or flow rates are too high. It is suggested that the resin also adsorbs small but significant amounts of inorganic ions from seawater and that this effect makes the resin unsuitable for quantitative measurements of trace metal speciation.

Kiriyama and Kuroda [604] determined vanadium, cobalt, copper, zinc and cadmium in seawater by adsorption in an anion exchange resin.

Pre-concentration from seawater was achieved in thiocyanate medium. A strongly basic anion exchange resin in the thiocyanate form concentrated the five metals from seawater adjusted to $1 M$ thiocyanate and $0.1 M$ hydrochloric acid. Sorbed metals were recovered simultaneously by elution with $2 M$ nitric acid prior to determination by graphite furnace atomic absorption spectrometry.

Adsorption of metals on a single bead of ion-exchange resin has been used as a means of effecting pre-concentration [605,606].

Koide, Lee and Stallard [605] used a single anion exchange bead to isolate trace metals from seawater prior to their determination. Optimal conditions for the adsorption of cadmium-109, palladium-103, indium-192, gold-105, plutonium-237 and technetium-95 m on to a single bead were determined. Three types of applications of the techniques were investigated, with no prior concentration, with pre-concentration; increasing the yield of plutonium and technetium on to a single bead for improved sensitivity in mass spectrometric analysis. Two types of anion exchange resin (gel type and macroporous type) were tested. Hodge et al. [606] determined platinum and iridium in marine waters by pre-concentration by anion exchange, purification by uptake on a single anion exchange bead and determination by graphite furnace atomic absorption spectrometry. All steps were followed by radiotracers (platinum-191 and iridium-192). Yields varied between 35 and 90% for determination of platinum and iridium in sediments, manganese nodules, seawater and microalgae.

Koide, Lee and Stallard [607] showed that cadmium, palladium, iridium, gold, plutonium and technetium can be concentrated from seawater on to a single bead of anion exchange resin. This process eliminated salt interference. The beads acted as point sources during subsequent analytical determinations.

Chelation solvent extraction or resin-adsorption pre-concentration procedures coupled with graphite furnace atomic absorption spectrometry or neutron activation analysis are capable of determining many elements at the baseline concentrations occurring in deep seawater samples. All workers in this field emphasize the need for extreme precautions during sampling of seawater for analysis at these very low concentrations.

The iminodiacetate containing resin, Chelex-100, is the most commonly employed chelating resin for the removal and pre-concentration of trace heavy metals from seawater. Work on the use of Chelex-100 resin for the pre-concentration of metals from seawater is reviewed in Table 4.50. In each case metals are desorbed from the resin with nitric acid (2–2.5 M) and then determined in the extract by graphite furnace atomic absorption spectrometry.

Table 4.50 Application of Chelex-100 resin to the pre-concentration of metals in seawater prior to analysis by graphite furnace atomic absorption spectrometry

Elements	Concentration factor	Eluent	Detection limit		Reference
Cd, Co, Cu, Fe, Mn, Ni, Pb, Zn	100:1	2.5 M nitric acid	subnanogram μg l^{-1}		596
Cu, Cd, Zn, Ni	120:1	2 M nitric acid	Cu	0.006	597
			Cd	0.006	
			Zn	0.015	
			Ni	0.015	
Cd, Zn, Pb, Cu, Fe, Mn, Co, Cr, Ni	20:1	2.5 M nitric acid	Not stated μg l^{-1}		598
Cd, Pb, Ni, Cu, Zn	100:1	2.5 M nitric acid	Cd	0.01	599
			Pb	0.16–0.28	
			Ni	0.24–0.68	
			Cu	0.6	
			Zn	1.8	

Pre-concentration factors of up to 100–120 [596,598,599] have been reported by this technique enabling metals to be determined at the μg l^{-1} or ng l^{-1} level.

Early Chelex-100 procedures only partially separated the alkali and alkaline earth metal components of seawater prior to the analysis of the eluted elements of interest.

A more recent separation procedure utilizing the Chelex resin produced a sample devoid of alkali, alkaline earth and halogen elements, and left a dilute nitric acid/ammonium nitrate matrix containing only the trace elements of the seawater samples [596]. While this procedure produces a highly desirable and appropriate matrix for most spectroscopic methods of analysis, a solid sample would be more appropriate for other instrumental techniques such as X-ray fluorescence or neutron activation analysis. In addition, the above separation procedure also makes it difficult or impossible to analyse several elements which are held strongly by the resin but cannot be quantitatively eluted. Chromium and vanadium exhibit this type of behaviour and attempts to reproducibly elute these elements from Chelex-100 have not been successful.

Greenberg and Kingston [600] described a method to prepare 0.5 g solid samples from 100 ml estuarine or seawater using Chelex-100 resin, followed by the determination of trace elements in the solid resin by neutron activation analysis. Using this procedure, typical decontamination factors of $\geqslant 10^7$ for sodium, $\geqslant 10^5$ for chlorine and $\geqslant 10^3$ for bromine are observed. They used this procedure to

Table 4.51 Trace elements in water – SRM 1643a ($\mu g\ g^{-1}$)

Element	Neutron activation analysis*	Certified†
Co	19 ± 1	19 ± 2
Cr	16 ± 2	17 ± 2
Cu	19.1 ± 0.6	18 ± 2
Fe	88 ± 16	88 ± 4
Mn	30.9 ± 0.6	31 ± 2
Mo	97 ± 6	95 ± 6
Ni	56 ± 8	55 ± 3
V	52 ± 1	53 ± 3
Zn	68 ± 5	72 ± 4

*Uncertainties are 2 SD.
†Uncertainties are 95% confidence limit.
Reproduced from Greenberg, R.D. and Kingston, H.M. (1982) *Journal of Radioanalytical Chemistry*, **71**, 147, by courtesy of authors and publishers.

Table 4.52 Trace elements in seawater sample (ng ml^{-1})*

Element	Neutron activation analysis [600]	GFAAS [596]	XRF [608]
Co	0.044 ± 0.003	<0.1	
Cr	3.3 ± 0.2		
Cu	2.01 ± 0.05	2.0 ± 0.1	2.0 ± 0.2
Fe	2.1 ± 0.3	2.1 ± 0.5	
Mn	1.89 ± 0.03	2.0 ± 0.1	2.0 ± 0.1
Mo	5.3 ± 0.1		
Ni	1.3 ± 0.2	1.2 ± 0.1	1.3 ± 0.2
Sc	0.00095 ± 0.00005		
Th	$\leqslant 0.0002$		
U	1.90 ± 0.04		
V	0.45 ± 0.01		
Zn	4.9 ± 0.2	4.8 ± 0.3	4.5 ± 0.4

*All uncertainties are at the 1SD level.
GFAAS = Graphite furnace atomic absorption spectrometry. XRF = X-ray fluorescence spectrography.
Reproduced from Greenberg, R.R. and Kingston, H.M. (1982) *Journal of Radioanalytical Chemistry*, **71**, 147, by courtesy of authors and publishers.

determine cobalt, chromium, copper, iron, manganese, molybdenum, nickel, scandium, thorium, uranium, vanadium and zinc in NBS Standard Reference Material 1643a (Trace Elements in Water) (Table 4.51).

They also analysed a sample of seawater taken in Chesapeake Bay by neutron activation analysis and compared these results with those obtained by other techniques. The good agreement is evident in Table 4.52.

Boniforti *et al.* [609] compared several pre-concentration methods in the determination of metals in seawater. A comparison was made of ammonia pyrrolidine dithiocarbamate-8-quinolinol complexation followed by extraction with methylisobutyl ketone or Freon-113, co-precipitation with magnesium hydroxide or iron (II) hydroxide or chelating by batch treatment with Chelex-100 for the determination of manganese, iron, cobalt, zinc, nickel, copper and chromium. Atomic absorption spectrometry and inductively coupled plasma atomic emission spectrometry were used for analysis. Interferences, recovery, precision, accuracy and detection limits were compared. The Chelex-100 resin method was most suitable for the pre-concentration of all determinands except chromium, whereas pre-concentration of Cr (III) and Cr (VI) was achieved only by co-precipitation with iron (II) hydroxide.

Paulson [610] studied the effects of flow rate and pre-treatment on the extraction of manganese, cadmium and copper from estuarine and coastal seawater by Chelex-100 resin. Decreasing the flow rate for column extraction of estuarine samples by Chelex 100 to $0.2\,ml\,min^{-1}$ increases the yield of trace metals and improves the precision for determination of these elements.

Detection limits achievable by chelation-solvent extraction and ion-exchange separation on Chelex-100 resin are summarized in Table 4.53.

Table 4.53 Applicability of pre-concentration procedures to seawater analysis. Detection limits (based on 2 SD of blank)

| Element | Concentration in seawater consensus values ($\mu g\,l^{-1}$) | Best achievable values by pre-concentration |||
| | | Chelation solvent extraction – graphite furnace AAS | Ion-exchange separation Chelex-100 resins ||
			Graphite furnace AAS	Neutron activation analysis [600]
Cr	0.03	0.05	–	0.14
Mn	0.02	0.004	–	0.16
Fe	0.2	0.02	–	1.2
Co	0.005	0.04	–	0.006
Ni	0.17	0.012	0.015	–
Cu	0.05	0.006	0.006	0.08
Zn	0.49	0.016	0.015	0.20
Cd	–	0.0001	0.0006	–
Pb	–	0.016	–	–
Th	0.01			0.0004
V	2.5			0.06

Pre-concentration of metals on chemically modified supports other than chelex resins and ion-exchange resins

Metal chelating resins and immobilized (adsorbed or chemically bonded) chelates have found widespread application for the concentration and/or separation of trace metals from a variety of matrices. Dimethylglyoxime, alkylamines, diamines, xanthates, and dithiocarbamates, propylenediamenetetra-acetic acid, n-butylamides, N-substituted hydroxylamine, hexylthioglycolate, ferroin-type chelating agents, iminodiacetate, amidoxime, dithizone and 8-hydroxyquinoline have all been immobilized on various substrates. Tailoring chemically bonded chelating agents to specific needs allows the use of selective or general concentration schemes and also permits the 'recycling' of the chelating agent. Their major use lies in the pre-concentrations of trace metal ions from aqueous and saline media.

Other pre-concentration techniques

Early work in this field included adsorption of metals from seawater on to columns of modified carbon [611], chitosan [612,613], p-dimethylamino-benzylidenerhodamine on silica gel [614] and 5,7-dibromo-8-hydroxyquinoline [614].

Buono, Burns and Fasching [616] showed that the addition of poly-5-vinyl 8-hydroxyquinoline to seawater leads to the formation of insoluble metal chelates of aluminium, cobalt, copper, iron, lead, manganese, nickel, vanadium, and zinc within 2 minutes if the pH value is suitably adjusted.

Since this reagent does not precipitate alkali metal or alkaline earths it constitutes an excellent reagent for separating trace metals from saline matrices.

Sturgeon et al. [617] pre-concentrated cadmium, copper, zinc, lead, iron, manganese, nickel and cobalt from seawater on to silica immobilized 8-hydroxylquinoline prior to determination by graphite furnace atomic absorption spectrometry. Results for the analyses of two near-shore seawater samples are given in Table 4.54. Table 4.55 presents results for the open-ocean sample. Near-shore samples were concentrated 50-fold, the open-ocean samples 90-fold. Calibration was achieved by spiking an aliquot of the concentrate with the element of interest, thereby obtaining an exact matrix match.

Results obtained by using the immobilized 8-hydroxyquinoline concentration procedure were compared to 'accepted' values for these samples. Good agreement

Table 5.54 Analyses of near-shore seawater (salinity 29.5‰) (means ± SD; n = 3)

Element	Concentration ($\mu g\ ml^{-1}$)			
	Sample 1		Sample 2	
	I-8-HOQ	Accepted value	I-8-HOQ	Accepted value
Cd	0.020 ± 0.001	0.024 ± 0.004	0.025 ± 0.001	0.023 ± 0.001
Pb	0.22 ± 0.01	0.22 ± 0.06	0.014 ± 0.003	0.018 ± 0.001
Zn	0.44 ± 0.01	0.41 ± 0.05	0.29 ± 0.03	0.28 ± 0.01
Cu	1.03 ± 0.06	0.96 ± 0.04	0.17 ± 0.01	0.22 ± 0.02
Fe	1.0 ± 0.1	1.03 ± 0.04	7.2 ± 0.9	6.9 ± 0.02
Mn	0.71 ± 0.02	0.68 ± 0.05	1.06 ± 0.01	1.13 ± 0.02
Ni	0.33 ± 0.01	0.31 ± 0.04	0.39 ± 0.01	0.34 ± 0.01
Co	0.018 ± 0.002	0.015 ± 0.007	0.017 ± 0.001	ND

ND = Not determined.
Reproduced from Sturgeon, R.E., et al. (1981) Analytical Chemistry, **53**, 2337, American Chemical Society, by courtesy of authors and publishers.

Table 4.55 Analysis of open-ocean seawater (salinity 35‰) (means ± SD; n = 3)

Element	Concentration ($\mu g\ ml^{-1}$)	
	I-8-HOQ	Accepted value
Cd	0.030 ± 0.002	0.033 ± 0.002
Pb	0.095 ± 0.009	0.09 ± 0.01
Zn	0.28 ± 0.01	0.30 ± 0.03
Cu	0.121 ± 0.008	0.11 ± 0.02
Fe	0.20 ± 0.01	0.18 ± 0.04
Mn	0.018 ± 0.001	0.023 ± 0.003
Ni	0.27 ± 0.01	0.27 ± 0.02
Co	0.003 ± 0.001	ND

ND = Not determined.
Reproduced from Sturgeon, R.E., et al. (1981) Analytical Chemistry, **53**, 2337, American Chemical Society, by courtesy of authors and publishers.

Table 4.56 Recovery (as %) of added metal ions from sea- and tap water. Results are given as means ± SD with number of replicates in parentheses

	Co (0.5)*	Cd (1)	Cu (5)	Hg (1)	Pb (10)
Seawater	55 ± 18 (6)	98 ± 9 (6)	93 ± 7 (6)	99 ± 5 (6)	93 ± 5 (4)
Seawater†	63	89	95	95	93
Tap water	54 ± 12 (4)	90 ± 8 (4)	95 ± 8 (4)	95 ± 8 (4)	92 ± 8 (4)

*The numbers in parentheses are the amounts added in µg l^{-1}.
†By passing solution through 50 mg DTC-cellulose spread on filter paper.
Reproduced from Murthy, R.S.S. and Ryan, D.E. (1982) *Analytica Chimica Acta*, **140**, 163, Elsevier Science Publishers, Amsterdam, by courtesy of authors and publishers.

Table 4.57 Results for seawater samples

	Volume used (litres)	Concentration (µg l^{-1})*			
		Cu	Cd	Hg	Pb
Seawater	1	0.66 ± 0.06	–	–	0.20 ± 0.03
	5	0.72 ± 0.07	<0.03†	0.026 ± 0.007	0.25 ± 0.03
				0.018 ± 0.008	0.30 ± 0.03

*Concentration in µg l^{-1} ± concentration equal to 2 (background)$^{1/2}$.
†Concentration less than 2 (background)$^{1/2}$.
Reproduced from Murthy, R.S.S. and Ryan, D.E. (1982) *Analytica Chimica Acta*, **140**, 163, Elsevier Science Publishers, Amsterdam, by courtesy of authors and publishers.

with accepted values is evident for all three samples. The precision of analysis, expressed as a relative standard deviation, averages 6% (range of RSD 2–12%) over all elements, concentrations and samples.

Murthy and Ryan [618] pre-concentrated copper, cadmium, mercury and lead from seawater on a column of a dithiocarbamate cellulose derivative. Metal concentrations on the adsorbent material were determined by neutron activation analysis. The recovery of added spikes to sea- and tap water shown in Table 4.56 suggests that Cu^{II}, Hg^{II} and Pb^{II} can be quantitatively collected.

Some typical results obtained by this procedure are listed in Table 4.57.

Sturgeon, Willie and Berman [619] pre-concentrated selenium and antimony from seawater on C-18 bonded silica gel prior to determination by graphite furnace atomic absorption spectrometry. The method was based on the complexation of selenium (IV), antimony (III) and antimony (V) with ammonium pyrrolidine dithiocarbamate. These complexes were adsorbed on to a column of C-18 bonded silica gel, then eluted with methanol, followed by evaporation to near dryness. The residue was taken up in 1% nitric acid. Concentration factors of 200 could be obtained. Detection limits for selenium (IV), antimony (III) and antimony (V) were 7, 50 and 50 ng l^{-1} respectively, based on a 300 ml sample volume.

Co-precipitation methods

Work in this field has been mainly concerned with co-precipitatives in seawater metals on to ferric hydroxide [620–623] cupric sulphide [624] and zirconium hydroxide [625]. In a typical procedure the iron salt was added to the seawater samples, the pH was adjusted and the iron plus trace metal precipitate was separated by filtration on paper. The paper and precipitate were dissolved in nitric acid and the acid solution was analysed directly by graphite furnace atomic absorption spectrometry. A 200-fold concentration was achieved with a recovery in excess of 90% of the metals in seawater and, of course, iron could not be determined. Manganese did not co-precipitate.

Weisel, Duke and Fasching [621] determined aluminium, lead and vanadium in Atlantic seawater after co-precipitation on to ferric hydroxide.

Patin and Morozov [624] developed a procedure for simultaneous concentration of mercury, lead and cadmium from seawater by co-precipitation with copper sulphide. The isolation yield is 99% for mercury and lead and 89% for cadmium. Mercury was determined by flameless atomic absorption spectrophotometry and lead and cadmium by flame atomic absorption spectrophotometry.

Siu and Berman [623] determined selenium in seawater by gas chromatography after co-precipitation with hydrous iron (III) oxide. The method used rapid co-precipitation of selenium with hydrous iron (III) oxide. Following a brief stirring and settling period the co-precipitate was filtered and dissolved in hydrochloric acid, derivatized to 5-nitropiazselenol and extracted into toluene. The selenium was determined by gas chromatography–electron capture detection. The detection limit was 1 pg injected or 5 ng selenium per litre of seawater using a 200 ml sample. Precision was 6% at 25 pg selenium per litre.

Akagi et al. [625] used zirconium co-precipitation for simultaneous multi-element determinations of trace metals in seawater by inductively coupled plasma atomic emission spectrometry. The co-precipitation procedure, ageing and washing of co-precipitates and optimal pH conditions are described, together with spectral interferences. Recoveries of most metals increased with increase in pH except for hexavalent chromium and hexavalent molybdenum. Improved detection limits for 17 metals are reported including aluminium, arsenic, cadmium, cobalt, chromium, copper, iron, lanthanum, manganese, molybdenum, nickel, lead, antimony, titanium, vanadium, yttrium and zinc.

Co-flotation with octadecylamine and ferric hydroxide as collectors has been used to separate copper, cadmium and cobalt from seawater [622]. The method

Table 4.58 Cu^{II}, Cd^{II} and Co^{II} recovery from spiked seawater samples by the co-flotation method (V = 2000 ml) and the APDC/MIBK method (V = 400 ml)

Method	Element addition (ng ml^{-1})	Species (ng ml^{-1})					
		Cu^{II}		Cd^{II}		Co^{II}	
		Found	RSD (%)	Found	RSD (%)	Found	RSD (%)
Co-flotation	0.0	2.5	13.3	0.8	6.0	5.8	7.0
	5.0	5.6	9.5	4.3	6.0	11.4	5.6
	10.0	10.1	2.0	8.4	3.4	15.9	4.7
	Mean recovery (%)	83 ± 6		79 ± 4		98 ± 5	
APDC/MIBK	0.0	4.1	29.4	2.8	14.5	3.2	22.0
	5.0	7.0	4.4	7.6	7.2	7.5	2.7
	10.0	13.8	8.2	12.0	4.9	11.7	1.7
	Mean recovery (%)	87 ± 6		94 ± 2		92 ± 2	

Reproduced from Cabazon, L.M., et al. (1984) Talanta, 31, 597, by courtesy of authors and publishers.

was based on the co-flotation or adsorbing colloid flotation technique. The substrates were dissolved in an acidified mixture of ethanol, water and methyl isobutyl ketone to increase the sensitivity of the determination of these elements by flame atomic absorption spectrophotometry. The results were compared with those of the usual ammonium pyrrolidine dithiocarbamate-methyl isobutyl ketone extraction method. While the mean recoveries were lower, they were nevertheless considered satisfactory (Table 4.58).

Copper, nickel and cadmium have been determined at the $ng\,l^{-1}$ level by co-precipitation with cobalt pyrrolidionedithiocarbamate followed by dissolution of the precipitate in an organic solvent and analysis by graphite furnace atomic absorption spectrometry (Boyle and Edmond [626]). Excellent results for the distribution of nickel and cadmium in the ocean were obtained by this technique.

Pre-concentration by flow injection analysis

Flow injection analysis has been recently applied as an automatic sample injection technique for atomic absorption spectrometry [627–630].

Ion-exchange pre-concentration has proved to be an effective means of increasing the sensitivity of atomic absorption spectrometry as well as a means of removing interferences. The concept has been utilized for many years and its practicality has recently been shown in the work of, for example, Kingston *et al.* [631]; Danielsson, Magnusson and Zhang [632]; and others [633,634]. However, the conventional column mode of ion-exchange pre-concentration is tedious and incompatible with the final rapid determination by atomic absorption spectrometry. Therefore, attempts have been made recently to apply an on-line flow injection pre-concentration technique by ion-exchange to atomic absorption spectrometry determinations with the aim of speeding up and simplifying the pre-concentration step.

Olsen *et al.* [635] gave details of equipment and procedure developed for pre-concentrating and determining traces of cadmium, lead, copper and zinc in seawater by atomic absorption spectrophotometry combined with flow injection analysis.

Pre-concentration was achieved by passing 2 ml of sample through a microcolumn packed with Chelex 100 resin. Metals were desorbed with 180 µl 2 M aqueous nitric acid which was passed to an atomic absorption spectrometer, thereby achieving a concentration factor of about 10. Seen from a practical viewpoint this combination results in timesaving because it allows an unprecedented sample throughput at the $\mu g\,l^{-1}$ level. As the analytical readout is available within 5 s for the direct assay and the latest within 110 s for the system including pre-concentration, smaller sample series can be treated expediently by manual injection.

Since the original work of Olsen many improvements have been proposed to flow injection analysis equipment. Valve designs have been improved; in addition to Chelex 100 other ion-exchange resins have been tested, and different flow systems have been proposed to increase the efficiency of the process [636–639]. These developments indicate that the new approach not only increases the speed of the pre-concentration process, but could also ultimately rival the sensitivity and speed of graphite-furnace atomic absorption spectrometry.

Fang, Ruzicka and Hansen [640] have described a flow-injection system (Figure 4.37) with on-line ion-exchange preconcentration on dual columns for the

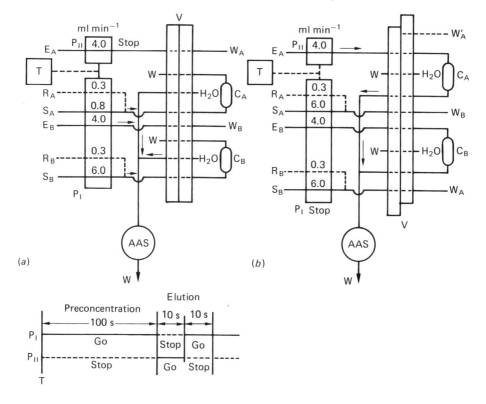

Figure 4.37 Manifold for dual-column on line ion-exchange pre-concentration system with flame atomic absorption detection. (*a*) Sampling and pre-concentration mode. (*b*) Elution mode for column A. PI, PII = Pumps I and II; T = timer; V = Valve; because of the circular arrangements of the valve channels, W_A' represents the same channel as W_A. (*c*) Time-sequencing programme for valve operation. The points marked T indicate turn of valve. Note that the elution of both columns A and B takes place in position (*b*) of the valve, but with pumps PI and PII sequenced stop-go and go-stop, respectively

determination of trace amounts of heavy metals at $\mu g\,l^{-1}$ and sub-$\mu g\,l^{-1}$ levels by flame atomic absorption spectrometry. The degree of pre-concentration ranges from 50- to 105-fold for different elements at a sampling frequency of 60 samples hourly. The detection limits for copper, zinc, lead and cadmium are 0.07, 0.03, 0.5 and 0.05 $\mu g\,l^{-1}$ respectively. Relative standard deviations were 1.2–3.2% at $\mu g\,l^{-1}$ levels. These workers studied the behaviour of the different chelating exchangers used with respect to their pre-concentration characteristics, with special emphasis on interferences encountered in the analysis of seawater.

Fang studied closely the swelling properties of Chelex 100 Resin. They showed that, depending on the pH of the element, the resin swelled by a factor up to two. To limit the maximum change in volume of a resin in a column to 25% during a single cycle of operation (which is imperative to avoid excessive pressure and void volume variations), the sampling period (during which chelation takes place) at a flow rate of 6 ml min^{-1} should not exceed 50 s for a 0.5 M ammonium acetate buffer, 75 s for a 0.1 M buffer or 100 s for a 0.05 M buffer. Secondly, the column should be packed about three-quarters full with resin in the H$^+$-form (washed with water) after conversion from the NH$_4^+$ form (in which state it should be equilibrated with buffer of the same concentration as that used in the ensuing procedure). Packing in the NH$_4^+$-form would otherwise result in an excessively loose column packing giving rise to large dispersion and degrading the sensitivity. Finally, resin columns in the NH$_4^+$-form should never be washed with water, lest excessive pressure develop in the column, causing blockages, leakages or dislodgement of the nylon retaining gauzes. When not in use, the columns should be converted to the H$^+$-form by normal elution with acid and washed thoroughly with water.

Fang concluded that the flow-injection atomic absorption spectrometry system with on-line pre-concentration will challenge the position of the graphite-furnace technique, because it yields comparable sensitivity for much lower cost by using simpler apparatus and separation mode. The method offers unusual advantages when matrices with high salt contents such as seawater are analysed because the matrix components do not reach the nebulizer.

Marshall and Mottola [641] have used silica immobilized 8-quinolol as a means of pre-concentrating metals for analysis by flow injection–atomic absorption spectrometry. This has proved to be a particularly useful material for sample preparation, matrix isolation and pre-concentration of trace metal ions [642–645].

Table 4.59 Results of copper determination in water samples (Environmental Protection Agency, USA)

Sample	Cu concentration (ng ml^{-1})	
	Determined	Reported
EPA-1	51 ± 4.6	50
EPA-2	252 ± 2.6	250
EPA-3	38 ± 3.0	40
EPA-4	8.9 ± 1.6	8.3
Tap*	26 ± 2.0	NA

Other elements known to be present (concentration in ng ml^{-1} in parentheses): EPA-1, Al (450), As (60), Be (250), Cd (13), Cr (80), Co (80), Fe (80), Pb (120), Mn (75), Hg (3.5), Ni (80), Se (30), V (250), Zn (80); EPA-2, Al (700), As (200), Be (750), Cd (50), Cr (150), Co (500), Fe (600), Pb (250), Mn (350), Hg (7.5), Ni (250), Se (40), V (750), Zn (200); EPA-3, Al (350), As (40), Be (150), Cd (10), Cr (60), Co (70), Fe (50), Pb (80), Mn (55), Hg (3.0), Ni (70), Se (20), V (200), Zn (60); EPA-4, Al (106), As (27), Be (29), Cd (9.1), Cr (7.1), Co (43), Fe (22), Pb (43), Mn (13), Hg (0.7), Ni (17), Se (11), V (130), Zn (10); tap water, unknown*. Stillwater, OK. NA = Not available.
Reproduced from Marshall, M.A. and Mottola, H.A. (1985) *Analytical Chemistry*, **57**, 729, American Chemical Society, by courtesy of authors and publishers.

The selectivity of silica immobilized 8-quinolol for transition-metal ions over the alkali and alkaline-earth metal ions makes it useful for samples containing large quantities of the latter such as seawater.

These workers evaluated breakthrough capacities under different flow, temperature and geometric characteristics of the pre-concentration column. The columns have relatively high capacities for metals and do not suffer from complications due to swelling. Excellent agreement was obtained in determinations of copper on standard environmental samples (Table 4.59).

Hydride generation pre-concentration methods

A number of elements in the fourth, fifth and sixth group of the periodic system form hydrides upon reduction with sodium borohydride [647], which are stable enough to be of use for chemical analysis (Ge, Sn, Pb, As, Sb, Se, Te). Of these elements, Andreae [646] have investigated in detail arsenic, antimony, germanium and tin. The inorganic and organometallic hydrides are separated by a type of temperature-programmed gas-chromatography. In most cases it is optimal to combine the functions of the cold trap and the chromatographic column in one device. The hydrides are quantified by a variety of detection systems which take into account the specific analytical chemical properties of the elements under investigation. For arsenic, excellent detection limits (about 40 pg) can be obtained with a quartz tube cuvette burner which is positioned in the beam of an atomic absorption spectrophotometer. For some of the methylarsines, similar sensitivity is available by an electron capture detector. The quartz-burner atomic absorption spectroscopy system has a detection limit of 90 pg for tin; for this element much lower limits (about 10 pg) are possible with a flame photometric detection system, which uses the extremely intense emission of the SnH molecule at 609.5 nm. The formation of GeO at the temperatures of the quartz tube furnace makes this device quite insensitive for the determination of germanium. Excellent detection limits (about 140 pg) can be reached for this element by the combination of the hydride generation system with a modified graphite furnace atomic absorption spectroscopy.

Many of the recent methods make use of the condensation of the hydrides in a cold trap at liquid nitrogen temperature. Braman and Foreback [648] pioneered the use of a packed cold trap to serve both as a substrate to collect the hydrides at liquid nitrogen temperature, and to separate arsine and the methylarsines gas chromatographically by controlled heating of the trap. In the same paper, they described the differentiation between As^{III} and As^{V} by a pre-reduction step and by control of the pH at which the reduction takes place.

Amaukwah and Fasching [654] have discussed the determination of arsenic (V) and arsenic (III) in seawater by solvent extraction-absorption spectrometry using the hydride generation technique.

Most of the detectors commonly used for gas chromatography have been applied to the detection of the hydrides, among them the thermal conductivity, flame ionization and the electron capture detector [649]. A molecular emission detector has been used for tin [650]. Atomic emission spectrometric detectors based on d.c. discharges [648,651] and microwave induced plasmas [652] were applied to the speciation of arsenic in environmental samples. The currently most popular detection system is atomic absorption spectrometry in one of its numerous variants. The hydrides were at first introduced into a normal atomic absorption flame, but it was soon recognized that better detection limits could be achieved with enclosed atom reservoirs and with very small flames or with flameless systems. A number of heated quartz furnace devices without internal flames are now on the commercial market. The lowest detection limits were achieved by cold-trapping of the hydrides and subsequent introduction into either a quartz cuvette furnace [649] or into a commercial graphite furnace [654].

The determination of the hydride element species consists of five steps: (1) the reduction of the element species to the hydrides; (2) the removal of interferent volatiles from the gas stream; (3) the cold-trapping of the hydrides; (4) the

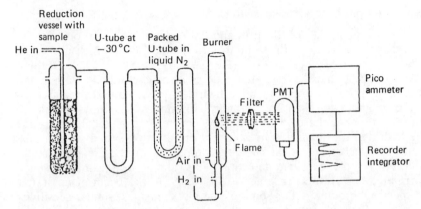

Figure 4.38 Schematic design of the $NaBH_4$–reduction/flame photometric detection system for the determination of tin species in natural waters

separation of the substituted and unsubstituted hydrides from each other and from interfering compounds; and (5) the quantitative detection of the hydrides. A typical instrumental configuration to accomplish these steps is shown in Figure 4.38 for the borohydride reduction/flame photometric detection system for tin speciation analysis.

Reduction of the element species to the hydrides
Most of the hydride elements occur in a number of different species. The optimum reduction conditions vary from element to element and between different species of the same element.

The conditions under which the element species are being reduced have been optimized as shown in Table 4.60.

Table 4.60 Reaction conditions for the reduction of various 'hydride element' species to the corresponding hydrides

Species	pK_a	pH	Composition of reaction medium	$NaBH_4^{2*}$
As^{III}	9.2	6–7	0.05 M Tris-HCl	1
As^{V}	2.3			
MMAA	3.6	~1	0.12 M HCl	3
DMAA	6.2			
Sn^{II}	9.5			
Sn^{IV}	~10	2–8†	0.01 M HNO_3	1
Me_xSn	11.7‡			
Ge^{IV}	9.3	~6	0.095 M Tris-HCl	3
Sb^{III}	11.0	~6	0.095 M Tris-HCl	2
Sb^{V}	2.7	~1	0.18 M HCl, 0.15 M KI	3
MMSA	–	1.5–2	0.06 M HCl	2
DMSA	–			

MMAA = monomethylarsonic acid $[(CH_3)AsO(OH)_2]$. DMAA = dimethylarsenic acid $[(CH_3)_2AsO(OH)]$. $Me_xSn = Me_xSn^{3+}$. Me_2Sn^{2+}, Me_3Sn^+. MMSA = monomethylstibonic acid $[(CH_3)SbO(OH)_2]$. DMSA = dimethylstibinic acid $[(CH_3)_2SbO(OH)]$.
*ml of 4% $NaBH_4$ solution per 100 ml sample.
†Increases during the reaction.
‡Data available only for $Me_2Sn(OH)_2$.
Reproduced from Andreae. M.O.P. (1981) *Trace Metals in Seawater*. Plenum Press. New York. by courtesy of authors and publishers.

With the exception of Sb^V which requires the presence of iodide for its reduction, all species can be reduced in an acid medium at pH of 1–2. However, the reduction of some species, including Sb^{III} and As^{III} and all tin species, will also proceed at higher pH, where As^V and Sb^V are not converted to their hydrides. This effect permits the selective determination of the different oxidation states of these elements [648,655,656].

In contrast to the findings of Foreback [652] and Tompkins [653], Andreae [646] was not able to reduce antimony (V) or antimony (III) quantitatively at pH 1.5–2.0 without the addition of potassium iodide. A concentration of at least 0.15 M potassium iodide in the final solution at a pH less than 1.0 was necessary to achieve complete reduction [650]. This is in agreement with the work of Fleming and Ide [655] who suggested an addition of about 0.2 M potassium iodide to ensure the reduction of Sb^V.

Germanium can be reduced through a wide range of pH [650]. The optimum pH is in the near-neutral region, as the efficiency of germanium reduction decreases at lower pH, probably due to the competitive acid-catalysed hydrolysis of the borohydride ion. At a pH above 8, the yield also decreases.

Removal of interferent volatiles from the gas stream
Depending on the detector used, some volatile compounds which are formed or released during the hydride generation step may interfere with the detection of the hydrides of interest. Most prominent among them are water, carbon dioxide and, in the case of anoxic water samples, hydrogen sulphide. The atomic absorption detector is insensitive toward these compounds; thus no precautions need to be taken when this detector is used. It has been found convenient in some applications, however, to remove most of the water before it enters the cold-trap/ column which serves to condense and separate the hydrides. This can be accomplished by passing the gas stream through a larger cold trap cooled by a dry-ice/alcohol mixture or by an immersion cooling system [652]. This method was also used with water-sensitive detectors, e.g. the electron-capture detector for methylarsines [653], or with plasma discharge detectors (e.g. Crecelius [656]).

Cold-trapping of the hydrides
Only when the very contamination-sensitive electron-capture detector is used is it necessary to provide separate gas streams, one for the reaction and stripping part of the system, the other for the carrier gas stream of the column and detector. Otherwise, the same gas stream can be used to strip the hydrides from solution and to carry them into the detector, which greatly simplifies the apparatus.

Initially, column packings of glass beads or glass wool [649] were used. These packings produce poor separation of the methylated species from each other and badly tailing peaks, however. Andreae [646] therefore used a standard gas chromatographic packing (15% OV-3 on Chromosorb W/AWDMCS, 60–80 mesh) in U-tubes for the separation of the inorganic and alkyl-species of arsenic, antimony and tin. This packing is quite insensitive to water and produces sharp and well-separated peaks, as demonstrated in Figure 4.39, for stibine, methylstibine and dimethylstibine in a standard and a seawater sample.

The most versatile system is the combination of hydride generation with atomic absorption spectroscopy. Here, the objective is to introduce the hydrides into an atom reservoir aligned in the beam of the instrument and to dissociate them to produce a population of the atoms of interest. This can be either achieved in a

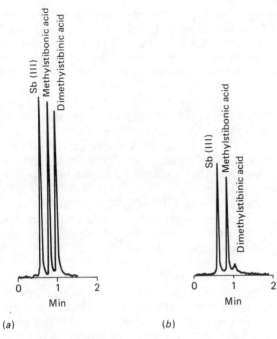

Figure 4.39 (*a*) Chromatogram of stibine, methylstibine and dimethylstibine as separated by the CV-3 trap/column with the quartz furnace/AA detector. The stibines were prepared by the NaBH$_4$ reduction of 2 ng Sb each as Sb^{III} methylstibonic and dimethylstibinic acid. (*b*) Analysis of a sample of seawater (100 ml) from the open Gulf of Mexico (Station SN4-3-1, 12 March 1981, latitude 27° 15.16′N, longitude 96° 29.88′W)

fuel-rich hydrogen/air flame in a quartz tube (cuvette) as described by Andreae [649] or in a standard graphite furnace by electrothermal atomization [653]. The quartz cuvette has a higher sensitivity than the graphite furnace for arsenic and antimony; it is therefore preferred for the determination of these two elements. When organotin compounds are analysed using the quartz cuvette system, spurious peaks are sometimes seen eluting with the methyl tins. The origin of these peaks is not clear. This interference can be avoided by using the graphite furnace system. Here, the hydrides are introduced with the carrier gas stream, to which some argon has been added, into the internal purge inlet of the graphite furnace. They have to pass through the graphite tube and leave through the internal purge outlet. The heating cycle of the furnace is timed so it reaches the required atomization temperature shortly before the arrival of the un-substituted hydride and is held at temperature until the last alkyl-substituted hydride has eluted. With this system, probably due to the higher operating temperature, no spurious tin peaks are present.

Bertine and Lee [658] have described hydride generation techniques for determining Sb^V and Sb^{III} Sb-S species and organoantimony species in frozen seawater samples.

The full potential of the hydride method has not yet been realized. Cutter [659] is studying its application to the determination of selenium in seawater. Other elements which might be amenable to the technique are bismuth, tellurium and lead.

Willis, Sturgeon and Berman [660] used the hydride generation-graphite furnace atomic absorption spectrometry technique to determine selenium in seawater. A Pyrex cell was used to generate selenium hydride which was carried to a quartz tube and then to a preheated furnace operated at 400°C. Pyrolytic graphite tubes were used. Selenium could be determined in seawater down to $20 \, \text{ng} \, \text{l}^{-1}$. No interference was found due to iron, copper, nickel and arsenic.

Yamamoto, Yasuda and Yamamoto [661] combined the technique of flow injection analysis with a gas segmentation procedure [662] to the hydride generation-atomic absorption spectrometry of arsenic, antimony, bismuth, selenium and tellurium in water. Standard reaction conditions included sodium borohydride (0.25%) hydrochloric acid ($8 \, \text{mol} \, \text{l}^{-1}$). A hydride generation tube length of 10 cm was used. Under these conditions, the hydrides were separated from the sample in less than 0.1 s after generation. In this short reaction period, most of the metal ions were not reduced to metal, which is preferable to minimize the interference from transition metal ions. One hundred-fold excesses of iron, cobalt, nickel and copper were without appreciable interference in the method.

A mixing coil was necessary for the reduction of As^V and Sb^V to tervalent states. To obtain the same sensitivity from pentavalent and tervalent species of arsenic and antimony, a coil length of 2 m (a reaction time of 20 s) was enough. Under the conditions, As^V and Sb^V were recovered quantitatively if the concentration of potassium iodide is higher than 4.2% and 0.4% in the mixture for respective ions, As^{III} and Sb^{III} could be determined selectively in the ranges of pH 4.5–5.5 and pH 5–7 respectively. Sensitivities of As^{III} and As^V were the same and the selectivity of As^{III} in the presence of As^V was satisfactory. The same was found for Sb^{III} and Sb^V.

The detection limits of the element for the species of As, Sn, Ge and Sb with different detection systems are shown in Table 4.61.

Table 4.61 Detection limits in nanograms of the element for the species of As, Sn, Ge and Sb with different detection systems

Species	QCAA	GFAA	FPD	ECD
As^{III}, As^V	0.03	0.09	–	–
MMAA	0.03	0.09	–	0.4
DMAA	0.05	0.15	–	0.2
Sn^{II}, Sn^{IV}	0.05	0.05	0.02	–
Me_xSn	0.06	0.06	0.015	–
Ge^{IV}	3	0.14	–	–
Sb^V	0.05	0.15	–	–
Sb^{III}				
MMSA	0.04	0.12	–	–
DMSA				

QCAA = quartz cuvette/atomic absorption; GFAA = graphite furnace/atomic absorption; FPD = flame photometric detector; ECD = electron capture detector.
Reproduced from Andreae, M.O.P. (1981) *Trace Metals in Seawater*, Plenum Press, New York, by courtesy of author and publishers.

Table 4.62 Selective determination of arsenic and antimony in saline water (μg l^{-1})*

As^{III+V}	As^{III}	Sb^{III+V}	Sb^{III}
1720 ± 20	90.0 ± 1.8	53.5 ± 1.2	1.50 ± 0.03
(1720)	(87.5)	(54.0)	(1.6)

*Onikoube, June 1982, Akita Prefecture, Japan. Average values of 10 determinations. Values in parentheses are obtained by HG AAS with batch method.
Reproduced from Yamamoto, Y., et al. (1985) Analytical Chemistry, 57, 1382. American Chemical Society, by courtesy of authors and publishers.

Comparison of the results by the flow injection analysis technique with those by the batch method [663] for the differential determination of arsenic and antimony are given in Table 4.62. Both results are in good agreement within experimental errors. The standard deviations were satisfactory.

References

1. Blake, W.E., Bryant, M.W.R. and Waters, A. *Analyst (London)*, **94**, 49 (1969)
2. Jagner, D. *Analytica Chimica Acta*, **68**, 83 (1974)
3. Schmid, R.W. and Reilley, C.N. *Analytical Chemistry*, **29**, 264 (1957)
4. Ringbom, A.G., Pensar, G. and Wanninen, E. *Analytica Chimica Acta*, **19**, 525 (1958)
5. Sadek, F.S., Schmid, R.W. and Reilley, C.N. *Talanta*, **2**, 38 (1959)
6. Wanninen, E. *Talanta*, **8**, 355 (1961)
7. Date, Y. and Toel, K. *Bulletin of the Chemical Society of Japan*, **36**, 518 (1963)
8. Culkin, F. and Cox, R.A. *Deep-Sea Research*, **13**, 789 (1966)
9. Tsunogai, S., Nishimura, M. and Nakaya, S. *Talanta*, **15**, 385 (1968)
10. Schwarzenbach, G. and Flaschka, H. *Complexometric Titrations*, 2nd English edn. Methuen, London (1969)
11. Horibe, Y., Endo, K. and Tsuboto, H. *Earth Planet Science Letters*, **23**, 136 (1974)
12. Jagner, D. *Analytica Chimica Acta*, **68**, 83 (1974)
13. Lebel, J. and Poisson, A. *Marine Chemistry*, **4**, 321 (1976)
14. Krumgalz, D. and Holzer, R. *Limnology and Oceanography*, **25**, 367 (1980)
15. Kanamori, S. and Ikegami, H. *Journal of the Oceanographic Society, Japan*, **36**, 177 (1980)
16. Olson, E.J. and Chen, C.T.A. *Limnology and Oceanography*, **27**, 375 (1982)
17. Smith, M.R. and Cockran, H.B. *Atomic Spectroscopy*, **2**, 97 (1981)
18. Pybus, J. *Clinica Chimica Acta*, **23**, 309 (1969)
19. Anfalt, T. and Graneli, A. *Analytica Chimica Acta*, **86**, 13 (1976)
20. Carr, R.A. *Limnology and Oceanography*, **15**, 318 (1970)
21. Epstein, M.S. and Zander, A.T. *Analytical Chemistry*, **51**, 915 (1979)
22. Roe, K.K. and Froelich, P.N. *Analytical Chemistry*, **56**, 2724 (1984)
23. Jagner, D. and Kerstein, A. *Analytica Chimica Acta*, **57**, 185 (1971)
24. Jagner, D. and Ferstein, A. *Analytica Abstracts*, **23**, 1190 (1972)
25. Kulev, I.I. and Bakalov, V.D. *C.r. Acad. Bulgarian Science*, **26**, 787 (1973)
26. Kulev, I.I., Stanev, D.S. and Bakalov, V.D. *Khimiya Indus.*, **46**, 65 (1974)
27. Marczak, M. and Ziaja, E. *Chemica Analytica*, **18**, 99 (1973)
28. Torbjoern, A. and Jaguer, D. *Analytica Chimica Acta*, **66**, 152 (1973)
29. Torbjoern, A. and Jaguer, D. *Analytical Abstracts*, **23**, 994 (1972)
30. Marquis, G. and Lebel, J. *Analytical Letters*, **14**, 913 (1981)
31. Culkin, F. and Cox, R.A. *Deep Sea Research*, **13**, 789 (1966)
32. Ward, G.K. *Report 29966*, National Technical Information Service, Springfield, VA (1979) (32696)
33. Benzwi, N. *Israel Journal of Chemistry*, **10**, 967 (1972)
34. Rona, M. and Schmuckler, G. *Talanta*, **20**, 237 (1973)
35. Wiernik, M. and Amiel, S. *Journal of Radioanalytical Chemistry*, **5**, 123 (1970)
36. Schoenfeld, I. and Held, S. *Israel Journal of Chemistry*, **7**, 831 (1969)

37. Frigieri, P., Trucco, R., Ciaccolini, I. and Pampurini, G. *Analyst (London)*, **105**, 651 (1980)
38. Frigieri, P., Trucco, R., Ciaccolini, I. and Pampurini, G. *Analyst (London)*, **105**, 651 (1980)
39. Ishibashi, M. and Kawai, T. *Nippon Kagaku Zasshi*, **73**, 380 (1952)
40. Ishibashi, M. and Motojima, K. *Nippon Kagaku Zasshi*, **73**, 491 (1952)
41. Motojimi, K. *Nippon Kagaku Zasshi*. **76**, 902 (1955)
42. *Japan Industrial Standard (JIS) (1974) K0102*
43. Shull, K.E. and Guthan, G.R. *Journal of American Water Works Association*, **59**, 1456 (1967)
44. Titkova, N.F. *Nauchn Doklady Vyssn Shk, Biol. Nauki*, **126**, 127 (1963)
45. Simons, L.H., Monaghan, P.H. and Taggart, M.S. *Analytical Chemistry*, **25**, 989 (1953)
46. Sackett, W. and Arrhenius, G. *Geochimica Cosmochimica Acta*, **26**, 955 (1962)
47. Nishikawa, Y., Hiraki, K., Morishige, K., Tsuchiyama, A. and Shigematsu, T. *Bunseki Kagaku*, **17**, 1092 (1968)
48. Shigematus, T., Nishikawa, Y., Hiraki, K. and Nagano, N. *Bunseki Kagaku*, **19**, 551 (1970)
49. Hydes, D.J. and Liss, P.S. *Analyst (London)*, **101**, 922 (1976)
50. Morishige, K., Hiraki, K., Nishikawa, K. and Shigematsu, T. *Bunseki Kagaku*, **27**, 109 (1978)
51. Ishibashi, M. and Fujinaga, T. *Nippon Kagaku Zasshi*, **73**, 783 (1952)
52. Hsu, D.Y. and Pipes, W.O. *Environmental Science and Technology*, **6**, 645 (1972)
53. Kahn, H.L. *Environmental Analytical Chemistry*, **3**, 121 (1973)
54. Barnes, R.A. *Chemical Geology*, **15**, 177 (1975)
55. Lee, M.L. and Burrell, D.C. *Analytica Chimica Acta*, **66**, 245 (1973)
56. Gosink, T.A. *Analytical Chemistry*, **47**, 165 (1975)
57. Pavlenko, L.I. and Safronova, N.S. *Zhur Analytical Khim*, **30**, 775 (1975)
58. Fujinaga, T., Kusaka, R., Koyama, M., *et al. Journal of Radioanalytical Chemistry*, **13**, 301 (1973)
59. Buono, J.A., Karin, R.W. and Fasching, J.L. *Analytica Chimica Acta*, **80**, 327 (1975)
60. Dougan, W.K. and Wilson, A.L. *Analyst (London)*, **99**, 413 (1974)
61. Dale, T. and Henriksen, A. *Vatten*, **31**, 91 (1975)
62. Henriksen, A. and Bergmann-Paulsen, L.M. *Vatten*, **31**, 339 (1975)
63. Spencer, D.W. and Sachs, P.L. *Atomic Absorption Newsletter*, **8**, 65 (1969)
64. Weisel, C.P., Duce, R.A. and Fasching, J.L. *Analytical Chemistry*, **56**, 1050 (1984)
65. Korenaga, T., Motomizum, S. and Toei, K. *Analyst (London)*, **105**, 328 (1980)
66. Van der Berg, C.H.G., Murphy, K. and Riley, J.P. *Analytica Chimica Acta*, **188**, 177 (1986)
67. Sturgeon, R.E., Willie, S.N. and Berman, S.S. *Analytical Chemistry*, **57**, 6 (1985)
68. Sturgeon, R.E., Willie, S.N. and Berman, S.S. *Analytical Chemistry*, **57**, 2311 (1985)
69. Afansev, Y.A., Ryabinin, A. and Azhipa, L. *Zhur Analytical Khim.*, **30**, 1830 (1975)
70. Ryabinin, A.I., Romanov, A.S., Khatawov, S.L., Kist, A.A. and Khamidova, R. *Zhur Analytical Chim*, **27**, 94 (1972)
71. Burton, J.D. In Trace metals in seawater. *Proceedings of a NATO Advanced Research Institute on Trace Metals in Seawater, 30/3–3/4/81, Sicily, Italy*, (eds C.S. Wong, *et al.*) Plenum Press, New York, p. 145 (1981)
72. Gilbert, T.R. and Hume, D.N. *Analytica Chimica Acta*, **65**, 451 (1973)
73. Florence, T.M. *Journal of Electroanalytical Chemistry*, **49**, 255 (1974)
74. Eskilsson, H. and Jaguer, D. *Analytica Chimica Acta*, **138**, 27 (1982)
75. Gillain, G., Duychaerts, G. and Distech, A. *Analytica Chimica Acta*, **106**, 23 (1979)
76. Jagner, D. and Aren, K. *Analytica Chimica Acta*, **100**, 375 (1978)
77. Danielsson, L.G., Jagner, D., Josefson, M. and Westerlund, S. *Analytica Chimica Acta*, **127**, 147 (1981)
78. Jagner, D., Josefson, M. and Westerlund, S. *Analytica Chimica Acta*, **128**, 155 (1981)
79. Portman, J.E. and Riley, J.P. *Analytica Chimica Acta*, **34**, 201 (1966)
80. Koroleff, F. *Acta Chimica Scandinavica*, **1**, 503 (1947)
81. Koroleff, F. (ed.) *Proceedings of the 12th Conference of the Baltic Oceanographers, Leningrad* (1980)
82. Uppstroem, L.R. *Analytica Chimica Acta*, **43**, 475 (1968)
83. Hulthe, P., Uppstroem, L.R. and Oestling, G. *Analytica Chimica Acta*, **51**, 31 (1970)
84. Nicolson, R.A. *Analytica Chimica Acta*, **56**, 147 (1971)
85. Oshima, M., Motomizu, S. and Toei, K. *Analytical Chemistry*, **56**, 948 (1974)
86. Marcantoncetos, M., Gamba, G. and Mannier, D. *Analytica Chimica Acta*, **67**, 220 (1973)
87. Horta, T.M.T.C. and Curtins, A.J. *Analytica Chimica Acta*, **95**, 207 (1977)
88. Horta, A.M.T.C. and Curtins, A.J. *Analytica Chimica Acta*, **96**, 209 (1978)
89. Tsaikov, S.P. *Comptes Rendus De l'Academic Bulgare Des Science*, **35**, 61 (1982)
90. Campbell, W.C. and Ottaway, J.H. *Analyst (London)*, **102**, 495 (1977)
91. Guevremont, R. *Analytical Chemistry*, **52**, 1574 (1980)

92. Guevremont, R., Sturgeon, R.E. and Berman, S.S. *Analytica Chimica Acta*, **115**, 1633 (1980)
93. Gardner, J. and Yates, J. *International Conference Management and Control of Heavy Metals in the Environment*, Water Research Centre, Stevenage Laboratory, Stevenage; CEP Consultants, Edinburgh, pp. 427–430 (1979)
94. Danielsson, L.G., Magnusson, B. and Westerlund, S. *Analytica Chimica Acta*, **98**, 47 (1978)
95. Batley, G.E. and Farrah, Y.J. *Analytica Chimica Acta* **99**, 283 (1978)
96. Danielson, L.G., Magnusson, B. and Westerlund, S. *Analytica Chimica Acta*, **98**, 47 (1978)
97. Sturgeon, R.E., Berman, S.S., Desauiniers, J.A.H. and Russell, D.S. *Talanta*, **27**, 85 (1980)
98. Bruland, K.W., Franks, R.P., Knauer, G.A. and Martin, J.H. *Analytica Chimica Acta*, **105**, 233 (1979)
99. Smith, R.G. Jr. and Windom, H.L. *Analytica Chimica Acta*, **113**, 39 (1980)
100. Rasmussen, L. *Analytica Chimica Acta*, **125**, 117 (1981)
101. Sturgeon, R.E., Berman, S.S., Desaulniers, J.A.H., *et al. Analytical Chemistry*, **52**, 1585 (1980)
102. Sperling, K.R. *Analytical Chemistry*, **292**, 113 (1978)
103. Bengtsson, M., Danielsson, L.G. and Magnusson, B. *Analytical Letters*, **12**, 1367 (1979)
104. Sperling, K.R. *Analytical Chemistry*, **301**, 294 (1980)
105. Kingston, H.M., Barnes, I.L., Brady, T.J., *et al. Analytical Chemistry*, **50**, 2084 (1978)
106. Lund, W. and Larsen, B.V. *Analytica Chimica Acta*, **72**, 57 (1974)
107. Bately, G.E. *Analytica Chimica Acta*, **124**, 121 (1981)
108. Lundgren, G., Lundmark, L. and Johansson, G. *Analytical Chemistry*, **46**, 1028 (1974)
109. Sperling, K.R. *Zeitschrift für Analytische Chemie*, **287**, 23 (1977)
110. Sperling, K.R. *Zeitschrift für Analytische Chemie*, **301**, 294 (1980)
111. Pruszkowska, E., Carnrick, G.R. and Slavin, W. *Analytical Chemistry*, **55**, 182 (1983)
112. Campbell, W.C. and Ottaway, J.M. *Analyst (London)*, **102**, 495 (1977)
113. Nakahara, T. and Chakrabarti, C.L. *Analytica Chimica Acta*, **104**, 99 (1979)
114. Brewer, S.W. *Analytical Chemistry*, **57**, 724 (1985)
115. Kounaves, S.P. and Zirino, A. *Analytica Chimica Acta*, **109**, 327 (1979)
116. Turner, D.R., Robinson, S.G. and Whitfield, M. *Analytical Chemistry*, **56**, 2387 (1984)
117. Nakayama, E., Kuwamoto, T., Tokoro, H. and Fujinaca, T. *Analytica Chimica Acta*, **131**, 247 (1981)
118. Cranston, R.E. and Murray, J.W. *Analytica Chimica Acta*, **99**, 275 (1978)
119. Dejong, G.J. and Brinkman, U.A.Th *Analytica Chimica Acta*, **98**, 243 (1978)
120. Gilbert, T.R. and Clay, A.M. *Analytica Chimica Acta*, **67**, 289 (1973)
121. Crosmun, S.T. and Mueller, T.R. *Analytica Chimica Acta*, **75**, 199 (1975)
122. Batley, G.E. and Matouseir, J.P. *Analytical Chemistry*, **52**, 1570 (1980)
123. Pankow, J.F. and Junauer, G.E. *Analytica Chimica Acta*, **69**, 97 (1974)
124. Miyazaki, A. and Barnes, R.M. *Analytical Chemistry*, **53**, 364 (1981)
125. Sandell, E.B. *Colorimetric Determination of Traces of Metals*, Interscience, New York (1959)
126. Marchart, H. *Analytica Chimica Acta*, **30**, 11 (1984)
127. Willems, G.J. and de Ranter, C.J. *Analytica Chimica Acta*, **68**, 111 (1974)
128. Myasoedova, G.V. and Savvin, S.G. *Zhur Analytica Khim*, **37**, 499 (1982)
129. Leyden, D.E. and Wegscheider, W. *Analytical Chemistry*, **63**, 1059A (1981)
130. Willie, S.N., Sturgeon, R.E. and Berman, S.S. *Analytical Chemistry*, **55**, 981 (1983)
131. Siu, W.M., Bednas, H.E. and Berman, S.S. *Analytical Chemistry*, **55**, 473 (1983)
132. Heumann, K.G. *Toxicological Environmental Chemical Review*, **3**, 111 (1980)
133. Colby, B.N., Rosecrance, A.E. and Colby, M.E. *Analytical Chemistry*, **53**, 1907 (1981)
134. Mykytiuk, A.P., Russell, D.S. and Sturgeon, R.E. *Analytical Chemistry*, **52**, 1281 (1980)
135. Sturgeon, R.E., Berman, S.S., Willie, S.N. and Desauiniers, J.A.H. *Analytical Chemistry*, **53**, 2337 (1981)
136. Chang, C.A., Patterson, H.H., Mayer, L.M. and Bause, D.E. *Analytical Chemistry*, **52**, 1264 (1980)
137. De Jong, G.J. and Brinkman, U.A.T. *Analytica Chimica Acta*, **98**, 243 (1978)
138. Cranston, R.E. and Murray, J.W. *Analytica Chimica Acta*, **99**, 275 (1978)
139. Cranston, R.E. and Murray, J.W. *Limnology and Oceanography*, **25**, 1104 (1980)
140. Mullins, R.L. *Analytica Chimica Acta*, **165**, 97 (1984)
141. Nakayama, E., Toroko, H. and Fujuiaga, T. *Analytica Chimica Acta*, **131**, 247 (1981)
142. Nakayama, E., Kuwamoto, H., Tokoro, H., Tsurubo, S. and Fujinaga, T. *Analytica Chimica Acta*, **130**, 289 (1981)
143. Grimaud, D. and Michard, G. *Marine Chemistry*, **2**, 229 (1974)
144. Cranston, R.E. and Murray, J.M. *Analytica Chimica Acta*, **99**, 275 (1978)

145. Nakayama, E., Kuwamoto, T., Tokoro, H. and Fiyinaka, T. *Analytica Chimica Acta*, **130**, 289 (1981)
146. Nakayama, E., Kuwamoto, T., Tokoro, H. and Fujinaka, T. *Analytica Chimica Acta*, **130**, 401 (1981)
147. Nakayama, E., Kuwamoto, T., Tokoro, H. and Fujinaka, T. *Analytica Chimica Acta*, **131**, 247 (1981)
148. Nakayama, E., Kumamoto, T., Tokoro, H. and Fujinaka, T. *Analytica Chimica Acta*, **130**, 401 (1981)
149. Takayama, E., Kuwamoto, T., Tokoro, H. and Fujinaka, T. *Analytica Chimica Acta*, **130**, 289 (1981)
150. Ishibashi, M. and Shigematsu, T. *Bulletin of the Institute of Chemical Research, Kyoto University*, **23**, 59 (1950)
151. Cheucas, L. and Riley, J.P. *Analytica Chimica Acta*, **35**, 240 (1966)
152. Fukai, R. *Nature (London)*, **213**, 901 (1967)
153. Fukai, R. and Vas, D. *Journal of the Oceanographic Society, Japan*, **23**, 298 (1967)
154. Kuwamoto, T. and Murai, S. *Preliminary Report of the Hakuho-maru Cruise KH-68-4*, Ocean Research Institute University, Tokyo, p. 72 (1970)
155. Grimaud, D. and Michard, G. *Marine Chemistry*, **2**, 229 (1974)
156. Yamamoto, T., Kadowski, S. and Carpenter, J.H. *Geochemical Journal*, **8**, 123 (1974)
157. Cranston, R.E. and Murray, J.M. *Analytica Chimica Acta*, **99**, 275 (1978)
158. Mullins, T.L. *Analytica Chimica Acta*, **165**, 97 (1984)
159. Ahern, F., Eckert, J.M., Payne, N.C. and Williams, K.L. *Analytica Chimica Acta*, **175**, 147 (1985)
160. Kentner, E. and Zeitlin, H. *Analytica Chimica Acta*, **49**, 587 (1970)
161. Motomizu, S. *Analytica Chimica Acta*, **64**, 217 (1973)
162. Forster, W. and Zeitlin, H. *Analytica Chimica Acta*, **34**, 211 (1966)
163. Riley, J. and Topping, G. *Analytica Chimica Acta*, **44**, 234 (1969)
164. Armitage, B. and Zeitlin, H. *Analytica Chimica Acta*, **53**, 47 (1971)
165. Going, J., Wesenberg, G. and Andrejat, G. *Analytica Chimica Acta*, **81**, 349 (1976)
166. Korkisch, J. and Sorio, A. *Analytica Chimica Acta*, **79**, 207 (1975)
167. Gurtler, O. *Fresenius Zeitschrift für Analytische Chemie*, **284**, 206 (1977)
168. Harvey, B.R. and Dutton, J.W.R. *Analytica Chimica Acta*, **67**, 377 (1973)
169. Kouimtzis, T.A., Apostolopoulou, C. and Staphilakis, I. *Analytica Chimica Acta*, **113**, 185 (1980)
170. Vasilikiotis, G., Kouimtzis, Th., Apostolopoulou, C. and Voulgaropoulos, A. *Analytica Chimica Acta*, **70**, 319 (1974)
171. Isshiki, K. and Nakayama, E. *Analytical Chemistry*, **59**, 291 (1987)
172. Spencer, C.P. *Journal of General Microbiology*, **16**, 282 (1957)
173. Johnston, R. *Journal of the Marine Biological Association, UK*, **43**, 427 (1962)
174. Johnston, R. *Journal of the Marine Biological Association, UK*, **44**, 87 (1964)
175. Barber, R.T. and Ryther, J.H. *Journal of Exploratory Marine Biology and Ecology*, **3**, 191 (1969)
176. Lewin, J. and Chen, C.H. *Limnology and Oceanography*, **16**, 670 (1971)
177. Lewis, A.G., Whitfield, R.H. and Ramnarine, A. *Marine Biology*, **17**, 215 (1972)
178. Davey, E.W., Morgan, M.J. and Erickson, S.J. *Limnology and Oceanography*, **18**, 993 (1973)
179. Prakash, A., Rshid, M.A., Jensen, A. and Subba Rao, D.V. *Limnology and Oceanography*, **18**, 516 (1973)
180. Gnassia-Barelli, M., Romero, M., Laumond, F. and Pesando, D. *Marine Biology*, **47**, 15 (1978)
181. Hongve, D., Skogheim, O.K., Hindar, A. and Abrahamsen, H. *Bulletin of Environmental Contamination and Toxicology*, **25**, 594 (1980)
182. Fisher, N.S. and Frood, D. *Marine Biology*, **59**, 85 (1980)
183. Sunda, W. and Guillard, R.R.L. *Journal of Marine Research*, **34**, 511 (1976)
184. Anderson, D.M. and Morel, F.M. *Limnology and Oceanography*, **23**, 283 (1978)
185. Murphy, L.S., Guillard, R.R.L. and Gavis, J. Evolution of resistant phytoplankton strains through exposure to marine pollutants. In: *Ecological Stress and the New York Blight; Science and Management* (ed. G.F. Mayer), Estuarine Research Federation, Columbia, SC, pp. 401–412 (1980)
186. Jackson, G.A. and Morgan, J.J. *Limnology and Oceanography*, **23**, 268 (1978)
187. Chester, R. and Stoner, J.H. *Marine Chemistry*, **2**, 17 (1974)
188. Alberts, J.J., Leyden, D.E. and Patterson, T.A. *Marine Chemistry*, **4**, 51 (1976)
189. Boyle, F.A., Sclater, F.R. and Edmond, J.M. *Science Letters*, **37**, 38 (1977)
190. Winlom, H.L. and Smith, R.G. *Marine Chemistry*, **7**, 157 (1979)
191. Zirino, A. and Yamamoto, S. *Limnology and Oceanography*, **17**, 661 (1972)

192. Turner, D.R., Whitfield, M. and Dickson, A.G. *Geochemica Cosmochimica Acta*, **45**, 855 (1981)
193. Stumm, W. and Morgan, J.J. *An Introduction Emphasizing Chemical Equilibria in Natural Waters*, 2nd edn. Wiley-Interscience, New York (1981)
194. Sibley, T.H. and Morgan, J.J. Equilibrium speciation of trace metals in freshwater; seawater mixtures. In *Proceedings of International Conference on Heavy Metals in the Environment* (ed. H.C. Hutchinson), University of Toronto, Toronto, Ont., pp. 310–338 (1975)
195. Pagenkopf, G.K., Russo, R.C. and Thurston, R.V. *Journal of the Fisheries Research Board, Canada*, **31**, 462 (1974)
196. Shuman, M.S. and Woodward, G.P. *Analytical Chemistry*, **45**, 2032 (1973)
197. Chau, R. and Lum-Shue-Chan, K. *Water Research*, **8**, 383 (1974)
198. Campbell, P.G.C., Blisson, M., Gagne, R. and Tessior, A. *Analytical Chemistry*, **49**, 2358 (1977)
199. Stiff, M.J. *Water Research*, **5**, 585 (1971)
200. Batley, G.E. and Florence, T.M. *Analytical Letters*, **9**, 379 (1976)
201. Plasvic, M., Krznacic, D. and Branica, M. *Marine Chemistry*, **11**, 17 (1982)
202. Van Den Berg, C.M.G. *Marine Chemistry*, **11**, 323 (1982)
203. Florence, T.M. and Batley, G.E. *Talanta*, **23**, 179 (1976)
204. Figura, P. and McDuffie, B. *Analytical Chemistry*, **51**, 120 (1979)
205. Abraham, J., Winpe, M. and Ryan, D.E. *Analytica Chimica Acta*, **48**, 431 (1969)
206. Muzzarelli, R.A.A. and Rocchetti, R. *Analytica Chimica Acta*, **69**, 35 (1974)
207. Fairless, C. and Bard, A.J. *Analytical Chemistry*, **45**, 2289 (1973)
208. Tasinski, R., Trachtenberg, I. and Andrychuk, D. *Analytical Chemistry*, **46**, 364 (1974)
209. Virmani, Y.P. and Zeller, E.J. *Analytical Chemistry*, **46**, 324 (1974)
210. Zharikov, V.F. and Senyavin, M.K. *Trudy gos okeanogr. Inst. (1970) (101) Ref. Zhur Khim 19GD (7)* Abstract No. 7G189
211. Marvin, K.T., Proctor, R.R. and Neal, R.A. *Limnology and Oceanography*, **15**, 320 (1970)
212. Wood, A.M., Evans, D.W. and Alberts, J.J. *Marine Chemistry*, **13**, 305 (1983)
213. Stolzberg, R.J. and Rosin, D. *Analytical Chemistry*, **49**, 226 (1977)
214. Giesy, J.P. Cadmium interactions with naturally occuring ligands. In *Biogeochemistry of Cadmium, Part 1* (ed. J.C. Nriagu), Wiley, New York, pp. 237–256 (1980)
215. Zorkin, N.G., Grill, E.V. and Lewis, A.G. *Analytica Chimica Acta*, **183**, 163 (1986)
216. Zirino, A. and Healy, M.L. *Limnology and Oceanography*, **15**, 956 (1970)
217. Bradford, W.L. *Limnology and Oceanography*, **18**, 757 (1978)
218. Bilinski, H., Huston, R. and Stumm, W. *Analytica Chimica Acta*, **84**, 157 (1976)
219. O'Shea, T.A. and Mancy, K.H. *Analytical Chemistry*, **48**, 1603 (1976)
220. Duinker, J.C. and Kramer, C.J.M. *Marine Chemistry*, **5**, 207 (1977)
221. Piro, A., Bernhard, M., Branica, M. and Verzi, M. *Radioactive Contamination of the Marine Environment; Proceedings of the Symposium, Seattle, July 1972*, IAFA edn., IAFA, Vienna, pp. 287–304 (1972)
222. Ernst, R., Allen, H.E. and Mancy, K.H. *Water Research*, **9**, 969 (1975)
223. Shuman, M.S. and Woodward, G.P. *Analytical Chemistry*, **45**, 2032 (1973)
224. Chau, Y.K. and Lum-Shue-Chan, K. *Water Research*, **8**, 383 (1974)
225. Florence, T.M. and Batley, G.B. *Journal of Electroanalytical Chemistry*, **75**, 791 (1977)
226. Zirino, A. and Yamamoto, S. *Limnology and Oceanography*, **17**, 661 (1972)
227. Stumm, W. and Bilinski, H. *Advances in Water Pollution Research, 6th International Conference, Jerusalem, June 8–23, 1972*, Pergamon, Oxford, p. 39 (1973)
228. Bubic, S. and Branica, M. *Thalassia Jugoslavia*, **9**, 47 (1973)
229. Zirino, A. and Kounaves, S.P. *Analytical Chemistry*, **49**, 56 (1977)
230. Zirino, A. and Kounaves, S.P. *Analytical Chemistry*, **51**, 592 (1979)
231. Zirino, A. and Kounaves, S.P. *Analytica Chimica Acta*, **113**, 79
232. Sheffrin, N. and Williams, E.E. *Analytical Proceedings*, **19**, 483 (1982)
233. Jagner, D. and Graneli, A. *Analytica Chimica Acta*, **83**, 19 (1976)
234. Jagner, D. and Aren, K. *Analytica Chimica Acta*, **100**, 375 (1978)
235. Jagner, D. *Analytical Chemistry*, **50**, 1924 (1978)
236. Jagner, D. and Aren, K. *Analytica Chimica Acta*, **107**, 29 (1979)
237. Brooks, R.R., Presley, B.J. and Kaplan, I.R. *Analytica Chimica Acta*, **38**, 321 (1967)
238. Brooks, R.R., Presley, B.J. and Kaplan, I.R. *Talanta*, **14**, 809 (1967)
239. Shuman, M.S. and Michael, L.C. In *Proceedings of the International Conference on Heavy Metals in the Environment*, Vol. 1, p. 227, John Wiley, New York (1975)
240. Shuman, M.S. and Michael, L.C. *Environmental Science and Technology*, **12**, 1069 (1978)
241. Odier, M. and Plichon, V. *Analytica Chimica Acta*, **55**, 209 (1971)

242. Weiss, H.V., Kenis, P.R., Korkisch, J. and Steffan, I. *Analytica Chimica Actda*, **104**, 337 (1979)
243. Van Der Berg, C.M.G. *Analytica Chimica Acta*, **164**, 195 (1984)
244. Nelson, A. *Analytica Chimica Acta*, **169**, 287 (1985)
245. Pilipenko, A.T. and Pavlova, V.K. *Zhur Analytical Khim*, **27**, 1253 (1972)
246. Matthews, A.D. and Riley, J.P. *Analytica Chimica Acta*, **51**, 287 (1970)
247. Hiiro, K., Tanaka, T. and Sawada, T. *Japan Analyst*, **21**, 635 (1972)
248. Moore, R.M. In *Trace Metals in Seawater. Proceedings of a Nato-Advanced Research Institute on Trace Metals in Seawater 30/3–3/4/81, Sicily, Italy*, (eds C.S. Wong *et al.*), Plenum Press, New York (1981)
249. Danielson, L., Magnusson, G.B. and Westerlund, S. *Analytica Chimica Acta*, **98**, 47 (1978)
250. Sharma, G.M. and DuBois, H.R. *Analytical Chemistry*, **50**, 516 (1978)
251. Shriadah, M.M.A. and Ohzeki, K. *Analyst (London)*, **111**, 555 (1986)
252. Frech, W. and Cedergreu, A. *Analytica Chimica Acta*, **88**, 57 (1977)
253. Halliday, M.C., Houghton, C. and Ottaway, J.M. *Analytica Chimica Acta*, **119**, 67 (1986)
254. Schaule, B.K. and Patterson, C.C. *Earth and Planetary Science Letters*, **54**, 97 (1981)
255. Hirao, Y., Fukomoto, K., Sugisaki, H. and Kimura, K. *Analytical Chemistry*, **51**, 651 (1979)
256. Torsi, G., Desimoni, E., Palmisano, F. and Sabbatini, L. *Analytica Chimica Acta*, **124**, 143 (1981)
257. McLaren, J.W. and Wheeler, R.C. *Analyst (London)*, **102**, 542 (1977)
258. Adams, M.J., Kirkbright, G.F. and Rienvatana, P. *Atomic Absorption Newsletter*, **14**, 105 (1977)
259. Culver, B.R. and Surles, T. *Analytical Chemistry*, **47**, 920 (1975)
260. Pritchard, M.W. and Reeves, R.D. *Analytica Chimica Acta*, **82**, 103 (1976)
261. Yasuda, S. and Kakihama, H. *Analytica Chimica Acta*, **84**, 291 (1976)
262. Tessari, G. and Torsi, G. *Analytical Chemistry*, **47**, 842 (1975)
263. Sturgeon, R.E., Chakrabarti, C.L. and Langford, C.N. *Analytical Chemistry*, **48**, 1792 (1976)
264. Fernandez, F.J. and Manning, D.C. *Atomic Absorption Newsletter*, **10**, 3 (1971)
265. Ediger, R.D., Peterson, G. and Kerber, J.D. *Atomic Absorption Newsletter*, **13**, 61 (1974)
266. Ediger, R.D. *Atomic Absorption Newsletter*, **14**, 127 (1975)
267. Regan, J.G.T. and Warren, J. *Analyst (London)*, **101**, 220 (1976)
268. Frech, W. and Cedergren, A. *Analytica Chimica Acta*, **88**, 57 (1977)
269. Manning, D.C. and Slavin, W. *Analytical Chemistry*, **50**, 1234 (1978)
270. Lundgren, G., Lundmark, L. and Johansson, G. *Analytical Chemistry*, **46**, 1028 (1974)
271. Scobbie, R. *Technical Topics*, Varian Techtron, Springvale, Vic., Australia (1973)
272. Donnelly, T.H., Ferguson, J. and Eccleston, A.J. *Applied Spectroscopy*, **29**, 2 (1975)
273. Shigematsu, T., Matsui, M., Fujino, O. and Kinoshita, K. *Analytica Chimica Acta*, **76**, 329 (1975)
274. Branderberger, H. and Bader, H. *Atomic Absorption Newsletter*, **6**, 101 (1967)
275. Lund, W. and Larsen, B.V. *Analytica Chimica Acta*, **70**, 299 (1974)
276. Lund, W. and Larsen, B.V. *Analytica Chimica Acta*, **72**, 57 (1974)
277. Lund, W., Larsen, B.V. and Gundersen, N. *Analytica Chimica Acta*, **81**, 319 (1976)
278. Lund, W., Thomassen, Y. and Doyle, P. *Analytica Chimica Acta*, **93**, 53 (1977)
279. Fairless, L. and Bard, A.J. *Analytical Letters*, **5**, 433 (1972)
280. Fairless, L. and Bard, A.J. *Analytical Chemistry*, **45**, 2289 (1973)
281. Batley, G.E. and Matousek, J.P. *Analytical Chemistry*, **49**, 2031 (1977)
282. Jensen, F.O., Dolezal, J. and Langmyhr, F.J. *Analytica Chimica Acta*, **72**, 245 (1975)
283. Newton, M.P., Chauvin, J.V. and Davis, D.G. *Analytical Letters*, **6**, 89 (1973)
284. Newton, M.P. and Davis, D.G. *Analytical Chemistry*, **47**, 2003 (1975)
285. Dawson, J.B., Ellis, D.J., Hartley, T.F., Evans, M.E.A. and Metcalf, K.W. *Analyst (London)*, **99**, 602 (1974)
286. Torsi, G. *Analytica Chimica (Rome)*, **67**, 557 (1977)
287. Torsi, G., Oesimoni, E., Palimisano, F. and Sabbatini, L. *Analytica Chimica Acta*, **124**, 143 (1981)
288. Halliday, M.C., Houghton, C. and Ottaway, J.M. *Analytica Chimica Acta*, **119**, 67 (1980)
289. Hirao, Y., Fukomoto, K., Sugisaki, H. and Kimura, K. *Analytical Chemistry*, **51**, 651 (1979)
290. Petric, L.M. and Baier, R.W. *Analytical Chemistry*, **50**, 351 (1978)
291. Acebal, S.A., DeLuca and Rebello, A. *Analytica Chimica Acta*, **148**, 71 (1983)
292. Shurnik Sarig, S., Zidon, M. and Zak, I. *Israel Journal of Chemistry*, **8**, 545 (1970)
293. Byrne, R.H. *Nature (London)*, **290**, 487 (1981)
294. Scaule, D.M. and Patterson, C.C. Trace metals in seawater. *In Proceedings of a Nato Advanced Research Institute on Trace Metals in Seawater, 30/3–3/4/81, Sicily, Italy* (eds C.S. Wong, *et al.*), Plenum Press, New York, p. 488 (1981)
295. Quentel, F., Madoc, C. and Courtot-Coupez, J. *Water Research*, **20**, 325 (1986)
296. Stumm, W. and Morgan, J.J. *Aquatic Chemistry*, Wiley-Interscience, New York (1970)

297. Murray, J.W. and Brewer, P.G. *Marine Manganese Deposits* (ed. G.P. Glasby), Elsevier, Amsterdam, pp. 291–325 (1977)
298. Morgan, J.J. Chemical equilibria and kinetic properties of manganese in natural waters. In *Principles and Applications of Water Chemistry* (eds S.D. Faust and J.V. Hunter), Wiley, New York, pp. 561–624 (1967)
299. Kessick, M.A. and Morgan, J.J. *Environmental Science and Technology*, **9**, 157 (1975)
300. Stumm, W. and Giovanoli, R. *Chimia*, **30**, 423 (1976)
301. Bryan, G.W. and Hummerstone, L.G. *Journal of the Marine Biological Association of the UK*, **53**, 705 (1973)
302. Holliday, L.M. and Liss, P.S. *Estuarine and Coastal Marine Science*, **4**, 349 (1976)
303. Evans, D.W., Cutshall, N.H., Cross, F.A. and Wolfe, D.A. *Estuarine and Coastal Marine Science*, **5**, 71 (1977)
304. Sholkovitz, E.R. *Earth and Planetary Science Letters*, **41**, 77 (1978)
305. Davison, W. *Journal of Electroanalytical Chemistry*, **72**, 229 (1976)
306. Davison, W. *Limnology and Oceanography*, **22**, 746 (1977)
307. Sturgeon, R.E., Berman, S.S., Desauiniers, A. and Russell, D.S. *Analytical Chemistry*, **51**, 2364 (1979)
308. Segar, D.A. and Gonzalez, J.G. *Analytica Chimica Acta*, **58**, 7 (1972)
309. Klinkhammer, G.P. *Analytical Chemistry*, **52**, 117 (1980)
310. McArthur, J.M. *Analytica Chimica Acta*, **93**, 77 (1977)
311. Hydes, D.J. *Analytical Chemistry*, **52**, 959 (1980)
312. Segar, D.A. *Advances in Chemistry Services*, **147**, 56 (1975)
313. Kingston, H.M., Barnes, I.L., Brady, T.J., *et al. Analytical Chemistry*, **50**, 2064 (1978)
314. Segar, D.A. and Cantillo, A.Y. *Analytical Chemistry*, **52**, 1766 (1980)
315. Montgomery, J.R. and Peterson, G.N. *Analytica Chimica Acta*, **117**, 397 (1980)
316. Manning, D.C. and Slavin, W. *Analytical Chemistry*, **50**, 1234 (1978)
317. L'vov, B.V. *Spectrochimica Acta, Part B*, **33B**, 153 (1978)
318. Segar, D.A. and Cantillo, A.Y. *Spectrochimica Symposium, American Society of Limnology and Oceanography*, **2**, 171 (1976)
319. Slavin, W. and Manning, D.C. *Analytical Chemistry*, **51**, 261 (1979)
320. Manning, D.C. and Slavin, W. *Analytica Chimica Acta*, **118**, 301 (1980)
321. Slavin, W. and Manning, D.C. *Spectrochimica Acta, Part B*, **35B**, 701 (1980)
322. Ediger, R.D., Peterson, G.E. and Kerber, J.D. *Atomic Absorption Newsletter*, **13**, 61 (1974)
323. Klinkhammer, G.P. *Analytical Chemistry*, **52**, 117 (1980)
324. Landing, W.M. and Bruland, K.W. *Earth Planet Science Letters*, **49**, 45 (1980)
325. Moore, R.M., Burton, J.D., Williams, P.J. le B. and Young, M.L. *Cosmochimica Acta*, **43**, 919 (1979)
326. Subramanian, K.S. and Meranger, J.C. *International Journal of Environmental Analytical Chemistry*, **7**, 25 (1979)
327. Sugimae, A. *Analytica Chimica Acta*, **121**, 331 (1980)
328. Statham, P.J. *Analytica Chimica Acta*, **169**, 149 (1985)
329. International Council for Exploration of the Sea (Charlottenhind, Denmark, Marine Environment Quality Committee Report CM1983/E24) (1983)
330. Klinkhammer, G.P. *Analytical Chemistry*, **42**, 117 (1980)
331. Burton, J.D. In *Trace Metals in Seawater. Proceedings of a Nato Advanced Research Institute on Trace Metals in Seawater, 30/3–3/4/81, Sicily, Italy.* (eds. C.S. Wong, *et al.*) Plenum Press, New York, p. 419 (1981)
332. Bender, M.L., Klinkhammer, G.P. and Spencer, D.W. *Deep Sea Research*, **24**, 799 (1977)
333. Klinkhammer, G.P. and Bender, M.L. *Earth Planet Science Letters*, **46**, 361 (1980)
334. Kingston, H.M., Barnes, I.L. and Rains, T.C. *Analytical Chemistry*, **50**, 2064 (1978)
335. Carnrick, G.R., Slavin, W. and Manning, D.C. *Analytical Chemistry*, **53**, 1866 (1981)
336. Knox, S. and Turner, D.R. *Estuarine and Coastal Marine Science*, **10**, 317 (1980)
337. Department of the Environment and the National Water Council, Standing Committee of Analysts *Manganese in Raw and Potable Waters by Spectrophotometry (using formaldoxime) 1977; Methods for the Examination of Waters and Associated Materials.* HMSO, London (1978)
338. Clem, R.G., Mitton, G. and Ornelas, L.D. *Analytical Chemistry*, **45**, 1306 (1973)
339. Biddle, V.L. and Wehry, E.L. *Analytical Chemistry*, **50**, 867 (1978)
340. Olafsson, J. *Science of the Total Environment*, **49**, 101 (1986)
341. Wiggins, P.F., Duffey, D. and El Kady, A.A. *Analytica Chimica Acta*, **61**, 421 (1972)
342. Hatch, W.R. and Ott, W.L. *Analytical Chemistry*, **40**, 2085 (1968)
343. Hinckle, M.E. and Learned, R.E. *U.S. Geological Survey, Professional Papers*, **650-D**, 251 (1969)

344. Fishman, M.J. *Analytical Chemistry*, **42**, 1462 (1970)
345. Lindsteat, G. *Analyst (London)*, **95**, 264 (1970)
346. Chau, Y.K. and Saitoh, H. *Environmental Science and Technology*, **4**, 839 (1970)
347. Omang, S.H. *Analytica Chimica Acta*, **53**, 415 (1971)
348. Omang, S.H. and Paus, P.E. *Analytica Chimica Acta*, **56**, 393 (1971)
349. Hwang, J.H., Ullucci, P.A. and Malenfant, A.L. *Canadian Spectroscopy*, **16**, 1 (1971)
350. Aston, S.R. and Riley, J.P. *Analytica Chimica Acta*, **59**, 349 (1972)
351. Muscat, V.I., Vickers, T.J. and Andren, A. *Analytical Chemistry*, **44**, 218 (1972)
352. Topping, G. and Pirie, J.M. *Analytica Chimica Acta*, **62**, 2001 (1972)
353. Carr, R.A., Hoover, J.B. and Wilkniss, P.E. *Deep Sea Research*, **19**, 747 (1972)
354. Omang, S.H. *Analytica Chimica Acta*, **63**, 247 (1973)
355. Gardner, D. and Riley, J.P. *Nature (London)*, **241**, 526 (1973)
356. Olafsson, J. *Analytica Chimica Acta,* **68**, 207 (1974)
357. Muscat, V.I. and Vickers, T.J. *Analytica Chimica Acta*, **18**, 2275 (1970)
358. Leatherland, T.M., Burton, J.D., McCartney, M.J. and Culkin, F. *Nature (London)*, **232**, 112 (1971)
359. Chester, R., Gardner, D., Riley, J.P. and Stoner, J. *Marine Pollution Bulletin*, **28**, 4 (1973)
360. Fitzgerald, W.F. and Hunt, C.D. *Abstract of the Bulletin of the French Oceanographic Union*, **A-5**, (1973)
361. Voyce, D. and Zeitlin, H. *Analytica Chimica Acta*, **69**, 27 (1974)
362. Fitzgerald, W.F., Lyons, W.B. and Hunt, C.D. *Analytical Chemistry*, **46**, 1882 (1974)
363. Coyne, R.V. and Collins, J.A. *Analytical Chemistry*, **44**, 1093 (1972)
364. Carr, R.A. and Wilkniss, P.E. *Environmental Science and Technology*, **7**, 62 (1973)
365. Feldman, C. *Analytical Chemistry*, **46**, 99 (1974)
366. Fitzgerald, W.F. Distribution of mercury in natural waters. In *The Biogeochemistry of Mercury in the Environment*, (ed. J.O. Nriagu), Elsevier, Amsterdam (1979)
367. Olafsson, J. Trace metals in seawater. *In Proceedings of a Nato Advanced Research Institute on Trace Metals in Seawater, 30/3–3/4/81, Sicily, Italy*, (eds C.S. Wong, *et al.*) Plenum Press, New York, p. 476 (1981)
368. Olafsson, J. *Analytica Chimica Acta*, **68**, 207 (1974)
369. Olafsson, J. *Marine Chemistry*, **11**, 129 (1982)
370. Hatch, W.R. and Ott, W.L. *Analytical Chemistry*, **40**, 2085 (1968)
371. Jensen, K.D. and Carlsen, V. *Journal of Radioanalytical Chemistry*, **47**, 121 (1978)
372. Dogan, S. and Haerdy, W. *International Journal of Environmental Analytical Chemistry*, **5**, 157 (1978)
373. Olafsson, J. *Analytica Chimica Acta*, **68**, 207 (1974)
374. Thibaud, Y. *Peche Bulletin Institute Peches Maritime*, **250**, 1 (1975)
375. Thompson, K.C. and Godden, R.G. *Analyst (London)*, **100**, 544 (1975)
376. Nelson, L.A. *Analytical Chemistry*, **51**, 2289 (1979)
377. Nishimura, M., Matsunaga, K. and Konishi, S. *Japan Analyst*, **24**, 655 (1975)
378. Ambe, M. and Suwabe, R. *Analytica Chimica Acta*, **92**, 55 (1977)
379. Filippelli, M. *Analyst (London)*, **109**, 515 (1984)
380. Nurnberg, H.W., Valenta, P., Sipos, L. and Branica, M. *Analytica Chimica Acta*, **115**, 25 (1980)
381. Hiyai, Y. and Murao, Y. *Japan Analyst*, **21**, 608 (1972)
382. Weiss, H.V. and Crozier, T.E. *Analytica Chimica Acta*, **58**, 231 (1972)
383. Fujita, M. and Iwashima, K. *Environmental Science and Technology*, **15**, 929 (1981)
384. Agemian, H. and Da Silva, J.A. *Analytica Chimica Acta*, **104**, 285 (1979)
385. Jaya, S., Rao, T.P. and Raog, P. *Analyst (London)*, **110**, 1361 (1985)
386. Davies, I.M. and Pirie, J.M. *Marine Pollution Bulletin*, **9**, 128 (1978)
387. Topping, G., Pirie, J.M., Graham, W.C. and Shepherd, R.J. *An Examination of the Heavy Metal Levels in Muscle, Kidney and Liver of Saithe in Relation to Year, Class, Area of Sampling, and Season, ICE.5.CM, E:37* (1975)
388. Topping, G. and Pirie, J.M. *Analytica Chimica Acta*, **62**, 200 (1972)
389. Nakahara, T. and Chakrabarti, C.L. *Analytica Chimica Acta*, **104**, 99 (1979)
390. Van den Sloot, H.A., Wals, G.D. and Das, H.A. *Progress in Water Technology*, **8**, 193 (1977)
391. Var der Sloot, H.A. and Das, H.A. *Progress in Water Technology*, **8**, 193 (1977)
392. Muzzarelli, R.A.A. and Rochetti, R. *Analytica Chimica Acta*, **64**, 371 (1973)
393. Emerick, R.J. *Atomic Spectroscopy*, **8**, 69 (1987)
394. Zeitlin, H. *Analytica Chimica Acta*, **51**, 516 (1970)
395. Shriadah, H.M.A., Katoeka, M. and Ohzebi, K. *Analyst (London)*, **110**, 125 (1985)
396. Riley, J.P. and Taylor, D. *Analytica Chimica Acta*, **40**, 479 (1968)

397. Kawabuchi, K. and Kuroda, R. *Analytica Chimica Acta*, **46**, 23 (1969)
398. Kawabuchi, K. and Kuroda, R. *Analytica Chimica Acta*, **46**, 23 (1969)
399. Kuroda, R. and Tarui, T. *Fresenius Zeitschrift fur Analytische Chemie*, **269**, 22 (1974)
400. Korkisch, J., Godl, L. and Gross, H. *Talanta*, **22**, 669 (1975)
401. Chan, J.K.M. and Riley, J.O. *Analytica Chimica Acta*, **36**, 220 (1966)
402. Ohta, N., Fujita, M. and Tomura, K. *Bunseki Kagaku*, **28**, 277 (1979)
403. Prabhu, V.G., Zarapkar, L.R. and Das, M.S. *Mikrochimica Acta*, **11**, 67 (1980)
404. Kim, Y.S. and Zeitlin, H. *Analytica Chimica Acta*, **46**, 1 (1969)
405. Head, P. and Burton, J.D. *Journal of The Marine Biological Association, UK*, **50**, 439 (1970)
406. Butler, L.R.P. and Matthews, P.M. *Analytica Chimica Acta*, **36**, 319 (1966)
407. Chau, Y.K. and Lum-Shue-Chan, K. *Analytica Chimica Acta*, **48**, 205 (1969)
408. Akama, Y., Nakai, T. and Kawamura, F. *Nippon Kaisui Gakkai-shi*, **33**, 180 (1979)
409. Fujinaga, T., Kusaka, Y., Koyama, M., *et al. Journal of Radioanalytical Chemistry*, **13**, 301 (1973)
410. Kulathilake, A.I. and Chatt, A. *Analytical Chemistry*, **52**, 828 (1980)
411. Weiss, H.V. and Lai, M.G. *Talanta*, **8**, 72 (1961)
412. Muzzarelli, R.A.A. and Rochetti, R. *Analytica Chimica Acta*, **64**, 371 (1973)
413. Muzzarelli, R.A.A. *Analytica Chimica Acta*, **54**, 133 (1971)
414. Vanderborght, B.M. and Van Grieken, R.E. *Analytical Chemistry*, **49**, 311 (1977)
415. Riley, J.P. and Taylor, D. *Analytical Chemistry*, **41**, 175 (1968)
416. Ficklin, W.H. *Analytical Letters*, **15**, 865 (1982)
417. Yoshimura, K., Hiraoka, S. and Tarutani, T. *Analytica Chimica Acta*, **142**, 101 (1982)
418. Van der Shoot, H.A., Wals, G.D. and Das, H.A. *Analytica Chimica Acta*, **90**, 193 (1977)
419. Chau, Y.K. and Lum-Shue-Chan, K. *Analytica Chimica Acta*, **48**, 205 (1969)
420. Parker, C.R. *Water Analysis by Atomic Absorption Spectroscopy*, Varian Techtron (1972)
421. Kiriyama, T. and Kuroda, R. *Talanta*, **31**, 472 (1984)
422. Shriadah, H.M.A., Katoaka, M. and Ohzeki, K. *Analyst (London)*, **110**, 125 (1985)
423. Kim, Y.S. and Zeitlin, H. *Separation Science*, **6**, 505 (1971)
424. Kuroda, R. and Tarui, T. *Fresenius Zeitschrift für Analytische Chemie*, **269**, 22 (1974)
425. Van Der Berg, C.H.G. *Analytical Chemistry*, **57**, 1532 (1985)
426. Hanson, G., Szalo, A. and Chasteen, N.D. *Analytical Chemistry*, **49**, 461 (1977)
427. Mok, W.M. and Wai, C.W. *Analytical Chemistry*, **56**, 27 (1984)
428. Kulathilake, A.I. and Chatt, A. *Analytical Chemistry*, **52**, 828 (1980)
429. Monien, H., Bovenkerk, R., Kringe, K.P. and Rath, D., *Fresenius Zeitschrift für Analytische Chemie*, **300**, 363 (1980)
430. Van der Beng, C.M.G. *Analytical Chemistry*, **57**, 1532 (1985)
431. Bruland, K.W., Franks, R.P., Knauer, G.A. and Martin, J.H. *Analytical Chemistry*, **105**, 233 (1979)
432. Boyle, E.A. and Edmond, J.A. *Analytica Chimica Acta*, **91**, 189 (1977)
433. Bruland, K.W. *Science Letters*, **47**, 176 (1980)
434. Boyle, E.A., Huested, S.S. and Jones, S.P. *Journal of Geographical Research*, **86**, 8048 (1981)
435. Kentner, E., Armitage, D.B. and Zeitlin, H. *Analytica Chimica Acta*, **45**, 343 (1969)
436. Yatsimirskii, K.B., Ewel'Yakov, E.M., Pavlova, V.K. and Savichenko, Ya.S. *Okeanologiya IIII*, **10**, Ref; Zhur Khim. 19GD Abstract No. 11G, 203 (11)
437. Rampon, H. and Cavelier, R. *Analytica Chimica Acta*, **60**, 226 (1972)
438. Lee, D.S. *Analytical Chemistry*, **54**, 1182 (1982)
439. Shigematsu, T., Nishikawa, Y., Hiraki, K., Goda, S. and Tsujimatu, Y. *Japan Analyst*, **20**, 575 (1971)
440. Elderfield, H. and Greaves, H.J. Trace metals in seawater. In *Proceedings of a Nato Advanced Research Institute on Trace Metals in Seawater, 30/3–3/4/81, Sicily, Italy* (eds C.S. Wong, *et al.*) Plenum Press, New York, p. 427 (1981)
441. Hooker, P.J. *BA Thesis*, University of Oxford (1974)
442. Hooker, P.J., O'Nions, R.K. and Pankhurst, R.J. *Chemical Geology*, **16**, 189 (1975)
443. Korkisch, J. and Arrhenius, G. *Analytical Chemistry*, **36**, 850 (1964)
444. Faris, J.P. and Warton, J.W. *Analytical Chemistry*, **34**, 1077 (1962)
445. Desai, H.B., Krishnamoorthy, I.R. and Sandar Das, M. *Talanta*, **11**, 1249 (1964)
446. Brunfelt, A.O. and Steinnes, E. *Analyst (London)*, **94**, 979 (1969)
447. Matthews, A.D. and Riley, J.P. *Analytica Chimica Acta*, **51**, 455 (1970)
448. Committee on Medical and Biologic Effects of Environmental Pollutants, Selenium, National Academy of Sciences, Washington, D.C. (1976)
449. Luckey, T.D. and Venegopal, B. *Chemical Engineering News*, **54**, 2 (1976)

450. Riley, J.P. and Skirrow, G. *Chemical Oceanography*, 2nd edn., Vol. 1, Academic Press, New York, p. 418 (1975)
451. Stein, V.B., Canelli, E. and Richards, A.H. *Atomic Spectroscopy*, **1**, 61 (1980)
452. Sturgeon, R.F., Willie, S.N. and Berman, S.S. *Analytical Chemistry*, **57**, 6 (1985)
453. Neve, J., Hanocq, M. and Molle, L. *International Journal of Environmental Analytical Chemistry*, **8**, 177 (1980)
454. Shimoishi, Y. *Analytica Chimica Acta*, **64**, 465 (1973)
455. Measures, C.I. and Burton, J.D. *Analytica Chimica Acta*, **120**, 177 (1980)
456. Sui, K.W.M. and Berman, S.S. *Analytical Chemistry*, **56**, 1806 (1984)
457. Tzeng, J.H. and Zeitlin, H. *Analytica Chimica Acta*, **101**, 71 (1978)
458. Kim, Y.S. and Zeitlin, H. *Separation Science*, **6**, 505 (1971)
459. Goldschmidt, V.M. and Strock, L.W. *Nachr. Akad. Wiss. Goettingen, Math-Physik, KI, IV, N.F.*, **1**, 123 (1935)
460. Wattenberg, H. *Zeitung Anorganic Chemise*, **236**, 339 (1938)
461. Ishibashi, M., Shigematsu, T. and Nakagawa, Y. *Recent Oceanographic Works, Japan, Special Number*, **1**, 44 (1953)
462. Chau, Y.K. and Riley, J.P. *Analytica Chimica Acta*, **33**, 36 (1965)
463. Schutz, D.F. and Turekian, K.K. *Abstracts of Papers XIII, General Assembly I.U.G.G., Berkeley*, Paper 23 (1963)
464. Schutz, D.F. *PhD Thesis, Department of Geology, Yale University* (1964)
465. Kawabuchi, K. and Riley, J.P. *Analytica Chimica Acta*, **65**, 271 (1973)
466. Petit, L. *Revue Internationale d'Oceanographic Medicale*, **19**, 79/80 (1985)
467. Kodama, Y. and Tsubota, H. *Japan Analyst*, **20**, 1554 (1971)
468. Brinckmann, F.E. Trace metals in seawater. *In Proceedings of a Nato Advanced Research Institute on Trace Metals in Seawater, 30/3–3/4/81, Sicily, Italy* (eds C.S. Wong, *et al.*), Plenum Press, New York (1981)
469. Aue, W.A. and Flinn, C.S. *Journal of Chromatography*, **142**, 145 (1977)
470. Jackson, J.A., Blair, W.R., Brinckmann, F.E. and Iverson, W.P. *Environmental Science and Technology*, **16**, 110 (1982)
471. Jewett, K.L. and Brinckmann, F.E. *Journal of Chromatographic Science*, **19**, 583 (1981)
472. Brinckmann, F.E., Blair, W.R., Jewett, K.L. and Iverson, W.P. *Journal of Chromatographic Science*, **15**, 493 (1977)
473. Brinckmann, F.E. *Journal of Organometallic Chemistry*, **12**, 343 (1981)
474. Jewett, K.L., Brinckman, F.E. and Bellama, J.M. Influence of environmental parameters on transmethylation between aquated metal ions. In *Organometals and Organometalloids: Occurrence and Fate in the Environment* (eds F.B. Brinckmann and J.M. Bellama), eds. American Chemical Society, Washington D.C., pp. 158–187 (1978)
475. Jackson, J.A., Blair, W.R., Brinckmann, F.E. and Iverson, W.P. *Environmental Science and Technology*, **16**, 110 (1982)
476. Michael, L.C., Erickson, M.D., Parks, S.P. and Pellizzari, B.D. *Analytical Chemistry*, **52**, 1836 (1980)
477. Braman, R.S. and Tompkins, M.A. *Analytical Chemistry*, **51**, 12 (1979)
478. Schaeffer, G.W. and Emilius, M. *Journal of the American Chemical Society*, **76**, 1203 (1954)
479. Hodge, V.F., Seidel, S.L. and Goldberg, F.D. *Analytical Chemistry*, **51**, 1256 (1979)
480. Hallas, L.E. *PhD Dissertation, University of Maryland* (1978)
481. Parris, G.E., Blair, W.E. and Brinckmann, F.E. *Analytical Chemistry*, **49**, 378 (1977)
482. Evans, W.H., Jackson, F.J. and Dellar, D. *Analyst (London)*, **104**, 16 (1979)
483. Dogan, S. and Haerdi, W. *International Journal of Environmental Analytical Chemistry*, **8**, 249 (1980)
484. Bergerioux, C. and Haerdi, W. *Analusis*, **8**, 169 (1980)
485. Van der Sloot, H.A. and Das, H.A. *Progress in Water Technology*, **8**, 193 (1977)
486. Yang, C.Y., Shih, J.S. and Yeh, Y.C. *Analyst (London)*, **106**, 385 (1981)
487. Milner, G.W.C., Wilson, J.D. and Barnett, G.A. *Journal of Electroanalytical Chemistry*, **2**, 25 (1961)
488. Van der Berg, C.M.G. and Huang, Z.Q. *Analytica Chimica Acta*, **164**, 209 (1984)
489. Nishimura, M., Matsunaga, K., Kudo, T. and Obara, F. *Analytica Chimica Acta*, **65**, 466 (1973)
490. Kirkyama, T. and Kuroda, R. *Analytica Chimica Acta*, **62**, 464 (1972)
491. Monien, H. and Stangel, R. *Fresenius Zeitschrift für Analytische Chemie*, **311**, 209 (1982)
492. Van der Berg, C.M.G. and Huang, Q. *Analytical Chemistry*, **56**, 2383 (1984)
493. Weiss, H.V., Guttman, H.A., Korkisch, J. and Steffan, I. *Talanta*, **24**, 509 (1977)

494. Korkisch, J. and Gross, H. *Talanta*, **20**, 1153 (1973)
495. Muzzarrelli, R.A.A. and Sipos, L.G. *Talanta*, **18**, 853 (1971)
496. Lieberman, S.H. and Zirino, A. *Analytical Chemistry*, **46**, 20 (1974)
497. Burrell, D.C. and Guener, G. *Analytica Chimica Acta*, **48**, 45 (1969)
498. Guevremont, R. *Analytical Chemistry*, **53**, 911 (1981)
499. Kim, Y.S. and Zeitlin, H. *Separation Science*, **7**, 1 (1972)
500. Van der Berg, C.M.G. *Marine Chemistry*, **16**, 121 (1985)
501. Van der Berg, C.M.G. *Talanta*, **31**, 1069 (1984)
502. Sastry, V.N., Krishnamoorthy, T.M. and Sarma, T.P. *Current Science (Bombay)*, **38**, 279 (1969)
503. Allain, P. and Mauras, Y. *Analytica Chimica Acta*, **165**, 141 (1984)
504. Doidge, P.S. *Spectrochimica Acta*, **40B**, 569 (1985)
505. Manning, D.C., Slavin, W. and Myers, S. *Analytical Chemistry*, **51**, 2375 (1979)
506. Bengtsson, M., Danielsson, L.G. and Magnusson, B. *Analytical Letters*, **12**, A13 (1979)
507. Campbell, W.C. and Ottaway, J.H. *Analyst (London)*, **102**, 495 (1977)
508. Batley, G.B. and Matousek, J.P. *Analytical Chemistry*, **49**, 2031 (1977)
509. Stein, V.B., Canelli, G. and Richards, A.H. *International Journal of Environmental Analytical Chemistry*, **8**, 99 (1980)
510. Cachetti, G., Cvelic-Lazzari, N. and Rolle, E. *Metodi Analitica per le Aeque*, **2**, 1 (1981)
511. Sturgeon, R.E., Berman, S.S., Desaneniers and Russel, D.G. *Analytical Chemistry*, **51**, 2364 (1979)
512. Segar, D.A. and Cantillo, A.Y. *Analytical Chemistry*, **52**, 1766 (1980)
513. Montgomery, J.R. and Peterson, G.N. *Analytica Chimica Acta*, **117**, 397 (1980)
514. Lyons, M.B. and Fitzgerald, W.F. Trace metals in seawater. *In Proceedings of a Nato Advanced Research Institute on Trace Metals in Seawater, Sicily, Italy* (eds C.S. Wong, *et al.*) Plenum Press, New York, p. 625 (1981)
515. Brooks, R.R., Presley, B.J. and Kaplan, I.R. *Talanta*, **14**, 809 (1967)
516. Brewer, P.G., Spencer, D.W. and Smith, C.L. *American Society of Testing Materials*, **443**, 70 (1969)
517. Tominaga, M., Bansho, K. and Umezaki, Y. *Analytica Chimica Acta*, **169**, 171 (1985)
518. Tominaga, M., Bansho, K. and Umezaki, Y. *Analytica Chimica Acta*, **169**, 176 (1985)
519. Segar, D.A. and Gonzales, J.G. *Analytica Chimica Acta*, **58**, 7 (1972)
520. Hayase, K., Shitashima, K. and Tsuibota, H. *Talanta*, **33**, 754 (1986)
521. Hadeishi, T. and McLaughlin, R.D. *Science*, **174**, 404 (1971)
522. Hadeishi, T., Church, D.A., McLaughlin, R.D., Zak, B.D., Nakamura, M. and Chang, B. *Science*, **187**, 348 (1975)
523. Koizumi, H. and Yasuda, K. *Analytical Chemistry*, **47**, 1679 (1975)
524. Stephens, R. and Ryan, D.E. *Talanta*, **22**, 655 (1975)
525. Stephens, R. and Ryan, D.E. *Talanta*, **22**, 659 (1975)
526. Veinot, D.E. and Stephens, R. *Talanta*, **23**, 849 (1976)
527. Koizumi, H. and Yasuda, K. *Spectrochimica Acta*, **31B**, 237 (1976)
528. Koizumi, H. and Yasuda, K. *Analytical Chemistry*, **48**, 1178 (1976)
529. Stephens, R. *Talanta*, **25**, 435 (1978)
530. Stephens, R. *Talanta*, **26**, 57 (1979)
531. Lawson, J.B., Grassam, E., Ellis, D.J. and Keir, M.J. *Analyst (London)*, **101**, 315 (1976)
532. Grassam, E., Dawson, J.B. and Ellis, D.J. *Analyst (London)*, **102**, 804 (1977)
533. Koizumi, H. and Yasuda, K. *Spectrochimica Acta*, **31B**, 523
534. Koizumi, H., Yasuda, K. and Katayama, M. *Analytical Chemistry*, **49**, 1106 (1977)
535. Pernandez, F.J., Bohler, W., Beatty, N.M. and Barnett, W.B. *Atomic Spectroscopy*, **2**, 73 (1981)
536. Carnrick, S.R., Slavin, W. and Manning, D.C. *Analytical Chemistry*, **53**, 1866 (1981)
537. Slavin, W., Manning, D.C. and Carrick, G.R. *Atomic Spectroscopy*, **2**, 137 (1981)
538. Grabenski, Z., Lehmann, R., Radzuik, B. and Voellkopf, U. *Atomic Spectroscopy*, **5**, 87 (1984)
539. De Kersabiec, A.M., Blanc, G. and Pinta, M. *Fresenius Zeitschrift für Analytische Chemie*, **322**, 731 (1985)
540. Margoshes, M. and Scribner, B.F. *Spectrochimica Acta*, **15**, 138 (1959)
541. Korolev, V.V. and Vainshtein, E.E. *Zeitschrift für Analytische Chemie*, **15**, 658 (1959)
542. Owen, L.W. *Applied Spectroscopy*, **15**, 150 (1961)
543. Valente, S.E. and Schrenk, W.G. *Applied Spectroscopy*, **24**, 197 (1970)
544. Marinkovic, M. and Vickers, E.J. *Applied Spectroscopy*, **25**, 319 (1971)
545. Chapman, J.F., Dale, L.S. and Whittem, R.N. *Analyst (London)*, **98**, 529 (1973)
546. Merchant, P. and Veillon, C. *Analytica Chimica Acta*, **70**, 17 (1974)
547. Murdick, D.A. and Plenmeier, E.H. *Analytical Chemistry*, **46**, 678 (1974)
548. Rippetoe, W.E., Johnson, E.R. and Vickers, T.J. *Analytical Chemistry*, **47**, 436 (1975)

549. Winge, R.K., Fassel, V.A., Kniseley, R.N., De Kalb, E. and Haas, W.J. *Spectrochimica Acta*, **32B**, 327 (1977)
550. Nygaard, D.D. *Analytical Chemistry*, **51**, 881 (1979)
551. McLeod, C.W., Otsuki, K., Okamoto, H., Fuwa, K. and Haraguchi, H. *Analyst (London)*, **106**, 419 (1981)
552. Sugimae, A. *Analytica Chimica Acta*, **121**, 331 (1980)
553. Mujazaki, A., Kimura, A., Bansho, K. and Amezaki, Y. *Analytica Chimica Acta*, **144**, 213 (1981)
554. Blockaert, J.A.C., Leis, F. and Lagua, K. *Talanta*, **28**, 745 (1981)
555. Berman, S.S., McLaren, J.W. and Willie, S.N. *Analytical Chemistry*, **52**, 488 (1980)
556. Sturgeon, R.E., Berman, S.S., Desaulniers, J.A.H., *et al. Analytical Chemistry*, **52**, 1585 (1980)
557. Hervath, Z. and Barnes, R.M. *Analytical Chemistry*, **58**, 1352 (1986)
558. Hiraide, M., Ito, T., Baba, M., *et al. Analytical Chemistry*, **52**, 804 (1980)
559. McLaren, J.W., Mykytiuk, A.P., Willie, S.N. and Berman, S.S. *Analytical Chemistry*, **57**, 2907 (1985)
560. Kingston, H.M., Barnes, I.L., Brady, T.J., *et al. Analytical Chemistry*, **50**, 2064 (1978)
561. Fujino, O., Matsui, M. and Shigematsu, Mizu. *Shori Gijutsu*, **20**, 201 (1979)
562. Jones, J.S., Harrington, D.E., Leone, B.A. and Branstedt, W.R. *Atomic Spectroscopy*, **4**, 49 (1983)
563. Buchanan, A.S. and Hannaker, P. *Analytical Chemistry*, **56**, 1379 (1984)
564. Brooks, R.R., Presley, B.J. and Kaplan, I.R. *Talanta*, **14**, 809 (1967)
565. Kremling, K. and Peterson, H. *Analytica Chimica Acta*, **70**, 35 (1974)
566. Kinrade, J.D. and Van Loon, J.C. *Analytical Chemistry*, **46**, 1894 (1974)
567. Jan, T.K. and Young, D.R. *Analytical Chemistry*, **50**, 1250 (1978)
568. Stolzberg, R.J. In *Analytical Methods in Oceanography* (ed. T.R.P. Gibb, Jr.) *Advanced Chemistry*, no. 147, American Chemical Society, Washington, DC, p. 30 (1975)
569. Danielsson, L., Magnusson, B. and Westerlund, S. *Analytica Chimica Acta*, **98**, 45 (1978)
570. Sturgeon, R.E., Berman, S.S., Desauiniers, A. and Russell, D.S. *Talanta*, **27**, 85 (1980)
571. Magnusson, B. and Westerlund, S. *Analytica Chimica Acta*, **131**, 63 (1981)
572. Armansson, H. *Analytica Chimica Acta*, **88**, 89 (1977)
573. Bruland, K.W., Franks, R.P., Knauer, G.A. and Martin, J.H. *Analytica Chimica Acta*, **105**, 233 (1979)
574. Lee, M.G. and Burrell, D.C. *Analytica Chimica Acta*, **62**, 153 (1972)
575. Tsalev, D.L., Alimarin, I.P. and Neiman, S.I. *Zhur Analytical Khim*, **27**, 1223 (1972)
576. El-Enamy, F.F., Mahmond, K.F. and Varma, M.M. *Journal of the Water Pollution Control Federation*, **51**, 2545 (1979)
577. Pellenbarg, R.E. and Church, T.M. *Analytica Chimica Acta*, **97**, 81 (1978)
578. Armannsson, H. *Analytica Chimica Acta*, **110**, 21 (1979)
579. Brugmann, L., Danielsson, L.R., Magnusson, B. and Westerlund, S. *Marine Chemistry*, **13**, 327 (1983)
580. Boyle, E.A. and Edmond, J.M. *Analytica Chimica Acta*, **91**, 189 (1977)
581. Jan, T.K. and Young, D.R. *Analytical Chemistry*, **50**, 1250 (1978)
582. Bruland, K.W., Franks, R.B., Knawer, G.A. and Martin, J.H. *Analytica Chimica Acta*, **105**, 233 (1979)
583. Rasmussen, L. *Analytica Chimica Acta*, **125**, 117 (1981)
584. Ediger, R.D., Peterson, G.E. and Kerber, J.D. *Atomic Absorption Newsletter*, **13**, 61 (1974)
585. Smith, R.G. and Windom, H.L. *Analytica Chimica Acta*, **113**, 39 (1980)
586. Danielsson, L.G., Magnusson, B. and Westerlund, S. *Analytica Chimica Acta*, **98**, 47 (1978)
587. Sturgeon, R.E., Berman, S.S., Desauliniers, A. and Russell, D.S. *Talanta*, **27**, 85 (1980)
588. Sperling, K.R. *Fresenius Zeitschrift für Analytische Chemie*, **301**, 294 (1980)
589. Sturgeon, R.E., Berman, S.S., Desauliniers, J.A.H., Mykytink, A.P., McLaren, J.W. and Russell, D.M. *Analytical Chemistry*, **52**, 1585 (1980)
590. Sperling, K.R. *Fresenius Zeitschrift für Analytische Chemie*, **310**, 254 (1982)
591. Lo, J.M., Hutchison, J.C. and Wal, C.M. *Analytical Chemistry*, **54**, 2536 (1982)
592. Lo, J.M., Yu, J.C., Hutchinson, F.I. and Wal, C.M. *Analytical Chemistry*, **54**, 2536 (1982)
593. Statham, P.J. *Analytica Chimica Acta*, **169**, 149 (1985)
594. Danielsson, L.G., Magnusson, B. and Westerlund, S. *Analytica Chimica Acta*, **144**, 183 (1982)
595. Florence, T.M. and Batley, G.E. *Talanta*, **23**, 179 (1976)
596. Kingston, H.M., Barnes, I.L., Brady, T.J., Rains, T.C. and Champ, M.A. *Analytical Chemistry*, **50**, 2064 (1978)
597. Bruland, K.W., Franks, R.P., Knauer, G.A. and Martin, J.H. *Analytica Chimica Acta*, **105**, 233 (1979)
598. Sturgeon, R.E., Bermann, S.S., Desauliniers, A. and Russell, D.S. *Talanta*, **27**, 85 (1980)

599. Rasmussen, L. *Analytica Chimica Acta*, **125**, 117 (1981)
600. Greenberg, R.R. and Kingston, H.M. *Journal of Radioanalytical Chemistry*, **71**, 147 (1982)
601. Colella, M.B., Siggia, S. and Barnes, R.M. *Analytical Chemistry*, **52**, 2347 (1980)
602. Wan, C.C., Chaing, S. and Corsini, A. *Analytical Chemistry*, **57**, 719 (1985)
603. Mackay, D.J. *Marine Chemistry*, **11**, 169 (1982)
604. Kiriyama, T. and Kuroda, R. *Mikrochimica Acta*, **1**, 405 (1985)
605. Koide, M., Lee, D.S. and Stallard, M.O. *Analytical Chemistry*, **56**, 1956 (1984)
606. Hodge, V., Stallard, M., Koide, M. and Goldberg, E.D. *Analytical Chemistry*, **58**, 616 (1986)
607. Koide, M., Lee, D.S. and Stallard, M.O. *Analytical Chemistry*, **56**, 1956 (1984)
608. Kingston, H. and Pella, P.A. *Analytical Chemistry*, **53**, 223 (1981)
609. Boniforti, R., Ferraroli, R., Frigieri, P., Heltai, D. and Queirazza, G. *Analytica Chimica Acta*, **162**, 33 (1984)
610. Paulson, A.J. *Analytical Chemistry*, **58**, 183 (1986)
611. Zharikov, V.F. and Senyavin, T.M. *Trudy gos Okeanogr. Inst. 9710 (101)* 128 Ref: Zhur. Khim. 195D (7) Abstract No. 7G189 (1971)
612. Riccardo, A.A., Muzzarrelli, R.A.A. and Tubertini, O. *Journal of Chromatography*, **47**, 414 (1970)
613. Muzzarrelli, R.A.A. and Sipos, L. *Talanta*, **18**, 853 (1971)
614. Terada, K., Morimoto, K. and Kiba, T. *Analytica Chimica Acta*, **116**, 127 (1980)
615. Riley, J.P. and Topping, G. *Analytica Chimica Acta*, **44**, 234 (1969)
616. Buono, J.A., Buono, J.C. and Fasching, J.L. *Analytical Chemistry*, **47**, 1926 (1975)
617. Sturgeon, R.E., Berman, S.S., Willie, S.N. and Desauliniers, J.A.H. *Analytical Chemistry*, **53**, 2337 (1981)
618. Murthy, R.S.S. and Ryan, D.E. *Analytica Chimica Acta*, **140**, 163 (1982)
619. Sturgeon, R.E., Willie, S.N. and Berman, S.S. *Analytical Chemistry*, **57**, 6 (1985)
620. Matsuzaki, C. and Zeitlin, H. *Separation Science*, **8**, 185 (1973)
621. Weisel, C.P., Duce, R.A. and Fasching, J.L. *Analytical Chemistry*, **56**, 1050 (1984)
622. Cabezon, L.M., Cabellero, M., Cela, M. and Perez-Bustamante, J.A. *Talanta*, **31**, 597 (1984)
623. Siu, K.W.H. and Berman, S.S. *Analytical Chemistry*, **56**, 1806 (1984)
624. Patin, A.A. and Morozov, N.P. *Zhurnal Analiticheskoi Khimii*, **31**, 282 (1976)
625. Akagi, T., Nojiri, Y., Matsui, M. and Haraguchi, H. *Applied Spectroscopy*, **39**, 662 (1985)
626. Boyle, I.A. and Edmond, J.M. *Analytica Chimica Acta*, **91**, 189 (1977)
627. Tyson, J.F., Applten, J.M.H. and Idris, A.B. *Analyst (London)*, **108**, 153 (1983)
628. Kimura, H., Oguma, K. and Kuroda, R. *Bunseki Kagaku*, **32**, T79 (1983)
629. Zhou, N., Frech, W. and Lundberg, E. *Analytica Chimica Acta*, **23**, 153 (1983)
630. Tyson, J.F. and Idris, A.B. *Analyst (London)*, **109**, 26 (1984)
631. Kingston, H.M., Barnes, I.L., Brady, T.J. and Rains, T.C. *Analytical Chemistry*, **50**, 2064 (1978)
632. Danielsson, L.G., Magnusson, B. and Zhang, K. *Atomic Spectroscopy*, **3**, 39 (1982)
633. Komson, O.F. and Townshend, A. *Analytica Chimica Acta*, **155**, 253 (1983)
634. Nord, L. and Karlberg, B. *Analytica Chimica Acta*, **145**, 151 (1983)
635. Olsen, S., Pessanda, L.C.R., Ruzicka, J. and Hansen, E.H. *Analyst (London)*, **108**, 905 (1983)
636. Jorgensen, S.S. and Petersen, K. Paper presented at *9th Nordic Atomic Spectroscopy and Trace Element Conference, Reykjavik, Iceland*, (1983)
637. Fang, S. Xu and Zhang, S. *Analytica Chimica Acta*, **164**, 41 (1984)
638. Malamas, F., Bengtsson, M. and Johansson, G. *Analytica Chimica Acta*, **160**, 1 (1984)
639. Krug, F.J., Reis, B.F. and Jorgensen, S.S. *Proceedings of the Workshop on Locally Produced Laboratory Equipment for Chemical Education*, Copenhagen, Denmark, p. 121 (1983)
640. Fang, Z., Ruzicka, J. and Hansen, E.H. *Analytica Chimica Acta*, **164**, 23 (1984)
641. Marshall, H.A. and Mottola, H.A. *Analytical Chemistry*, **57**, 729 (1985)
642. Sugawara, K.R., Weetall, H.H. and Schucker, G.D. *Analytical Chemistry*, **46**, 489 (1974)
643. Gudes da Mota, M.M., Romer, F.G. and Griepink, B. *Fresenius Zeitschrift für Analytische Chemie*, **287**, 19 (1977)
644. Moorhead, E.D. and Davis, P.H. *Analytical Chemistry*, **46**, 1879 (1974)
645. Sturgeon, R.E., Berman, S.S., Willie, S.N. and Desaulniers, J.A.H. *Analytical Chemistry*, **53**, 2337 (1981)
646. Andreae, M.O.P.I. Trace metals in seawater. In *Proceedings of a Nato Advanced Research Institute on Trace Metals in Seawater, 30/3/81–3/4/81, Sicily, Italy* (eds C.S. Wong, *et al.*), Plenum Press, New York (1981)
647. Braman, R.S., Justen, L.C. and Foreback, C.C. *Analytical Chemistry*, **44**, 2195 (1972)
648. Braman, R.S. and Foreback, C.C. *Science*, **182**, 1247 (1973)
649. Andreae, M.O. and Froelich, P.N. *Analytical Chemistry*, **53**, 287 (1981)

650. Andreae, M.O., Asmode, J.R., Foster, P. and Vant'dack, L. *Analytical Chemistry*, **53**, 1766 (1981)
651. Andreae, M.O. *Analytical Chemistry*, **49**, 820 (1977)
652. Foreback, C.C. *Some Studies on the Detection and Determination of Mercury Arsenic and Antimony in Gas Discharges, Thesis, University of South Florida, Tampa* (1973)
653. Tompkins, M.A. *Environmental Analytical Studies in Antimony, Germanium and Tin, Thesis, University of South Florida, Tampa* (1977)
654. Amaukwah, S.A. and Fasching, J.L. *Talanta*, **32**, 111 (1985)
655. Fleming, H.H. and Ide, R.G. *Analytica Chimica Acta*, **83**, 67 (1976)
656. Crecelius, E.A. *Analytical Chemistry*, **50**, 826 (1978)
657. Braman, R.S. and Tompkins, M.A. *Analytical Chemistry*, **51**, 12 (1979)
658. Bertine, K.K. and Lee, D.S. Trace metals in seawater. *In Proceedings of a Nato Advanced Research Institute on Trace Metals in Seawater, 30/3–3/4/81, Sicily, Italy*, (eds C.S. Wong *et al.*), Plenum Press, New York (1981)
659. Cutter, G.A. *Analytica Chimica Acta*, **98**, 59 (1978)
660. Willie, S.N., Sturgeon, R.E. and Berman, S.S. *Analytical Chemistry*, **58**, 1140 (1986)
661. Yamamoto, M., Yasuda, M. and Yamamoto, Y. *Analytical Chemistry*, **57**, 1382 (1985)
662. Skeggs, L.T. *American Journal of Clinical Pathology*, **28**, 311 (1957)
663. Yamamoto, M., Urato, K. and Murashige, T. *Analyst (London)*, **109**, 1461 (1984)
664. Whitnack, G.C. and Basselli, R. *Analytica Chimica Acta*, **47**, 267 (1969)
665. Komatsu, M., Matsueda, T. and Kakiyami, H. *Japan Analyst*, **20**, 987 (1971)
666. David, J. and Redmond, J.D. *Journal of Electroanalytical Chemistry*, **33**, 169 (1971)
667. Florence, T.M. *Journal of Electroanalytical Chemistry*, **35**, 237 (1972)
668. Rojohn, T. *Analytica Chimica Acta*, **62**, 438 (1972)
669. Seitz, W.R., Jones, R., Klatt, L.N. and Mason, W.D. *Analytical Chemistry*, **45**, 810 (1973)
670. Clem, R.G., Litton, G. and Ornelas, L.D. *Analytical Chemistry*, **45**, 1306 (1973)
671. Clem, R.G. *MPI Application Notes*, **8**, 1 (1973)
672. Bubic, S., Sipos, L. and Branica, M. *Thalass Jugoslavia*, **9**, 55 (1973)
673. Chau, Y.K., Gachter, R. and Lum-Shue-Chan, K. *Journal of the Fisheries Research Board, Canada*, **31**, 1515 (1974)
674. Nurnberg, H.W., Valenta, P., Mart, L., Raspor, B. and Sipos, L. *Fresenius Zeitschrift für Analytische Chemie*, **282**, 357 (1976)
675. O'Shea, T.A. and Mancy, K.H. *Analytical Chemistry*, **48**, 1603 (1976)
676. Florence, T.M. and Batley, G.E. *Talanta*, **24**, 151 (1977)
677. Whitnak, G.C. and Sanelli, R. *Analytica Chimica Acta*, **47**, 267 (1969)
678. Smith, J.D. and Redmond, J.D. *Journal of Electroanalytical Chemistry*, **33**, 169 (1971)
679. Sinko, I. and Dolezal, J. *Journal of Electroanalytical Chemistry*, **25**, 299 (1970)
680. Cambon, J.P., Cauw, G. and Monaco, A. *Review of International Oceanographic Medicine T. XLVIII*, **2**, 73 (1977)
681. Welghe, N. and Claeys, A. *Journal of Electroanalytical Chemistry*, **35**, 229 (1972)
682. Brugmann, L. *Acta Hydrochimica Hydrobiologica*, **2**, 123 (1974)
683. Florence, T.M. *Journal of Electroanalytical Chemistry*, **35**, 237 (1972)
684. Lund, W. and Salberg, M. *Analytica Chimica Acta*, **76**, 131 (1975)
685. Gardner, J. and Stiff, M.J. *Water Research*, **9**, 517 (1975)
686. Abdullah, M.I., Reusch Berg, B. and Klimek, R. *Analytica Chimica Acta*, **84**, 307 (1976)
687. Duyckaerts, G. and Gillain, G. *Essays in Memory of Anders Ringbom*, Pergamon Press, Oxford (1977)
688. Florence, T.M. and Batley, G.E. *Talanta*, **23**, 179 (1977)
689. Florence, T.M. and Batley, G.E. *Talanta*, **24**, 151 (1977)
690. Brezonik, P.L. In *Aqueous Environmental Chemistry of Metals* (ed. A.J. Rubin), Ann Arbor Science Publishers, Michigan, p. 167 (1974)
691. Odier, M. and Plichon, V. *Analytica Chimica Acta*, **55**, 209 (1971)
692. Elder, J.F. *Limnology and Oceanography*, **20**, 96 (1975)
693. O'Shea, T.A. and Maney, K.H. *Analytical Chemistry*, **48**, 1603 (1976)
694. Liebermann, S.H. and Zirino, A. *Analytical Chemistry*, **46**, 20 (1974)
695. Batley, G.E. and Matousek, J.P. *Analytical Chemistry*, **49**, 2031 (1977)
696. Batley, G.E. and Matousek, J.P. *Analytical Chemistry*, **52**, 1570 (1980)
697. Young, J., Gurtisen, J.M., Apts, C.W. and Crecelius, E.A. *Marine Environmental Research*, **2**, 265 (1979)
698. Batley, G.E. *Analytica Chimica Acta*, **124**, 121 (1981)
699. Scarponi, G., Capoaglio, G., Cescon, P., *et al.* *Analytica Chimica Acta*, **135**, 263 (1982)

700. Brugman, L., Magnusson, B. and Westerlund, S. *Marine Chemistry*, **13**, 327 (1983)
701. Clem, R.G. and Hodgson, A.T. *Analytical Chemistry*, **50**, 102 (1978)
702. Van der Berg, C.M.G. *Science of the Total Environment*, **49**, 89 (1986)
703. Nurnberg, H.W. *The Science of the Total Environment*, **37**, 9 (1984)
704. Gillain, G., Daychaerts, G. and Disteche, A. *Analytica Chimica Acta*, **106**, 23 (1979)
705. Mart, L., Nurnberg, W.H. and Dryssen, D. Trace metals in seawater. *In Proceedings of a Nato Advanced Research Institute on Trace Metals in Seawater, 30/3–3/4/81, Sicily, Italy*, (eds C.S. Wong, *et al.*) Plenum Press, New York (1981)
706. Valenta, P., Mart, L. and Rutzel, H. *Journal of Electroanalytical Chemistry*, **82**, 327 (1977)
707. Pihlar, B., Valenta, P. and Nurnberg, H.W. *Fresenius Zeitschrift für Analytische Chemie*, **307**, 337 (1981)
708. Pietrowicz, S.R., Springer-Young, M., Puig, J.A. and Spencer, M.J. *Analytical Chemistry*, **54**, 1367 (1982)
709. Lund, W. and Onshus, D. *Analytica Chimica Acta*, **86**, 109 (1976)
710. Abdullah, B., Berg, R. and Klimek, R. *Analytica Chimica Acta*, **84**, 307 (1976)
711. Van der Berg, C.M.G. *Geochimica et Cosmochimica Acta*, **48**, 2613 (1984)
712. Krznaric, D. *Marine Chemistry*, **15**, 117 (1984)
713. Bond, A.M., Heritage, I.D. and Thormann, W. *Analytical Chemistry*, **58**, 1063 (1986)
714. Turner, D.R., Robinson, S.G. and Whitfield, M. *Analytical Chemistry*, **56**, 2387 (1984)
715. Jagner, D. and Granelli, A. *Analytica Chimica Acta*, **83**, 19 (1976)
716. Jagner, D. *Analytical Chemistry*, **50**, 1924 (1978)
717. Jagner, D., Josefson, M. and Westerlund, S. *Analytica Chimica Acta*, **129**, 153 (1981)
718. Drabek, I., Pheiffer, O., Madsen, P. and Sorensen, J. *International Journal of Environmental Analytical Chemistry*, **15**, 153
719. Robertson, D.E. and Carpenter, R. *Report No. BNWL-SA-4455 Access 22569* National Information Service, Springfield, Va. (1972)
720. Robertson, D.R. and Carpenter, R. *Report No. NAS-NS-3114 Access 26418* National Technical Information Centre, Springfield, Va. (1974)
721. Piper, D.S. and Goles, G.G. *Analytica Chimica Acta*, **47**, 560 (1969)
722. Ghoda, S. *Bulletin of the Chemical Society of Japan*, **45**, 1704 (1972)
723. Brine, E. *Report Aktiebolaget Atomenerai, AE-466* (1972)
724. Fujinaga, T., Kusaka, R., Koyama, M., *et al. Journal of Radioanalytical Chemistry*, **13**, 301 (1973)
725. Lieser, K.H., Calmano, W., Heuss, E. and Neitzert, V. *Journal of Radioanalytical Chemistry*, **37**, 717 (1977)
726. Lee, C., Kim, N.B., Lee, I.C. and Chung, K.S. *Talanta*, **24**, 241 (1977)
727. Lo, J.M., Wei, J.C. and Yeh, S.J. *Analytical Chemistry*, **49**, 1146 (1977)
728. Greenberg, R.R. and Kingston, H.M. *Journal of Radioanalytical Chemistry*, **71**, 147 (1982)
729. Greenberg, R.R. and Kingston, H.M. *Analytical Chemistry*, **55**, 1160 (1983)
730. Murthey, R.S.S. and Ryan, D.E. *Analytical Chemistry*, **55**, 682 (1983)
731. Huh, C.A. and Bacon, M.P. *Analytical Chemistry*, **57**, 2065 (1985)
732. Stiller, M., Mantel, M. and Rappaport, M.S. *Journal of Radioanalytical and Nuclear Chemistry*, **83**, 345 (1984)
733. Armitage, B. and Zeitlin, H. *Analytica Chimica Acta*, **53**, 47 (1971)
734. Morris, A.W. *Analytica Chimica Acta*, **42**, 397 (1968)
735. Knockel, A. and Prange, A. *Fresenius Zeitschrift für Analytische Chemie*, **306**, 252 (1981)
736. Murata, M., Omatsu, M. and Muskimoto, S. *X-ray Spectrometry*, **13**, 83 (1984)
737. Prange, A., Knockel, A. and Michaelis, W. *Analytica Chimica Acta*, **172**, 79 (1985)
738. Chow, T.J. and Goldberg, E.D. *Journal of Marine Research*, **20**, 163
739. Chow, T.J. and Goldberg, E.D. *Geochimica Cosmochimica Acta*, **20**, 192 (1961)
740. Chow, T.J. and Patterson, C.C. *Earth Planet Science Letters*, **1**, 397 (1966)
741. Smith, R.C., Pillai, K.C., Chow, T.J. and Folson, T.R. *Limnology and Oceanography*, **10**, 226 (1965)
742. Wilson, J.D., Webster, R.K., Milner, G.W.C., *et al. Analytica Chimica Acta*, **23**, 505 (1960)
743. Abe, T., Itoh, K. and Murozumi, M. *Bunseki Kiki*, **15**, 65 (1977)
744. Harvey, B.R. *Analytical Chemistry*, **50**, 1866 (1978)
745. Leipziger, F.D. *Analytical Chemistry*, **37**, 171 (1965)
746. Paulsen, P., Alvarez, R. and Kelleher, D. *Spectrochimica Acta*, **24**, 535 (1969)
747. Chow, T.J. *Journal of the Water Pollution Control Federation*, 399 (1968)
748. Patterson, C.C., Settle, D.M. and Glover, B. *Marine Chemistry*, **4**, 305 (1976)
749. Schaule, B. and Patterson, C.C. The occurrence of lead in the Northeast Pacific, and the effects

of anthropogenic inputs. In *Lead in the Marine Environment* (eds M. Brancia and Z. Konrad), Pergamon Press, Oxford, pp. 31–43 (1980)

750. Stukas, V.J. and Wong, C.S. *Science*, **211**, 1424 (1981)
751. Murozumi, M. Isotope dilution mass spectrometry of copper, cadmium, thallium and lead in marine environments. Presented at *The American Chemical Society Chemical Congress, Hawaii.*
752. Mykytiuk, A.P., Russell, D.S. and Sturgeon, R.E. *Analytical Chemistry*, **52**, 1281 (1980)
753. Stuckas, V.J. and Wong, C.S. Trace metals in seawater. *In Proceedings of a Nato Advanced Research Institute on Trace Metals in Seawater, 30/3–3/4/81, Sicily, Italy* (eds C.S. Wong, *et al.*), Plenum Press, New York, p. 513 (1981)
754. Mykytiuk, A.P., Russell, D.S. and Sturgeon, R.E. *Analytical Chemistry*, **52**, 1281 (1980)
755. Sturgeon, R.E., Berman, S.S., Desauliniers, A. and Russell, D.S. *Talanta*, **27**, 85 (1980)

Chapter 5

Radioactive compounds

5.1 Naturally occurring isotopes

Uranium

Spencer [1] has reviewed the determination of uranium in seawater.

Bertine, Chan and Turekian [2] have discussed the determination of uranium in deep sea sediments and water utilizing the fission track technique. In this technique a weighed aliquot (50–100 mg) of the powdered sample is made into a pellet with sufficient cellulose (as binder). The pellet is placed in a high-purity aluminium capsule and covered by polycarbonate plastic film (Lexan: 10 μm thick).

The capsule is then wrapped in high purity aluminium foil and irradiated in an integrated neutron flux (as in neutron activation analysis) adjusted according to the anticipated level of uranium. Each disintegration of a uranium nucleus induced by neutron bombardment produces a track on the Lexan sheet. The Lexan sheets are then removed from the pellets, suspended in 6.25 M potassium hydroxide at 60 ± 0.2°C, rinsed in water and dried, and the centre areas that were adjacent to the pellet surfaces are cut out. The etched tracks on the Lexan circles are then counted on a digital discharge counter. The track pattern is transferred to aluminium foil by exposing the Lexan film, in contact with a piece of aluminium-backed Mylar, to high-voltage sparks. The pulses generated during this process are counted by the scaler. The counting, at 500 V, is repeated several times to give an average track count. The number of counts varies rectilinearly with uranium content up to track densities of 5000 cm^{-2}. For replicate determinations from two irradiated samples the coefficient of variation was 7%.

Adsorbing colloid flotation has been used to separate uranium from seawater [4].

To the filtered seawater (500 ml; about 1.5 μg U) is added 0.05 M ferric chloride (3 ml), the pH is adjusted to 6.7 ± 0.1 and the uranium present as $(UO_2(CO_3)_3)^{4-}$ is adsorbed on the colloidal ferric hydroxide which is floated to the surface as a stable froth by the addition of 0.05% ethanolic sodium dodecyl sulphate (2 ml) with an air-flow (about 10 ml min^{-1}) through the mixture for 5 min. The froth is removed and dissolved in 12 M hydrochloric acid–16 M nitric acid (4 : 1) and the uranium is salted out with a solution of calcium nitrate containing EDTA, and determined spectrophotometrically at 555 nm by a modification of a Rhodamine B method. The average recovery of uranium is 82%; co-adsorbed WO_4^{2-} and MoO_4^{2-} do not interfere.

Adsorbing colloid flotation has also been used by Williams and Gillam [5]. The fusion track method has also been used by Hashimoto [3]. In this method, the uranium is first co-coprecipitated with aluminium phosphate [6], the precipitate is dissolved in dilute nitric acid and an aliquot of the solution is transferred to a silica ampoule into which small pieces of muscovite are inserted before sealing. The uranium is then determined by measuring the density of fission tracks formed on the muscovite during irradiation of the ampoule for 15 min at 80°C in a neutron reactor. The muscovite is etched with hydrofluoric acid for 1 hour before the photomicrography; the density is referred to that obtained with standard solution of uranium. There is no interference from thorium, and no chemical separations are required. An average concentration of $3–40 \pm 0.12 \, \mu g$ uranium per litre was obtained, in good agreement with the normally accepted value.

Leung, Kim and Zeitlin [7] and Kim and Zeitlin [8] describe a method for the separation and determination of uranium in seawater. Thoric hydroxide $(Th(OH)_4)$ was used as a collector. The final uranium concentration was measured via the fluorescence (at 575 nm) of its Rhodamine B. complex. The detection limit was about $200 \, \mu g \, l^{-1}$.

Korkisch and Koch [9,10] determined low concentrations of uranium in seawater by extraction and ion-exchange in a solvent system containing trioctyl phosphine oxide. Uranium is extracted from the sample solution (adjusted to be 1 M in hydrochloric acid and to contain 0.5% of ascorbic acid) with 0.1 M trioctylphosphine oxide in ethyl ether. The extract is treated with sufficient 2-methoxyethanol and 12 M hydrochloric acid to make the solvent composition 2-methoxyethanol–0.1 M ethereal trioctylphosphine acid–12 M hydrochloric acid (9 : 10 : 1); this solution is applied to a column of Dowex 1-X8 resin (Cl^- form). Excess of trioctylphosphine oxide is removed by washing the column with the same solvent mixture. Molybdenum is removed by elution with 2-methoxyethanol–30% aqueous hydrogen peroxide–12 M hydrochloric acid (18 : 1 : 1); the column is washed with 6 M hydrochloric acid and uranium is eluted with molar hydrochloric acid and determined fluorimetrically or spectrophotometrically with ammonium thiocyanate. Large amounts of molybdenum should be removed by a preliminary extraction of the sample solution (made 6 M in hydrochloric acid) with ether.

Spectrophotometric analysis following extraction with Aliquot 336 has been used to determine uranium in seawater [11]. Kim and Burnett [12] used X-ray spectrometry to determine the uranium series nucleides including ^{238}U, ^{226}Ra and ^{210}Pb in marine phosphorites.

Bowie and Clayton [13] used γ-ray spectrometry to determine uranium, thorium and potassium in sea bottom surveys.

Polonium and lead

Various workers [14–18] have discussed the determination of polonium and lead in seawater.

Similar affinity of polonium and plutonium for marine surfaces implies that studies of the more easily measured polonium might be valuable in predicting some consequences of plutonium disposal in the oceans [19–22]. Rates at which plutonium and polonium deposit out of seawater onto surfaces of giant brown algae and 'inert' surfaces, such as glass and cellulose, suggest that both nuclides are associated in coastal seawater with colloidal sized species having diffusivities of about $3 \times 10^{-7} \, cm^2 \, s^{-1}$. The parallel behaviour possibly represents an initial step in the incorporation of both α-radioactive heavy elements into marine food chains and/or their transport by the greater activity concentrations found on marine surfaces and in seawater, about 200 times that of plutonium.

Tsunogai and Nozaki [17] analysed Pacific Oceans surface water by consecutive co-precipitations of polonium with calcium carbonate and bismuth oxychloride after addition of lead and bismuth carriers to acidified seawater samples. After concentration, polonium was spontaneously deposited onto silver planchets. Quantitative recoveries of polonium were assumed at the extraction steps and plating step. Shannon, Cherry and Orren [18], who analysed surface water from the Atlantic Ocean near the tip of South Africa, extracted polonium from acidified samples as the ammonium pyrrolidine dithiocarbamate complex into methyl isobutyl ketone. They also autoplated polonium onto silver counting disks. An average efficiency of 92% was assigned to their procedure after calibration with ^{210}Po-^{210}Pb tracer experiments.

Shannon [14] determined polonium 210 and lead 210 in seawater. These two elements are extracted from seawater (at pH 2) with a solution of ammonium pyrrolidine dithiocarbamate in isobutyl methyl ketone (20 ml organic phase to 1.5 litres of sample). The two elements are back-extracted into hydrochloric acid and plated out of solution by the technique of Flynn [23], but with use of a PTFE holder in place of the Perspex one, and the α-activity deposited is measured. The solution from the plating-out process is stored for 2–4 months, then the plating-out and counting are repeated to measure the build-up of polonium 210 from lead 210 decay and hence to estimate the original ^{210}Pb activity.

Nozaki and Tsunogai [15,25] determined lead 210 and polonium 210 in seawater. The lead 210 and polonium 210 in a 30–50 litre sample are co-precipitated with calcium carbonate together with lead and bismuth and are then separated from calcium by precipitation as hydroxides. The precipitate is dissolved in 0.5 M hydrochloric acid, and polonium 210 is deposited spontaneously from this solution on to a silver disc and is determined by α-spectrometry. Chemical yields of lead and bismuth are determined in a portion of the solution from which the polonium has been deposited; hydroxides of lead and other metals are precipitated from the remainder of this solution and after a period exceeding 3 months, the polonium 210 produced by decay of lead 210 is determined as before. The activity of lead 210 is calculated from the activity of polonium 210. The method was used to determine the vertical distribution of lead 210 and polonium 210 activities in surface layers of the Pacific Ocean.

Cowen, Hodge and Folson [16] showed that polonium can be electrodeposited onto carbon rods directly from acidified seawater, stripped from the rods and auto-plated onto silver counting disks with an overall recovery of tracer of 85 ± 4% for an electrodeposition time of 16 h [24].

These workers compared two procedures for concentrating polonium 210 from seawater:

1. Co-precipitation upon partial precipitation of the natural calcium and magnesium with sodium hydroxide.
2. Electrodeposition of polonium directly from acidified seawater onto carbon rods.

Polonium thus concentrated was autoplated onto silver counting disks held in spinning Teflon holders.

Table 5.1 Comparison of sodium hydroxide precipitation and carbon red plating methods for concentrating ^{210}Po from aliquots of acidified seawater removed from 50 litre parent samples

Method	Date sampled	^{210}Po activity (pCl l^{-1})*	Recovery (%)
Sample 1 (0.5 M HCl)†			
NaOH	5/12/75	0.115 ± 0.009	71
NaOH	5/12/75	0.113 ± 0.009	66
Carbon rod	5/12/75	0.116 ± 0.007	81‡
Carbon rod	5/12/75	0.110 ± 0.005	89‡
NaOH	5/17/75	0.104 ± 0.008	74
NaOH	5/17/75	0.097 ± 0.008	80
Sample 2 (0.5 M HCl)‖			
NaOH	5/20/75	0.031 ± 0.002	86
NaOH	5/20/75	0.025 ± 0.002	74
Carbon rod	5/20/75	0.034 ± 0.002	63§
Carbon rod	5/20/75	0.035 ± 0.003	65§
Carbon rod	5/27/75	0.034 ± 0.003	38¶
Carbon rod	5/27/75	0.028 ± 0.002	41¶
NaOH	5/27/75	0.040 ± 0.003	73
NaOH	5/27/75	0.034 ± 0.002	90

*± 1 standard counting error.
†Collected on 4/25/75 at Scripps Pier.
‡Electroplating time, 48 h.
‖Collected on 5/20/75 at Scripps Pier.
§Electroplating time, 24 h.
¶Electroplating time, 16 h.
Reproduced from Cowen, J.P., *et al.* (1977) *Analytical Chemistry*, **49**, 444, American Chemical Society, by courtesy of authors and publishers.

A comparison of results obtained by the two methods is shown in Table 5.1. Recoveries of polonium 208 tracer in the precipitation method were 77 ± 7% (n = 8) compared with 40 ± 2% (n = 2) for the electrodeposition method with 16 h plating time, 64 ± 1% (n = 2) in 24 h and 85 ± 4% (n = 2) in 48 h. Even though the electrode-position method requires less attention, it requires long plating times for high recoveries. Thus the recovery of polonium 210 by direct plating appears to be rate limited by diffusion of polonium to the cathodes since the applied potential difference is far in excess of that required to reduce Po^{IV} to polonium.

Recoveries are based on the added polonium 208 tracer, presumably PoCl$_6^{2-}$. Equilibrium was assumed to have occurred during the 2 h mixing of spike and sample which was always 0.5 m in acid at this stage.

Table 5.2 shows that recovery of polonium 208 from seawater was insensitive to the amount of 1 M sodium hydroxide used in precipitation. Sodium hydroxide 6 ml 1 M was chosen because it gave an easily manipulated volume of precipitate.

Table 5.2 Effect of added sodium hydroxide on the recovery of ^{208}Po tracer from seawater*

Volume of 1 N NaOH (ml l^{-1})	Recovery (%)
3	86
6	83
10	86

*After neutralization of acid used to stabilize polonium in solution (0.5 M HCl).
Reproduced from Cowen, J.P., *et al.* (1977) *Analytical Chemistry*, **49**, 444, American Chemical Society, by courtesy of authors and publishers.

The need to acidify seawater to be used for polonium analysis was investigated by periodically sub-sampling two parent samples over a week. Two 50 litre polyethylene carboys were filled with seawater on the same date at the same location. One was acidified with 12 MHCl to 0.5 M. The 50 litre carboys were shaken thoroughly before water was removed. The polonium 210 concentration in the unacidified seawater showed a dramatic decrease to less than half in 7 days the time sometimes lapsing between field sampling and analysis in the laboratory. Since the unacidified seawater sample was acidified to 0.5 M on addition of the polonium 208 tracer, polonium 210 was either lost to the walls of the carboy or converted to a suspended species not leachable in the 0.5 M acid.

Another approach to the effect of acidification was to subsample a large bottle of seawater, first untreated with acid and then progressively made more acidic, allowing 2 or more days between treatments. The results of such a test are summarized in Table 5.3. The need for acidification on collection is again demonstrated and the large variations encountered in analysing raw seawater by the sodium hydroxide precipitation method are evident.

Table 5.3 Effect of acidity and time on ^{210}Po concentrations in aliquots removed from single 100 litre parent sample*

Method	Date sampled	pCi l^{-1}	Recovery (%)
Raw seawater			
NaOH	5/29/75	0.097 ± 0.006	62
Carbon rod‡	5/29/75	0.044 ± 0.003	74
NaOH	6/2/75	0.028 ± 0.002	86
Carbon rod‡	6/2/75	0.047 ± 0.003	53
0.1 M HCl			
Carbon rod‡	6/2/75	0.053 ± 0.003	64
Carbon rod‡	6/4/75	0.050 ± 0.004	50
0.5 M HCl			
NaOH	6/10/75	0.052 ± 0.005	72
Carbon rod‡	6/10/75	0.060 ± 0.006	55
NaOH	6/23/75	0.050 ± 0.002	82
Carbon rod‡	6/23/75	0.058 ± 0.002	60

*Collected on 5/28/75 at Scripps Pier.
†±1 standard counting error.
‡Electroplating time, approximately 24 h.
Reproduced from Cowen, J.P., et al. (1977) Analytical Chemistry, **49**, 494, American Chemical Society, by courtesy of authors and publishers.

Thorium

Huh [26] has used neutron activation analysis to determine thorium in seawater. The method used preirradiation and postirradiation radiochemical separations.

Bowie and Clayton [13] used gamma-ray spectrometry to determine thorium, uranium and potassium in sea-bottom surveys.

Bacon and Anderson [27] determined ^{234}Th, ^{230}Th and ^{228}Th concentrations, in both dissolved and particulate forms, in seawater samples from the eastern equatorial Pacific. The results indicate that the thorium isotopes in the deep ocean are continuously exchanged between seawater and particle surfaces. The estimated rate of exchange is fast compared with the removal rate of the particulate matter, suggesting that the particle surfaces are nearly in equilibrium with respect to the exchange of metals with seawater.

Because of the large volumes of water that were required, an *in situ* sampling procedure was used. Submersible, battery powered pumping systems [28,29] were used to force the water first through filters ($62\,\mu m$ mesh Nitex followed by $1.0\,\mu m$ pore-size Nuclepore) then through an adsorber cartridge packed with Nitex netting that was coated with manganese dioxide to scavenge the dissolved thorium isotopes, and finally through a flow meter to record the volume of water that was filtered. Natural ^{234}Th served as the tracer for monitoring the efficiency of the adsorber cartridges. Standard radiochemical counting techniques were used [30]. On average 4% of the ^{234}Th, 15% of the ^{228}Th and 17% of the ^{230}Th were found in the particulate form, i.e. the percentage increases with increasing radioactive half-life. However, the percentages varied considerably from sample to sample and were found to be strongly dependent on total suspended matter concentration.

Radium, barium and radon

Perkins [31] carried out radium and radiobarium measurements in seawater by sorption and direct multi-dimensional gamma-ray spectrometry. The procedure described includes the removal of radium and barium from water samples on sorption beds of barium sulphate impregnated alumina (0.5–1 cm thick) and direct counting of these beds on a multi-dimensional γ-ray spectrometer. The radioisotopes can be removed at linear flow rates of sample of up to $1\,m\,min^{-1}$.

Oceanographers have developed methods to measure the ^{228}Ra content of seawater as it is a useful tracer of mixing in the ocean. These procedures are based on concentrating radium from a large volume of seawater, removing all ^{228}Th from the sample and ageing the sample while a new generation of ^{228}Th partially equilibrates with ^{228}Ra. After storage periods of 6–12 months, the sample is spiked with ^{230}Th and after ion-exchange and solvent exchange separations, the thorium isotopes are measured in a γ-ray spectrometer system utilizing a silicon surface barrier detector. Early work was based on concentrating the radium from the seawater sample by adding barium and co-precipitating with borium sulphate. This concentration procedure has been replaced by one involving the extraction of radium from seawater on acrylic fibre coated with manganese dioxide [32,33] (Mn fibres). By use of this technique, volumes of 200–2000 litres may be sampled routinely.

Radium is extracted from ground waters with even better efficiency on the manganese fibres [34]. Radium is removed from the manganese fibre by reducing the manganese dioxide with hot hydrochloric acid or desorbing the radium into cold dilute nitric acid. The hydrochloric acid treatment removes the radium quantitatively from the fibres but generates a considerable quantity of chlorine. The nitric acid treatment is much easier and safer but only removes about 70% of the radium [33].

Measurements of ^{226}Ra are simpler than those for ^{228}Ra and are more precise. These measurements are generally made by concentrating the radium from up to a few litres via barium sulphate precipitation followed by thick source α counting or by ^{222}Rn extraction following dissolution of barium sulphate [35].

Oceanographers use different techniques for measuring ^{226}Ra in seawater. Some workers store the sample in a 20 litre glass bottle and extract successive generations of ^{222}Rn [36,37]. Others quantitatively extract the radium onto manganese fibre and measure ^{222}Rn directly emanating from the manganese fibre [38] or in a hydrochloric acid extract from the fibres [39]. The ^{222}Rn activity is then determined by α-scintillation counting. All of these techniques give high levels of reproducibility and accuracy as determined by the oceanographic consistency of the results [36,37].

The introduction of high-resolution, high-efficiency γ-ray detectors composed of lithium-drifted germanium crystals has revolutionized γ- measurement techniques. Thus, γ-spectrometry allows the rapid measurement of relatively low-activity samples without complex analytical preparations. A technique described by Michel, Moore and King [40] uses Ge(Li) γ-ray detectors for the simultaneous measurements of ^{228}Ra and ^{226}Ra in natural waters. This method simplifies the analytical procedures and reduces the labour while improving the precision, accuracy and detection limits.

In this method the radium isotopes are pre-concentrated in the field from 100 to 1000 litre water sample onto manganese impregnated acrylic fibre cartridges, leached from the fibre and co-precipitated with barium sulphate. Lower limits of detection are controlled by the volume of water processed through the manganese fibres. In a 1 day count, samples as low as 10 dpm are measured to ±10% uncertainty. This manganese fibre–γ-ray technique is shown to be more accurate than the actinium-228 methods recommended by the Environmental Protection Agency [41] and as accurate but more rapid than the ^{228}Th ingrowth procedure.

Table 5.4 shows the results for samples from the Environmental Protection Agency Radium in Water Crosscheck Program using both the ^{228}Th ingrowth and the Ge(Li) γ-ray techniques. Both types of analyses were run on 3.8 litre samples. The ^{228}Th ingrowth samples were stored for 3–11 months to accumulate sufficient activity and then counted by α spectrometry for 2 days. The Ge(Li) samples were stored for 3 weeks and counted for 400–1000 min. Both techniques required a separate analysis for the absolute ^{226}Ra activity whereas the ^{228}Th ingrowth method requires an additional analysis for the total ^{226}Ra in the large volume sample to obtain the ^{228}Ra/^{226}Ra activity ratio.

Table 5.4 Comparison of the known values of the EPA radium in water cross-check samples with both the ^{228}Th ingrowth/^{222}Rn emanation and Ge(Li) γ-ray methods*

EPA No.	Ra/^{228}Ra activity ratio		^{228}Ra, decays/min	
	Th ingrowth	*Known*	*Th ingrowth*	*Known*
Dec 78	1.04 ± 0.06	0.96 ± 0.20	21.2 ± 0.8	19.8 ± 2.9
Mar 79	1.59 ± 0.31	1.66 ± 0.34	28.6 ± 5.5	30.2 ± 4.4
Apr 79	1.05 ± 0.05	1.05 ± 0.22	13.6 ± 0.5	13.8 ± 2.0
	Ge(Li)	*Known*	*Ge(Li)*	*Known*
Sept 80	1.07 ± 0.05	1.10 ± 0.24	29.1 ± 1.5	31.9 ± 1.8
Oct 80	0.66 ± 0.03	0.72 ± 0.09	19.8 ± 1.0	20.4 ± 1.8
Dec 80	0.76 ± 0.06	0.79 ± 0.17	22.0 ± 1.1	23.3 ± 3.6

*Measurements are reported as weighted means with weighted standard deviations. The known values are reported by EPA with expected laboratory standard deviations.
Reproduced from Michael, J., et al. (1981) *Analytical Chemistry*, **53**, 1885, American Chemical Society, by courtesy of authors and publishers.

Table 5.5 Comparison of ^{226}Ra, ^{228}Ra and ^{228}Ra/^{226}Ra activity ratio in seawater of the same sampling using two different methods

Sample	^{226}Ra (decays/min)*	^{228}Ra (decays/min)*	^{228}Ra/^{226}Ra	Method*
451–99	260 ± 7	210 ± 10	0.81 ± 0.04	1
	305 ± 2	223 ± 5	0.73 ± 0.02	2
452–99	235 ± 10	125 ± 13	0.53 ± 0.06	1
	229 ± 12	153 ± 3	0.67 ± 0.04	2
453–99	154 ± 1	56 ± 7	0.36 ± 0.04	1
	161 ± 9	53 ± 2	0.33 ± 0.02	2
454–99	246 ± 13	94 ± 11	0.38 ± 0.05	1
	291 ± 8	94 ± 3	0.32 ± 0.01	2

*Method 1, Ge(Li) γ-ray spectrometry. Results reported as weighted means and weighted standard deviations of both 1SD counting and efficiency uncertainties. Method 2, ^{226}Ra by Rn emanation; ^{228}Ra by ^{228}Th ingrowth. Results reported as 1SD counting uncertainties. Reproduced from Michael, J., *et al.* (1981) *Analytical Chemistry*, **53**, 1885, American Chemical Society, by courtesy of authors and publishers.

Table 5.5 shows the results for seawater samples collected by the Mn fibre technique and analysed by both the Ge(Li) and ^{228}Th ingrowth techniques. The ground-water samples were analysed first by γ-ray counting followed by fusion of the barium sulphate and the ^{228}Th ingrowth and radon emanation techniques. The seawater samples were analysed in the reverse order. Although the bulk activities of each isotope vary for the different techniques, the ^{228}Ra/^{226}Ra activity ratios are in close agreement.

Key *et al.* [42] have described improved methods for the measurement of radon and radium in seawater and marine sediments. The basic method that these workers used was that of Broecker [43]. Seawater samples were taken in 30 litre Niskin bottles. The analysis of seawater samples is discussed below in further detail.

Figure 5.1 shows the system used for helium stripping of radon from the water sample. After bubbling through the sample, the gas stream passes over Drierite and Ascarite to remove carbon dioxide and water. The gas streams then flow into primary sample traps kept at liquid-nitrogen temperature. These traps are made of 1/4 inch stainless steel tubing, packed with fine (00) bronze wool on activated charcoal at dry-ice acetone temperature to improve trapping efficiency by increasing thermal contact and surface area. After passing through the primary trap to the volume reduction trap, the carrier helium is routed through the 'keyblers' rather than through a bypass (Figure 5.2). When degassing is complete (75 minutes) the bypass valve is opened and the valves on the wash bottles closed; the inlet and exit valves to the primary sample traps are closed, the helium flow stopped and the transfer portion of the radon board evacuated.

The next step is to transfer the radon to a secondary, smaller trap; this volume reduction ensures quantitative sample transfer to the counting cell.

Radium is extracted with 99% efficiency from the seawater after radon degassing is completed by draining the water from the keybler through a column (1 inch diameter × 6 inch 1 PVC pipe) packed with manganese-impregnated acrylic fibre [44,45].

After all the seawater in a sample has passed through the column, the fibre is bagged and returned to the laboratory for analysis. Radium is leached from the fibre by boiling in about 300 ml concentrated hydrochloric acid for about 30 min.

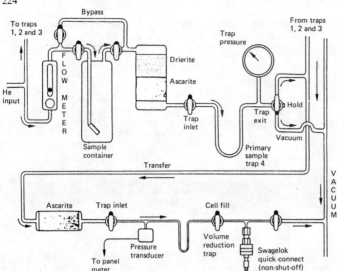

Figure 5.1 Schematic of radon stripping and transfer system. The upper section is one of four channels used in this system; the lower (transfer) section is shared by all four. In this upper section and the three others like it, four samples are degassed, purified of H_2O and CO_2 and trapped. The samples are then moved individually via the transfer manifold, to the sample reduction trap, then on to counting cells (see Figure 5.3). All plumbing is 1/4 inch copper tubing except (1) the connections between flowmeters and Drierite – Ascarite columns (1/4 inch polyethylene with pinch-clamp valves). 2 = The primary sample traps (1/4 inch 316 SS tubing packed with fine brass wool) and 3 = the volume-reduction trap (1/8 inch 316 SS tubing packed with fine bronze wool). All valves are made by Whitey; rotometers are Brooks Model 2 – 65A with needle valve inlets (Reproduced from Kay, R.M., *et al.* (1979) *Marine Chemistry*, **7**, 251, Elsevier Science Publishers, by courtesy of authors and publishers.)

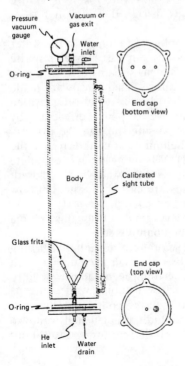

Figure 5.2 Diagram illustrating keybler construction. The bodies are 8 inch o.d. schedule 40 polyvinyl chloride pipe; endplates are milled from 1½ inch PVC stock. Samples are sucked into the evacuated keyblers through the nipple on the top plate. The volume of each keybler was calibrated by weighing. Volume is read from the sight tube attached to the side of the containers. Under normal conditions volume accuracy is about 50 ml. Radon is degassed by helium bubbles formed at two glass frits mounted to the bottom plate. The sample and helium flow out of the keybler through the quick-connect fitting on the top plate. After degassing the water is drained out of the bottom and into PVC columns packed with manganese impregnated acylic fibre to quantitatively extract radium from the sample (Reproduced from Key, R.M., *et al.* (1979) *Marine Chemistry*, **7**, 251, Elsevier Science Publishers, by courtesy of the authors and publishers.)

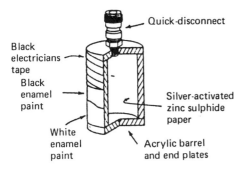

Quick-disconnect

Black
electricians
tape

Black
enamel
paint

White
enamel
paint

Silver-activated
zinc sulphide
paper

Acrylic barrel
and end plates

Figure 5.3 Radon counting cell. See text for
construction details (Reproduced from Key, R.M.,
et al. (1979) *Marine Chemistry*, **7**, 251, Elsevier
Science Publishers, by courtesy of authors and
publishers.)

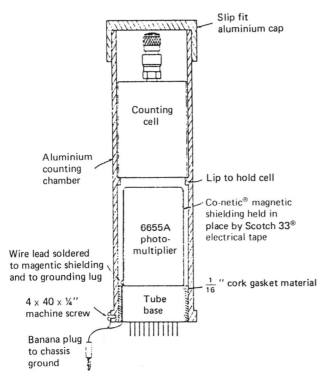

Slip fit
aluminium cap

Counting
cell

Aluminium
counting
chamber

Lip to hold cell

Co-netic® magnetic
shielding held in
place by Scotch 33®
electrical tape

6655A
photo-
multiplier

Wire lead soldered
to magentic shielding
and to grounding lug

$\frac{1}{16}$ " cork gasket material

4 x 40 x ¼"
machine screw

Tube
base

Banana plug
to chassis
ground

Figure 5.4 Detail of light-tight photomultiplier tube assembly showing positioning of the counting cell
(Reproduced from Key, R.H., *et al.* (1979) *Marine Chemistry*, **7**, 251, Elsevier Science Publishers, by
courtesy of authors and publishers.)

Leachate and wash are combined and the resulting solution is transferred to a gas
wash bottle, degassed for 40 min and sealed to allow radon growth.

The radon counting system has been described by Schink *et al.* [46] and Chung
[47].

Radon is counted in a Lucas-type cell [48] constructed of acrylic plastic (Figure
5.3). Cells are fabricated from 2 1/4 inch o.d. 1/4 inch wall tubing with end caps
made from 3/8 inch plate. Scintillations from the α-particles striking the zinc
sulphide coated paper are detected by a photomultiplier in a light-tight cylinder
shown in Figure 5.4. The cell rests on a machined lip so that it does not touch the
photomultiplier face.

Table 5.6 Mean radium concentrations determined on samples assumed to be uniform in radium content

Station	Average Ra (dpm/100 litres)	Number of samples	SD (%)
76–G–3–6	9.8 ± 0.2	6	2.5
76–G–4–14	19.4 ± 0.6	7	3.1
76–G–11–14	12.9 ± 0.3	8	2.3
76–G–11–23	13.5 ± 0.3	8	2.6
			Av. 2.6 ± 0.3

Reproduced from Key, R.M., *et al.* (1979) *Marine Chemistry*, **7**, 251, Elsevier Science Publishers BV, by courtesy of authors and publishers.

Table 5.6 summarizes the results for analytical radon precision in radium determinations from four near-bottom casts.

Potassium

Bowie and Clayton [13] used γ-ray spectrometry to determine potassium, uranium and thorium in sea-bottom surveys.

5.2 Fall-out products and nuclear plant emissions

Plutonium

The plutonium concentration in marine samples is principally due to environmental pollution caused by fall-out from nuclear explosions and is generally at very low levels [49]. Environmental samples also contain microtraces of natural α emitters (uranium, thorium and their decay products) which complicate the plutonium determinations [50]. Methods for the determination of plutonium in marine samples must therefore be very sensitive and selective. The methods reported for the chemical separation of plutonium are based on ion-exchange resins [50–54] or liquid–liquid extraction with tertiary amines [55], organophosphorus compounds [56,57] and ketones [58,59].

Wong [51] has described a method for the radiochemical determination of plutonium in seawater, sediments and marine organisms. This procedure permits routine determinations of ^{239}Pu activities (d.p.m.) down to 0.004 d.p.m. per 100 litres of seawater (50 litre sample). 0.02 d.p.m. per kg sediments (100 g samples) and 0.002 d.p.m. per kg of organisms (1 kg sample). The plutonium is separated from seawater by co-precipitation with ferric hydroxide and from dried sediments or ashed organisms by leaching with nitric acid and hydrochloric acid [60]. After further treatment and purification by ion-exchange, the plutonium is electro-deposited on to stainless-steel discs for counting and resolution of the activity by α-spectrometry. For 30 samples the average deviation was generally well within the 1 SD counting error. For seawater the average recovery was $52 \pm 18\%$ and for sediments and organisms it was $63 \pm 20\%$. The most serious interference is from ^{228}Th, which is present in most samples and is also a decay product of the ^{236}Pu tracer.

Livingstone, Mann and Bowen [61] carried out double tracer studies to optimize conditions for the radiochemical separation of plutonium from large volumes of seawater.

In this procedure ^{242}Pu is added to determine the overall recovery of plutonium from the sample and the recovery of ^{242}Pu at any point in the procedure is measured by the addition of a similar amount of ^{236}Pu at that point, the final recovery of ^{236}Pu being used to calculate the recovery of ^{242}Pu at the time of the addition of ^{236}Pu. Experience with this double-tracer experiment has permitted improvement in the ability to recover plutonium from 50 litre samples for α-spectrophotometric analysis of ^{239}Pu, ^{240}Pu and, sometimes, ^{238}Pu.

Delle Site, Marchionni and Testa [62] have used extraction chromatography to determine plutonium in seawater, sediments and marine organisms.

These workers used double extraction chromatography with Microthene-210 (microporous polyethylene) supporting tri-n-octylphosphine oxide (TOPO); a technique that has been used previously to isolate plutonium from other biological and environmental samples [63]. Plutonium 236 and plutonium 242 were tested as the internal standards to determine the overall plutonium recovery, but plutonium 242 was generally preferred because plutonium 236 has a shorter half-life and an α-emission (5.77 MeV) which interferes strongly with the 5.68 MeV (95%) α-line of ^{224}Ra, the daughter of ^{228}Th. However, the 5.42–MeV α-lines of ^{228}Th interfere with those of ^{238}Pu (5.50 MeV) and so a complete purification from thorium isotopes is required.

Add the extraction slurry A to the nitric acid solution obtained from the pretreatment of the samples (about 4 M in nitric acid) and stir magnetically for 1 h.

Plutonium sources were counted by an α-spectrometer with good resolution, background and counting yield. The counting apparatus used had a resolution of 40 keV. The mean (±SD) background value was 0.0004 ± 0.0003 cpm in the ^{239,240}Pu energy range and 0.0001 ± 0.0001 cpm in the ^{238}Pu energy range. The mean (±SD) counting yield, obtained with ^{239}Pu, ^{240}Pu reference sources counted in the same geometry, was found to be 25.08 ± 0.72%.

To determine the overall recovery obtained by this procedure (chemical recovery and electrodeposition yield) a known activity of ^{242}Pu was added to the different samples; the plutonium sources were counted by α-spectrometry for 3000 min and the percentage overall recovery was calculated from the area of ^{242}Pu peak. The percentage overall recovery for the different samples is reported in Table 5.7.

Owing to the very low activity of the samples, the determination of ^{239}Pu, ^{240}Pu and ^{238}Pu in the reagents is very important in calculating the net activity of the

Table 5.7 Overall recovery

Sample	Size	No. of samples considered	Mean recovery (% ± SD)*
Seawater	200 litres	5	62.6 ± 9.7
Sediments	100 g†	7	45.4 ± 9.6
Marine organisms	300 g†	5	81.7 ± 4.5

*Standard deviation calculated on the mean value of the analyses.
†Sample dried at 105°C.
Reproduced from Delle Site, A., et al. (1980) Analytica Chimica Acta, **117**, 217, Elsevier Science Publishers BV, by courtesy of authors and publishers.

Table 5.8 Reagent and process blank activity

Sample		$^{239,240}Pu$ (fCi ± SD)*	^{238}Pu (fCi ± SD)*
Seawater	50 litres	4.3 ± 1.3	0.8 ± 0.5
	200 litres	7.1 ± 1.5	2.1 ± 0.8
Sediments	100 g†	7.5 ± 2.5	0.8 ± 0.8
Marine organisms	300 g†	5.4 ± 0.8	1.1 ± 0.6

*Standard deviation of a single source α-counting.
†Sample dried at 105°C.
Reproduced from Delle Site, A., et al. (1980) Analytica Chimica Acta, **117**, 217, Elsevier Science Publishers BV, by courtesy of authors and publishers.

Table 5.9 Experimental check of the method with IAEA samples

Sample	Size (g)	IAEA activity (fCi ± SD)* $^{239,240}Pu$ ^{238}Pu		Present activity (fCi ± SD)† $^{239,240}Pu$	Mean value	^{238}Pu	Mean value
Seawater	1000	103 ± 7	18 ± 1	113 ± 7 96 ± 7 119 ± 8	109 ± 7	17 ± 4 17 ± 3 15 ± 4	16 ± 4
Sediment	1‡	501–585	14–31	629 ± 19 516 ± 20 575 ± 19	574 ± 19	17 ± 3 23 ± 4 21 ± 4	20 ± 4

*Standard deviation for mean of 5 (^{238}Pu) and 10 (^{239,240}Pu) results.
†Standard deviation of a single source α-counting.
‡Sample dried at 105°C.
Reproduced from Delle Site, A., et al. (1980) Analytica Chimica Acta, **117**, 217, Elsevier Science Publishers BV, by courtesy of authors and publishers.

radionuclides. The ^{239}Pu, ^{240}Pu and ^{238}Pu activity found in the reagent blanks is reported in Table 5.8.

The method proposed was checked by analysing some seawater and sediment reference samples prepared by the IAEA Marine Radioactivity Laboratory (Monaco) for intercomparison programmes. The values reported by IAEA and the experimental values obtained here are compared in Table 5.9; the agreement is fairly good.

Anderson and Fleer [64] determined the natural actinides ^{227}Ac, ^{228}Th, ^{230}Th, ^{232}Th, ^{234}Th, ^{231}Pa, ^{238}U and ^{234}U and the α-emitting plutonium isotopes in samples of suspended marine particulate material and sediments. Analysis involves total dissolution of the samples to allow equilibration of the natural isotopes with added isotope yield monitors followed by co-precipitation of hydrolysable metals at pH 7 with natural iron and aluminium acting as carriers to remove alkali and alkaline earth metals. Final purification is by ion-exchange chromatography (Dowex AG1-X3) and solvent extraction for palladium. Overall chemical yields generally range from 50% to 90%. The method has been successfully interfaced with methods to include the determination of the fall-out elements of ^{55}Fe, ^{137}Cs, ^{90}Sr and ^{241}Am on the same samples.

Table 5.10 illustrates results obtained by this procedure when applied to oceanic sediments.

Table 5.10 Radiochemical results for sediment trap samples (given as disintegrations min⁻¹ g⁻¹ dried sediment)

Depth (m)	^{227}Ac	^{238}U	^{230}Th	^{231}Pa	^{232}Th	^{228}Th	$^{239,240}Pu$	Activity ratio	
								$^{234}U/^{238}U$	$^{238}Pu/^{239,240}Pu$
Site E, 13.5°N, 54.0°W, 5288 m									
389-A	ND	0.72 ± 0.01	0.36 ± 0.02	0.026 ± 0.003	0.21 ± 0.01	13.9 ± 0.2	0.18 ± 0.03	1.15 ± 0.04	ND
389-B	ND	0.67 ± 0.04	0.37 ± 0.02	0.023 ± 0.005	0.18 ± 0.02	12.1 ± 0.7	0.20 ± 0.03	1.06 ± 0.08	ND
988	ND	0.76 ± 0.03	1.22 ± 0.03	0.087 ± 0.006	0.60 ± 0.02	13.8 ± 0.1	0.14 ± 0.02	1.01 ± 0.07	ND
3755	ND	0.81 ± 0.03	4.80 ± 0.08	0.221 ± 0.010	0.99 ± 0.03	14.1 ± 0.2	0.15 ± 0.02	0.98 ± 0.07	ND
5086	ND	0.83 ± 0.03	5.69 ± 0.15	0.200 ± 0.009	1.26 ± 0.06	21.7 ± 0.5	0.12 ± 0.02	1.06 ± 0.06	ND
Site P, 15.3°N, 151.5°W, 5792 m									
378	ND	3.78 ± 0.14	0.38 ± 0.03	<0.03	0.06 ± 0.02	0.7 ± 0.1	0.28 ± 0.03	1.20 ± 0.07	0.054 ± 0.036
978	<0.012	0.62 ± 0.04	1.93 ± 0.06	0.105 ± 0.009	0.07 ± 0.01	5.5 ± 0.1	1.61 ± 0.09	1.20 ± 0.12	0.033 ± 0.007
2778	ND	0.41 ± 0.02	10.4 ± 0.2	0.345 ± 0.007	0.11 ± 0.01	5.9 ± 0.1	2.0 ± 0.1	1.29 ± 0.13	0.038 ± 0.005
4280	0.204 ± 0.018	0.34 ± 0.02	16.6 ± 0.3	0.48 ± 0.02	0.12 ± 0.01	5.4 ± 0.1	2.4 ± 0.1	1.28 ± 0.13	0.036 ± 0.005
5582	0.519 ± 0.035	0.44 ± 0.04	27.4 ± 0.5	0.91 ± 0.04	0.23 ± 0.02	10.0 ± 0.2	2.2 ± 0.1	1.10 ± 0.19	0.049 ± 0.007

ND = Not determined.
Reproduced from Anderson, R.F. and Fleer, A. (1982) *Analytical Chemistry*, **54**. American Chemical Society, by courtesy of authors and publishers.

Testa and Staccioli [65] have pointed out that Microlthene-710 (a microporous polyethylene) as a support material for triphenylphosphine oxide in cyclohexane medium) has a potential application for the determination of plutonium in fall-out samples.

Hirose and Sugimura [66] investigated the speciation of plutonium in seawater using adsorption of plutonium(IV)-xylenol orange and plutonium-arsenazo (III) complexes on the macroreticular synthetic resin XAD-2. Xylenol orange was selective for plutonium (IV) and arsenazo (III) for total plutonium. Plutonium levels were determined by α-ray spectrometry.

Strontium-90

Silant'ev, Chumichev and Vakulowski [67] have described a procedure for the determination of strontium-90 in small volumes of seawater. This method is based on the determination of the daughter isotope ^{90}Y. The sample is acidified with hydrochloric acid, heated and, after addition of iron, interfering isotopes are separated by double co-precipitation with ferric hydroxide. The filtrate is acidified with hydrochloric acid, yttrium carrier added, the solution set aside for 14 days for ingrowth of ^{90}Y, and $Y(OH)_3$ precipitated from the hot solution with carbon dioxide free aqueous ammonia. Then $Y(OH)_3$ is re-precipitated from a small volume in the presence of hold-back carrier for strontium, the precipitate dissolved in the minimum amount of nitric acid, the solution heated and yttrium oxalate precipitated by adding precipitated oxalic acid solution. The precipitate is collected and ignited at $800°$ to $850°$ to Y_2O_3. The cooled residue is weighed to determine the chemical yield, then sealed in a polyethylene bag and the radioactivity of the saturated yttrium measured on a low-background β-spectrometer. If the short-lived nuclides ^{140}Ba and ^{140}La are thought to be present in the seawater sample, lanthanum carrier is introduced after the first $Y(OH)_3$ separation, and the system is freed from ^{140}La by precipitation of the double sulphate of lanthanum and potassium from a solution saturated with potassium sulphate.

Gordon and Larson [68] used photon activation analysis to determine ^{87}Sr in seawater. Samples (2 ml, acidified to pH 1.67 or 2.54 for storage) were filtered and freeze-dried. The residues, together with strontium standards, were wrapped in polyethylene and aluminium foil and irradiated in a 30 MeV bremsstrahlung flux of γ-radiation. After irradiation, the samples were dissolved in 50 ml of acidified water and ^{87m}Sr was operated by precipitation as strontium carbonate for counting (γ-ray spectrometer, Ge(Li) detector and multi-channel pulse-height analyser). The standard deviation at the 7 ppm strontium level was ± 0.47.

Caesium-137

Dutton [69] has described a procedure for the determination of caesium-137 in water. This procedure comprises a simple one-step separation of the radio-caesium from the sample using ammonium dodecamolybdophosphate or potassium cobaltihexacyanoferrate and ^{137}Cs and ^{134}Cs are measured by γ-ray counting of the dried adsorbent with a NaI(T1) crystal coupled to a γ-ray spectrometer. Levels of ^{137}Cs activity down to about 1 pCi per litre can be determined in seawater and lake, rain and river waters without sophisticated chemical processing.

Low-level γ-ray spectrometry with lithium drifted germanium detectors has been used to determine strontium-90 in seawater, and sediment samples [70].

The system is capable of unambiguous qualitative and quantitative analyses for minute amounts of γ-emitting nuclides in complex mixtures. It incorporates a $42\,cm^3$ Ge(Li) crystal as the main detector; this is operated in anti-coincidence with a $40 \times 40\,cm$ plastic scintillator device and the whole assembly is enclosed in a lead shield (10 cm walls). When the instrument is operated in the anti-coincidence mode, the background continuum is reduced to 0.3 c.p.m. per keV at 100 keV and to less than 0.005 c.p.m. per keV at 1000 keV for an average reduction of normal background of 99.5%. The resolution of the system is 3.0–3.5 keV (f.w.h.m.) depending on the γ-energy. Many radionuclides can be detected in environmental samples at levels of less than 0.02 pCi per g without preliminary chemical separation. Accuracy of counting is improved up to three-fold by using the anti-coincidence mode, depending on the initial peak-to-background ratio. Results of measurements of ^{137}Cs on several samples of biota, sediment and seawater are given.

A further method for the determination of caesium isotopes in saline waters [71] is based on the high selectivity of ammonium cobalt ferrocyanide for caesium. The sample (100–500 ml) is made 1 M in hydrochloric acid and 0.5 M in hydrofluoric acid, then stirred for 5–10 min with 100 mg of the ferrocyanide. When the material has settled, it is collected on a filter (pore size 0.45 μm), washed with water, drained dried under an infra-red lamp, covered with plastic film and β-counted for ^{137}Cs. If ^{131}Cs is also present, the γ-spectrometric method of Yamamoto [72] must be used. Caesium can be determined at levels down to $10\,pCi\,l^{-1}$.

Mason [73] has described a rapid method for the separation of caesium-137 from a large volume of seawater. In this procedure the sample (50 litres) is adjusted to pH 1 with nitric acid and ammonium nitrate (100 g) and caesium chloride (30 mg) added as carrier. A slurry is prepared of ammonium molybdophosphate (7.5 g) and Gooch-crucible asbestos (715 g) with 0.01 M ammonium nitrate and deposited by centrifugation on a filter paper fitted in the basket of a continuous-flow centrifuge. The sample is centrifuged at 600–3000 r.p.m. and the deposit washed on the filter with 1 M ammonium nitrate (60–70 ml) and 0.01 M nitric acid (30–40 ml). The caesium collected on the filter is then prepared for counting by the method of Morgan and Arkell [74]. With this method the caesium can be extracted in less than 1 h.

Iron-55

Testa and Staccioli [65] used Microthene-710 (microporous polyethylene) as a support material for bis-(2-ethylhexyl) hydrogen phosphate in the determination of ^{55}Fe in environmental samples.

Ruthenium-106

Kiba *et al.* [75] has described a method for determining this element in marine sediments. The sample is heated with a mixture of potassium dichromate and condensed phosphoric acid (prepared by dehydrating phosphoric acid at 300°C). The ruthenium is distilled off as RuO_4, collected in 6 M hydrochloric acid–ethanol and determined spectrophotometrically (with thiourea) or radiometrically. Osmium is separated by prior distillation with a mixture of condensed phosphoric acid and $Ce(SO_4)_2$. In the separation of ruthenium-osmium mixtures, recovery of each element ranged from 96.8 to 105.0%.

Cobalt-60

Hiraide, Sakurai and Mizuike [76] used continuous flow co-precipitation–flotation for the radiochemical separation of cobalt-60 from seawater. The cobalt-60 activity was measured by liquid scintillation counting with greater than 90% yield and a detection limit of 5 fCi per litre seawater.

Tseng, Hsieh and Yong [77] determined cobalt-60 in seawater by successive extractions with tris(pyrrolidine dithiocarbamate) bismuth (III) and ammonium pyrrolidine dithiocarbamate and back-extraction with bismuth (III). Filtered seawater adjusted to pH 1.0–1.5 was extracted with chloroform and 0.01 M tris(pyrrolidine dithiocarbamate) bismuth (III) to remove certain metallic contaminants. The aqueous residue was adjusted to pH 4.5 and re-extracted with chloroform and 2% ammonium pyrrolidine thiocarbamate, to remove cobalt. Back-extraction with bismuth (III) solution removed further trace elements. The organic phase was dried under infrared and counted in a germanium/lithium detector coupled to a 4096 channel pulse height analyser. Indicated recovery was 96%, and the analysis time excluding counting was 50 minutes per sample.

Manganese-54

Flynn [78] has described a solvent extraction procedure for the determination of manganese-54 in seawater in which the sample with bismuth, cerium and chromium carriers, is extracted with a heptane solution of bis(2-ethylhexyl) phosphate and the manganese back-extracted with 1 M hydrochloric acid. After oxidation with nitric acid and potassium chlorate, manganese is determined spectrophotometrically as permanganate ion.

Manganese-54 and zinc-65

Manganese-54 has also been determined by a method [79] using co-precipitation with ferric hydroxide. The precipitate is boiled with hydrogen peroxide and the iron is removed by extraction with isobutyl methyl ketone. Zinc is separated on an anion-exchange column and manganese is separated by oxidizing it to permanganate in the presence of tetraphenylarsonium chloride and extracting the resulting complex with chloroform. Both ^{65}Zn and ^{54}Mn are counted with a 512-channel analyser with a well-type NaI(T1) crystal (3 × 3 in). Recoveries of known amounts of ^{65}Zn and ^{54}Mn were between 74% and 84% and between 69% and 74% respectively.

References

1. Spencer, R. *Talanta*, **15**, 1307 (1968)
2. Bertine, K.K., Chan, L.H. and Turekian, K.K. *Geochimica Cosmochimica Acta*, **34**, 641 (1970)
3. Hashimoto, T. *Analytica Chimica Acta*, **56**, 347 (1971)
4. Kim, Y.S. and Zeitlin, H. *Analytical Chemistry*, **43**, 1390 (1971)
5. Williams, W.J. and Gillam, A.H. *Analyst (London)*, **103**, 1239 (1979)
6. Smith, J. and Grimaldi, O. *Bulletin of the U.S. Geological Survey*, **1006**, 125 (1957)
7. Leung, G., Kim, Y.S. and Zeitlin, H. *Analytica Chimica Acta*, **60**, 229 (1972)
8. Kim, Y.S. and Zeitlin, H. *Analytical Abstracts*, **22**, 4571 (1972)
9. Korkisch, J. and Koch, W. *Mikrochimica Acta*, **1**, 157 (1973)
10. Korkisch, J. *Mikrochimica Acta*, 687 (1972)

11. Barbano, P.G. and Rigoli, L. *Analytica Chimica Acta*, **96**, 199 (1978)
12. Kim, K.H. and Burnett, W.C. *Analytical Chemistry*, **55**, 796 (1983)
13. Bowie, S.H.U. and Clayton, C.G. *Translations of the Institution of Minerals and Metals B*, **81**, 215 (1972)
14. Shannon, L.V. and Orren, M.J. *Analytica Chimica Acta*, **52**, 166 (1970)
15. Nozaki, Y. and Tsunogai, S. *Analytica Chimica Acta*, **64**, 209 (1973)
16. Cowen, J.P., Hodge, V.F. and Folson, T.R. *Analytical Chemistry*, **49**, 494 (1977)
17. Tsunogai, S. and Nozaki, Y. *Geochemical Journal*, **5**, 165 (1971)
18. Shannon, L.V., Cherry, R.D. and Orren, M.J. *Geochimica and Cosmochimica Acta*, **34**, 701 (1970)
19. Hodge, V.F., Hoffman, F.L. and Folsom, T.R. *Health Physics*, **27**, 29 (1974)
20. Folsom, T.R. and Hodge, V.R. *Marine Science Communications*, **1**, 213 (1975)
21. Folsom, T.R., Hodge, V.F. and Gurney, M. *Marine Science Communications*, **1**, 39 (1975)
22. Goldberg, E.D., Koide, M. and Hodge, V.F. Scripps Institution of Oceanography, La Jolla, CA, USA, (1976)
23. Flynn, A. *Analytical Abstracts*, **18**, 1624 (1970)
24. Reid, D.F., Key, R.M. and Schink, D.R. *Earth Planet Science Letters*, **43**, 223 (1979)
25. Nozaki, Y. and Tsunogai, S. *Analytica Chimica Acta*, **64**, 209 (1973)
26. Huh, C.A. *Analytical Chemistry*, **57**, 2138 (1985)
27. Bacon, M.P. and Anderson, R.F. Trace metals in sea water. *In Proceedings of a NATO Advanced Research Institute on trace Metals in Sea water, 30/3–3/4/81, Sicily, Italy,* (eds C.S. Wong, *et al.*) Plenum Press, New York, p. 368 (1981)
28. Spencer, D.W. and Sachs, P.L. *Marine Geology*, **9**, 117 (1970)
29. Krishnaswami, S., Lal, D., Somayajulu, B.L.K., Weiss, R.F. and Craig, H. *Earth Planet Science Letters*, **32**, 420 (1976)
30. Anderson, R.F. *The Marine Geochemistry of Thorium and Protactinium*, PhD dissertation, Massachusetts Institute of Technology/Woods Hole Oceanographic Institution, WH01-81-1 (1981)
31. Perkins, R.W. *Report of The Atomic Energy Commission, US*, BNWL 1051 (Pt. 2) pp. 23–27 (1969)
32. Moore, W.S. and Reid, D.F. *Journal of Geophysical Research*, **78**, 8880 (1973)
33. Moore, W.S. *Deep Sea Research*, **23**, 647 (1976)
34. Moore, W.S. and Cook, L.M. *Nature (London)*, **253**, 262 (1975)
35. US Environmental Protection Agency *Radiochemical Methodology for Drinking Water Regulations*, EPA 600/4-75-005 (1975)
36. Ku, T.L., Huh, C.A. and Chen, P.S. *Earth Planet Science Letters*, **49**, 293 (1980)
37. Chung, Y. *Earth Planet Science Letters*, **49**, 319 (1980)
38. Moore, W.S. *Estuarine, Coastal Shelf Science*, **12**, 713 (1982)
39. Reid, D.F., Key, R.M. and Schink, D.R. *Earth Planet Science Letters*, **43**, 223 (1979)
40. Michel, J., Moore, W.S. and King, P.T. *Analytical Chemistry*, **53**, 1885 (1981)
41. US Environmental Protection Agency. *National Interim Primary Drinking Water Regulations*, EPA-57019-79-003 (1976)
42. Key, R.M., Brewer, R.L., Stockwell, J.H., Guinasso, N.L. and Schink, R.D. *Marine Chemistry*, **7**, 251 (1979)
43. Broecker, W.S. An application of natural radon to problems in oceanic circulations. In *Proceedings of the Symposium on Diffusion in the Oceans and Fresh Waters*, Lamont Geological Observatory, New York, pp. 116–145 (1965)
44. Moore, W.S. Sampling ^{228}Ra in the deep ocean. *Deep-Sea Research*, **23**, 647 (1976)
45. Reid, D.F., Key, R.M. and Schink, D.R. Radium extraction from sea water; efficiency of manganese impregnated fibers. *EOS Transactions of The American Geophysical Union, December* (1974)
46. Schink, D., Guinasso, N., Jr., Charnell, R. and Sigalove, J. Radon profiles in the sea–a measure of air-sea exchange. *IEEE Transactions in Nuclear Science*, **NS-17**, 184–190 (1970)
47. Chung, Y. *Pacific Deep and Bottom Water Studies Based on Temperature, Radium and Excess-radon Measurements*, Dissertation, University of California, San Diego (1971)
48. Lucas, H.F. *Review Scientific Instruments*, **28**, 680 (1957)
49. Comar, C.L. *Plutonium; Facts and Interferences*, EPRI EA-43-SR (1976)
50. Livingston, H.D., Mann, D.R. and Bowen, V.T. *Analytical Methods in Oceanography. Advances in Chemistry*, Series No. 147, A.C.S., p. 124 (1975)
51. Wong, K.M. *Analytica Chimica Acta*, **56**, 355 (1971)
52. Pillai, K.C., Smith, R.C. and Folsom, T.R. *Nature (London)*, **203**, 568 (1964)
53. Ballestra, S., Holm, E. and Fukai, R. Presented at *The Symposium on the Determination of Radionuclides in Environmental and Biological Materials*, Central Electricity Generating Board, London, October 1978

54. Holm, E. and Fukai, R. *Talanta*, **24**, 659 (1977)
55. Sakanous, M., Nakamura, M. and Imai, T. *Rapid Methods for Measuring Radioactivity in the Environment*. Proceedings of the Symposium, Neuherberg, IAEA, Vienna, p. 171 (1971)
56. Statham, C. and Murray, C.N. *Report of the International Committee of the Mediterranean Ocean*, **23**, 163 (1976)
57. Hampson, B.L. and Tennant, D. *Analyst (London)*, **98**, 873 (1973)
58. Levine, H. and Lamanna, A. *Health Physics*, **11**, 117 (1965)
59. Aakrog, A. *Reference Methods of Marine Radioactivity Studies II, Technical Report Services*, No. 169, IAEA, Vienna (1975)
60. Chu, A. *Analytical Abstracts*, **22**, 427 (1972)
61. Livingston, H.D., Mann, D.R. and Bowen, U.J. *Report of The Atomic Energy Commission, US*. COO-3563-12 Woods Hole Oceanographic Institute, Massachusetts, U.S.A. (1972)
62. Delle Site, A., Marchionni, V. and Testa, C. *Analytica Chimica Acta*, **117**, 217 (1980)
63. Testa, C. and Delle Site, A. *Journal of Radioanalytical Chemistry*, **34**, 121 (1976)
64. Anderson, R.F. and Floor, A.I. *Analytical Chemistry*, **54**, 1142 (1982)
65. Testa, C. and Staccioli, L. *Analyst (London)*, **97**, 527 (1972)
66. Hirose, K. and Sugimura, Y.J. *Radioanalytical and Nuclear Chemistry Articles*, **92**, 363 (1985)
67. Silant'ev, A.N., Chumichev, U.B. and Vakulouski, S.M. *Trudy Inst. eksp. Met. glav. uprav, gidromet, Sluzhty Sov. Minist. SSSR*, **15**, (2) (1970) Ref: Zhur Khim, 19GD (1) Abstr. No. 1 G209 (1971)
68. Gordon, C.M. and Larson, R.E. *Radiochemical and Radioanalytical Letters*, **5**, 369 (1970)
69. Dutton, J.W.R. *Report of The Fisheries and Radiobiological Laboratory*, FRL 6 Ministry of Agriculture, Fish and Food, UK (1970)
70. Lewis, S.R. and Shafrir, H.N. *Nuclear Instrumental Methods*, **93**, 317 (1971)
71. Janzer, V.J. *Journal of Research of the U.S. Geological Survey*, **1**, 113 (1973)
72. Yamamoto, O. *Analytical Abstracts*, **14**, 6669 (1967)
73. Mason, W.J. *Radiochemical and Radioanalytical Letters*, **16**, 237 (1974)
74. Morgan, A. and Arkell, O. *Health Physics*, **9**, 857 (1963)
75. Kiba, T., Terada, K., Kiba, T. and Suzuki, K. *Talanta*, **19**, 451 (1972)
76. Hiraide, M., Sakurai, K. and Mizuike, A. *Analytical Chemistry*, **56**, 2851 (1984)
77. Tseng, C.L., Hsieh, Y.S. and Yong, M.H. *Journal of Radioanalytical and Nuclear Chemistry Letters*, **95**, 359 (1985)
78. Flynn, W.W. *Analytica Chimica Acta*, **67**, 129 (1973)
79. Stah, S.M. and Rao, S.R. *Current Science (Bombay)*, **41**, 659 (1972)

Chapter 6

Sample preparation prior to analysis for organics

Methods of identifying and measuring organic compounds have improved greatly in the past few decades. The development of the various kinds of instrumental chromatography has made identification of components of complex organic mixtures a matter of routine laboratory procedure, and the coupling of the gas chromatograph to the mass spectrometer has produced a tool of almost unbelievable power and versatility, one that will, in the long run, enable us to unravel even the most complex mixtures.

The concentration of organic materials in seawater is too low to merit direct utilization of many of the modern analytical instruments; concentration by a factor of a hundred or more is necessary in many instances. Furthermore, the water and inorganic salts interfere with many of the analytical procedures. Separation of the organic components from seawater therefore accomplishes two purposes; it removes interfering substances, and at the same time concentrates enough organic matter to make analysis possible. It is not surprising that considerable effort has been put into methods of separation and concentration.

Even after the organic compounds have been concentrated and separated from seawater, the resulting mixture may be too complex for the analytical method we wish to employ. If we have collected a large enough sample of organic material, we may then resort to any of a number of fractionation schemes, based perhaps on functional groups or molecular weights, in order to simplify the mixtures to the point where analysis could be relatively straightforward. Alternatively, if we wish to measure only one compound or class of compounds, we may try to design a concentration method which will be specific for the compounds of interest, thus achieving concentration and fractionation in one step. Both of these approaches have been followed with some success.

There is an extensive literature on concentration, separation and fractionation methods. References [1–5] give general reviews.

The simplest approach to the collection and subdivision of organic materials in seawater is to use some physical or chemical means of removing one fraction from solution or suspension. The techniques vary, from simple filtration to collect particulate matter, to chemical methods, such as solvent extraction and co-precipitation. With each of these methods, the analyst must know the efficiency of collection and exactly which fraction is being collected. Very often the fraction is defined by the method of collection; two methods that purport to collect the same fraction may in fact be sampling very different universes. Comparisons between the results of different investigators are usually difficult because we do not know whether differences in results stem from real differences in the areas or times of collection, or from differences in the methods used. Intercalibration of analytical methods is generally agreed to be a good thing, once a field has become sufficiently stabilized so that certain methods are more or less recognized as 'standard' methods. It is not so often realized that the intercalibration should start not with the analytical methods, but with the sampling methods, if true comparisons are to be made.

6.1 The particulate fraction

Seemingly, one of the simplest organic fractions to separate and measure should be the particulate fraction. The amount of material that can be collected is limited primarily by the clogging of the filter used for collection, so there should be no limitation on analytical methods due simply to sample size. One would normally consider, then, that once some agreement had been reached on what the minimum size of a particle would be, agreement would quickly be reached on filter sizes, and results from the various laboratories would be strictly comparable. However, this has not proved to be the case.

In the first place, it is generally agreed that the distribution of particle sizes in the oceans is continuous, from the whale to the simple single molecule [6, 7]. The size at which one calls an aggregate of molecules a particle is therefore arbitrary. In the case of seawater, the dividing line between dissolved and particulate has been chosen as $0.45\,\mu m$, largely because the first commercially available membrane filters had that as their pore size.

(a) Filtration

The acceptance of $0.45\,\mu m$ as the dividing line is purely nominal, since few workers in the field actually use filters with this pure size in making determinations of particulate organic carbon (POC) content. The glass-fibre filters used by many workers have pore sizes which are considerably larger, ranging from $0.7\,\mu m$ for Whatman GF/F to $1.32\,\mu m$ for GF/C. With these filters, all particles larger than the nominal pore size are retained, but many smaller particles are also trapped. The silver filters also in common use have a more uniform pore size, but the pore sizes as quoted are largely illusory. The filters, and more particularly the $0.45\,\mu m$ size, contain relatively large and variable amounts of carbon, which must be removed by combustion. After this combustion, the pore sizes are considerably enlarged, with the $0.45\,\mu m$ filter approaching $0.8\,\mu m$ in pore size. The nominal $0.8\,\mu m$ pore size filter is used by many investigators because the pore size changes very little under heat treatment. Thus although $0.45\,\mu m$ has been accepted as the minimum size for particulate matter by definition, the filters actually used have a somewhat larger pore size, and retain particles which are considerably smaller than the nominal cut-off size [8].

The choice of filter can determine the amount of material considered as particulate, sometimes with unexpected results. Thus the Whatman GF/C filter, with its larger pore size, actually retains about three times as much particulate organic carbon as does the $0.8\,\mu m$ silver filter. Presumably the difference results from the larger number of small particles retained by the glass-fibre filters.

The method of calculation of the blank can also influence the particulate organic carbon values. If surface seawater is filtered through a pad consisting of two or more filters, either glass fibre or silver, the bottom filter will often contain a small amount of organic carbon over and above the blank value. Some workers have maintained that the presence of this carbon is due to the adsorption of dissolved organic matter to the filter and that this value should therefore be subtracted from the particulate organic carbon value for the top filter [9, 10]. Other workers feel that the material caught in the second filter is largely composed of smaller particles passing through the first filter. Depending upon the way in which the particulate fraction is defined, the material caught by the second filter should either be added to that caught on the first [11] or ignored [12–14].

The choice of blank calculation can cause a considerable difference in the final values given for particulate organic carbon, at least in the surface waters. In as much as the particulate fraction is defined not in terms of particle size, but in terms of material caught on a specific filter, it is recommended that only one filter, rather than a pad of two or more, be used, since the material caught on subsequent filters is irrelevant, by definition.

When uniform methods of collection and analysis are used, the deeper layers of the oceans give remarkable consistent results. Replicate samples, taken with a Niskin rosette sampler rigged to close six 5 litre bottles simultaneously, displayed a standard deviation of $\pm$ 1.3 µg C/per litre [14].

Methods that collect a greater proportion of the smaller particles have also been employed. For example, a layer of fine inorganic particulate matter deposited on a filter of coarser porosity has been used to separate the particulate from the dissolved fraction. Thus Fox, Oppenheimer and Kittredge [15] used layers of $Ca(OH)_2$ and $Mg(OH)_2$, while Ostapenya and Kovalevskaya [16] used powdered glass. These filters suffered from three disadvantages: they were troublesome to construct, the nominal pore size was irreproducible, and adsorption of truly dissolved material was possible. The techniques were abandoned with the advent of the first membrane filters having graduated pore sizes.

Some work has been done on size fractionation of particulate matter by the use of graduated filters. Since the filters in common use do not display a sharp cut-off in particle size retention, interpretation of the results is difficult. Repeated filtration of a single sample through filters of different pore size does not divide the particulate matter into definite size classes, since each filter retains particles smaller than the nominal pore size. The results of the filtration of separate aliquots through a series of filters can only be reported in terms of 'particles smaller than' the nominal pore size and are equally difficult to interpret. Although such size fractionation has been reported [17] the conclusions can only be accepted in the broadest possible sense. Particle size distributions based on filtration should be supported by Coulter counter data before any strong conclusions are drawn.

We must also consider that the collection methods normally used are biased towards those particles falling very slowly. If the residence time of a particle in the water column is only a few days, the probability of being caught in a 5 litre Niskin bottle is vanishingly small. This has been pointed out by the work of Bishop *et al.* [18]. These investigators used an *in situ* pump and filtration apparatus to filter very large quantities ($5–30 m^3$) of seawater and caught many classes of particles never seen in Niskin bottle samples. Their results are not comparable with any other filtration work.

Mercury concentrations currently measured in some oceanic waters (North Sea)

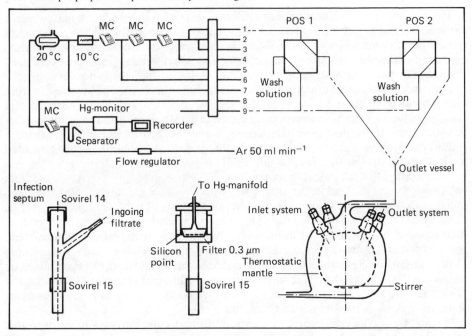

Figure 6.1 Schematic of system for determination of adsorption rates. 1 = Sample, 3.9 ml min^{-1}; 2 = H$_2$O, 3.9 ml min^{-1}; 3 = air, 2 ml min^{-1}; 4 = HNO$_3$ (1 + 2) 0.42 ml min^{-1}; 5 = H$_2$SO4 (1 + 1), 0.42 ml min^{-1}; 6 = KMnO$_4$ (1%), 0.1 ml min^{-1}; 7 = NH$_4$OCl (1%), 0.23 ml min^{-1}; 8 = NaBH$_4$ (1%), 0.10 ml min^{-1}; 9 = waste, 3.90 ml min^{-1}; MC = mixing coil. POS 1, 2 = position 1, 2 of the switching valve (Reproduced from Balyens, W., Recadt, G, Dehairs, F, *Goeyeuse Oceanologica Acta,* **51**, 261, 1982.)

appeared much lower than those permitted by thermodynamical solid-dissolved equilibrium calculations [19]. The reason that seawater and estuarine water are so depleted in mercury, and according to Turekian [20] also in several other trace metals, is the role particulates play as sequestering agents for these elements [21–26]. Kinetics are, however, still poorly known. This is mainly due to the fact that adsorption rate measurements require the elaboration of an appropriate experimental device. Indeed, Lockwood and Chen [22] reported that in many of their runs, the rate was too fast to measure even by the most rapid solids separation methods available. Similar difficulties were encountered by Reimers and Krenkel [23]. They solved the problem by assuming the adsorption rates to be of zero order as a simple approximation.

Baeyens *et al.* [26] have described an automated auto analyser method for the assessment of mercury adsorption rates on particulate suspended matter in seawater (Figure 6.1).

(b) Centrifugation

A method for removing particles which is not limited in volume sampled and which suffers less from problems of overlapping classification is continuous-flow centrifugation. Separation into density classes can be achieved by choice of speed of centrifugation. At least theoretically a cascade system could be constructed which would collect colloidal material separately from the larger particles; practically, this has not been attempted, and batch processing has been used instead.

Centrifugation methods are better known for their potential than for their actual performance. Jacobs and Ewing [27] used continuous centrifugation to collect total suspended matter in the oceans, and Lammers [28] discussed the possible uses of the method. The biggest drawback seems to be the separation by density, rather than the more usual separation by particle size; no one has yet been interested in such a separation. However, the method holds considerable promise for the collection of colloidal material as a separate fraction.

(c) Dialysis and ultrafiltration

Variations on the theme of filtration, what might be called filtration at the molecular level have been used for the concentration of the larger organic molecules in seawater; these techniques include dialysis and ultrafiltration. Dialysis as a method for desalting is a common biochemical technique. What has made the technique attractive for the concentration of organic materials is the advent of pressure filtration. In this technique, both water and the organic and inorganic smaller compounds, are forced through the dialysis membrane under pressure. With the judicious addition of distilled water, the solution can be both concentrated and desalted. In the past few years, a large number of devices designed to simplify and automate this process have been marketed. They have not yet been appreciably exploited for oceanographic studies.

The membranes available for these devices range from hollow fibres which permit fixed gasses to escape from solutions [29] to what are essentially very fine filters, capable of trapping particles of colloidal size. To date, the greatest use of these techniques has been in fractionation, rather than in concentration, although their greatest potential is probably in concentration from solution.

6.2 The soluble fraction

(a) Reverse osmosis

This technique has been applied to the concentration of organochlorine and organophosphorus insecticide [30, 31] and various ethers, glycols, amines, nitriles, hydrocarbons and chlorinated hydrocarbons. Although this work was concerned with drinking water, it is a useful technique which may have applications in seawater analysis. Cellulose acetate [33], ethyl cellulose acetate [30] and cross-linked polyethyleneinine [32] were used as semi-permeable membranes.

(b) Freeze drying

This is another technique which has applications in seawater analysis. Approximately 100% recovery of glucose and lindane at the 0.1 and 0.15 mg l^{-1} level have been obtained from water by this technique.

Pocklington [34] has separated amino acids in seawater using this technique. The first step in his concentration of free amino acids from seawater was the freeze drying of the seawater sample. To reduce interferences in the later steps of the procedure, the sea salts were packed into a chromatographic column and washed with diethyl ether to remove non-polar compounds. The diethyl ether extract, particularly from surface water samples, quite often contained coloured materials as well as other organics. If a series of solvents of graduated polarity were passed through a sea salt column, a fractionation by polarity should be obtained. With the proper choice of solvents, a form of gradient elution could be devised which would result in a continuous, rather than a batch fractionation.

(c) Freezing out methods

Slow freezing, with constant stirring, results in a concentration of organic materials in the solution remaining. The technique is most effective in water of low salinity. It has been applied to lake water with some success but marine applications do not seem to have been developed [35–38].

(d) Froth flotation

In its passage through a water column, a bubble acts as an interface between the liquid and vapour phases, and as such collects surface-active dissolved materials as well as colloidal micelles on its surface. Thus in a well-aerated layer of water, the upper levels will become progressively enriched in surface-active materials. In the open ocean, an equilibrium undoubtedly exists between the materials carried downward by bubble injection from breaking waves and those carried upward by rising bubbles. In the laboratory, however, this effect may behold to enrich the surface layer with organic materials.

When the concentrations of surface-active materials are high, the injection of bubbles into the solution from well below the surface may result in the formation of a foam at the surface. The foam can be as much as 200 times as concentrated in organic material as the body of the solution [39]. Natural foams of this kind can often be seen along beaches during periods of strong winds and violent wave action. Even when the surface-active materials are not present at levels high enough for foam formation, a considerable enrichment of organic material can be found in the upper part of the water column [40, 41].

When bubbles break at the sea surface, some portion of the organic materials collected on the bubble surface is ejected into the air. The phenomenon is well known [42] but has only recently been considered as a possible method for the collection of surface-active organic materials. MacIntyre [43] suggested that adjustment of the bubble size and the depth at which the bubbles were formed might be used to control the thickness of the surface layer sampled. The collection of the charged particles ejected by bursting bubbles can be enhanced by the use of a charged glass plate as a collector [44]. While most of the bubble ejection work has been done in the laboratory, a surface microlayer sampler using this principle has been devised for work in protected inlets [45]. The literature on bubble ejection and collection has been reviewed by Blanchard [46].

(e) Solvent extraction

Solvent extraction is another attractive method for concentrating a particular fraction of the dissolved organic matter, the fraction concentrated being determined by the choice of solvent. The most obvious limitation on the method is that set by solvent choice; the solvent should have only limited solubility in water, which limits the materials removed to the less polar compounds. Another limitation is that set by contamination. The solvent used must be purified carefully, since the amounts of the various organic compounds collected from seawater will be about as large as the trace impurities in the solvents. Because of the relatively large amount of time and equipment required for the processing of each sample, these methods must be used to characterize the organic compounds at a few selected stations and depths, rather than in mass surveys. Once the separation into the organic solvent has been accomplished, any of a number of techniques of fractionation and analysis can be applied.

The first step in the solvent extraction is the actual sampling of the water column. All of the problems associated with sampling can occur in this step, and may be aggravated by the large volumes of seawater customarily employed. The cleverest approach to this problem is to avoid sampling in the normal manner altogether. Ahnoff and Josefsson [47] have described an *in situ* apparatus for solvent extraction. This apparatus is buoyed at the sampling depth, anywhere between the surface and 50 m, and water is pumped through a series of extraction chambers. The capacity of the unit is 50 litres per 48 hrs. The use of an *in situ* pumping system on the far side of the extraction chambers eliminates the pump and hose contamination, as well as much of the contamination coming from the passage through the surface film. Since the apparatus is battery powered, the unit may be suspended from a free-floating surface buoy; no ship time is required, except for placement and recovery of the samplers. Thus while each sample may require up to 48 h to collect, a number of depths and areas may be sampled in the same time period.

Since the non-polar organic content of seawater is fairly low, identification of unknown compounds requires the processing of large quantities of seawater. Where possible, continuous extraction should be favoured over batch extraction. Several workers have developed such systems; Goldberg, Delong and Sinclair [48] designed and evaluated a system for fresh water. This apparatus seems fragile, and would require considerable re-design before it could be used routinely on shipboard. Ahnoff and Josefsson [49] built a solvent extraction apparatus for river work which was later modified into their *in situ* extractor [50]. The unit as described in the earlier work could easily be adapted for seawater analysis. A unit based on a Teflon helix liquid–liquid extractor, some 332 feet (101.5 m) in length, was constructed by Wu and Suffet [51]. The extractor was optimized for the removal of organophosphorus compounds, specifically pesticides, with an efficiency of around 80%. For some compounds, these continuous extraction methods should be the methods of choice, and should be further explored. These compounds are the ones that are not seriously affected by hose or pump contamination, are present in low concentrations, and are important enough, for either scientific or practical reasons, to warrant the extra time spent on station.

The conventional approach to solvent extraction is the batch method. Early work with this method was hampered by the low concentrations of the compounds present and the relative insensitivity of the methods of characterization. Thus lipids and hydrocarbons have been separated from seawater by extraction with petroleum ether and with ethyl acetate. The fractionation techniques included column and thin-layer chromatography with final characterization by the in-layer chromatography, infrared and ultra-violet spectroscopy and gas chromatography. Of these techniques, only gas chromatography is really useful at the levels of organic matter present in seawater. With techniques available today such as glass capillary gas chromatography and mass spectrometry, much more information could be extracted from such samples [52].

The information could be restricted to a tractable amount by performing some preliminary fractionation before the gas chromatography–mass spectrometry step.

This type of separation and fractionation has been proposed by Copin and Barbier [53].

The Oil Companies International Study Group for Conservation of Clear Air and Water–Europe (Concawe) [54] have made a detailed study of the application of solvent extraction to the determination of organics in water. In this procedure, one portion of the aqueous sample is adjusted to pH 11 and extracted with pure methylene chloride. The methylene dichloride extract containing basic and neutral substances is examined by gas chromatography coupled to a mass spectrometer. The alkaline aqueous phase is acidified and extracted with methylene dichloride to provide an extract containing phenols which, again, is examined by gas chromatography–mass spectrometry. A second portion of the water sample is acidified and extracted with methylene dichloride to provide a further phenol-containing extract for examination by high performance liquid chromatography.

A third portion of the water sample is extracted with methylene dichloride and examined for polynuclear aromatic hydrocarbons by high performance liquid chromatography using an ultraviolet/fluorescence detector and by gas chromatography using a flame ionization detector.

Solvent extraction has proved to be most useful when applied to the concentration of particular compounds for which there exists an analytical method of great sensitivity. The major application of the method has been for the determination of hydrocarbons in seawater.

In general, solvent extraction is an excellent method for the concentration and determination of specific compounds, chiefly non-polar, in seawater [55]. Special precautions must be taken to prevent contamination from trace materials in the solvents used. *In situ* methods offer many advantages, not the least being the elimination of lengthy processing in the shipboard laboratory; these methods should be investigated more thoroughly and, if possible, extended to greater depths. When coupled to modern separation and detection systems, the methods may offer us the simplest and most direct approach to the measurement of certain classes of compounds. In most cases, we have little or no estimate of the efficiency of the solvent extraction techniques. Because of the great variety of compounds present in any one sample, a true efficiency of extraction may be impossible to obtain. Working efficiencies, using model compounds, may be the only approach in trying to make the analysis truly quantitative.

(f) Co-precipitation techniques

Another method of segregating and removing a portion of the dissolved organic matter includes incorporation into a solid phase, which can then be removed by filtration or centrifugation. This incorporation can result either from co-precipitation with a solid phase formed in a reaction in the solution or, as discussed in the next section, from adsorption of the organic material on to a pre-existing solid phase.

The adsorption of organic matter on any surface presented to seawater has been well documented. Neihof and Loeb [56] have demonstrated this adsorbance by following the change in surface charge of newly immersed surfaces. There has even been an attempt to use this phenomenon as a means of measuring dissolved organic carbon. Chave [57] found an association between calcite and dissolved organic materials in seawater, and Meyers and Quinn [58] tried to use the effect as a method for the collection of fatty acids. As a collection technique, adsorption on calcite has several advantages. The pH of the sample is not greatly altered by the addition of small amounts of calcite; the precipitate is dense and should settle quickly; and after filtration the inorganic support can be removed by acidification. Unfortunately, the recovery of added fatty acids was inefficient; of the order of 18%. Meyers and Quinn [59] achieved a somewhat greater efficiency of collection of fatty acids with clays, but the insolubility of the clays nullified one of the advantages of this concentration technique.

The precipitant most commonly used for the collection of organics has been what is loosely called 'ferric hydroxide'. It is formed by the *in situ* formation of hydrated ferric oxides, usually by the addition of ferrous iron, followed by potassium hydroxide. The technique was first used for the precipitation of organic matter from an aged algal culture [60]. They recovered 79–95% of the ^{14}C-labelled material from such cultures. Williams and Zirono [61] measuring efficiencies of removal of DOC, found that such scavenging collected between 38% and 43% of the organic carbon measurable by wet oxidation with persulphate. Chapman and Rae [62] examined the effect of this precipitation on specific compounds. They found co-precipitation to be more complete with copper hydroxides, but still far from satisfactory. Only certain compounds were removed effectively by this treatment and the efficiency of removal varied with the water type and the organic compound involved.

A limited amount of work has been carried out using zirconium phosphates, compounds with well-defined coagulation and adsorption properties. The efficiency of co-precipitation was about 70% for free amino acids and albumin.

These methods may prove useful in the qualitative analysis of organic compounds, once the selectivities of the precipitants are understood. The metallic oxides suffer from the disadvantage of producing a precipitate which is difficult to filter, while calcite and zirconium phosphates produce relatively well-mannered precipitates. Even when the efficiences of collection of various model compounds in seawater will be known, the immense variety of organic compounds in seawater will keep this technique largely qualitative.

(g) Adsorption techniques

Co-precipitation techniques, as discussed in the previous section, are basically batch processes operating on a limited amount of organic matter. When really large concentrations of organic matter are required, it is much more efficient, as discussed below, to hold the adsorptive material in a column, through which the seawater is passed. After the proper amount of water has passed through the column, the organic matter can be desorbed with the proper choice of solvents. If the proper sequence of solvents is employed, a rough fractionation of the adsorbed material can also result.

Table 6.1 Application of macroreticular resins to the analysis of non-saline waters

Resin	Type of water	Substances adsorbed	Reference
Macroreticular	Fresh water	Miscellaneous	73, 74, 75
Macroreticular	Natural water	Benzo(a)pyrene	76
XAD-2, XAD-4	Natural water	Benzoic acid p-methylbenzyl-alcohol, aniline, fulvic acid, acetic acid, sucrose, ethylene diamine, citric acid, quinoline, glycine, quinaldic acid	77
XAD-4	Waste water Drinking water	Alcohols, esters, ketones, aldehydes, alkylbenzenes, polynuclear hydrocarbons, phenols, chlorocompounds	78
XAD-2, XAD-7, Tenax GC	Water	Hydrocarbons	79
XAD-2	Drinking water	Alkanes Alkylbenzenes, haloforms, polynuclear hydrocarbons, ketones, alcohols, phenols, carboxylic acids	80
XAD-4	Drinking water Waste water	Alkanes, alkylbenzenes Chlorocompounds, alcohols, ketones, esters, bromocompounds	81
XAD-2, XAD-7		Alkylbenzenes, alkanes, phenols, carboxylic acids, chlorocompounds	82
Sulphonic acid type (Styrene-divinyl benzene matrix) Aminex A-5	Waste water	Phenols, carboxylic acids, aldehydes, alcohols	83
XAD-2	Water	Organochlorine and organophosphorus pesticides, triazines, chlrophenoxyacids, phthalate esters	84

One of the earliest choices of adsorbent for seawater organics was activated charcoal [60, 63]. The technique has been refined for use in both fresh and seawater by a number of workers [64, 67]. A major problem in this technique has been the unknown efficiences of collection and desorption. Jeffrey [68] using ^{14}C-labelled material, found that 80% of the organic material in the seawater was adsorbed by the charcoal. Of that 80% approximately 80% again was desorbed by the solvents used. The overall efficiency of the method was therefore about 64%. This method of collection is one of the few that permits the accumulation of gram amounts of organic materials from seawater and therefore also permits the application of many of the standard techniques of organic analysis. However, we do not know how the distribution of compounds is changed in the process of adsorption and desorption; certain classes of compounds are probably entirely removed from the mixture, while others may be retained only in part. When so active a surface is used, there is also a possibility that chemical changes may take place during adsorption. The mixture as released from the charcoal may be considerably different from that originally present in seawater. This collection method, although attractive for its simplicity and speed, is limited to qualitative results.

Macroreticular resins have also been used for the collection of trace organics. An excellent early review of the properties of the various XAD resins, along with comparisons with EXP-500 and activated carbon, can be found in Gustafson and Paleos [69].

In fact, few references have been found concerning the application of this type of resin (XAD-1) to the determination of dissolved organic materials in seawater. Riley and Taylor [70] have studied the uptake of about 30 organics from seawater on to the resin at pH 2–9. At the 2–5 μg l^{-1} level none of the carbohydrates, amino acids, proteins or phenols investigated were adsorbed in any detectable amounts. Various carboxylic acids, surfactants, insecticides, dyestuffs and especially humic acids are adsorbed. The humic acids retained on the XAD-1 resin were fractionated by elution with water at pH 7, 1M aqueous ammonia and 0.2 M potassium hydroxide.

Osterroht [70, 72] studied the retention of non-polar organics from seawater on to macroreticular resins.

More applications have been found for the use of macroreticular resins in non-saline water (Table 6.1).

Each of the XAD resins has slightly different properties and should collect a slightly different organic fraction from seawater. The major differences between the resins is in the degree of their polarity.

The macroreticular resins should be useful in the analysis of particular classes of compounds; the stumbling block will be the determination of efficiences. Earlier experience, admittedly with early versions of the resins, was that the collection of organic materials from water was far from complete. Once the properties of the resins are well understood, the analysis of at least some classes of compounds may quickly become routine.

6.3 Volatile components

(a) Gas stripping

Whilst much of the literature on this subject is concerned with non-saline water samples, it is believed that many of these procedures will also work satisfactorily with seawater; indeed, the presence of salts in the sample may assist in the removal of volatiles.

In the collection of organic materials, separation of the more volatile materials can be achieved by transferring them into the vapour phase for collection and concentration. The usual method for effecting such a transfer is to bubble some inert gas through the liquid. The effluent gas is then passed through an adsorbent to collect the organic materials [85, 90]. The desorbed material is then analysed by gas chromatography or by linked gas chromatography–mass spectroscopy.

Material removed from the water by stripping in this manner can also be concentrated by trapping in a loop immersed in a cooling bath. The usual cooling baths are liquid nitrogen or solid carbon dioxide with or without an organic solvent. The major disadvantage of the liquid nitrogen bath, along with the cost and the limited availability, particularly aboard ship, is that it condenses carbon dioxide and water, along with the organic materials actually desired. These trapping techniques have been used by many workers and are usually described in conjunction with a gas chromatographic determination of some fraction of the organic materials. An example of this kind of separation is the work of Novak *et al.* [91]. They improved the yield of volatile organics by salting out the volatiles with sodium sulphate.

While the gases used in strippig are usually air, nitrogen or helium, electrolyti-cally evolved hydrogen has been used as a collector for hydrocarbons [92]. In this technique, the gas is not passed through a column of adsorbent, but instead collects in the headspace of the container. Since the volume of seawater and of hydrogen are known, the hydrocarbon concentration in the headspace can be used to calculate the partition coefficients and the concentration of hydrocarbon in the seawater.

As in most of the separation methods, the stripping and collection techniques have not been investigated for their recovery efficiences except in isolated instances. Kuo *et al.* [93] have found recovery efficiences for volatile polar organics ranging from 9.5% for ethanol to 88.8% for acetone. Most of the model compounds displayed recovery efficiencies in the 60–80% region under the conditions of their experiment.

The Oil Companies International Study Group for Conservation of Clean Air and Water–Europe (Concawe) has made a detailed study [54] of the application of gas stripping to the determination of hydrocarbons in amounts down to parts per billion in water. In this procedure the water sample is purged with nitrogen and helium and the volatiles trapped on a solid adsorbent such as activated carbon. The organics are then released from the carbon by heating and purging directly into a gas chromotograph linked to a mass spectrometer. For chlorine and bromine containing impurities a halide selective detector such as the Hall electrolytic conductivity detector is used on the gas chromatograph. Alternatively, an alkali flame ionization detector or an electron capture detector could be used.

Colenut and Thorburn [94, 95] have also described the procedure using gas stripping of the aqueous sample followed by adsorption to active carbon from which surface they are taken up in an organic solvent for gas chromatographic analysis. They optimized conditions for the determination of parts per billion of pesticides and polychlorinated biphenyls.

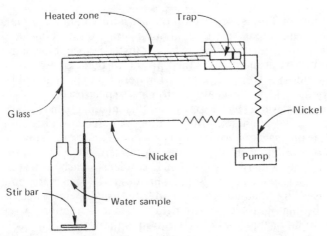

Figure 6.2 CLSA flow scheme (Reproduced from Stephen, S.I., *et al.* (1978) *Water Research* **12**, 447, Pergamon Publications, by courtesy of authors and publishers.)

The closed loop gas stripping system has been discussed by various workers [96–100]. In this technique organic compounds are removed from water by purging with a gas saturated with water vapour (Figure 6.2). Volatile and semi-volatile compounds will partition out into the headspace and are swept to an activated charcoal trap. The charcoal will retain organics while allowing the purge gas to pass through. The purge gas is then returned to repurge the sample via a pump. At the end of the purge time, typically 2 h, the trap is removed and fitted with a glass collection vial. Organic compounds are extracted from the charcoal with a small volume of a suitable solvent such as carbon disulphide or dichloromethane which is then collected and injected into a capillary gas chromatograph or a capillary gas chromatograph coupled with a mass spectrometer.

Grob and Zurcher [96–99] have carried out very detailed and systematic studies of the closed loop gas stripping procedure and applied it to the determination of parts per billion of 1-chloroalkanes in water. Westerdorf [100] applied the technique to chlorinated organics and aromatic and aliphatic hydrocarbons.

Waggott [101] reported that a factor of major concern in adapting the technique to more polluted samples is the capacity of the carbon filter which usually contains only 1.5–2 mg carbon. He showed that the absolute capacity of such a filter for a homologous series of l-chloro-n-alkanes was 6 μg for complete recovery. Maximum recovery was dependent on carbon number being at a maximum between C_8 and C_{12} for the 1-chloro-n-alkane series. It is important, therefore, to balance the amount of sample stripped with the capacity of the carbon filter to obtain better than 90% recoveries.

(b) Headspace analysis

Techniques of collection and separation of volatile organics using headspace gas analysis have been in the literature for many years; they have long been employed in the analysis of commercial products, particularly in the brewing industry. An application to water samples was described by Bassett, Ozeris and Whitnah [102]. Their application was relatively general, for trace volatiles in biological fluids. Cowen [103] used the method for the analysis of ketones and aldehydes in seawater. Halocarbons were similarly separated from environmental samples by Kaiser and Oliver [104]. There have been many other applications of the technique [105–114]. The major advantage of the headspace method is simplicity in handling the materials. At most, only one chemical, the salt used in the salting-out procedure, needs to be added and in most cases the headspace gas can be injected directly into a gas chromatograph or carbon analyser. On the other hand, the concentration of organic materials present is limited by the volume of seawater in the sample bottle. This is very much a batch process.

Equilibration between the headspace gas and the solution can take a considerable time. This is not a problem when the salting-out material is added at sea, and the samples are then brought into the laboratory for analysis some time later. When the salting-out is done in the laboratory, equilibration can be hastened by recirculating the headspace gas through the solution. A system could be devised which would permit the accumulation of volatiles from a large volume of water into a relatively small headspace, perhaps by recirculating both water and headspace gas through a bubbling and collection chamber, but much of the simplicity and freedom from possible contamination would be lost in the process.

Volatile organic materials can also be removed from solution by distillation, either at normal or at elevated pressures. While the amounts to be collected in this fashion are small, if headspace samples are taken at elevated temperatures and pressures trace quantities of organics can be detected [115]. It should be emphasized, however, that whenever extreme conditions are employed to free an organic fraction, that fraction is defined by the conditions of the separation and cannot profitably be compared with fractions defined by different sampling conditions. The use of elevated temperatures and pressures may also alter the compounds separated, limiting the amount of information that can be extracted from the analyses.

Friant and Suffet [111] have investigated in detail the interactive effects of temperature, salt concentration and pH on headspace analysis for isolating volatile trace organics in aqueous samples. Optimal conditions were derived from a statistical evaluation of the effect of parameter variation on the partition coefficient. These were a pH of 7.1, a sample temperature of 50°C and a salt concentration equivalent to 3.35 M sodium sulphate. Dowty, Green and Laseter [18] passed the headspace purge gas through a column of Tenax GC (poly (p-2,6-diphenyl-phenylene)oxide) adsorbent to trap the organics. The organics are then released from the Tenax GC and swept into a gas chromatograph for analysis in the parts per billion range. Bellar and Lichtenberg [110] also used the principle of adsorption of the organic components of the purge gas on a solid adsorbent material.

Chau *et al.* [114] point out that the Bellar and Lichtenbeng [110] procedure of gas stripping followed by adsorption on to a suitable medium and subsequent thermal desorption on to a gas chromatograph–mass spectrometer is not very successful for trace determinations of volatile polar organic compounds such as the low molecular weight alcohols, ketones and aldehydes. To achieve their required sensitivity of parts per billion, Chian *et al.* [114] carried out a simple distillation of several hundred ml of sample to produce a few ml of distillate. This achieved a concentration factor of between 10 and 100. The headspace gas injection–gas chromatographic method was then applied to the concentrate obtained by distillation.

6.4 Fractionation

Once a sample of dissolved organic matter has been isolated, it is still seldom in a form that permits simple analysis. In most cases, there are far too many compounds present and some form of fractionation must take place to remove interferences and simplify analytical procedures.

One could devise many different bases for the fractionation of organic materials, functional groups, degree of saturation, presence or absence of aromatic groups, and degree of polarity have all been used. The approach most often used is a fractionation by size. At the upper end of the size range we are dealing with particles consisting of many discrete molecules. Fractionation is accomplished by differential filtration, using filters and screens of decreasing pore size. A good example of the results of such fractionation is found in Sheldon [116].

Particles of smaller sizes, from the colloidal to the macromolecular, are separated by membrane filters. The most familiar of these is the Amicon Diflo filter, although several other companies now manufacture similar products.

Separations in the same size range can also be achieved with hollow polymeric fibres. At the upper end of their size range, these filters can be used to separate different size classes of material normally considered as colloidal. At the smaller end, the separation is made on the basis of molecular size. The results are presented in terms of molecular weight, but the molecular weight calibration is done with spherical molecules. The results are therefore given as equivalent spheres, rather than as true molecular weights. The techniques have been applied to coastal seawater. Ultrafiltration as a fractionation method gave recoveries of 80–100% when the carbon present in each fraction is summed.

Ultrafiltration techniques employing membrane filters and those using hollow fibres both worked well for the concentration and desalting of humic and fulvic acids, but the high priming volume needed for the hollow fibre apparatus restricts it to large volume applications. This is not likely to be a problem in marine work, where large volumes are required because of the low concentrations of organic materials. Both membranes and fibres retained material well below the expected molecular weight cut-off.

These techniques are only just coming into use in marine organic chemistry. The apparatus is now available for processing large quantities of seawater, at pilot plant levels, to yield gram quantities of dissolved organic materials in specified molecular size classes. This should be one of the most fruitful methods for accumulation, separation and rough fractionation of dissolved organic materials.

Separation into molecular size classes by ultra-filtration is necessarily discontinuous; the fractions resulting are composed of mixtures of compounds within a given band of molecular sizes. We would often prefer a continuous separation by molecular size, particularly if we suspect that the material in question might naturally fall into only a few fractions, each consisting of a tight grouping of molecular size or weights. This kind of fractionation is best carried out by some form of column chromatography. If molecular size is to be the criterion for separation, then materials such as Sephadex can be used as column packing. Sephadex separates compounds by exclusion, holding the smaller molecules within the particles and rejecting those that will not fit within the pores of the resin. Thus with a Sephadex column, the large components come off the column first. The system is not perfect; some charged compounds, such as phenols, can be bound irreversibly to some of the resins. The procedure has been used in the analysis of natural waters [117, 118] but it has not been developed to its full usefulness.

XAD resins have been used to collect and concentrate organic materials from seawater. They can also be used as packings for fractionation by column chromatography. While they have been used in simple gravity flow column chromatography, high pressure liquid chromatography has also been used [119, 120]. This technique is in its infancy; in marine chemistry, it offers many possibilities and a few major drawbacks. The first drawback is the lack of sensitivity of the detectors. If the compounds sought happen to be fluorescent, or can be made into fluorescent derivatives, the inherent sensitivity of fluorescence can be used. Otherwise, we are largely limited to the much less sensitive refractive index and ultraviolet absorbance detectors. There have been recent attempts to couple the sensitive gas chromatography detectors to the liquid chromatograph.

Ultimately the possibility is seen on combining liquid chromatography with mass spectrometry as a standard technique for identification of at least the compounds of lower molecular weight.

Workers who have examined the applicability of high performance liquid chromatography to water analysis include Waggott [121], Scott and Kucera [122], Snyder [123] and Engelhardt [124].

Reversed phase chromatography is a variant of high performance liquid chromatography where a non-polar organic phase is immobilized and a polar solvent is used as eluent. This variant may also be applied when non-polar compounds are to be sorbed from a polar solvent. This very situation is encountered in the attempt to accumulate non-polar organic substances from seawater by liquid–solid adsorption.

Activated charcoal was one of the first adsorbants used to accumulate dissolved organic material from water [125–126]. Its excellent adsorption properties are reflected by its extensive use for water purification. For more specific applications activated charcoal was later replaced by synthetic adsorbants such as macroreticular resins, e.g. Amberlite XAD [127–128] or polyurethane foam [129]. These adsorbants suffer from the drawback of being difficult to clean and of retaining traces of the material collected rather tenaciously. Therefore, these adsorbants have to be soxhlet-extracted in order to remove the sorbate which for all practical purposes eliminates the possibility of fractionated desorption.

In applying the principle of reversed-phase chromatography to the accumulation of dissolved organic material from water, Ahling and Jensen [130] used a mixture of n-undecane and Carbowax 4000 monostearate on chromosorb W as the collecting medium. Uthe and Reinke [131] tested porous polyurethane coated with liquid phases such as SE3, DC 200, QF-1, DEGS, OV-25, OV-225 for the same purposes. In each case the coating is achieved easily and may be modified to the desired adsorption properties. However, the coating is not chemically bonded to the support and may thus be removed together with the sorbate. Aue, Kapila and Hastings [132] finally demonstrated the potential of support-bonded polysiloxanes for a simple, fast and sensitive analysis of organochlorine compounds in natural aqueous systems.

Derenbach et al. [133] tested a technique for the accumulation of certain fractions of dissolved organic material from seawater; and subsequently for the fractionated desorption of the collected material.

The handling of water extracts and possible sources of contamination would thus be reduced to a minimum. Furthermore, fractionated desorption of the accumulated material under mild conditions should result in less complex mixtures with little risk of denaturation.

These workers investigated the suitability of numerous support materials for use in reversed-phase high performance liquid chromatography for the recovery of non-polar organic compounds from seawater. Porous glass treated with trichloro-n-octadecyl silane was found to permit at least a semi-quantitative recovery of test compounds. This silanized glass support was found to be easy to keep free from contamination and in addition, had a relatively high adsorption capacity, permitting fractional desorption of the test compounds. Results obtained with this column were compared with those obtained using Amberlite XAD-2. Darenbach et al. [133] give full details of the preparation of this support material. They used ^{14}C labelled spike compounds, $1-^{14}$C n-hexadecane and di(2-ethylhexyl) (carboxyl-^{14}C)phthalate) and also non-labelled compounds (C n.C_{16}–n C_{24} alkanes, diethyl, di-isobutyl, di-n-butyl, butylbenzyl dichyclohexyl, bisethylhexyl phthalic acid esters, p,p'DDE,DDMU,Dieldrin, Endrin pesticides) in recovery experiments.

Twenty-five litre samples of natural seawater were spiked and 5 litre subsamples of these were extracted with 20, then 10 ml pure n-hexane. The hexane phases were allowed to separate for 30–60 min. The combined extracts were dried over anhydrous sodium sulphate, reduced in volume with a rotary evaporator at 40°C and tap-water vacuum, and taken for silica gel clean-up followed by gas chromatography. The remaining 20 litres of the sample were drawn through the adsorption system at a pumping rate of 2–5 bed volumes per minute. The system consisted of a precombusted glass fibre filter (same type as above: diameter 140 mm) and the adsorption column (length 90 mm; i.d. 23 mm) either packed with silanized porous glass, silanized glass beads, or Amberlite XAD-2. Columns and filters were then soxhlet extracted for approximately 8 hours with methanol-water or acetone-water (v:v 3:2). After evaporating a major portion of the organic solvent (as above), the remaining extract was partitioned into hexane and dried over anhydrous sodium sulphate.

Table 6.2 Average spike recovery from 25 litre samples given in percentages (spike concentration added to the sample is taken as 100%)

Spike compound	Recovery by liquid–liquid extraction	Recovery by liquid–solid adsorption onto:			Recovery by glass-fibre filters
		XAD-2	Silanized porous glass	Silanized glass beads	
n-alkanes	71 ± 8	7 ± 2	6 ± 2	~5	13 ± 2
Phthalates	103 ± 5	67 ± 28	73 ± 36	14 ± 12	5 ± 4
DDT, DDE, DDMU	74 ± 30	47 ± 16	50 ± 21	47 ± 35	8 ± 12
Dieldrin, Endrin	71 ± 13	65 ± 4	68 ± 15	26 ± 15	1 ± 2

Reproduced from Derenback, J.B., et al. (1978) Marine Chemistry, **6**, 351, Elsevier Science Publishers, by courtesy of authors and publishers.

Table 6.2 shows spiking recoveries from 25 litre seawater samples obtained by liquid–liquid extraction followed by (a) gas chromatography and (b) by liquid–solid adsorption on to silanized porous glass, silanized glass beads and XAD-2 resin. The recovery was measured from 25 litre samples for a range of spike compounds: n-alkanes and phthalates at concentrations of 0.5 µg l^{-1} and pesticides at concentration of 40 and 20 ng l^{-1}. The liquid–liquid extraction is superior to any liquid–solid adsorption technique. The adsorption materials XAD-2 and silanized porous glass gave poor recoveries for alkanes varying between 3d and 20% for different sets of water samples. Both adsorption materials were equally inefficient. However, the recovery of phthalates is much better. On the other hand, the recovery of phthalates drops with increasing alkane character, e.g. diisobutyl-, di-n-butyl-, benzylbutyl-, dicyclohexyl-phthalate acid esters are still recovered at about the average value (diethyl-phthalate was detected at below-average values due to evaporation in the work-up procedure); while for bis-(ethylhexyl)-phthalate 28% was found on XAD-2 and 29% on silanized porous glass. The same effect is less pronounced for alkanes with increasing chain lengths, which can be taken as a further indication of a michelle formation for non-aromatic hydrocarbons. The recoveries of pesticides are in the range 30–100% and are distinguished by a scattering of the results over a wide range, sometimes exceeding 100% by more than 50%. On average, DDE and DDMU give the highest values; this might be explained by the degradation of DDT. Dieldrin and Eldrin were equally well

Table 6.3 Fractionated desorption of spike material from a silanized porous glass column

Solvent	Colour of extract	Spike material desorbed (%) total recovery is taken as 100%)	
		Phthalates	*n-Alkanes*
Water	Slightly green	Not measured	Not measured
Ethanol–water (1:10)	Slightly green	1	0
Ethanol–water (1:1)	Slightly green-yellow	92	6–12
Ethanol	Slightly yellow	3–4	84
n-pentane	Colourless	0–4	0–4

Reproduced from Derenback, J.B., *et al.* (1978) *Marine Chemistry*, **6**, 351, Elsevier Science Publishers, by courtesy of authors and publishers.

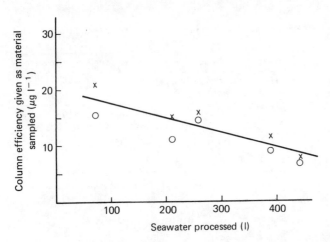

Figure 6.3 Sampling efficiency of a 36 ml adsorption column when sampling different sized water volumes. ○ = contained by neutral fraction. × = contained by the acidic fraction (Reproduced from Derenbach, J.B., *et al.* (1978) *Marine Chemistry*, **6**, 351, Elsevier Science Publishers, by courtesy of authors and publishers.)

recovered by all sampling techniques. When the column capacity was increased by a factor of two (two columns connected in series, either XAD-2 or silanized porous glass), an additional 3–8% of alkanes and phthalates were recovered. No pesticides could be eluted from the second column.

Spiked phthalates and *n*-alkanes are easily desorbed from silanized porous glass columns in fractions according to the polarity of solvents and compounds. In Table 6.3 the results are given for the fractionated elution from an 18 ml column using a total of 18 ml solvent. The eluate may even be free from most of the humic substances also sampled from seawater, if the column is pre-washed in a pH gradient ranging from pH 5 to 8.5.

An indication of the actual sampling capacity of silanized porous glass may be derived from Figure 6.3. The efficiency of a 36 ml column is plotted against the volume of processed natural seawater. The column extracts were separated into neutral, acidic and basic fractions, the latter being approximately one fifth of the neutral fraction. Best recoveries of neutral compounds were found to be just above $15\,\mu gl^{-1}$. The use of larger sampling columns only resulted in a slightly better recovery for this group of compounds (a 110 ml column bed sampled an average of $18\,\mu g$ neutral compounds per litre out of 50, 80 and 100 litre volumes). The acidic fraction, however, increased by more than a factor of two.

Compromising on the column volume for sampling neutral compounds from 100 litre seawater samples, a 40 ml column bed would be sufficient with 1 ml column material extracting around $40\,\mu g$ neutral compounds out of 2500 bed volumes. This is equal to $2\text{--}10^{-3}$ g column load per gram n-C_{18} phase bonded to the support, which is the order of magnitude for a maximum load of an analytical reversed phase-column, thus indicating the necessity for a considerable excess of binding sites for the sampling purpose in seawater.

Using 110 ml columns, 100 litre seawater samples were processed at different pH values. As can be seen from Figure 6.4, the recovery of neutral compounds drops

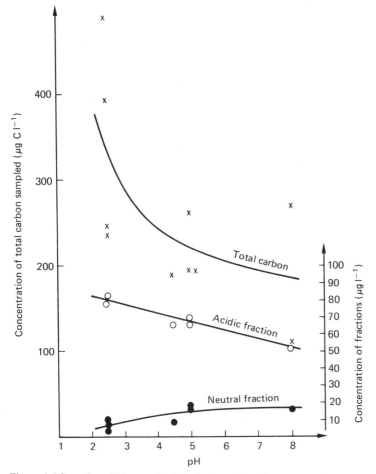

Figure 6.4 Sampling efficiency of a 110 ml adsorption column processing 100 litre water samples at different pH values (Reproduced from Derenbach, J.B., *et al.* (1978) *Marine Chemistry*, **6**, 351, Elsevier Science Publishers, by courtesy of authors and publishers.)

with decreasing pH, which would indicate an overloaded column. This also might explain the linearity in the recovery of acidic compounds, which one would expect to increase significantly at a lower pH. The recovery of both fractions is most probably suppressed in favour of a fraction of the bulk of uncharacterized organic compounds now sampled (the scattering of results could be due to changes in the concentration of total dissolved organic carbon in the water, which varied between 4 and 1 mg C per litre). Here, the limit of a good column efficiency seems to be reached with an average column load of about 10^{-3} g DOMC per gram of n-C_{18} phase.

Thus if the sampling unit of a liquid–solid adsorption system is kept within certain proportions to the concentration of compounds to be sampled, it will help to extract at least a major portion of the more hydrophobic compounds from seawater.

A disadvantage of the reversed-phase hplc technique or Derenbach et al. [133] is that it is only semi-quantitative. Outweighted against this is the fact that the technique using silanized porous glass is easy to keep free of contamination, has a comparatively high adsorption capacity and permits the fractionated desorption of accumulated material.

Hayase et al. [135] applied reversed phase liquid chromatography with double detectors (fluorescence and absorption) to the determination of dissolved organic matter in estuarine seawater. The dissolved organic matter was extracted into chloroform at pH 3 or 8. The hydrophilic–hydrophobic balance and aromatic character of the seawater dissolved organic matter was represented on the chromatograms. The results indicated that reversed phase liquid chromatography with double detectors was an effective technique in the characterization of dissolved organic matter in seawater.

6.5 Chemical pre-treatment of organics

In some cases, the forms in which the organic materials are present are not easily separated or identified. Some chemical pre-treatment is necessary before the analytical procedures can be used. These pre-treatments may include the preparation of volatile derivatives for use in gas chromatography, or the splitting of a large and otherwise intractable molecule, as in the hydrolysis of proteins to their constituent amino acids.

For many of the organic materials in seawater, some form of chemical pre-treatment is necessary before analysis is possible. The obvious cases are the hydrolysis of polysaccharides and proteins before the analysis for monomeric constituents, and the formation of volatile derivatives to permit analysis by gas chromatography. These methods will be discussed in later sections on the analysis of specific compounds.

There are few more general reactions which are useful in determining the skeletal structure of unknown organic compounds. One such reaction is a treatment with hydrogen, resulting in the replacement of halogens, oxygen, sulphur and nitrogen and the saturation of double bonds [134]. The system is attached on line in a gas chromatograph, the hydrogen being supplied by the carrier gas, the conversion taking place in a catalyst containing tube. The resultant chromatographic pattern is greatly simplified, making identification of chemical structures somewhat easier.

Carbohydrazide can be used in a similar manner as a reducing agent, converting azo and nitro compounds to the corresponding amines for better volatilization [136]. A considerable literature exists on methods of this kind; it has scarcely been used by marine chemists, perhaps because they are still concerned with concentration and isolation of compounds in seawater.

One form of chemical pre-treatment that has been used more extensively is the oxidation of dissolved organic matter to carbon dioxide, with high intensity to ultraviolet light. This is the basis for at least one method of measurement of dissolved organic carbon; it has also been used for the measurement of stable carbon isotope ratios in dissolved and particulate organic matter [137, 138]. This measurement rests on the assumption that oxidation of all organics in seawater by ultraviolet light goes to completion, an assumption that is by no means proven.

References

1. Koyama, T. *Journal of the Oceanographic Society of Japan*, **20**, 563 (1963)
2. Andelmen, J.B. and Caruso, S.C. Concentration and separation techniques. In *Water and Water Pollution Handbook* (ed. L.L. Ciaccio), Vol. 2, Marcel Dekker, New York, pp. 483–591 (1971)
3. Webb, R.G. *Isolating Organic Water Pollutants, XAD Resins, Urethane Foams, Solvent Extraction*, U.S. National Technical Information Service, PB Rep. No. 245647. (1975)
4. Whitby, F.J. *Proceeding of the Analytical Division of the Chemical Society*, **12**, 110 (1975)
5. Fritz, J.S. *Accounts of Chemical Research*, **10**, 67 (1977)
6. Sharp, J.H. *Marine Chemistry*, **1**, 211 (1973)
7. Sharp, J.H. *Limnology and Oceanography*, **18**, 441 (1976)
8. Sheldon, R.W. and Sutcliffe, W.H., Jr. *Limnology and Oceanography*, **14**, 441 (1969)
9. Banoub, M.W. and Williams, P.J. *Deep Sea Research*, **19**, 433 (1972)
10. Menzel, D.W. *Deep Sea Research*, **14**, 229 (1974)
11. Bishop, J.K.B. and Edmond, J.M. *Journal of Marine Research*, **34**, 181 (1976)
12. Gordon, D.C., Jr. and Sutcliffe, W.H., Jr. *Limnology and Oceanography*, **19**, 989 (1974)
13. Sharp, J.H. *Limnology and Oceanography*, **19**, 984 (1974)
14. Wangersky, P.J. *Limnology and Oceanography*, **19**, 980 (1974)
15. Fox, D.L., Oppenheimer, C.H. and Kittredge, J.S. *Journal of Marine Research*, **12**, 233 (1953)
16. Ostapenya, A.P. and Kovalevskaya, R.Z. *Okeanologiya* **4**, 649 (1965)
17. Mullin, M.M. *Limnology and Oceanography*, **10**, 459 (1965)
18. Bishop, J.K.B., Edmond, J.M., Ketten, D.R., Bacon, M.P. and Silker, W.B. *Deep Sea Research*, **24**, 511 (1977)
19. Baeyens, W., Decadt, G. and Elskens, L. *Oceanologica Acta*, **2**, 447 (1979)
20. Turekian, K.K. *Geochimica and Cosmochimica Acta*, **41**, 139 (1977)
21. Bothner, M.H. and Carpenter, R. *Sorption–Desorption Reactions of Mercury with Suspended Matter in The Columbia River*, IAEASM-158/5, Washington, July 1972, 73–87
22. Lockwood, R.A. and Chen, K.Y. *Environmental Science and Technology*, **7**, 1028 (1973)
23. Reimers, R.S. and Krenkel, P.A. *Journal of The Water Pollution Control Federation*, 352 (1974)
24. Lindberg, S.E. and Harriss, R.C. *Journal of the Water Pollution Control Federation*, 2479 (1977)
25. Frenet-Robin, M. and Ottmann, F. *Estuarine Coastal Marine Science*, **7**, 425 (1978)
26. Baeryens, W., Decadt, G., Dehairs, F. and Geoyeus, L. *Oceanologica Acta*, **5**, 261 (1982)
27. Jacobs, M.B.M. and Ewing, M. *Science*, **163**, 380 (1969)
28. Lammers, W.G. *Environmental Science and Technology*, **1**, 52 (1967)
29. Westover, L.B., Tou, J.C. and Mark, J.H. *Analytical Chemistry*, **46**, 568 (1974)
30. Klein, E., Eichelberger, J., Eyer, C. and Smith, J. *Water Research*, **9**, 807 (1975)
31. Chian, E.S.K., Bruce, W.N. and Fang, H.H. *Experimental Science and Technology*, **9**, 52 (1975)
32. Deinzer, H., Melton, R. and Mitchell, D. *Water Research*, **9**, 799 (1975)
33. Kammerer, P.A. and Lee, G.F. *Environmental Science and Technology*, **3**, 276 (1969)
34. Pocklington, R. *A New Method for the Determination of Amino-Acids in Sea Water and an Investigation of the Dissolved Free Amino-Acids of North Atlantic Ocean Waters*, PhD thesis, Dalhousie University, pp.1–102 (1970)

35. Shapiro, J. *Science*, **133**, 2063 (1961)
36. Shapiro, J. *Analytical Chemistry*, **39**, 280 (1967)
37. Habermann, H.M. *Science*, **140**, 292 (1963)
38. Kammerer, P.A., Jr. and Lee, G.F. *Environmental Science and Technology*, **3**, 276 (1969)
39. Wallace, G.T., Jr. and Wilson, D.F. *Foam Separations as a Tool in Chemical Oceanography*. U.S. Naval Research Laboratory Report, No. 6958, p.1 (1969)
40. Dorman, D.C. and Lemlich, R. *Nature (London)*, **207**, 145 (1965)
41. Karger, B.L. and DeVivo, D.G. *Separation Science*, **3**, 393 (1968)
42. Blanchard, D.C. *Science*, **146**, 396 (1964)
43. MacIntyre, F. *Journal Chemical Physics*, **72**, 589 (1968)
44. Blanchard, D.C. and Syzdek, L.D. *Limnology and Oceanography*, **20**, 762 (1975)
45. Fasching, J.L., Courant, R.A., Duce, R.A. and Piotrowicz, S.R. *Journal Rech. Atmosphere*, **8**, 649 (1974)
46. Blanchard, D.C. Applied chemistry at protein interfaces. In *Advances in Chemistry Series*, No. 145 (ed. R.E. Baier), American Chemical Society, Washington DC, pp. 360–387 (1975)
47. Annoff, M. and Josefsson, B. *Analytical Chemistry*, **48**, 1268 (1976)
48. Goldberg, M.C., Delong, L. and Sinclair, M. *Analytical Chemistry*, **45**, 89 (1973)
49. Ahnoff, M. and Josefsson, N. *Analytical Chemistry*, **46**, 658 (1974)
50. Ahnoff, M. and Josefsson, N. *Analytical Chemistry*, **48**, 1268 (1976)
51. Wu, C. and Suffet, I.H. *Analytical Chemistry*, **49**, 231 (1977)
52. Fielding, W., Gibson, T.M., James, H.A., McLoughlin, K. and Steep, C.P. *Organic Micropollutants in Drinking Water*, Technical Report TR 159, Water Research Centre, Medmenham, U.K. (1981)
53. Copin, G. and Barbier, M. *Oceanography*, **23**, 455 (1971)
54. The Oil Companies International Study Group for Conservation of Clean Air and Water, Europe. *Analysis of Trace Substances in Aqueous Effluents from Petroleum Refineries*. Concave Report No. 6/82 (1982)
55. Gomella, C., Belle, J.P. and Auvray, J. *Techniques et Sciences Municipales*, **71**, 439 (1976)
56. Neihof, R. and Loeb, G. *Journal of Marine Research*, **32**, 5 (1974)
57. Chave, K.E. *Science*, **148**, 1723 (1965)
58. Meyers, P.A. and Quinn, J.G. *Limnology and Oceanography*, **16**, 992 (1971)
59. Meyers, P.A. and Quinn, J.G. *Geochimica and Cosmochimica Acta*, **37**, 1745 (1973)
60. Jeffrey, L.M. and Hood, D.W. *Journal of Marine Research*, **17**, 247 (1958)
61. Williams, P.M. and Zirino, A. *Nature (London)*, **204**, 462 (1964)
62. Chapman, G. and Rae, A.C. *Nature (London)*, **214**, 627 (1967)
63. Wangersky, P.J. *Science*, **115**, 685 (1952)
64. Vaccaro, R.F. *Environmental Science and Technology*, **5**, 134 (1971)
65. Grob, K. *Journal of Chromatography*, **84**, 255 (1973)
66. Kerr, R.A. and Quinn, J.G. *Deep Sea Research*, **22**, 107 (1975)
67. Grob, K. and Zuercher, F. *Journal of Chromatography*, **117**, 285 (1976)
68. Jeffrey, L.M. *Development of a Method for Isolating Gram Quantities of Dissolved Organic Matter from Sea Water and Some Chemical and Isotopic Characteristics of the Isolated Material*, PhD thesis, A. & M. University, Texas, p. 152 (1969)
69. Gustafson, R.L. and Paleos, J. Interactions responsible for the selective adsorption of organics on organic surfaces in *Organic Compounds in Aquatic Environments* (eds S.J. Faust and J.V. Hunter), Marcel Dekker, New Yok, pp. 213–237 (1971)
70. Riley, J.P. and Taylor, D. *Analytica Chimica Acta*, **46**, 307 (1969)
71. Osterroht, C. *Kiel. Meeresforsch*, **28**, 48 (1972)
72. Osterroht, C. *Journal of Chromatography*, **101**, 289 (1974)
73. Burnham, A.K., Calder, G.V., Fritz, J.S., *et al. Analytical Chemistry*, **44**, 139 (1972)
74. Junk, G.A., Richard, J.J., Grieser, M.D., *et al. Journal of Chromatography*, **99**, 745 (1974)
75. Fritz, J.S. *Industrial Engineering Product Development Research*, **14**, 94 (1975)
76. Saxena, J., Kozuchowski, J. and Basu, D.K. *Environmental Science and Technology*, **11**, 682 (1977)
77. Leenheer, J.A. and Huffman, E.W.D. *Journal of Research, US Geological Survey*, **4**, 737 (1976)
78 Tateda, A. and Fritz, J.S. *Journal of Chromatography*, **152**, 329 (1978)
79. Stephen, S.F., Smith, J.F., Flego, U. and Renkers, J. *Water Research*, **12**, 447 (1978)
80. Chang, R.C. and Fritz, J.S. *Talanta*, **25**, 659 (1978)
81. Ryan, J.P. and Fritz, J.S. *Journal of Chromatographic Science*, **16**, 488 (1978)
82. Stepan, S.I. and Smith, J.F. *Water Research*, **11**, 339 (1977)
83. Iahangir, L.M. and Samuelson, O. *Analytica Chimica Acta*, **100**, 53 (1978)

84. Rees, G.A.V. and Au, L. *Bulletin of Environmental Contamination and Toxicology*, **22**, 561 (1979)
85. Dravnieks, A., Krotoszynski, S.K., Whitfield, J., O'Donnell, A. and Burgwald, T. *Environmental Science and Technology*, **5**, 1220 (1971)
86. Kaiser, R. *Haus Tech., Essen, Vortragsveroeff.* **1972**, 13 (1972)
87. Zlatkis, A., Lichtenstein, H.A. and Tichbee, A. *Chromatographia*, **6**, 67 (1973)
88. Bergert, K.H., Betz. V. and Pruggmayer, D. *Chromatographia*, **7**, 115 (1974)
89. Bertsch, W., Chang, R.C. and Zlatkis, A. *Journal of Chromatographic Science*, **12**, 175 (1974)
90. Bertsch, W., Anderson, E. and Hylser, G. *Journal of Chromatography*, **112**, 701 (1975)
91. Novak, J., Zluticky, J., Kubelka, V. and Mostecky, J. *Journal of Chromatography*, **76**, 45 (1973)
92. Wasik, S.P. *Journal of Chromatographic Science*, **12**, 845 (1974)
93. Kuo, P.R.K., Chian, E.S.K., DeWalle, F.B. and Kim, J.H. *Analytical Chemistry*, **49**, 1023 (1977)
94. Colenut, B.A. and Thorburn, S. *International Journal of Environmental Analytical Chemistry*, **7**, 231 (1980)
95. Colenut, B.A. and Thorburn, S. *International Journal of Environmental Analytical Chemistry*, **15**, 25 (1980)
96. Grob, K. and Zurcher, F. *Journal of Chromatography*, **117**, 285 (1976)
97. Grob, K. *Journal of Chromatography*, **84**, 255 (1973)
98. Grob, K. and Grob, G. *Journal of Chromatography*, **90**, 303 (1974)
99. Grob, K. and Grob, G. *Journal of Chromatography*, **106**, 299 (1975)
100. Westerdorf, R.G. *International Laboratory*, **September**, 32 (1982)
101. Waggott, A. and Reid, W.J. *Separation of Non-Volatile Organic Carbon in Sewage Effluent Using High Performance Liquid Chromatography*, Part 1, Technical Report No. 52, Water Research Centre, Stevenage, Herts (1977)
102. Bassett, R., Ozeris, S. and Whitnah, C.H. *Analytical Chemistry*, **34**, 1540 (1962)
103. Corwen, J.F. Volatile organic materials in sea water. In *Organic Matter in Natural Waters* (ed. W. Hood), Institute of Marine Science, University of Alaska Publication no. 1, pp. 169–180 (1970)
104. Kaiser, K.L.E. and Oliver, B.G. *Analytical Chemistry*, **48**, 2207 (1976)
105. Gjavotchanoff, S., Luessen, H. and Schlimme, E. *Gas and Wasser*, **112**, 448 (1971)
106. Chesler, S.N., Gump, B.H., Hertz, H.P., *et al. Analytical Chemistry*, **50**, 805 (1978)
107. Heyndrickx, A. and Peteghem, C. *European Journal of Toxicology*, **8**, 275 (1975)
108. Dowty, B., Green, L. and Laseter, J.L. *Journal of Chromatographic Science*, **14**, 187 (1976)
109. Waggott, A. *Proceedings of the Analytical Division of Chemical Society, London*, **15**, 232 (1978)
110. Bellar, T.A. and Lichtenberg, J.J. *Journal of American Water Works Asociation*, **566**, 739 (1974)
111. Friant, S.L. and Suffet, I.H. *Analytical Chemistry*, **51**, 2161 (1979)
112. Malter, L. and Vreden, V. *Forum Stadte-Hygiene*, **29**, 37 (1978)
113. Cowen, W.F., Cooper, W.J. and Hisfill, J.W. *Analytical Chemistry*, **47**, 2483 (1975)
114. Chian, E.S.K., Kuo, P.P.K., Cooper, W.J., Cowen, W.F. and Fuentes, R.C. *Environmental Science and Technology*, **11**, 282 (1977)
115. Luessem, H. Detection of trace materials by concentration by pressure distillation. *Haus Tech., Essen, Vortragsveroeff.*, **1972**, 69 (1972)
116. Sheldon, R.W. *Limnology and Oceanography*, **17**, 494 (1972)
117. Gjessing, E. and Lee, F.G. *Environmental Science and Technology*, **1**, 631 (1967)
118. Sirotkina, I.S., Varshal, G.M., Lur'e, Y.Y. and Stepanova, N.P. *Zhur Analytical Khim*, **29**, 1626 (1974)
119. Pietrzyk, D.I. and Chu, C.H. *Analytical Chemistry*, **49**, 757 (1977)
120. Pietrzyk, D.J. and Chu, C.H. *Analytical Chemistry*, **49**, 860 (1977)
121. Waggott, A. *Proceedings of the Analytical Division of the Chemical Society (London)*, **15**, 232 (1978)
122. Scott, R.P.W. and Kucera, P. *Analytical Chemistry*, **45**, 749 (1973)
123. Snyder, L.R. *Analytical Chemistry*, **46**, 1384 (1974)
124. Engelhardt, H. *Hochdruck Flussegkeits–Chromatographie*. Springer, Berlin (1975)
125. Braus, H., Middleton, F.M. and Walton, G. *Analytical Chemistry*, **23**, 1160 (1951)
126. Eichelberger, J.W. and Lichtenberg, J.J. *Journal of the American Water Works Association*, **63**, 25 (1971)
127. Riley, J.P. and Taylor, D. *Analytica Chimica Acta*, **46**, 307 (1969)
128. Harvey, G.R. *Adsorption of Chlorinated Hydrocarbons from Seawater by a Cross linked Polymer*, Woods Hole Oceanographic Institute Technical Report WHOK-72-86 (unpublished) (1972)
129. Gesser, H.D., Chow, A., Davis, F.C., Uthe, J.F. and Reinke, J. *Analytical Letters*, **4**, 833 (1971)
130. Ahling, B. and Jensen, S. *Analytical Chemistry*, **42**, 1483 (1970)
131. Uthe, J.F. and Reinke, J. *Environmental Letters*, **3**, 117 (1972)

132. Aue, W.A., Kapila, S. and Hastings, C.R. The use of support-bonded silicones for the extraction of organochlorines of interest from water. *Journal of Chromatography*, **73**, 99 (1972)
133. Derenbach, J.R., Ehrhardt, M., Osterroht, C. and Petrick, G. *Marine Chemistry*, **6**, 351 (1978)
134. Beroza, M. *Analytical Chemistry*, **34**, 1801 (1962)
135. Hayase, K., Shitashima, K. and Tsubota, H. *Journal of Chromatography*, **322**, 358 (1985)
136. Rahn, P.C. and Siggia, D. *Analytical Chemistry*, **45**, 2336 (1973)
137. Williams, P.M. *Nature (London)*, **219**, 152 (1968)
138. Williams, P.M. and Gordon, L.I. *Deep Sea Research*, **17**, 19 (1970)

Chapter 7

Organic compounds

While there has always been some interest in the nature of the organic compounds in seawater, identification of actual compounds has been slow because of the low concentrations. With a total organic carbon concentration of 0.5–1.5 mg C per litre, the total concentration of any single organic compound is likely to be less than 10^{-7} M. Therefore, in the past, identification of individual compounds has been limited to those few for which specific, sensitive chemical methods existed. These methods were usually spectrophotometric, and were often developments of methods originally used in clinical chemistry.

The advent of the newer physical methods of separation and identification, together with the impetus given to the field by the imposition of anti-pollution legislation, has resulted in a flood of new and often unproven, methods. While most of these methods were specifically designed to measure materials added to the environment by man's activities, in many cases they have added greatly to our knowledge of the naturally occurring compounds as well.

Until the advent of modern instrumental methods of analysis, the best we could hope to do was to measure the amounts of certain broad classes of compounds present, as, for example, the total protein or total carbohydrate.

Using the newer methods, such as gas chromatography, liquid–liquid chromatography, fluorimetry, and mass spectrometry, it is possible to measure many compounds at the parts-per-billion level, and a few selected compounds with special characteristics at the parts-per-trillion level. Even with these sensitivities, however, a considerable concentration must usually be undertaken to permit the chemical or physical fractionation necessary to render the final analyses interpretable. A major effort has therefore been expended on the study of methods of separation and concentration and this is discussed further in Chapter 6.

A problem which has been less well recognized is that of contamination in sampling and sample handling. The oceanographic vessel itself is a major source of contamination; from the moment that the ship stops on station, a surface film of oil, flakes of metal and rust spreads out in all directions. The means of sampling also often acts as a source of contamination. Once the sample is on board, along with the normal problems of contamination through sample handling and through high blank values from the reagents used, we must also attempt to cope with chemical and biological changes occurring during storage, since many of the modern instruments are not easily taken to sea. In environmental chemistry, the work of the analyst does not begin with the delivery of the sample to the laboratory, every aspect of sampling, storage and pre-treatment must be considered part of the analyst's domain. This is discussed further in Chapter 1.

7.1 Hydrocarbons

Probably the most studied group of organic compounds in seawater is the hydrocarbons, not because of the importance of the naturally occurring materials, but because of the continuing threat of large-scale pollution. However, the methods devised for the measurement of anthropogenic hydrocarbons will also measure the natural materials. A major problem, not altogether solved, is that of distinguishing between the two sources.

There have been many reviews of analytical methods for hydrocarbons particularly those given in references [1–11].

Separation methods

Separation methods for hydrocarbons are simple, largely because of their non-polar nature and, in many cases, their density. Thus tarry materials which have collected into tar balls can be harvested from the surface layers with a neuston net [12]. These tar balls are normally less dense than seawater and therefore are not sought in the water column. However, with time, the balls often acquire a considerable population of marine organisms, sometimes including barnacles. They may thus become heavier than seawater and sink to the ocean floor. The tar balls have not yet been sought in the upper layers of the sediments.

The hydrocarbons may be present in the water column in true solution, in suspension in particulate or droplet form, or in the poorly defined state called 'accommodation'. In all of these forms, they may be separated from seawater by extraction with organic solvents [13]. The solvents normally used include: methylene chloride [14–19], chloroform [20–24], carbon tetrachloride [25–31], trichlorotrifluoroethane [32–34], benzene [39], pentane [40, 41], petroleum ether [42], methanol–benzene [43], and ethyl acetate–petroleum ether [44].

It is obvious from these references that there is no single well-accepted extraction technique for hydrocarbons. To some extent, the choice of solvent may depend on the analytical technique to be used for subsequent measurements. The most commonly used solvents are carbon tetrachloride and hexane. In place of a simple solvent extraction, hydrocarbons can be concentrated from the surface layer, or from an oil slick, by adsorption on a sheet of polyurethane foam and from the water column by adsorption on a polyurethane foam plug. The foam is then extracted with an organic solvent and analysis is carried out in the usual manner.

Fingerprinting

In the early days of pollution research, many methods were investigated in the hope of finding a single technique which would infallibly link the deed and the doer, the oil spill and the leaking container. The research was aided by a number of cases in which the provenance was obvious; the *Torrey Canyon* and the *Arrow* incidents are two examples. Thus it was possible to study the changes brought about by weathering and to discover how these changes would hamper identification of the source of the spill. The difficulty of attempting such identification using only a single technique was clearly shown by the number of proposals to add radioactive or chemical labels to petroleum products as they were loaded into the tankers.

Of the methods developed for the identification of hydrocarbon mixtures, only coupled gas chromatography–mass spectrometry holds any real promise of certain identification and this only at a prohibitive cost in time spent characterizing minor peaks. It would be far more efficient to develop rapid screening procedures which would eliminate all but a few possibilities, and then use gas chromatography–mass spectrometry to isolate and identify a few key peaks to confirm the characterization. This is precisely the scheme adopted independently by a number of laboratories.

The methods included in fingerprinting a hydrocarbon mixture are infra-red and ultraviolet spectrophotometry, nuclear magnetic resonance, gas chromatography, coupled gas chromatography–mass spectrometry isotope ratios [45, 46], thin layer and liquid chromatography [47], microscopy and sulphur content [48], and gel filtration and charge transfer complexation [49].

Every additional piece of information, although not necessarily critical in its own right, helps to make the identification more certain. There is little doubt now that positive identification of the source of any simple oil spill can be made, as recent court cases indicate.

Gas chromatographic methods

In the marine environment gas chromatography [50–54] has been employed to identify petroleum products. Here the pollutants are crude oil, marine residual fuel oil and crude oil sludge consisting of a concentrated suspension of high-melting point paraffin wax in crude oil. Although weathering of marine oil pollutants can be such that the oil is rendered unrecognizable, the time required to achieve this was found to be so long as to be insignificant in regard to pollutant identification [54]. Ramsdale and Wilkinson [52] employed a specially constructed injection device which permitted introduction of samples contaminated with water, seaweed, sand or sediments. Pollutants were generally readily identified.

Occasionally the mound of unresolved compounds on the chromatogram supporting the superimposed n-alkane peaks confuse the true n-alkane profile. This has been overcome by separating off the n-alkanes using molecular sieves, prior to gas chromatography [53]. However, separation of n-alkanes in this way, or by urea complex formation [55] is reported as being more applicable to distillates rather than residual materials, and also the separation is not entirely specific [54]. In the case of marine pollutants it has been found advantageous to chromatograph a distilled residue, b.p. 343°C [53] or fraction, b.p. 245–370°C [56] which avoids problems caused by evaporation of lower ends by weathering.

Chromatographs of marine pollutants have been published; crude oils, 200 second fuel oil; 2000 second fuel oil, pollution samples [53]; sludge wax, crude oil sludge, Bahrein fuel oil, 3000 second diesel fuel, pollution samples [52]; marine diesel fuel, sludges, crude oil, residual fuel oil [56].

Freegarde, Hatchard and Parker [57] have discussed the identification, determination and ultimate fate of oil spilt at sea.

Erhardt and Blumer [58] have developed a method for the identification of the source of marine oil spills, by gas chromatographic analysis and results for eight different crude oils are given. Distinguishing compositional features are still recognizable after weathering for more than 8 months. The method was used for the tentative source identification of samples of beach tar.

Zafiron, Myers and Freestone [59] have shown that commercial oil spill emulsifiers can interfere with the gas chromatographic detection of the source of oil spills.

Boylan and Tripp [60] determined hydrocarbons in seawater extracts of crude oil and crude oil fractions. Samples of polluted seawater and the aqueous phases of simulated samples (prepared by agitation of oil-kerosene mixtures and unpolluted seawater to various degrees) were extracted with pentane. Each extract was subjected to gas chromatography on a column (8 ft × 0.06 in) packed with 0.2% of Apiezon L on glass beads (80–100 mesh) and temperatures programmed from 60°C to 220°C at 4°C per minute. The components were identified by means of ultraviolet and mass spectra. Polar aromatic compounds in the samples were extracted with methanol-dichloromethane (1:3).

Investigations on pelagic tar in the North West Atlantic have been carried out using gas chromatography [61]. Their report collects together the results of various preliminary investigations. It is in the Sargasso Sea, where the highest concentrations ($2–40\,\mathrm{mg\,m^{-2}}$) occur, and on beaches of isolated islands, such as Bermuda. These workers discuss the occurrence, structure, possible sources, and possible fate of tar lumps found on the surface of the ocean.

Zafiron and Oliver [62] have developed a method for characterizing environmental hydrocarbons using gas chromatography. Solutions of samples containing oil were separated on an open-tubular column (50 ft × 0.02 in) coated with OV-101 and temperature programmed from 75°C to 275°C at 6°C per minute; helium ($50\,\mathrm{ml\,min^{-1}}$) was used as carrier gas and detection was by flame ionization. To prevent contamination of the columns from sample residues the sample was injected into a glass-lined injector assembly, operated at 175°C, from which gases passed into a splitter before entering the column. Analysis of an oil on three columns gave signal:intensity ratios similar enough for direct comparison or for comparison with a standard. The method was adequate for correlating artifically weathered oils with sources and for differentiating most of 30 oils found in a sea port.

Hertz et al. [63] have discussed the methodology for the quantitative and qualitative assessment of oil spills. They describe an integrated chromatographic technique for studies of oil spills. Dynamic headspace sampling and gas chromatography and coupled-column liquid chromatography are used to quantify petroleum-containing samples and the individual components in these samples are identified by gas chromatography and mass spectrometry.

Rasmussen [64] describes a gas chromatographic analysis and a method for data interpretation that he has successfully used to identify crude oil and bunker fuel spills. Samples were analysed using a Dexsil-300 support coated open tube (SCOT) column and a flame ionization detector. The high-resolution chromatogram was mathematically treated to give 'g.c. patterns' that were a characteristic of the oil and were relatively unaffected by moderate weathering. He compiled the 'g.c. patterns' of 20 crude oils. Rasmussen [64] uses metal and sulphur determinations and infra-red spectroscopy to complement the capillary gas chromatographic technique.

The gas chromatograms of most oil samples examined had similar basic features. All were dominated by the n-paraffins, with as many as 13 resolved but unidentified, smaller peaks appearing between the n-paraffin peaks of adjacent carbon numbers. Each oil had the same basic peaks, but their relative size within bands of one carbon number varied significantly with crude source. To aid in the

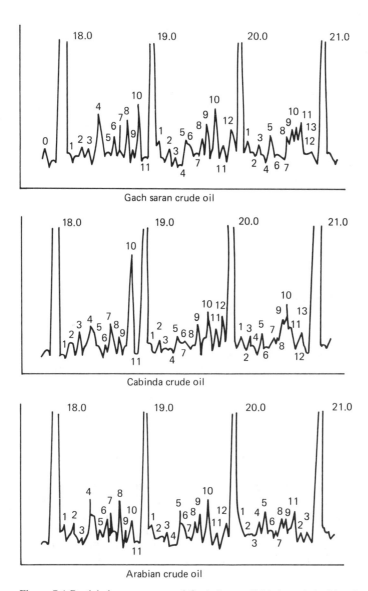

Figure 7.1 Partial chromatograms of Gach Saran, Cabinda and Arabian Crudes (Reproduced from Rasmussen, D.V. (1976) *Analytical Chemistry*, **48**, 1562, American Chemical Society, by courtesy of author and publishers.)

comparison of chromatograms, the peaks were labelled as shown in Figure 7.1 and the relative areas of the smaller peaks were expressed as a percentage of the preceding *n*-paraffin peak. In other words, n-$C_{19}H_{40}$ was assigned the peak number 19.0 and the 12 peaks eluting between n-$C_{19}H_{40}$ and n-$C_{20}H_{42}$ were assigned the peak numbers 19.1–19.12. The areas of the peaks 19.1–19.12 were then expressed as a percentage of peak 19.0 to form the C_{19} segment of the 'g.c. pattern' of the oil. Generally, the C_{14}–C_{35} portion of the chromatogram was treated in this manner.

The partial chromatograms in Figure 7.1 are typical of analyses performed on the Dexsil-300 SCOT column and demonstrate the peak numbering system. Differences between the three crude oils are evident from the chromatograms.

The preferred approach to an oil spill identification programme was to analyse the oil spill sample and all potential sources followed by the comparison of their gas chromatographic patterns. When it was not possible to analyse all potential sources, the spill sample was compared with a library of gas chromatographic patterns of known crude oils and products.

Various other workers have investigated methods for the identification of oil spills [65–68]. Wilson, Ferreto and Coloman [65] have reviewed approaches based on trace metal (Ni–V) ratio, nitrogen and sulphur content, infra-red spectroscopy and gas chromatography. The fingerprinting of oil spills by gas chromatography has been reported using flame ionization and flame photometric detection [66, 67]. Interpretation of the flame ionization chromatograms is normally based on the distribution of the n-paraffin peaks [65], however, the use of the pristane to phytane ratio has been reported by Blumer and Sass [68] and is thought to be more independent of weathering.

Ramsdale and Wilkinson [69] have identified petroleum sources of beach pollution by gas chromatography. Samples containing up to 90% of sand or up to 80% of emulsified water were identified, without pre-treatment by gas chromatography on one of a pair of matched stainless steel columns (750×3.2 mm i.d.) fitted with precolumns (100 mm) to retain material of high molecular weight, the second column being used as a blank. The column packing is 5% of silicone E301 on Celite (52–60 mesh), the temperature is programmed at 5°C per minute from 50°C to 300°C, nitrogen was used as carrier gas and twin flame ionization detectors were used.

Adlard, Creaser and Matthews [70] improved the method of Ramsdale and Wilkinson [69] by using a S-selective flame photometric detector in parallel with the flame ionization detector. Obtaining two independent chromatograms in this way greatly assists identification of a sample. Evaporative weathering of the oil samples has less effect on the information attainable by flame photometric detection than on that attainable by flame ionization detection. A stainless steel column (1 m $\times$ 3 mm i.d.) packed with 3% of OV-1 on AWDMCS Chromosorb G (85–100 mesh) was used, temperature programmed from 60°C to 295°C per minute with helium (35 ml min^{-1}) as carrier gas, but the utility of the two-detector system is enhanced if it is used in conjunction with a stainless steel capillary column (20×0.25 mm) coated with OV-101 and temperature programmed from 60°C to 300°C at 5° per minute because of the greater detail shown by the chromatograms.

Brunnock, Duckworth and Stephens [71] have also analysed beach pollutants. They showed that weathered crude oil, crude oil sludge and fuel oil can be differentiated by the n-paraffin profile as shown by gas chromatography, wax content, wax melting point and asphaltene content. The effects of weathering at sea on crude oil were studied; parameters unaffected by evaporation and exposure are the contents of vanadium, nickel and n-paraffins. The scheme developed for the identification of certain weathered crude oils includes the determination of these constituents, together with the sulphur content of the sample.

McKay [72] has investigated the use of automatic sample injection on to the chromatographic column in his investigations of hydrocarbon pollution of beach sands.

Adlard and Matthews [73] applied the flame photometric sulphur detector to pollution identification. A sample of the oil pollutant was submitted to gas chromatography on a stainless steel column (1 m × 3 mm) packed with 3% of OV-1 on AWDMCS Chromosorb G (85–100 mesh). Helium was used as carried gas (35 ml min^{-1}) and the column temperature was programmed from 60°C to 295°C at 5° per minute. The column effluent was split between a flame ionization and a flame photometric detector. Adlard and Matthews [73] claim that the origin of oil pollutants can be deduced from the two chromatograms. The method can also be used to measure the degree of weathering of oil samples.

Garra and Muth [74] and Wasik and Brown [75] characterized crude, semi-refined and refined oils by gas chromatography. Separation followed by dual-response detection (flame ionization for hydrocarbons and flame photometric detection for S-containing compounds) was used as a basis for identifying oil samples. By examination of chromatograms, it was shown that refinery oils can be artificially weathered so that the source of oil spills can be determined.

There are three main approaches to the problem of removing and concentrating the volatile hydrocarbons in seawater.

1. We can allow the solution to come to equilibrium with a known volume of gas, and sample this gas.
2. We can extract the volatile materials by pulling a vacuum on the solution and collect the volatiles either by adsorption or by cold-trapping.
3. We can sweep the volatile materials from the solution by bubbling with an inert gas, again collecting the volatiles by adsorption or by cold-trapping.

'Headspace' methods

The first method, the so-called 'headspace' method, is conceptually and operationally the simplest. However, it requires much more information before the calculation of concentration can be completed. This technique is discussed in more detail in section 6.3(a).

May et al. [76] have described a gas chromatographic method for analysing hydrocarbons in seawater which is sensitive at the submicrogram per kilogram level. Dynamic headspace sampling for volatile hydrocarbon components, followed by coupled-column liquid chromatography for analysing the non-volatile components, requires minimal sample handling, thus reducing the risk of sample component loss and/or sample contamination. The volatile components are concentrated on a Tenax gas chromatographic precolumn and determined by gas chromatography or gas chromatography–mass spectrometry.

Vacuum removal of volatiles

The second system, the removal of volatiles by vacuum, can be set up in two ways; either as a flow-through or as a batch process. As a flow-through process, the sample is drawn continuously through the system, and the gases taken off by the vacuum pass through a sampling loop. Periodically, the material in the loop is injected into the gas chromatograph. In this manner it is possible to derive almost continuous profiles of volatile hydrocarbon concentrations [77].

In the batch mode, a larger sample can be treated over a longer period, and the volatiles collected by cold-trapping or adsorption. These techniques are not as fast as flow-through sampling, nor do they permit semi-continuous profiling, but they result in a greater concentration of the hydrocarbons, and thus in greater sensitivity [78].

Gas stripping

The third technique, stripping-out, is by far the most common. In this technique, an inert gas is bubbled through the sample to remove the volatile materials (see section 6.3(a)). When the concentration of hydrocarbons is great enough, as, perhaps, after a petroleum spill, the emergent gas stream can be sampled directly [79]. This is seldom the case in true oceanic samples, however, and some form of concentration is needed.

It is possible to collect the volatiles in a cold trap [80]. A more favoured technique is the collection of the gases by adsorption on some support such as one of the Chromosorbs, or Tenax GC [76, 80, 81]. The volatiles are then desorbed by heating and injected into a gas chromatograph.

Of the three general methods, the last seems to be the most practical. Theoretically, with high enough concentrations of hydrocarbons the first method, the headspace analysis, should be both the most accurate and the easiest to calibrate. Operationally, it leaves much to be desired, both because of the problems of sensitivity and those of the accommodation of the larger molecules in water. The second method, vacuum degassing, requires much more equipment than the third method and requires that large amounts of water vapour be removed before the sample is injected into the gas chromatograph. The last method is so much less complicated that even with its calibration problems it has been adopted almost universally.

Gas chromatography–mass spectrometry

One of the most versatile of the detectors that can be used with a gas chromatograph is the mass spectrometer. Whereas most detectors furnish only a retention time for identification and a peak height or area which is proportional to the amount of the compound present, the mass spectrometer furnishes, in addition, some clues as to the nature of the compounds present in any peak. Identification is seldom absolute; there are usually several possible ways of assembling the fragments found into compounds. When the mixture of compounds is as complex as those found in natural water, the identifications are only as good as the reference library of mass spectra and the computer search program available. For the spectra to be reasonably intelligible, the separations must be as clean as possible. It is difficult enough to make sense out of the fragments of one large compound; it is next to impossible to do so if the peaks analysed consist of mixtures of several compounds in uncertain proportions. Therefore, the use of capillary columns in the gas chromatographic section of the instruments is becoming all but universal.

Albaiges and Albrecht [82] propose that a series of petroleum hydrocarbons of geochemical significance (biological, markers) such as C_{20}–C_{40} acyclic isoprenoids and C_{27} steranes and triterpenes are used as passive tags for the characterization of oils in the marine environment. They use mass fragmentography of samples to make evident these series of components without resorting to complex enrichment treatments. They point out that computerized gas chromatography–mass spectrometry permits multiple fingerprinting from the same gas chromatographic run. Hence rapid and effective comparisons between samples and long-term storage of the results for future examination can be carried out.

Usually, Albaiges and Albrecht [82] first deasphaltenized with n-pentane (40 volumes) prior to gas chromatography. When, however, the recovery of the branched plus cyclic alkanes was needed for subsequent analysis the saturated hydrocarbon fraction was isolated by conventional silica-gel adsorption chromatography (eluting solvent *n*-pentane) and refluxed in iso-octane with 5 Å (0.5 nm) molecular sieves.

The gas chromatograph (Perkin-Elmer 900) equipped with flame ionization and flame photometric detectors was operated either with 9 ft × ¼ inch packed columns (1% Dexsil 300 on Gas-Chrom Q 100– 120) from 150°C to 300°C at 6°C per minute or with 200 ft × 0.02 inch capillary columns (OV-101 or Apiezon L) from 120°C to 180°C at 6°C per minute.

Gas chromatographic profiles of petroleum residues exhibit several characteristic features that have been applied for identification or correlation purposes.

Generally, the most apparent is the *n*-paraffin distribution that has proved to be useful in differentiating the main types of pollutant samples (crude oils, fuel oils and tank washings [83] or even types of crude oil [84]) although in this case the method uses the quantification of the previously isolated *n*-paraffins, therefore lengthening the analysis time.

Another relevant feature of the gas chromatographic profile is the acyclic isoprenoid hydrocarbon pattern that is made evident with capillary columns or by the inclusion of the saturated fraction in 5 Å (0.5 nm) molecular sieves or in urea. The predominant peaks usually correspond to the C_{19} (pristane) and C_{20} (phytane) isomers, which ratio serves as an identification parameter [85], although the series extends to lower and higher homologues.

Finally, the sulphur compounds that are present in minor quantities in petroleum products also exhibit a typical gas chromatographic fingerprint easily obtained by flame photometric detection. This fingerprint has been introduced to complement the flame ionization detection chromatogram with the aim of resolving the ambiguities or increasing the reliability in the identification of the pollutants [86].

All the above fingerprints exhibit a different usefulness for characterizing oils. Their variation between crudes and their resistance to the sea weathering processes are not enough, in many cases, for providing the unequivocal identification of the pollutant. The *n*-paraffins can, apparently, be removed by biodegradation as well as the lower acyclic isoprenoids at respectively slower rates.

Boylan and Tripp [87] used gas chromatography–mass spectrometry to determine hydrocarbons in pentane and methanol dichloromethane (1:3) extracts of seawater.

Hertz *et al.* [88] developed an integrated gas chromatographic technique for the characterization of oil spill materials. Dynamic headspace sampling and the complementary analytical technique of gas chromatography and coupled-column liquid chromatography are utilized for quantitation of petroleum-containing samples. Gas chromatography–mass spectrometry was employed for identification of individual components in these samples.

Some typical results obtained by these procedures are given in Table 7.1.

Table 7.1 Hydrocarbon content of samples obtained from catastrophic oil spill. Numbers in parentheses indicate the number of determinations carried out

Sample	Type	Percent hydrocarbons by various methods			Weathering factor (ratio of $n - C_{15}/n - C_{11}$ peak heights)	% water by Karl Fischer
		Soxhlet extraction corrected for water content Karl Fischer (%)	Gas chromatography (%)	Coupled-column LC (%)		
M-1	Oil–water emulsion	67 (1)	50 ± 20 (4)	80 ± 12 (3)	7 ± 7	25
S-1	Sediment	0.2 (1)	0.2 ± 0.06 (3)	0.7 ± 0.07 (2)	60 ± 20	0.8
S-2	Sediment	16 (1)	9 ± 3 (3)	10 ± 1 (2)	30 ± 10	16
S-3	Control sediment	Trace (1)	0.0002 ± 0.00004 (3)	<0.0002 (2)	NA	5
T-1	Tissue	7.7 (1)	0.03 ± 0.01 (3)	1.8 ± 0.2 (2)	NA	76
Unweathered spill oil		–	–	–	0.6 ± 0.2	0.2

LC = Liquid chromatography.
NA = Not applicable.
Reproduced from Hertz, H.S., et al. (1978) Environmental Science and Technology, **10**, 900, American Chemical Society, by courtesy of authors and publishers.

High performance liquid chromatography

By its very nature, the gas chromatograph is only useful with those compounds which can be made volatile in some manner. Compounds that are non-volatile at the temperatures that can be achieved in the gas chromatograph injection port, or those that degrade and polymerize, will be left as a residue in the injection port or at the top of the column. For these compounds, high performance liquid chromatography is the natural technique. The weak point of this technique has been the sensitivity of the detectors; the common commercially available detectors measure refractive index, ultraviolet and visible light absorption and fluorescence. Of these, only the fluorescence detector can approach the sensitivity of the gas chromatograph detectors, and it is useful only for those few compounds that are naturally fluorescent. There have also been attempts to link the liquid chromatograph to flame ionization detectors and atomic absorption spectrometers.

HPLC has been used, with an ultraviolet absorption detector set for 254 nm, for the determination of aromatic hydrocarbons and with a flow calorimeter for the detection of all hydrocarbons. Increased sensitivity and decreased interference can be achieved with the ultraviolet absorption detector by measuring absorption at two wavelengths and using the ratios of the absorption at those wavelengths [89].

Gel permeation chromatography

Done and Reid [90] applied gel permeation chromatography to the identification of crude oils and products isolated from seawater. The technique, which appears more suited to the analysis of crude oils, is based on the separation of oil components in order of their molelcular size - for practical purposes, their molecular weight.

Oils were dissolved in tetrahydrofuran or toluene and each solvent resulted in a different output profile, and therefore more information for identification. The solutions were pumped through a 24 × 3/8 inch column packed with 6 nm (60 Å) Styragel and the eluent monitored by a differential refractometer. Elution time was about 1 hour, although this was improved by staggered injection, and 6 mg samples were required. 'Fingerprints' of the crude oils examined, although claimed to be unique for the 50 oils investigated, fall into several discrete groups. Crude oil residues (b.p. 525°C) were excluded by the Styragel employed. While crude oil and weathered samples were found to give similar traces, some changes were introduced by evaporation processes, and it could be difficult to differentiate between similar but weathered crude oils. Heavy fuel oils were more difficult to identify, owing to the variety of fuel oils in use.

Spectrophotometric methods

The methods chosen for analysis of hydrocarbons will depend upon whether a measure of the total amount of hydrocarbons present is wanted or whether it is required to identify the separate components present. If we are monitoring for pollution, we would search for a simple indicator of total hydrocarbon content. Since aromatic compounds display some absorbance in the ultraviolet, this absorbance can be used as a relative measure of the hydrocarbons present if other organic compounds absorbing in this region can be removed. One group of workers extract the hydrocarbons into chloroform, then determine the absorbance at 292 nm without further purification. On the other hand, Zurina, Stradomskaya and Semenov [91] separate the hydrocarbons from other organic compounds by thin layer chromatography on alumina, and then determine absorbance. The separation of aromatic hydrocarbons from other organic compounds can be achieved efficiently without the problem, common in gas chromatography, of the identification of many separate peaks, by a high pressure liquid chromatographic separation on a very short column. The use of an ultraviolet detector set at 254 nm then provides peak detection and quantitative analysis [92].

The ultraviolet methods, although relatively simple and convenient, are only semi-quantitative at best. Aromatic hydrocarbons vary greatly in their molar absorbance at any specific wavelength; to be truly quantitative, the method should separate the compounds present, then determine the concentration of each compound present by measuring the absorbance at the appropriate wavelength, using the appropriate standards. There are methods of separation and measurement which are both easier and more sensitive than those using ultraviolet absorption.

Fluorescence methods

Methods of considerably greater sensitivity are those using fluorescence. These methods can be either qualitative or, like the ultraviolet absorption methods, roughly quantitative. Fractionation into separate compounds is necessary to make the measurements qualitative. Thus, Van Duuren [93] examined the use of emission and excitation spectra in the identification of aromatic hydrocarbons. Contour diagrams of fluorescence activity at various excitation and emission wavelengths have been used as a means of identifying petroleum residues.

However, the main use of fluorescence has been in the semi-quantitative determination of aromatic hydrocarbons by extraction into an organic solvent, followed by excitation at a standard wavelength and comparison with the emission from a chosen standard. These techniques have been studied by many workers [94–98].

The difficulties in the use of fluorescence for quantitative measurement of hydrocarbons are much like those for the ultraviolet absorption methods. Each compound has its own excitation and emission maxima, with the fluorescence quantum yields varying sometimes by an order of magnitude. Thus the amount of hydrocarbon reported by an analysis will depend upon the emission and excitation wavelengths chosen, and upon the compound selected as the standard.

Petroleum products contain many fluorescing compounds, e.g. aromatic hydrocarbons, polycyclic aromatic hydrocarbons and various heterocyclic compounds. The use of fluorescence technique and instrumentation has led to the use of this technique for the identification of crude and residual oil pollutants in a marine environment [99, 100] and of motor oils and related petroleum products [101–104].

Maher [105] used fluorescence spectroscopy for monitoring petroleum hydrocarbon contamination in estuarine and ocean waters.

An ingenious variation on the standard fluorescence methods was proposed by Red'kin, Voitenko and Teplyakcv [106]. Water samples were extracted with non-polar solvents, transferred into hexane and the hexane solution frozen at 77K. At that temperature the normally diffuse luminescence emission bands are present as sharp emission lines, making identification of fluorescing compounds considerably simpler. In the case of a complex mixture, some separation by column or thin layer chromatography might be necessary.

In general, for routine semi-quantitative analysis of hydrocarbons, as in pollution monitoring, the fluorescence methods offer many advantages, including higher sensitivity than the absorption methods. The disadvantages of the methods, including the non-linearity of response and the non-coincidence of emission peaks, are shared with the absorption methods.

Infra-red spectroscopy

Infra-red absorption methods, like the fluorescence emission and ultraviolet absorption methods, can be used both qualitatively and quantitatively. Infra-red spectra can be used to characterize petroleum samples [107] and to evaluate the effects of weathering on oil spills [108]. The process of matching the spectrum of an unknown with that of a standard petroleum sample had been primarily a subject matter, since no method has existed for assigning a degree of match or mismatch. Mattson *et al.* [109] have attempted to use a multivariate normal probability density function for this purpose. The method should be useful when a large data base and a computer are available.

Electrolytic stripping

Wasik [110] has used an electrolytic stripping cell to determine hydrocarbons in seawater. Dissolved hydrocarbons in a known quantity of seawater were equilibrated with hydrogen bubbles, evolved electrolytically from a gold electrode, rising through a cylindrical cell. In an upper headspace compartment of the cell, the hydrocarbon concentration is determined by gas chromatography. The major advantages of this cell are that the hydrocarbons in the upper compartment are in equilibrium with the hydrocarbons in solution and that the hydrogen used as an extracting solvent does not introduce impurities into samples.

Wasik [110] used this method to successfully determine ppb of gasoline in seawater. He found that a convenient method for concentration of the hydrocarbons is to recycle the hydrogen stream containing the hydrocarbons many times over a small amount of charcoal (2.3 mg).

The charcoal, while still in the filter tube, was extracted three times with 5 μl carbon disulphide. A 2–3 μl aliquot of this solution was then injected into a SCOT capillary column.

Figure 7.2 shows a drawing of the stripping cell.

A 10 ml portion of the seawater extracted was diluted to 1000 ml with seawater. To this solution was added toluene-d, as an internal standard to give a concentration of 0.05 ppm. The diluted extract was poured into a 1000 ml stripping cell located in a constant temperature water bath. The temperature of the bath was raised to 80°C and a current of 0.3 A was passed through the cell. The upper chromatogram shown in Figure 7.3 was obtained by sampling 1 ml of the headspace after 30 ml hydrogen had bubbled through the stripping cell. The chromatogram shows the large number of aliphatic and olefinic hydrocarbon peaks that are eluted at the same time as the early aromatic hydrocarbon peaks. The lower chromatogram in Figure 7.3 was obtained after 46 ml ($V/V_L = 0.046$) of hydrogen had

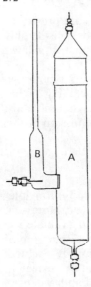

Figure 7.2 Electrolytic stripping cell (Reproduced from Wasik, S.P. (1974) *Journal of Chromatographic Science*, **12**, 845, Preston Publications, by courtesy of the author and publisher.)

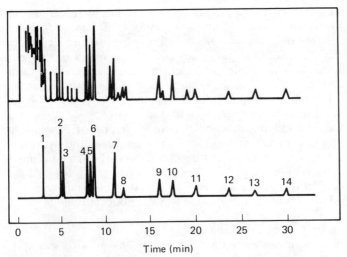

Figure 7.3 Hydrocarbons stripped from artificial seawater after 10 ml hydrogen had bubbled through the stripping cell (upper chromatogram) and after 100 ml hydrogen had bubbled through the stripping cell (lower chromatogram). A 1.0 ml volume of headspace gas was injected into a Scot column, 15 m × 0.5 mm i.d., coated with *m*-bis(*m*-phenoxyphenoxy) benzene; carrier gas He, 4 ml min⁻¹; temperature 80°C; attenuation 1 × 8. Peak identification from retention times of known compounds. 1 = Benzene; 2 = toluene; 3 = toluene-d_8; 4 = ethylbenzene; 5 = paraxylene; 6 = metaxylene; 7 = orthoxylene; 8 = *n*-propylbenzene; 9 = 1-methyl-4-ethylbenzene; 10 = 1,3,5-trimethylbenzene; 11 = 1-methyl-2-ethylbenzene; 12 = 1,2,4-trimethylbenzene; 13 = *n*-butylbenzene; 14 = 1,2,3-trimethylbenzene (Reproduced from Wasik, S.P. (1974) *Journal of Chromatographic Science*, **12**, 845, Preston Publications, by courtesy of author and publishers.)

bubbled through the stripping cell. All the aliphatic and olefinic hydrocarbon peaks are absent from the chromatogram. The aromatic hydrocarbon peaks were identified from retention times of known compounds.

Flow calorimetry

Zsolnay and Kiel [111] have used flow calorimetry to determine total hydrocarbons in seawater. In this method the seawater (1 litre) was extracted with trichlortri-fluoroethane (10 ml) and the extract was concentrated, first in a vacuum desiccator, then with a stream of nitrogen to 10 μl. A 50 μl portion of this solution was injected into a stainless steel column (5 cm × 1.8 mm) packed with silica gel (0.063–0.2 mm) deactivated with 10% of water. Elution was effected, under pressure of helium, with trichlorotrifluorethane at 5.2 ml per hour and the eluate passed through the calorimeter. In this the solution flowed over a reference thermistor and thence over a detector thermistor. The latter was embedded in porous glass beads on which the solutes were adsorbed with evolution of heat. The difference in temperature between the two thermistors was recorded.

Each solute first displayed the adsorbed solvent molecules, giving a desorption peak (negative); this was followed by an adsorption peak (positive). The area of the desorption peak was proportional to the amount of solute present. All hydrocarbons were eluted as one peak, the silica gel column removing the non-hydrocarbon solutes. Responses to unsaturated hydrocarbons were less than those to saturated ones; if the peak area were calculated as nonocane, an error of 25% could be introduced. The limit of detection was about $5 \mu g \, l^{-1}$.

Schuldiner [112] used paper chromatography to distinguish between crude oil, marine fuel oil and crude oil sludge.

Mass spectrometry

Brown and Huffman [113] reported an investigation of the concentration and composition of non-volatile hydrocarbons in Atlantic Ocean and nearby waters. Non-volatile hydrocarbons were identified by mass spectrometric technique. The results show that the non-volatile hydrocarbons in Atlantic and nearby waters contained aromatics at lower concentrations than would be expected if the source of the hydrocarbons were crude oil or petroleum refinery products. Hydrocarbons appeared to persist in the water to varying degrees with the most persistent being the cycloparaffins, then isoparaffins and finally the aromatics.

Miscellaneous

The Institute of Petroleum [114] in 1970 recommended analytical methods for the identification of the source of pollution by oil of seas, rivers and beaches. Methods are described for recovering the oil from the sample and for determination of boiling range, vanadium and nickel contents, and wax content. For the determination of boiling range, the hydrocarbon components are separated in the order of their boiling point by means of gas chromatography. For the determination of metals after incineration of the oil with sulphur and dissolution of the ash, vanadium and nickel are determined by reaction with tungstophosphate and dimethylglyoxime respectively, and use of X-ray fluorescence or spectrophotometry. For the determination of wax content, asphalt-free material is dissolved in hot dichloromethane, the solution is cooled to −32°C and the precipitated wax is separated and weighed.

Millard and Arvesen [115] discuss the airborne optical detection of oil on water. They undertook absolute radiometry, differential radiometry and polarimetry measurements utilizing reflected sunlight over the range 380–950 nm to evaluate methods for detecting oil spills on seawater. Maximum contrast between oil and water was observed at wavelength less than 400 nm and greater than 600 nm, minimum contrast being in the range 450–500 nm. Oil always appeared brighter than the water but it was not possible to distinguish one oil from another. These workers commented that differential polarization appeared to be a promising technique.

Freegarde, Hatchard and Parker [116] have reviewed methods of identifying and determining oils in seawater. They describe a method for determining down to $0.001\,mg\,l^{-1}$ of crude oil in seawater.

7.2 DDT derivatives

Work on the determination of chlorinated insecticides has been almost exclusively in the area of gas chromatography using different types of detection systems, although a limited amount of work has been carried out using liquid chromatography and thin-layer chromatography. Work on the determination of polychlorinated biphenyls and mixtures of these with chlorinated insecticides is reported at the end of this section.

Wilson and co-workers [117, 118] have discussed the determination of aldrin, chlordane, dieldrin, endrin, lindane, *o.p.* and *p.p'* isomers of DDT and its metabolites, mirex and toxaphene in seawater and molluscs. The US environmental Protection Agency has also published methods for organochorine pesticides in water and wastewater. The Food and Drug Administration (USA) [119] has conducted a collaborative study of a method for multiple organochlorine insecticides in fish. Earlier work by Wilson *et al.* [118, 120] in 1968 has indicated that organochlorine pesticides were not stable in seawater as indicated in Table 7.2. These conclusions were confirmed in the 1974 work.

Table 7.2 Stability of pesticides in natural seawater (salinity 29.8 ppt). Results expressed as ppb

Pesticide	Days after start of experiment					
	0	*6*	*17*	*24*	*31*	*38*
p,p'-DDT	2.9	0.75	1.0	0.27	0.18	0.16
p,p'-DDE*		0.096	0.95	0.065	0.034	0.037
p,p'-DDD*			0.081	0.041	0.038	0.037
Aldrin†	2.6	0.58	0.096	<0.01	<0.01	<0.01
Dieldrin*		0.74	1.0	1.0	0.75	0.56
Malathion	3.0	<0.2	<0.2			
Parathion	2.9	1.9	1.25	1.0	0.71	0.37

*Metabolites of parent compound.
†From the seventeenth day onward, two unidentified peaks appeared on the chromatographic charts after aldrin had eluted.
Reproduced from Wilson, A.J. and Forester, J. (1971) *Journal of the Association of Official Analytical Chemists*, **54**, 54, Association of Official Analytical Chemists by courtesy of authors and publishers.

Table 7.3 Percentage recovery of p,p'-DDT from seawater by different extraction solvents

Day	Extraction solvent		
	Petroleum ether	15% ethyl ether in hexane	Methylene chloride
0	93	93	93
6	67	66	76

Reproduced from Wilson, A.J. and Forester, J. (1971) *Journal of the Association of Official Analytical Chemists*, **54**, 54, Association of Official Analytical Chemists, by courtesy of authors and publishers.

Table 7.4 Percentage recovery of p,p'-DDT from seawater by petroleum ether and methylene chloride

0	90	95
4	67	85

Reproduced from Wilson, A.J. and Forester, J. (1971) *Journal of the Association of Official Analytical Chemists*, **54**, 54, Association of Official Analytical Chemists, by courtesy of authors and publishers.

Table 7.5 Percentage recovery of p,p'-DDT from seawater and distilled water by petroleum ether and methylene chloride

Day	Seawater		Distilled water	
	Petroleum ether	Methylene chloride	Petroleum ether	Methylene chloride
0	90	94	90	91
7	58	78	90	91
14	46	68	94	92

Reproduced from Wilson, A.J. and Forester, J. (1971) *Journal of the Association of Official Analytical Chemists*, **54**, 54, Association of Official Analytical Chemists, by courtesy of authors and publishers.

Table 7.6 Percentage recovery of p,p'-DDT from seawater incubated under different light and temperature conditions

Day	12 hours light and 12 hours dark at 20°C	Dark at 5°C
0	87	88
7	69	81
14	68	86

Reproduced from Wilson, A.J. and Forester, J. (1971) *Journal of the Association of Official Analytical Chemists*, **54**, 54, Association of Official Analytical Chemists, by courtesy of authors and publishers.

Since petroleum ether was the solvent used in these earlier studies for extracting the DDT from seawater, Wilson and Forester [117] initiated further studies to evaluate the extraction efficiencies of other solvent systems.

The recovery rates of o,p'-DDE in all tests were greater than 89% with petroleum ether, 15% diethylether in hexane followed by hexane or methylene dichloride, indicating no significant loss during analyses.

Tables 7.3–7.6 show the average percentage recovery of p,o'-DDT extracted from duplicate seawater (salinity 16–21 ppt) or distilled water samples up to 14 days after initiation of the experiment.

Table 7.3 shows that immediately after the seawater was fortified with 3.0 ppb of DDT all solvent systems removed 93% of the DDT.

Liquid scintillation counting of (^{14}C)DDT has been used to study the pick-up and metabolism of DDT by freshwater algae and to determine DDT in seawater [121].

Wilson [122] showed that liquid–liquid extraction of estuarine water immediately after addition of DDT gave acceptable recovery levels with all solvent systems tested but analyses carried out several days later gave only partial recovery owing to adsorption of DDT on suspended matter.

7.3 Parathion

Biochemical degradation of parathion has also been observed in seawater although the degradation products were not identified [123].

By the process shown in Table 7.7 aliquots of an ethanolic solution of parathion were separated into undecomposed insecticides and decomposed insecticide products. Among the products free p-nitrophenol was detected photometrically at 410 nm in alkaline solution. p-Nitrophenol, chemically bound in acidic phosphorus

Table 7.7 Procedure for parathion analysis (30 ml aliquot acidified with 2 M HCl to pH 1)

Extracted with diethylether (Et$_2$0)
(3 times with 20 ml) (consecutively)

Aqueous phase
(basic compound)
either discarded or
checked for amino
compounds

Organic phase
(neutral and acid compounds)
extracted with 0.1 M NaOH
at 5°C (once with 10 ml,
twice with 5 ml consecutively)

Aqueous phase
(acidic compounds)

photometric detection of
p-nitrophenolate:
FREE p-NITROPHENOL (B)

reacidified with 2 M HCl to pH 1

extracted with Et$_2$O (as above)

Organic phase
(acidic compounds)

Aqueous phase
(discard)

Organic phase
(neutral compounds)
checked for organophosphates
other than parathion

ether evaporated

ether evaporated

saponification of the residue

saponification of the residue

photometric detection of
p-nitrophenolate:
p-NITROPHENOL
EQUIVALENTS
IN ACIDIC COMPOUNDS
(C + B)

photometric detection of
p-nitrophenolate:
p-NITROPHENOL
EQUIVALENTS IN
NEUTRAL COMPOUNDS
(A)

Reproduced from Weber, K. (1976) *Water Research*, **10**, 4237, Pergamon Publications, by courtesy of author and publishers.

compounds and in non-hydrolysed neutral phosphorus compounds were detected in the same way after saponification. As a control an aliquot of 10 ml was taken at the same time as the 30 ml aliquots and extracted with ether in the way described in Table 7.7. Saponification after removal of ether without separation of neutral and acidic compounds yielded total *p*-nitrophenol equivalents. Saponifications were conducted in closed and calibrated reaction tubes; a 2 h reaction time at 100°C in 2 ml 3.5 M potassium hydroxide was used. Thereafter the solutions were diluted, if necessary, to ensure accurate measurement at 410 nm.

7.4 Other organochlorine insecticides

Aspila, Carron and Chau [124] reported the results of an interlaboratory quality control study in five laboratories on the electron capture gas chromatographic determination of ten chlorinated insecticides in standards and spiked and unspiked seawater samples (lindane, heptachlor, aldrin, δ-chlordane, α-chlordane, dieldrin, endrin, *p,p'*-DDT, methoxychlor and mirex). The methods of analyses used by these workers was not discussed, although it is mentioned that the methods were quite similar to those described in the water quality Branch Analytical Methods Manual [125]. Both hexane and benzene were used for the initial extraction of the water samples. In Table 7.8 results obtained by the five participating laboratories are presented for five water samples and the pair of standards (column 6). Samples 1A and 1B were unspiked natural water.

The standard deviation values are large because they contain all the errors, bias, and deviations accumulated from the interlaboratory comparisons [131]. Heptachlor revealed a poor recovery (Table 7.8) which confirmed its degradation [126–130]. The degradation product [126–128] is known to be 1-hydroxychlordene.

Table 7.8 Mean ($\pm$ SD) recovery of insecticides (n = 8)*

Insecticide	Sample				
	2 (as is)	3 (stripped)	4 (filtered)	5 (distilled water)	6 (standard)
Lindane	113 ± 31	83 ± 25	117 ± 34	130 ± 30	89 ± 25
Heptachlor	†	†	†	†	81 ± 8
Aldrin	49 ± 20	55 ± 23	83 ± 22	85 ± 15	93 ± 14
α-Chlordane	68 ± 18	59 ± 17	82 ± 25	94 ± 10	98 ± 10
γ-Chlordane	64 ± 17	53 ± 13	80 ± 21	87 ± 8	88 ± 8
Dieldrin	98 ± 22	85 ± 21	89 ± 18	102 ± 13	99 ± 6
Endrin	95 ± 20	114 ± 55	96 ± 19	105 ± 8	100 ± 13
p,p'-DDT	52 ± 36	46 ± 13	89 ± 30	108 ± 50	100 ± 4
Methoxychlor	82 ± 31	96 ± 25	105 ± 18	110 ± 14	101 ± 5
Mirex	56 ± 60	31 ± 23	54 ± 43	95 ± 13	100 ± 5
Mean recovery (%)	78	74	93	103	96
Mean SD	24	24	23	19	11

*Mean recoveries exclude data for mirex and heptachlor. All data from laboratory 2 were excluded in calculating the average per cent recoveries.
†No results were used because heptachlor is too unstable.
Reproduced from Aspila, K.I., *et al.* (1977) *Journal of the Association of Official Analytical Chemists*, **60**, 1097, Association of Official Analytical Chemists, by courtesy of authors and publishers.

Table 7.9 Mean (± SD) insecticide concentrations (as µg l^{-1}). Results for paired samples (2, 3, 4 and 5 in series A and B)

Insecticide	Lab 1		Lab 2		Lab 3		Lab 4		Lab 5	
	A	B	A	B	A	B	A	B	A	B
Lindane	9.0 ± 1.4	11.5 ± 2.4	5.25 ± 8.06	7.75 ± 9.67	12.5 ± 2.3	16.8 ± 3.3	7.37 ± 1.76	12.3 ± 1.7	8.0 ± 3.46	9.95 ± 2.87
B/A ratio		1.28		1.48		1.34		1.66		1.22
Aldrin	11.5 ± 3.5	15.0 ± 3.2	8.5 ± 10.6	11.0 ± 12.6	16.3 ± 3.8	22.8 ± 6.2	8.5 ± 3.2	14.0 ± 6.1	7.0 ± 3.6	12.0 ± 4.1
B/A ratio		1.30		1.29		1.4		1.65		1.71
α-Chlordane	17.5 ± 3.4	23.8 ± 5.5	16.8 ± 13.2	19.0 ± 13.2	21.8 ± 13.3	30.5 ± 4.0	13.7 ± 7.8	20.5 ± 25	15.3 ± 45	20.3 ± 5.0
B/A ratio		1.36		1.13		1.40		1.50		1.33
γ-Chlordane	19.8 ± 5.1	25.5 ± 5.7	21.5 ± 18.4	20.5 ± 3.4	30.3 ± 5.9	13.3 ± 2.1	21.0 ± 5.5	16.0 ± 7.3	20.0 ± 5.0	
B/A ratio		1.29		1.01		1.48		1.58		1.30
Dieldrin	23.3 ± 1.0	34.0 ± 2.5	19.0 ± 12.7	25.8 ± 17.8	28.3 ± 2.6	42.3 ± 4.6	19.7 ± 3.1	32.0 ± 1.4	23.0 ± 9.3	29.0 ± 6.2
B/A ratio		1.46		1.36		1.50		1.63		1.37
Endrin	30.0 ± 1.8	45.0 ± 4.7	83.0 ± 15.6	85.6 ± 15.8	34.0 ± 2.2	53.0 ± 2.0	36.3 ± 13.7	55.0 ± 24.7	23.3 ± 9.2	31.8 ± 9.0
B/A ratio		1.50		1.03		1.56		1.51		1.37
p,p'-DDT	32.2 ± 16.4	42.3 ± 15.8	19.3 ± 16.1	27.8 ± 23.0	41.0 ± 13.3	64.3 ± 24.7	*		34.5 ± 21.0	44.8 ± 21.3
B/A ratio		1.31		1.44		1.57				1.30
Methoxychlor	111 ± 11.7	160 ± 8.0	73.5 ± 35.4	104 ± 44	107 ± 9.4	158 ± 11	74.7 ± 30.0	115 ± 33.2	86.8 ± 36.3	110 ± 29
B/A ratio		1.44		1.42		1.47		1.54		1.26
Mean ratio (B/A)		1.368		1.270		1.465		1.581		1.344
SD for mean ratio		0.086		0.089		0.081		0.067		0.155

*Insufficient data.
Reproduced from Aspila, K.I., et al. (1977) Journal of the Association Of Official Analytical Chemists, **60**. 1097. Association of Official Analytical Chemists, by courtesy of authors and publishers.

Table 7.10 Comparison of total errors (%)

Insecticide	Lab 1	Lab 2	Lab 3	Lab 4	Lab 5	Mean % total error	Mean concentration insecticides ($\mu g\ l^{-1}$)
Lindane	46	216	104	41	77	66	0.010
Aldrin	68	190	53	90	92	76	0.020
α-Chlordane	56	133	46	63	84	63	0.0275
γ-Chlordane	61	145	65	73	82	70	0.030
Dieldrin	15	147	34	29	66	36	0.031
Endrin	17	223	27	100	61	51	0.038
p,p'-DDT	93	130	58	–	107	86	0.051
Methoxychlor	29	93	29	75	80	53	0.119
mean total for lab	48	160	52	67	81		

Reproduced from Aspila, K.I., *et al.* (1977) *Journal of the Association Of Official Analytical Chemists*, **60**, 1097, Association of Official Analytical Chemists, by courtesy of authors and publishers.

The average values for each insecticide, if it was indeed present, are shown in Tables 7.9 and 7.10 as well as the standard deviation for each series.

7.5 Polychlorinated biphenyls

In actual practice, environmental samples which are contaminated with PCBs are also highly likely to be contaminated with chlorinated insecticides. Many reports have appeared discussing cointerference effects of chlorinated insecticides in the determination of PCBs and vice versa and much of the more recent published work takes account of this fact by dealing with the analysis of both types of compounds. This work is discussed below.

PCBs have gas chromatographic retention times similar to the organochlorine insecticides and therefore complicate the analysis when both are present in a sample. Several techniques have been described for the separation of PCB from organochlorine insecticides. A review of these methods has been presented by Zitko and Choi [132]. These techniques are time consuming and, in general, semiquantitative. In addition, differential adsorption or metabolism of the Aroclor isomers in marine biota prevent accurate analysis of the PCBs. The gas chromatographic determination of chlorinated insecticides together with PCBs is difficult. Chlorinated insecticides and PCBs are extracted together in routine residue analysis, and the gas chromatographic retention times of several PCB peaks are almost identical with those of a number of peaks of chlorinated insecticides, notably of the DDT group. The PCB interference may vary, because the PCB mixtures used have different chlorine contents, but it is common for PCBs to be very similar to many chlorinated insecticides and the complete separation of chlorinated insecticides from PCBs is not possible by gas chromatography alone [133–137].

Figure 7.4 illustrates the possibility of the interference of DDT type compounds in the presence of PCBs. In an early paper on the determination of PCBs in water samples which also contain chlorinated insecticides, Ashling and Jensen [138] pass the sample through a filter containing a mixture of Carbowax 4000 monostearate on Chromosorb W. The adsorbed insecticides are eluted with light petroleum and

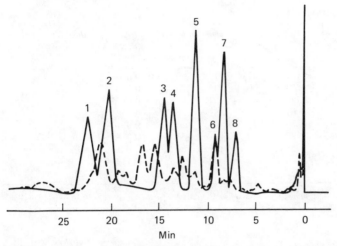

Figure 7.4 Interference of DDT-type compounds in the presence of PCBs. Gas chromatography: 5% QF-1 on Gas Chrom Q, 100–120 mesh. Solid time: 1 = *p,o'*-DDT; 2 = *p,p'*-DDD; 3 = *o,p'*-DDT; 4 = *o,p'*-DDD; 5 = *p,p'*-DDE; 6 = *p,p'*-DDMU; 7 = *o,p'*-DDE; 8 = *o,p'*-DDMU. Broken line = PCB Chiophen A 50 (Reproduced from Zikbo, V. and Choi, P (1971) Fisheries Research Board, Ottawa, Canada, by courtesy of authors and publishers.)

then determined by gas chromatography on a glass column (160 × 0.2 cm) containing either 4% SF-96 or 8% QF-1 on Chromosorb W pretreated with hexamethyldisilazane, with nitrogen as carrier gas (30 ml min^{-1}) and a column temperature of 190°C. When electron capture detector is used the sensitivity was 10 ng lindane per cubic metre with a sample size of 300 litres. The recoveries of added insecticides range from 50 to 100%; for DDT the recovery is 80% and for PCB, 93–100%.

Musty and Nickless [139] used Amberlite XAD-4 for the extraction and recovery of chlorinated insecticides and PCBs from water. In this method a glass column (20 × 1 cm) was packed with 2 g XAD-4 (60–85 mesh), and 1 litre of tap water (containing 1 part per 10^9 of insecticides) was passed through the column at 8 ml min^{-1}. The column was dried by drawing a stream of air through, then the insecticides were eluted with 100 ml ethyl ether-hexane (1:9). The eluate was concentrated to 5 ml and was subjected to gas chromatography on a glass column (5.5 ft (1.7 m) × 4 mm) packed with 1.5% OV-17 and 1.95% QG-1 on Gas-Chrom Q (100–120 mesh). The column was operated at 200°C, with argon (10 ml min^{-1}) as carrier gas and a ^{63}Ni electron capture detector (pulse mode). Recoveries of BHC isomers were 106–114%; of aldrin, 61%; of DDT isomers, 102–144%; and of polychlorinated biphenyls 76%.

Girenko, Klisenko and Dischcholka [140] noted that PCBs interfered with the electron capture gas chromatographic identification and determination of chlorinated insecticides in water and fish. Prior to gas chromatography, water samples and biological material were extracted with *n*-hexane. A mixture of organochlorine insecticides was completely separated (Figure 7.5(a)). Girenko, Klisenko and Dishcholka [140] noted that it was difficult to analyse samples of seawater because they are severely polluted by various co-extractive substances, chiefly chlorinated

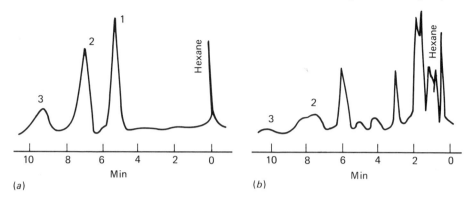

Figure 7.5 Chromatograms of a synthetic mixture of organochlorine insecticides (*a*) and of a water extract containing chlorinated biphenyls (*b*) 1 = *p,p'*-DDE; 2 = *p,p'*-DDD; 3 = *p,p'*-DDT (Reproduced from Musty, P.R. and Nickless, G. (1974) *Journal of Chromatography*, **89**, 185, Elsevier Science Publishers, by courtesy of authors and publishers.)

biphenyls. To determine organochlorine insecticide residues by gas chromatography with an electron capture detector, the chlorinated biphenyls were eluted from the column together with the insecticides. They produce unseparable peaks with equal retention time, thus interfering with the identification and quantitative determination of the organochlorine insecticides. The presence of chlorinated biphenyls is indicated by additional peaks on the chromatographs of the water samples and aquatic organic organisms (Figure 7.5(b)). Some of the peaks coincide with the peaks of the *o,p'* and *p.p'* isomers DDE, DDD and DDT and some of the constituents are eluted after *p.p'*-DDT.

Elder [141] determined PCBs in Mediterranean coastal waters by adsorption on to XAD-2 resin followed by electron capture gas chromatography. The overall average PCB concentration was $13 \, ng \, l^{-1}$.

Amberlite XAD-2 resin is a suitable adsorbent for polychlorinated biphenyl and chlorinated insecticides (DDT and metabolites, dieldrin) in seawater. These compounds can be suitably eluted from the resin prior to gas chromatography [142].

Picer and Picer [143] evaluated the application of XAD-2; XAD-4 and Tenax macroreticular resins for concentrations of chlorinated insecticides and polychlorinated biphenyls in seawater prior to analysis by electron capture gas chromatography. The solvents used eluted not only the chlorinated hydrocarbons of interest but also other electron capture sensitive materials, so that eluates had to be purified.

The extraction apparatus used by these workers is shown in Figure 7.6.

The eluates from the Tenax column were combined and the non-polar phase was separated from the polar phase in a glass separating funnel. Then the polar phase was extracted twice with *n*-pentane. The *n*-pentane extract was dried over anhydrous sodium sulphate, concentrated to 1 ml and cleaned on an alumina column using a modification of the method described by Holden and Marsden [144].

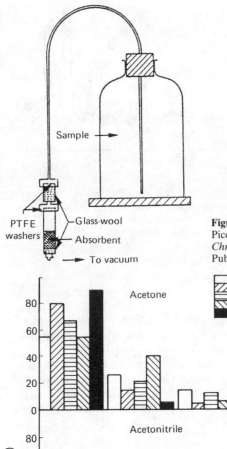

Sample →

PTFE washers

Glass-wool

Absorbent

→ To vacuum

Figure 7.6 Adsorption apparatus (Reproduced from Picer, N. and Picer, M. (1980) *Journal of Chromatography*, **193**, 357, Elsevier Science Publishers, by courtesy of authors and publishers.)

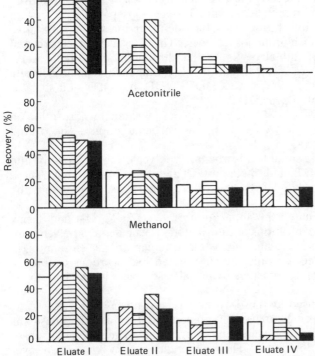

DDT
DDE
TDE
Dieldrin
Ar 1254

Acetone

Acetonitrile

Methanol

Recovery (%)

Eluate I   Eluate II   Eluate III   Eluate IV

Figure 7.7 Recovery of chlorinated hydrocarbons from 10 litres seawater percolated through XAD-2 resin. Seawater samples were spiked with 10 mg pesticides and 100 ng Aroclor 1254. Solvents for elution, acetone, acetonitrile, methanol. Eluates: I, 5 ml; II, 5 ml; III, 5 ml; IV, 10 ml (Reproduced from Picer, N. and Picer, M. (1980) *Journal of Chromatography*, **193**, 357, Elsevier Science Publications, by courtesy of authors and publishers.)

The eluates were placed on a silica gel column for the separation of PCBs from DDT, its metabolities and dieldrin using a procedure described by Snyder and Reinert [145] and Picer and Abel [146].

Picer and Picer [143] investigated the recovery from 10 litre samples of seawater of 0.1–1.0 μg l^{-1} chlorinated pesticides (DDT, DDE, TDE and Dieldrin), and 1–2 μg I per litre PCB (Aroclor 1254).

The results of the recovery tests on seawater samples are presented in Figure 7.7. Interestingly, for the elution of all chlorinated hydrocarbons 25 ml polar solvent were required, and 50 ml n-pentane was used as the re-extractant. Mirex was used as an internal standard, added to the eluate after the percolation of the polar solvent through the resin column. Hence this internal standard shows only the loss of chlorinated hydrocarbons during the re-extraction, alumina clean-up and silica gel separation. The recovery of Mirex during these steps varied between 80% and 90%. Losses of the investigated chlorinated hydrocarbons during these steps were 10–30% for about 10 ng pesticides.

Figure 7.8 shows gas chromatograms obtained after the percolation of a 10 litre seawater sample through XAD-2 resin and elution with 25 ml methanol; 10 ng of a pesticide mixture and 100 ng of Aroclor 1254 were added to the eluate, then eluate re-extraction with n-pentane and clean-up on an alumina column were performed. A chromatogram was also obtained without the addition of chlorinated hydrocarbons to the eluate. The experimental set-up is obviously capable of

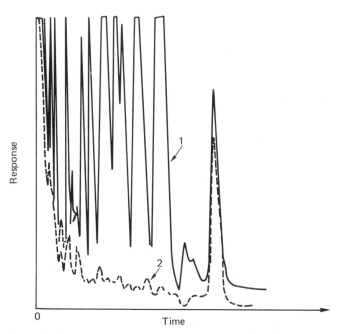

Figure 7.8 Comparison of chromatograms obtained after the percolation of 10 litres seawater through XAD-2 resin column. 1 = Eluate to which 10 nsg of pesticides and 100 ng Aroclor 1254 were added. 2 = Eluate with no addition of chlorinated hydrocarbons (Reproduced from Picer, N. and Picer, M. (1980) *Journal of Chromatography*, **193**, 357, Elsevier Science Publications, by courtesy of authors and publishers.)

determining chlorinated hydrocarbons in a 10 litre seawater sample at levels far below $1.0\,ng\,l^{-1}$ for pesticides and $10\,ng\,l^{-1}$ for PCBs.

Of the three macroreticular resins investigated XAD-2 was the best using methanol as elution solvent. Very unsatisfactory solvent blanks were obtained using Tenax resin. These workers conclude that application of macroreticular resins for the adsorption of chlorinated hydrocarbons from water samples and their determination after elution with different solvents has revealed several limitations. When water samples were spiked at levels close to the reported concentrations in seawater, the recovery of the investigated chlorinated hydrocarbons was low and unpredictable.

7.6 Chlorinated aliphatic compounds

Dawson, Riley and Tennent [147] have described samplers for large volume collection of seawater samples for chlorinated hydrocarbon analyses. The samplers use the macroreticular absorbent Amberlite XAD-2.

Lovelock and co-workers [148–149] determined methyl fluoride, methyl chloride, methyl bromide, methyl iodide and carbon tetrachloride in the Atlantic Ocean. This shows a global distribution of these compounds. Murray and Riley [150–151] confirmed the presence of carbon tetrachloride and also found low concentrations of chloroform and tri- and tetrachloroethylene in Atlantic surface waters.

Eklund, Josefsson and Ross [152, 153] have developed a capillary column method [154, 155] for the determination of down to $1\,\mu g\,l^{-1}$ volatile organohalides in waters which combine the resolving power of the glass capillary column with the sensitivity of the electron capture detector. The eluate from the column is mixed with purge gas of the detector to minimize band broadening due to dead volumes. This and low column bleeding give enhanced sensitivity. Ten different organohalides were quantified in seawater. Using this technique these workers detected bromoform in seawater for the first time.

Halogenated hydrocarbons in different waters were identified by comparison with a standard solution. A chromatogram of a standard pentane solution of various volatile organochlorine compounds including trihaloforms is shown in Figure 7.9. Retention times were measured on two columns with different stationary phases, i.e. SE-52 and Carbowax 400.

Extraction of 100 ml water with various amounts of pentane ranging from 1 to 15 ml showed that the extraction efficiency was increased when using a lower water to pentane ratio (Figure 7.10).

The approximate detection limits of the method for some compounds in water are given in Table 7.11, together with the gas chromatographic detection limits for eight organohalides.

Results obtained in the analysis of a seawater sample are shown in Table 7.12.

7.7 Organophosphorus compounds

The soluble organic phosphorus compounds arising from natural sources have been examined by a number of techniques, including concentration by freeze-drying and

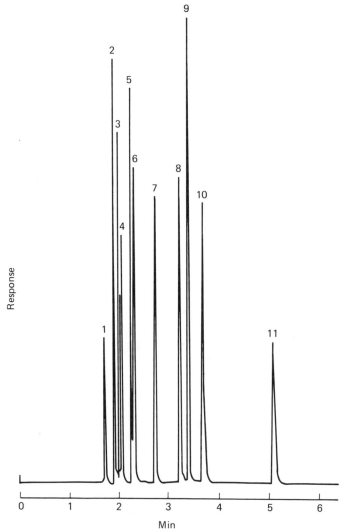

Figure 7.9 Chromatogram of a standard mixture containing 11 organohalides. Stationary phase, SE-52; 33 m × 0.3 mm i.d., injection temperature 200°C, column temperature 50°C, interface 250°C, detector temperature 250°C. Helium carrier gas flow rate 36 ml s^{-1}, scavenger gas flow 30 ml min^{-1}. Split ratio 1:20. 1 = CH_2Cl_2; 2 = $CHCl_3$; 3 = CH_3CCl_3; 4 = CCl_4; 5 = $CHClCCl_2$; 6 = $CHBrCl_2$; 7 = $CBrCl_3$; 8 = $CHBr_2Cl$; 9 = CCl_2CCl_2; 10 = $CHCl_2I$; 11 = $CHBr_3$. (Reproduced from Eklund, G, Josefson, B and Roos, C. *Journal of Chromatography*, **142**, 575 (1977).)

separation by molecular size on Sephadex gels, by Minear [156] and Minear and Walanski [157]. The most familiar of these compounds, adenosine triphosphate, is normally considered to be found in the particulate fraction, and is measured by the well-known luciferin–luciferinase reaction [158]. An adaptation of the method for estuarine waters and sediments was published by Wildish [159] and improvements on the usual method, including improvements in sampling and sample handling, were discussed by Hofer-Siegrist [160].

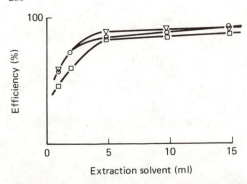

Figure 7.10 Extraction efficiency for CHCl₃ (□), CHBrCl₂ (▽) and CHBr₂Cl (○) in tap water (100 ml) as a function of amount extraction solvent. (Reproduced from Eklund, G, Josefson, B. and Roos, C. *Journal of Chromatography*, **142**, 575 (1977).)

Table 7.11 Electron capture detection limits and approximate detection limits of eight organohalides in water

	Electron capture detection limit (fg)	Approximate detection limits of organohalides in water (ng l⁻¹)
CH₂Cl₂	500	50
CHCl₃	10	1
CCl₄	1	0.1
CHClCCl₃	5	0.5
CHBrCl₂	2	0.2
CHBr₂Cl	2	0.2
CCl₂CCl₂	2	0.2
CHBr₃	10	1

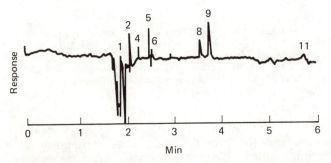

Figure 7.11 Chromatogram showing the minimum detectable amount for eight organohalides. 0.1 μl injected, split ratio 1:70, attenuation × 2. Approximate amounts on column: 1 = 200 fg; 2 = 20 fg; 4 = 1 fg; 5 = 10 fg; 6 = 3 fg; 8 = 4 fg; 9 = 4 fg; 11 = 8 fg; gas chromatography conditions and notations as in Figure 7.9. (Reproduced from Eklund, G., Josefson, B. and Roos, C. *Journal of Chromatography*, **146**, 575 (1977).)

Table 7.12 Concentrations of different organohalides in different waters

	Seawater ($\mu g \ l^{-1}$)
CH_2Cl_2	–
$CHCl_3$	0.026
CH_3CCl_3	0.046
CCl_4	<0.0005
$CHClCCl_2$	0.015
$CHBrCl_2$	–
$CHBr_2Cl$	–
CCl_2CCl_2	0.005
$CHCl_2I$	–
$CHBr_3$	0.027

Many of the organic phosphorus compounds now appearing in natural waters are the result of manufacturing or agricultural practices, rather than the degradation of aquatic organisms. Since the composition of these compounds is known, along with their chemistry, analytical methods can be designed specifically for these compounds. Thus the anticholinesterase nerve gases isopropylmethyl-phosphonofluoridate (GB) and O-ethyl S-diisopropylaminoethylmethylphos-phonothioate (VX) can be measured in seawater by an enzymatic technique [161]. The insecticide fenitrothion (O,O-dimethyl-O-4-nitro-3-methylphenyl thiophos-phate) can be measured in fish, water and sediments by gas chromatography, using a flame photometric detector to determine P and S [162]. The degradation products of the organophosphorus insecticides can be concentrated from large volumes of water by collection on Amberlite XAD-4 resin for subsequent analysis [163]. These are certainly not the only references to methods for the phosphorus-containing insecticides in natural waters; however, most of the work has been done in fresh water and at concentrations very much higher than those to be expected in seawater.

Cambella and Antia [164] determined phosphonates in seawater by fractionation of the total phosphorus.

The seawater sample was divided into two aliquots. The first was analysed for total phosphorus by the nitrate oxidation method capable of breaking down phosphonates, phosphate esters, nucleotides and polyphosphates. The second aliquot was added to a suspension of bacterial (*E. coli*) alkaline phosphatase enzyme, incubated for 2 h at 37°C and subjected to hot acid hydrolysis for 1 h. The resultant hot acid–enzyme sample was assayed for molybdate reactive phosphate which was estimated as the sum of enzyme hydrolysable phosphate and acid hydrolysable phosphate. Phosphate concentrations were calculated as the difference between total phosphorus and molybdate-reactive phosphorus (enzyme hydrolysable phosphate plus acid hydrolysable phosphate) thus exploiting the inert nature of the strong carbon–phosphorus bond. This bond was resistant to phosphatase action and acid hydrolysis.

Lores *et al.* [165] discussed the determination of the organophosphate insectide fenthion in seawater. The method comprises a solvent extraction followed by silica gel clean-up procedure prior to determination by gas chromatography with thermionic detection. Detection levels of $0.01\ \mu g\,l^{-1}$ were achieved.

Weber [123] has described a kinetic method for studying the degradation of parathion in seawater. Weber observed two pathways whereby parathion is hydrolysed. The first reaction proceeds via dearylation with loss of *p*-nitrophenol. This reaction is well known:

$$
\overset{\text{S}}{\underset{}{\|}} \\
(C_2H_5O)_2 - P - O - \langle C_6H_4 \rangle - NO_2 \quad \xrightarrow{+H_2O}
$$

$$
\overset{\text{S}}{\underset{}{\|}} \\
(C_2H_5O)_2 - P - OH \ + \ HO - \langle C_6H_4 \rangle - NO_2
$$

Additionally they observed a second main pathway, hydrolysis through dealkylation leading to a secondary ester of phosphoric acid which still contains the *p*-nitrophenyl moiety, i.e. de-ethylparathion (O-ethyl-O-p-nitrophenyl-monothiophosphoric acid):

$$
\overset{\text{S}}{\underset{}{\|}} \\
(C_2H_5O)_2 - P - O - \langle C_6H_4 \rangle - NO_2 \quad \xrightarrow{+H_2O}
$$

$$
C_2H_5O \overset{\text{S}}{\underset{}{\|}} \\
\diagdown P - OH \ + \ C_2H_5OH \\
O_2N - \langle C_6H_4 \rangle - O
$$

7.8 Surfactants

There are two different classes of surface-active materials in seawater, those that are naturally present and those that have been added to the oceans by man's activities. Most of the analytical methods proposed for use in seawater actually measure the anthropogenic input and attempt as much as possible to eliminate interferences from naturally occurring compounds. Yet sea foam was known to exist long before detergents. It is to be expected that both kinds of surfactants would be concentrated at the air–sea interface.

There has been at least one method suggested for the measurement of dissolved organic carbon which is actually a measurement of the carbon content of the surface-active materials. Formaro and Trasatti [166] dipped platinum electrodes into water, and measured the change in capacitance occurring as the organic materials were adsorbed onto the electrodes. This technique detected organic compounds down to 1 ppm C, which is about the level of organic carbon in oceanic surface water.

Both polarographic and spectrophotometric methods have been devised for the measurement of surface activity in seawater.

The presence of surface-active material depresses the oxygen reduction maximum so that if all other paraameters are held constant, this reduction will be a function of surfactant concentration only [167]. The result of the analysis is expressed in Triton-X-100 equivalents.

(a) Anionic detergents

The spectophotometric methylene blue method for anionic surfactants has been applied to seawater. In one version, the surfactants are collected in ethyl acetate. The solvent is then evaporated, the surfactants put back in solution in water and the standard spectrophotometric methylene blue method is applied to this solution. In this manner, the salt error introduced by seawater is eliminated [168]. A similar method with the methylene blue-surfactant complex extracted into chloroform and measured directly was proposed by Hagihara [169].

Titration methods are often used as alternatives to the spectrophotometric methods. Wang, Yang and Wang [170] determined anionic surfactants by adding an excess of cationic surfactant, acidifying then titrating with standard sodium tetraphenylboron. The method was adapted for seawater by the same authors [171], and a simplified field kit was described.

The methylene blue reaction can also be used in a fractionation procedure for surfactants. The complexes with methylene blue can be collected in an organic solvent, concentrated, dissolved in methanol and separated by high pressure liquid chromatography [172]. A variation of this method, permitting the collection of surfactant from large volumes of sample, should be workable in seawater.

Colour forming reagents other than methylene blue have been employed, particularly when specific surfactants were sought. Favretto and Tunis [173] extracted polyoxyethylene alkylphenyl ethers as picrates into an organic solvent, complexed the polymer with sodium ion and measured the absorption of the complex. This is one of the few specific methods available.

A rough estimate of the total amount of anionic surfactant present can be obtained by reacting the surfactant with a metal-containing material, such as bis(ethylene diamine) copper (II) [174, 178, 179], or o-phenanthroline-$CuSO_4$ [175–177], extracting the complex into an organic solvent and determining the metal by atomic absorption and this is the basis of more recent methods of analysis.

Cationic surfactants can be determined by adding an excess of anionic surfactant, then measuring this excess in the same manner.

Le Bihan and Courtot-Coupez [175, 176] analysed fresh water and seawater as follows. To 1 litre of filtered seawater was added, with shaking, 10 ml hydrochloric acid and 10 ml 0.023 M copper 1,10-phenanthroline sulphate. After 5 min 43 ml isobutyl methyl ketone were added, shaken vigorously for 1 min allowed to stand for 5 min and, after separating the phases, re-extracted with 25 ml of the ketone. The copper was determined by atomic absorption spectrometry using the 324.7 nm line. A blank was prepared from seawater containing about 1% of the amount of detergent to be determined. Calibration graphs must be prepared for each anionic detergent. The following species (weight per litre) do not interfere: CO^{II} and Ni^{II} (10 mg), HSO_4^- (3 g), $H_2PO_4^-$ (0.2 g), Bg^- (10 mg) and I^- (2 mg), Fe^{3+} (30 mg), CrO_4 (1 mg), SCN (20 mg) and NO_3^- (1 mg) produce a small constant increase in the atomic absorption, and hydrogen sulphide must be eliminated by oxidation. Non-ionic detergents do not interfere.

Le Bihan and Courtot-Coupez [177] used the same complex and flameless atomic absorption spectroscopy to determine anionic detergents. Crisp, Eckert and Gibson [180] were the first to use bis(ethylenediamine) Cu^{II} ion for the determination of anionic detergents. They determine the concentration of detergents by flame atomic absorption spectroscopy or by a colorimetric method. The colorimetric method was more sensitive with a limit of detection of $0.03 \, \mu g \, l^{-1}$ (as linear alkyl sulphonic acid) compared with $0.06 \, \mu g \, l^{-1}$ for atomic absorption spectroscopy. Crisp et al. [181] determined anionic detergents in fresh estuarine and seawater, at the ppb level. The detergent anions in a 750 ml water sample are extracted with chloroform as an ion association compound with the bis(ethylene-diamine) copper (II) cation and determined by atomic absorption spectrometry using a graphite furnace atomizer. The limit of detection (as linear alkyl sulphonic acids) is $2 \, \mu g \, l^{-1}$.

Gagnon [178] has described a rapid and sensitive atomic absorption spectrometric method developed from the work of Crisp, Eckert and Gibson [180] for the determination of anionic detergents at the ppb level in natural waters. The method is based on determination by atomic absorption spectrometry using the bis(ethylenediamine)copper (II) ion. The method is suitable for detergent concentrations up to $50 \, \mu g \, l^{-1}$ but it can be extended up to $15 \, mg \, l^{-1}$. The limit of detection is $0.3 \, \mu g \, l^{-1}$.

The recovery of different concentrations of detergents added to seawater was used to evaluate the accuracy of the method. Six duplicate determinations were carried out at different levels (Table 7.13). The recovery is 80% at $1 \, \mu g \, l^{-1}$ but reaches 90% at $10 \, \mu g \, l^{-1}$. The recovery is 97%, or better at higher concentrations. Precision was very good (Table 7.14).

Bhat, Eckert and Gibson [179] used complexation with the bis(ethylenediamine) copper (II) cation as the basis of a method for estimating anionic surfactants in fresh estuarine and seawater samples. The complex is extracted into chloroform and copper measured spectrophotometrically in the extract using 1,2(pyridyl azo)-2-naphthol.

Bhat, Eckert and Gibson [179] using the same extraction system were able to improve the detection limit of the method to $5 \, \mu g \, l^{-1}$ (as linear alkyl sulphonic acid) in fresh estuarine and seawater samples.

Hon Nami and Hanya [182, 183] have investigated the applicability of a combined gas chromatographic–mass spectrometric method to the determination of linear alkyl benzene sulphonates in chloroform extracts of 5 litre samples of

Table 7.13 Recovery of sodium dodeylsulphate (SDS) in seawater

SDS added (μg l^{-1})	Number of samples	Mean SDS found (μg l^{-1})	Mean recovery (%)
1.0	10	0.8	80
10.0	6	9.2	92
20.0	6	18.4	92
30.0	6	29.1	97
40.0	6	38.8	97
50.0	6	49.5	99

Reproduced from Gagnon, M.J. (1979) *Water Research*, **13**, 53, Pergamon Publications, by courtesy of author and publishers.

Table 7.14 Precision of method for determining detergent in seawater

Linear alkyl sulphuric acid taken*	s(μg)	s_1† (%)	LAS taken* (μg)	s(μg)	s_r (%)
2.5	0.3	12	50.0	1.2	2
10.0	0.3	3	250.0	2.0	1
25.0	0.5	2			

*Twelve determinations were carried out at each level. Between 84 and 97% recovery of linear alkyl sulphonate was obtained from seawater samples.
†s_r% = relative standard deviation.
Reproduced from Gagnon, M.J. (1979) *Water Research* **13**, 53, Pergamon Publications, by courtesy of authors and publishers.

estuary and bay water samples. They also determined the ratio of the concentration of linear alkyl benzene sulphonates to those of methylene blue active substances [184]. Linear alkyl benzene sulphonates were determined as their methyl sulphonate derivatives.

Hon Nami and Hanya [182] found that the LAS/MBAS ratio was less than 0.2, i.e. most of the methylene blue active substance was not linear alkyl benzene sulphonate. It has been reported that many organic and inorganic substances react with the methylene blue cation, for example, all sorts of anionic surfactants, chloride, nitrate and thiocyanate, etc. [185, 186]. A 3.3% of sodium chloride solution was found to give the same methylene blue active substance response as 0.06 ppm n-dodecylbenzene sulphonate. Therefore, the non-surfactant methylene blue active substance response except linear alkyl benzene sulphonate found in seawater appears to be largely chloride interference.

2-Phenyl, 3-phenyl and 4-7-phenyl C_{10} to C_{13} linear alkyl benzene sulphonates were determined by this procedure in a range of estuary and seawaters (Figure 7.12) and the procedure was used to measure the changes in concentration of these substances due to biodegradation reactions.

In cases where the information of interest is the total amount of surfactant present, methods like these seem the most useful. The results must be expressed in terms of equivalents of some standard surfactants; if identification of the components is required, some method such as the liquid chromatographic

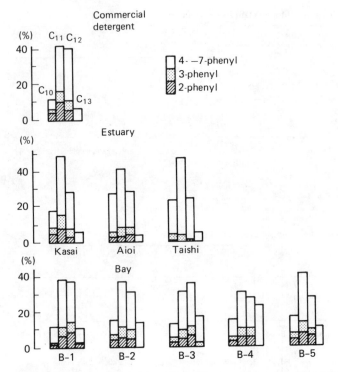

Figure 7.12 Percent composition of homologues and isomers of LAS in estuary and bay water (Reproduced from Hon-Nami, H. and Hanya, T. (1980) *Water Research*, **14**, 125, Pergamon Publications, by courtesy of authors and publishers.)

separation must be employed. If our interest is in naturally occurring surfactants, rather than the anthropogenic compounds, then either the polarographic [167] or the spectrophotometric [168] methods adapted to seawater should be used.

(b) Non-ionic detergents

Courtot-Coupez and Le Bihan [187, 188] determined non-ionic detergents in sea- and fresh-water samples at concentrations down to $2 \mu g l^{-1}$ ppm by benzene extraction of the tetrathiocyanatocobaltate (II) $(NH_4)_2 (Co(SCN)_41)$ [189] detergent ion-pair followed by atomic absorption spectrophotometric determination of cobalt [187].

Courtot-Coupez and Le Bihan [187] determined the optimum pH (7.4) for extraction of non-ionic surfactants with the above complex–benzene system. Cobalt in the extract is estimated by atomic absorption spectrometry after evaporation to dryness and dissolution of the residue in methyl isobutyl ketone. The method is applicable to surfactant concentrations in the range $0.02–0.5 \, mg \, l^{-1}$ and is not seriously affected by the presence of anionic surfactants.

Crisp, Eckert and Gibson [190] have described a method for the determination of non-ionic detergent concentrations between 0.05 and $2 \, mg \, l^{-1}$ in fresh, estuarine and seawater based on solvent extraction of the detergent–potassium tet-rathiocyanotozincate (II) complex followed by determination of extracted zinc by atomic absorption spectrometry. A method is described for the determination of non-ionic surfactants in the concentration range $0.05-2\,\text{mg}\,1^{-1}$. Surfactant molecu-les are extracted into 1,2-dichlorobenzene as a neutral adduct with potassium tetrathiocyantozincate (II) and the determination is completed by atomic absorp-tion spectrometry. With a 150 ml water sample the limit of detection is $0.03\,\text{mg}\,1^{-1}$ (as Triton X-100). The method is relatively free from interference by anionic surfactants; the presence of up to $5\,\text{mg}\,1^{-1}$ of anionic surfactant introduces an error of no more than $0.07\,\text{mg}\,1^{-1}$ (as Triton X-100) in the apparent non-ionic surfactant concentration.

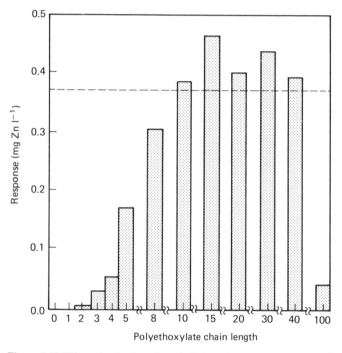

Figure 7.13 Effect of polyethoxylate chain length. Response of $1\,\text{mg}\,1^{-1}$ solutions of *n*-nonylphenol ethoxylates, containing 2–100 ethoxy units per molecule, is expressed as the concentration of zinc in the acid back-extract. The dashed line indicates the zinc concentration obtained from Triton X-100 $(1\,\text{mg}\,1^{-1})$ (Reproduced from Crisp, P.T., *et al.* (1979) *Analytica Chimica Acta*, **104**, 93, Elsevier Science Publishers, by courtesy of authors and publishers.)

Figure 7.13 shows the effect of polyethoxylate chain length. The response to the method of a range of *n*-nonylphenol ethoxylates containing up to 100 ethoxy units per molecule was studied. Response is greatest between 8 and approximately 40 ethoxy units. The effects of various ions on the recovery of Triton X-100 $(1\,\text{mg}\,1^{-1})$ are shown in Table 7.15. Only sulphide, iron (III) aluminium (III), and chromium (III) ions interfere at concentrations likely to be found in contaminated waters. There is, however, no interference from sulphide at up to $100\,\text{mg}\,1^{-1}$ if the water is

Table 7.15 Allowable concentrations of foreign ions*

Concentration	Ion
5 M	Cl^-, Na^+
0.5 M	NO_3^-, SO_4^{2-}
1000 mg l^{-1}	F^-, Br^-, I^-, NO_2^-, SCN^-, CH_3, CO_2^-, K^+, Mg^{2+}
100 mg l^{-1}	$P_3O_{10}^{5-}$, NH_4^+, Ca^{2+}, Ni^{2+}, Co^{2+}, Mn^{2+}, Zn^{2+}
10 mg l^{-1}	Cu^{2+}
0.1 mg l^{-1}	S^{2+}, Fe^{3+}, Al^{3+}, Cr^{3+}

*95–105% recovery of Triton X-100 (1 mg l^{-1}) was obtained in the presence of the stated condition.
Reproduced from Crisp, P.T., *et al.* (1979) *Analytica Chimica Acta*, **104**, 93, Elsevier Science Publishers, by courtesy of authors and publishers.

Table 7.16 Recovery of Triton X-100 from water and seawater in the presence of linear alkylbenzene sulphonates

Linear alkylbenzene sulphonates	Triton X-100 added (mg l^{-1})	Mean Triton X-100 found (mg l^{-1})	
		In water	*In seawater*
5	0	0.04	0.03
5	0.10	0.15	0.15
5	1.00	1.04	1.04
5	2.00	2.07	2.00
0.5	0	0.03	0.03
0.5	0.10	0.13	0.15
0.1	0	0.03	0.03
0.1	0.10	0.13	0.15

Reproduced from Crisp, P.T., *et al.* (1979) *Analytica Chimica Acta*, **104**, 93, Elsevier Science Publishers, by courtesy of authors and publishers.

treated with 1 ml 30% hydrogen peroxide, shaken, and allowed to stand for 10 min before addition of the zinc thiocyanate reagent. Interference from up to 100 mg l^{-1} concentrations of iron (III), aluminium (III) or chromium (III) is suppressed by addition of 1 g of EDTA (disodium salt dihydrate) prior to the addition of the zinc thiocyanate reagent.

The performance of this method in the presence of anionic surfactants is of special importance, since most natural or wastewater samples which contain non-ionic surfactants also contain anionic surfactants. Recovery data from Triton X-100 from water and seawater containing the anionic surfactants (linear alkylbenzene sulphonates) are given in Table 7.16. The presence of up to 5 mg l^{-1} linear alkylbenzene sulphonate increases the apparent concentration of non-ionic surfactant by a maximum of 0.07 mg l^{-1} (as Triton X-100). Soaps, such as sodium stearate, do not interfere with the recovery of Triton X-100 (1 mg l^{-1}) when present at the same concentration (i.e. 1 mg l^{-1}). Cationic surfactants, however, form extractable non-association compounds with the tetrathiocyanatozincate ion and interfere with the method.

7.9 Fatty acids and hydroxy acids

(a) Spectrophotometric method for total fatty acids

If a measurement of total fatty acid concentration is desired, short of attempting to sum the amounts of each compound found by gas chromatography, some indirect method must be employed. Harwood and Huyser [191] made iron (III) hydroxamates of the fatty acids and measured these colorimetrically.

Treguer, Le Corre and Courtot [192] determined total dissolved free fatty acids in seawater. The sample (1 litre) was shaken with chloroform (2 × 20 ml) to remove the free fatty acids and the extract evaporated to dryness under reduced pressure at 50°C. Chloroform–heptane (29:21) (2 ml) and fresh copper reagent (1M triethanolamine–1 M acetic acid–6.8% $CuSO_45H_20$ solution (9:1:10) (1.5 ml) was added to the residue. The solution was shaken vigorously for 3 min and centrifuged at 3000 rpm for 5 min. A portion (1.6 ml) of the organic phase was evaporated to dryness and 1% ammonium diethyldithiocarbamate solution in isobutylmethyl ketone (2 ml) was added to the residue to form a yellow copper complex. The copper in the solution was determined by atomic absorption spectrophotometry at 324.8 nm (air–acetylene flame). Palmitic acid was used to prepare a calibration graph. The standard deviation for samples containing $30 \, \mu g \, l^{-1}$ of free fatty acids (as palmitic acid) was $\pm 1 \, \mu g \, l^{-1}$.

A method that has not yet been applied to seawater, but which should work, has been suggested by Dunges [193]. He labelled monocarboxylic aliphatic acids with 4-bromomethyl-7-methoxycoumarin, forming a fluorescent derivative. The fluorescence can be measured directly or the derivative can be concentrated by extraction into an organic solvent. In this last variation, separation of the different fatty acid derivatives by liquid chromatography would seem an obvious next step.

Much of the effort that has been expended on the measurement of fatty acids has been devoted to glycollic acid, largely because of its prevalence in the medium in which axenic phytoplankton cultures have been grown. Antia et al. [194] proposed a colorimetric method comprising a concentration step and a reaction with 2,7-dihydroxynaphthalene. This method has a detection limit of $0.1 \, mg \, l^{-1}$. Methods comprising concentration by adsorption on alumina, followed by elution and colour development, yielded a detection limit of $5 \, ng \, l^{-1}$ [195, 196]. These methods have not been too successful in hands other than those of the original authors, possibly because of the amount of manipulation necessary in the analysis.

Stradomskaya and Goncharova [197] have developed an extraction and colour development method for formic acid which should work in seawater. The sample is acidified to pH 2 and extracted with diethyl ether. After removal of the ether, the formic acid is reduced to formaldehyde and determined spectrophotometrically with chromotropic acid. A sensitivity of $1 \, \mu g \, l^{-1}$, with a normal range of $0.9 \, \mu g \, l^{-1}$, is claimed.

(b) Chromatographic procedures–column chromatography

All of the various chromatographic methods have been used for the separation and subsequent determination of the individual fatty acids. The earlier work used column and paper chromatography for the concentration and separation of the acids, with spectrophotometric methods for the final measurement. Thus Mueller, Larson and Ferretti [198] concentrated the acids by evaporation and extraction, then separated them by column and paper chromatography. Koyama and Thompson [199] used vacuum distillation and extraction for the concentration step, and column chromatography for the separation. Makeimova and Pimenova [200], working with culture medium in which phytoplankton had been grown, separated derivatives of the fatty acids on paper chromatograms. These methods, although effective, are tedious; if they were to be attempted today, they would undoubtedly be adapted to thin-layer and liquid chromatography.

The fatty acids measured by these techniques have all been small monomeric molecules. Lamar and Goerlitz [201] studied the acidic materials in highly coloured water and found that most of the non-volatile material was composed of polymeric hydroxy carboxylic acids, with some aromatic and olefinic unsaturation. Their methods included gas, paper and column chromatography with infrared spectrophotometry as the major technique used for the actual characterization of the compounds.

Horikawa [202] has adapted a thermal detector for the determination of formic, acetic and propionic acids by liquid chromatography; little else has been done with liquid chromatography of the fatty acids.

Gorcharova and Khomenko [203] have described a column chromatographic method for the determination of acetic, propionic and butyric acids in seawater and thin-layer chromatographic methods for determining lactic, aconitic, malonic, oxalic, tartaric, citric and malic acids. The pH of the sample is adjusted to 8–9 with sodium hydroxide solution. It is then evaporated almost to dryness at 50–60°C and the residue washed on a filter paper with water acidified with hydrochloric acid. The pH of the resulting solution is adjusted to 2–3 with hydrochloric acid (1:1), the organic acids are extracted into butanol, then back-extracted into sodium hydroxide solution; this solution is concentrated to 0.5–0.7 ml, acidified and the acids separated on a chromatographic column.

Application of high performance liquid chromatography to the resolution of complex mixtures of fatty acids [204, 205] has provided an alternative to the high temperature separation obtained by gas chromatography. Both techniques have similar limits of detection, but lack the ability to analyse directly environmental samples. Analysis requires that the fatty acids be separated from the organic and inorganic carbon matrices followed by concentration. Typically, these processes can be accomplished simultaneously by the appropriate choice of methods. Initial isolation of the fatty acids is based on the relative solubility of the material of interest in an organic phase compared with the aqueous phase. Secondary separation is determined by the functional group content and affinity for a solid support.

(c) Gas chromatography

One of the earliest applications of gas chromatography to marine problems was in the measurement of fatty acids in seawater. In general, the gas chromatographic methods have used extraction into an organic solvent, followed by the formation of a volatile derivative, often the methyl ester. Garrett et al. [206] applied these techniques to materials collected from the surface layer with his screen finding fatty acids from C_8–C_{20} to be present. Slowey, Jeffery and Hood [207] applied similar analytical techniques to material isolated from the water column.

Earlier analytical results from the gas chromatographic analysis of fatty acids seemed to be very high. Williams [208] repeated much of the early work, using almost extreme care in the avoidance of contamination, and found very much smaller quantities. Papers concerned with the fatty acid content of the water column and sediment include references [209–215].

Analyses of the surface film have been performed [216–220] and analytical methods for the volatile members of the group have been reported [221–224].

Gas chromatography and gas chromatography–mass spectrometry have also been used to measure fatty alcohols [225], phytol [226] and the several isomers of inositol [227].

7.10 Aldehydes and ketones

Many different kinds of methods have been applied to the measurement of aldehydes and ketones in seawater. In most cases, the methods measure one particular aldehyde, usually formaldehyde, but some methods do differentiate between the various compounds.

A spectrophotometric method for aldehydes in either fresh or seawater was described by Kamata [228]. It used the colour-forming reaction between the aldehyde, 3-methyl-2-benzothiazolone hydrazone and ferric chloride and claimed a sensitivity of $0.01\,mg\,l^{-1}$ as formaldehyde equivalents. While Kamata clearly found evidence of the presence of aldehydes, the method appears to be not quite sensitive enough for the quantities to be found in seawater.

Automated colorimetric and fluorimetric methods for aldehydes were proposed by Afghan et al. [229]. The colorimetric method used chromatotropic acid, while the fluorimetric method was based on the reaction between formaldehyde, 2,4-pentadione and ammonia. The sensitivities of these methods were about the same as that of the Kamata method; they could be employed usefully in fresh water, but were marginal for seawater. A more sensitive version of this fluorimetric method for formaldehyde was developed by Zika [230] using the reaction as described by Belman [231]. This version had a sensitivity of 1×10^{-7} M in sea water.

A continuous flow system utilizing the oxidation of formaldehyde and gallic acid with alkaline hydrogen peroxide to produce a chemiluminescence was studied by Slawinska and Slawinski [232]. While the major peak of the chemiluminescence spectrum occurred at 635 nm, the photomultiplier used summed all of the available light between 560 and 850 nm. The intensity of the chemiluminescence was linearly proportional to formaldehyde concentration from 10^{-7} to 10^{-2} M, producing a detection limit of $1\,\mu g\,l^{-1}$. This method should be sensitive enough for use in seawater; however, it has not yet been tested in marine systems.

Another fresh-water method which holds some promise for seawater analysis is twin cell potential sweep voltammetry, as proposed by Afghan, Kulkarni and Ryan [233]. In this method, semicarbazones are formed by reaction with semicarbazide hydrochloride and the peak height of the semicarbazone polarogram is then measured. The shape of the polarographic peak is an indication of the type of compound present. The detection limit of this method is 1×10^{-8} M and the effective range runs from 10^{-8} to 10^{-4} M.

For those aldehydes and ketones which are volatile enough, gas chromatography of the headspace gases can be used and this has been used to measure acetone, butyraldehyde and 2-butanone in oceanic waters. In a gas chromatography–mass spectrometry analysis of a single sample of volatile materials concentrated from inshore waters onto Tenax GC, MacKinnon [234] reported tentative identification of methyl isopropyl ketone, bromoform, 4-methyl-2-pentanone, 2-hexanone and 2-hexanal.

Eberhardt and Sieburth [235] devised a spectrophotometric procedure for the determination of aldehydes in seawater. The method is based on the reaction of aldehydes with 3-methyl-2-benzothiazolinone hydrazone hydrochloride and ferric chloride to produce a coloured compound. A detection limit of $0.072 \mu M$ formaldehyde per litre was obtained using a 5 cm path-length.

7.11 Phthalate esters

Among the most ubiquitous of man's contributions to the environment are the phthalate esters. These compounds are used extensively in the plastics industry and have been found in both fresh and seawater at concentrations in the $\mu g \, l^{-1}$ range. The usual method of analysis at this level of concentration is gas chromatography; some form of concentration, usually adsorption on a column, is always needed, and great care is required to keep the background levels sufficiently low. Methods of collection, storage and analysis have been described [236–238]. The use of coupled gas chromatography–mass spectrometry as well as liquid chromatography was discussed by Hites [239].

7.12 Carbohydrates

Of all of the classes of organic materials to be found in seawater, the carbohydrates are probably the most widely investigated. This is partly because of their role in photosynthesis and their function as storage and structural compounds in algae. Concentrations as low as a few $\mu g \, l^{-1}$ are of significance.

The early methods were all spectrophotometric and were usually based on the condensation of carbohydrates with a concentrated acid, usually sulphuric, followed by reaction of the condensed product with some colour-forming compound including N-ethylcarbazole, anthrone [240]; phenol; orcinol and tryptophan [241].

Some comparative studies of these methods have been made; anthrone and N-ethylcarbazole were compared by Lewis and Rakestraw [240] and by Collier [242] and anthrone, phenol, orcinol and N-ethylcarbazole and L-tryptophan were examined by Josefsson, Uppstrom and Ostling [243]. In general, the comparative studies show anthrone to be more reliable than N-ethylcarbazole, although somewhat less sensitive. However, Josefsson and co-workers [241] found that of the five methods, the tryptophan method gave the best results when adapted to automatic analysis and was capable of analysis at concentrations of interest in seawater analysis, i.e. $\mu g \, l^{-1}$ levels.

The acid condensation methods do not distinguish between monosaccharides and polysaccharides as the various classes of carbohydrates each have different absorption maxima, which results in different molar absorptions at any chosen wavelength. Furthermore, when treated with concentrated sulphuric acid, some three and four-carbon compounds will condense into structures which will produce colours with those reagents. When the object of the analysis is to obtain some estimate of the total amount of carbohydrate or carbohydrate-like material present, the inclusiveness of these methods is useful. However, when the object is to distinguish between the easily metabolized simple sugars and the complex storage and structural materials, these methods give no information at all.

A spectrophotometric method that does not use condensation with sulphuric acid was proposed by Mopper and Gindler [244]. The method used the copper (I) complex with 2,2-bicinchoninate to form a colour with simple sugars, with a hydrolysis step which is included to make simple sugars from the polysaccharides. Thus the analysis of two aliquots, one of them hydrolysed, would yield values for both simple and combined sugars.

Another spectrophotometric method measuring both simple and combined sugars was described in papers by Johnson and Sieburth [245] and Burney and Sieburth [246]. The basic method comprised reduction of sugars to alditols with sodium borohydride and oxidation of the alditols to form free formaldehyde. The formaldehyde was then determined spectrophotometrically with 3-methyl-2-benzothiazolinone hydrazone hydrochloride.

In the determination of carbohydrates sensitivity can often be increased by using fluorescence rather than absorbance for the final determination. With compounds that are not normally fluorescent, it becomes necessary to find fluorescent derivatives. Hirayama [247] concentrated the carbohydrates in coastal water samples, using electrodialysis and evaporation and made fluorescent derivatives using anthrone and 5-hydroxyl-1-tetralone determining pentoses separately from hexoses in the process. While this method does seem to have the extra sensitivity expected from fluorescent methods, the extra manipulations render it unsatisfactory for routine use.

The methods discussed above are all class reactions, designed to estimate the total amount of carbohydrate present and usually actually furnishing some sort of weighted average, weighted by the unequal responses of the different classes of sugars. There are a few methods that are specific for a single class of carbohydrates or for a single sugar. It has long been suspected that uronic acids make up a considerable portion of the dissolved organic carbon in the ocean, but most of the carbohydrate methods do not measure these acids. Williams and Craigie [248] developed a modification of the acidic decarboxylation method of Lefevre and Tollens and found uronic acids in phytoplankton and in particulate organic carbon. A direct and somewhat simpler procedure was developed by Mopper [249].

Enzymatic methods are usually very specific and sensitive. Unfortunately the only methods in the literature for carbohydrates are all for glucose. Hicks and Carey [250] reported such a method, with a fluorimetric final measurement, which was useful down to 3×10^{-8} M. Andrews and Williams [251] used a pre-concentration step, sorption onto charcoal, elution and a final determination with glucose oxidase.

Chromatographic techniques are required to distinguish between the different classes of carbohydrates. Chromatographic methods for both fresh and seawater were described by Whittaker and Vallentyne [252]; Degens, Reudter and Shaw [253] and Starikova and Yablokova [254, 255]. These methods often include a hydrolysis step, to permit the measurement of the monomers held in polysaccharides. A discussion of these methods, together with a comparison between paper chromatographic, colorimetric and enzymatic methods, can be found in Geller [256].

The discovery of the usefulness of the trimethylsilyl derivatives for the gas chromatography of simple sugars was a major step forward in the analysis of complex mixtures of carbohydrates. The derivatives are not difficult to synthesize and are easily separated. These methods quickly became the standard carbohydrate methods in biochemical investigations, and attempts were made to adapt them to use in seawater. The reagents used in making the derivatives are easily hydrolysed; the samples must therefore be completely free of water. If simple freeze-drying is used to remove the water, the volume of sea salts left behind is so great that the volume of solvent necessary for the reaction results in too great a dilution of the carbohydrate derivatives. Therefore, a concentration and desalting step is usually necessary.

Such methods, usually also including a hydrolysis step to break down polysaccharides, have been described by Modzeleskie, Laurie and Nagy [257] and Tesarik [258].

Josefsson and Roos [259] developed a method for sensitive gas chromatographic analysis of monosaccharides in seawater, using trifluoroacetyl derivatization and electron capture detection. It is difficult to determine accurately the monosaccharide concentrations by this method because a number of chromatographic peaks result from each monosaccharide.

These separations suffer from an unexpected disability, they are, if anything, too good. They not only separate the various simple sugars, they also separate the anomers of each sugar. Thus the final chromatographic records become hopelessly confusing, with considerable overlapping of peaks, even when capillary columns are used [258]. This problem was also discussed by Josefsson [260], who recommended a separation by liquid chromatography after desalting by ion-exchange membrane electrodialysis. The electrodialysis cell used had a sample volume of 430 ml and an effective membrane-surface area of 52 cm^2. Perinaplex A-20 and C-20 ion-exchange membranes were used. The water-cooled carbon electrodes were operated at up to 250 mA and 500 V. The desalting procedure normally took less than 30 h. After the desalting, the samples were evaporated nearly to dryness at 40°C in vacuo, then taken up in 2 ml of 85% ethanol and the solution was subjected to chromatography on anion-exchange resins (sulphate form) with 85% ethanol as mobile phase. By this procedure, it was possible to determine eight monosaccharides in the range 0.15–46.5 µg l^{-1} with errors of less than 10% and to detect traces of sorbose, fucose, sucrose, diethylene glycol and glycerol in seawater.

Larsson and Samuelson [261] devised an automated system for determining carbohydrates in biological samples using partition chromatography for the separation and the orcinol colorimetric method for the final analysis. Later versions of this kind of autoanalyser, using tetrazolium blue or a Cu^I complex of bicinchoninate for the final measurement, have been reported [262].

Williams [263] has studied the rate of oxidation of ^{14}C-labelled glucose in seawater by persulphate. After the oxidation, carbon dioxide was blown off and residual activity was measured. For glucose concentrations of 2000, 200 and 20 µg l^{-1} residual radioactivities (as percentage of total original radioactivity) were 0.04, 0.05 and 0.025 respectively, showing that biochemical compounds are extensively oxidized by persulphate. With the exception of change of temperature, modifications of conditions had little or no effect. Oxidation for 2.5 h at 100°C was the most efficient.

While there appears to be a profusion of methods for the analysis of carbohydrates in seawater, a study of the actual capabilities of the methods soon reveals that few of them furnish us with much useful information. At the moment, only the methods of Johnson and Sieburth [245] and Burney and Sieburth [246] and the bicinochoniate method of Mopper and Gindler [244] furnish any real estimate of the total amount of carbohydrate present in seawater. For the analysis of the separate sugars, liquid chromatography of carbohydrate derivatives would seem to be the obvious choice.

Several methods of determining carbohydrates have been described [272–279].

7.13 Phenols

While the major emphasis in the analysis of phenols in seawater has been on those compounds introduced by industrial processes, as much phenolic material is probably added by the disintegration of fixed algae in the intertidal regions. A high value for total phenols, particularly in coastal waters, cannot be interpreted simply as a high degree of industrial pollution; the kinds of phenols present must also be ascertained.

A number of colorimetric methods for phenols in seawater have been reported. Alekseeva [264] oxidized the phenols to antipyrin dyes, then extracted the dyes into an organic solvent of considerably smaller volume. The reported detection limit of the method was $20 \mu g \, l^{-1}$ with a linear range of $20-80 \mu g \, l^{-1}$. Goulden [265] used an automated system with steam distillation of the phenols from acidic solution, followed by formation of a coloured derivative and extraction of the derivative into an organic solvent for the final determination. They evaluated both 4-amino-antipyrine and 3-methyl-2-benzothiazolinone as reagents, with a detection limit of $0.2 \mu g \, l^{-1}$ for either reagent. Nitroaniline was used as the colorimetric reagent by Schlungbaum and Behling [266], while Stilinovic et al. [267] used both nitroaniline and 4-aminoantipyrine and Gales [268] modified the 3-methylbenzothiazolinone method to lower the detection limit to $1 \mu g \, l^{-1}$.

Phenolic substances can be measured directly, without colour development, by the difference in their ultraviolet absorption in acidic and basic solution [269]. Interference due to the ultraviolet absorption of non-ionizing nonphenolic organic species are cancelled out by this difference method. The method was adapted for natural fresh waters by Fountaine et al. [270], with the use of two sealed hollow cathode lamps to monitor the difference spectrum. Comparison with the 4-aminoantipyrine colorimetric method generally showed higher values for the ultraviolet difference method, probably due to the presence of some para-blocked phenols, which will not react in the colorimetric procedure. A simple photometric difference instrument has been developed with a range of $100 \mu g \, l^{-1}$ full scale.

7.14 Nitrogen compounds

Simply on the basis of the normal composition of marine organisms, we would expect proteins and peptides to be normal constituents of the dissolved organic carbon in seawater. While free amino acids might be expected as products of enzymatic hydrolysis of proteins, the rapid uptake of these compounds by bacteria would lead us to expect that free amino acids would normally constitute a minor part of the dissolved organic pool. This is precisely what we do find; the concentration of free amino acids seldom exceeds $150 \mu g \, l^{-1}$ in the open ocean. It would be expected that the concentration of combined amino acids would be many times as great. There have been relatively few measurements of proteins and peptides and most of the measurements were obtained by measuring the free amino acids before and after a hydrolysis step. Representative methods of this type have been described [280–285]. Since these methods are basically free amino acid methods, they will be discussed in conjunction with those methods.

(a) Proteins and peptides

The older methods for the measurement of protein in natural waters usually depended upon the presence of aromatic amino acids in the protein, and calculated total protein on the basis of an average tyrosine, tryptophan or phenylalanine content. A method representative of this type was the Folin reagent method published by Debeika, Fyabov and Nabivanets [286]. While these methods were useful in fresh water and in some coastal regions, they were not sensitive enough for the lower concentrations to be found in oceanic waters.

A fluorescence method which would measure either free or combined amino acids, depending upon the pH of the solution, was originally proposed by Udeufriend et al. [287] and adapted for seawater by North [288], and Packard and Dortch [289]. In this Fluran method, peptides normally yield maximum fluorescence at pH 7, while amino acids fluoresce best at pH 9; with the proper choice of buffers, the fluorescence of peptides and proteins can be differentiated from that due to free amino acids.

An attempt to use the infra-red spectrum of materials collected at the sea surface for a quantitative measure of composition has been made by Baier et al. [290]. They dipped a germanium crystal through the surface film, then ran an internal reflectance spectrum on the material clinging to the crystal. From the spectrum, they concluded that the bulk of the material present in the surface film was there as glycoproteins and proteoglycans.

(b) Free amino acids

Although the free amino acids are present only at very low concentrations in oceanic waters, their importance in most biological systems has led to an inordinate amount of effort toward their determination in seawater. A sensitive, simple, and easily automated method of analysis, the colorimetric ninhydrin reaction, has been known in biochemical research for many years. In order for the method to be useful in seawater, the amino acids had to be concentrated. This concentration was usually achieved by some form of ion-exchange [291]. An automated method not using a concentration step was developed by Coughenower and Curl [292]. While the method was used successfully in Lake Washington, its limit of detection ($0.5 \, \mu mol \, l^{-1}$) is just too great for most oceanic samples.

As we have seen with other organic compounds, a shift from colorimetric to fluorimetric methods often brings about an increase in sensitivity of an order of magnitude or more. The Fluran or fluorescamine method already mentioned [289] offers almost two orders of magnitude greater sensitivity, and can be used in seawater without a concentration step. A further increase in sensitivity can be gained by the addition of dimethylsulphoxide to the final solution [293]. These methods, like all fluorescent class reactions, are only semi-quantitative, since they really measure amine groups and report the results in terms of the concentration of some model compound, perhaps as glycine equivalents. This particular group of methods is also sensitive to all primary amines, and not just to amino acids.

Creatine has been measured by a fluorescence method based on the reaction with ninhydrin [294]. This method has also been adapted for automated analysis. The values found ranged around $0.1 \, \mu g \, l^{-1}$.

Chromatographic methods
Several of the earlier methods used paper chromatography for the final separation and determination, after some concentration step. Starikova and Korzhikova [283] concentrated the free amino acids by taking the acidified solutions to dryness, then extracting the sea salts with 80% ethanol. Riley and Segar [281] dropped the detection limit to $0.1 \, \mu g \, l^{-1}$, using TLC. Litchfield and Prescott [295] made dansyl derivatives, extracted these into an organic solvent, then separated the amino acids by TLC. Ligand exchange was used as a concentrating mechanism by Clark, Jackson and North [296] followed by TLC for the final separation. The formation of 2,4-dinitro-1-fluorobenzene derivatives, followed by solvent extraction of these derivatives and circular TLC was suggested by Palmork [297].

Ligand exchange has been a favoured method for the concentration of amino acids from solution because of its selectivity [298]. Ion exchange has often then been used for the final separation of the acids [299–302].

Other methods of concentration have also been used to bring the levels of amino acids into a range suitable for ion-exchange chromatography. Bohling [303, 304] and Garrasi and Degens [305] used evaporation or freeze drying, followed by extraction of the dried salts with organic solvents. Tatsumota *et al.* [306] used co-precipitation with ferric hydroxide as the method of concentration prior to ion-exchange chromatography.

Ion-exchange chromatography coupled with fluorimetric detector. Amino acid analyses based on ion-exchange resins are now available commercially. These achieve good separations of amino acid mixtures. Fluorescent derivatives of separated amino acids constitute a very sensitive means of detecting these compounds in seawater [294, 295]. Fluorescent derivatives that have been studied include *o*-phthalaldehyde [307], dansyl [308], fluorescamine [309] and ninhydrin [309].

The amino acid analyser using fluorescamine as the detecting reagent, has been used to measure 250 picomoles of individual amino acids routinely [310] and dansyl derivatives have been detected fluorimetrically at the 10^{-15} M level [308]. Where the amounts of amino acid are high enough, the fluorescamine method, with no concentration step, can be recommended for its simplicity. At lower concentrations, the dansyl method, with an extraction of the fluorescent derivatives into a non-polar solvent, should be more sensitive and less subject to interferences. For proteins and peptides, the fluorescamine method seems to be the most sensitive available method.

In this connection, it is interesting to note that Gardner [311] isolated free amino acids at the $20 \, nmol \, l^{-1}$ level in from as little as 5 ml of sample, by cation exchange, and measured concentrations on a sensitive amino acid analyser equipped with a fluorimetric detector.

The classical work of Dawson and Pertchard [312] on the determination of α-amino acids in seawater uses a standard amino acid analyser modified to incorporate a fluorimetric detection system. In this method the seawater samples

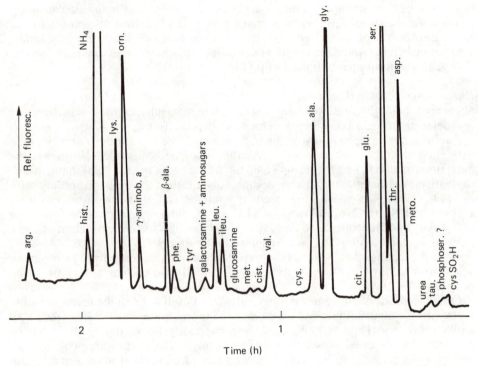

Figure 7.14 Chromatogram of a seawater extract (20 ml) sample for amino acids collected at 6 m in the Kiel Fjord. The concentrations of the individual acids were quantified as follows (in nmol l^{-1}): meto, 11; asp. 34.4; thr, 23.2; ser, 88; glu, 36; gly, 100; ala, 56; vol, 16; ileu, 9.6; leu, 12; galactosamine and aminosugars, 4; tyr, 6.8; phe, 7.2; B-ala, 20.8; α-aminoba, 14.4; orn, 44; lys, 12; hist, 7.2; arg, 8.6; cysSO$_2$H, 4; cit, trace; tan, cys, trace; glucoseamine, trace; met, trace; urea, trace; phosphoserine, trace; OH-lys, trace. The total concentration of amino acid in the sample lies around 51 µg l^{-1}, assuming a mean molecular weight of 100 (Reproduced from Dawson, R. and Pritchard, R.G. (1978) *Marine Chemistry*, **6**, 27, Elsevier Science Publishers, by courtesy of authors and publishers.)

are desalinated on cation-exchange resins and concentrated prior to analysis. The output of the fluorimeter is fed through a potential divider and low-pass filter to a compensation recorder.

Dawson and Pritchard [312] point out that all procedures used for concentrating organic components from seawater, however mild and uncontaminating, are open to criticism, simply because of the ignorance as to the nature of these components in seawater. It is, for instance, feasible that during the process of desalting on ion-exchange resins under weakly acidic metal chelates dissociated and thereby larger quantities of 'free' components are released and analysed.

An example of a chromatogram obtained from a seawater sample and the mole percentage of each amino acid in the sample is depicted in Figure 7.14.

Mopper [313] has recently discussed recent development in the reverse phase high performance liquid chromatographic determination of amino acids in seawater. He describes the development of a simple, highly sensitive procedure based on the conversion of dissolved free amino acids to highly fluorescent, moderately hydrophobic isoindoles by a derivatization reaction with excess o-phthalaldehyde and a thiol, directly in seawater. Reacted seawater samples were injected without further treatment into a reversed-phase high-performance liquid chromatography column, followed by a gradient elution. The eluted amino acid derivatives were detected fluorometrically. Detection limit for most amino acids was 0.1–0.2 nmol per 500 μl injection. Problems of inadequacies with the method itself, sample handling and whether chromatographically determined concentrations might be considered as biologically available concentrations in seawater, are discussed.

Gas chromatography.   Since the amino acids are not volatile, derivatives that are both volatile and easily synthesized had to be found. A silyl derivative was used by Pocklington [314] after the amino acids had been separated as a group and concentrated by extraction of the freeze-dried sea salts with an organic solvent, followed by evaporation of the solvent.

A still greater increase in sensitivity can often be achieved by using an electron capture detector instead of flame ionization, if the compound or its derivative contains a halogen atom. Derivatives of this sort had been found for the amino acid analyses; one procedure yielded *n*-butyl N-trifuluoroacetyl esters [315]. The method has been adapted for seawater and a similar method using the methyl ester was devised by Gardner and Lee [316]. While sufficient sensitivity could be achieved with the use of these techniques, there was far too much sample handling in the separation and derivative formation. With the development of easily synthesized fluorescent derivatives, liquid chromatography has largely supplanted gas chromatography for these analyses.

(c) Aliphatic and aromatic amines

Aliphatic amines have been determined by a number of methods. Batley, Florence and Kennedy [316] extracted the amines into chloroform as ion-associatioin complexes with chromate, then determined the chromium in the complex colorimetrically with diphenylcarbazide. The chromium might also be determined, with fewer steps, by atomic absorption. With the colorimetric method, the limit of detection of a commercial tertiary amine mixture was 15 ppb. The sensitivity was extended to 0.2 ppb by extracting into organic solvent the complex formed by the amine and Eosin Yellow. The concentration of the complex was measured fluorimetrically. Gas chromatography, with the separations taking place on a modified carbon black column, was used to measure aliphatic amines by Di Corcia and Samperi [318]. For aromatic amines, the reaction with 4-aminoantipyrine was used, with a colorimetric determination, by El-Dib, Abdel-Rahman and Aly [196].

Varney and Preston [320] discussed the measurement of trace aromatic amines in seawater using high performance liquid chromatography.

Aniline, methyl aniline, l-naphthylamine and diphenylamine at trace levels were determined using this technique and electrochemical detection. Two electrochemical detectors (a thin-layer, dual glassy-carbon electrode cell and a dual porous electrode system) were compared. The electrochemical behaviour of the compounds was investigated using hydrodynamic and cyclic voltammetry. Detection limits of 15 and 1.5 nmol l^{-1} were achieved using coulometric and amperometric cells respectively when using an in-line pre-concentration step.

(d) Urea

Urea has been of interest to the biological oceanographer because of its role as an excretion product of protein metabolism, its function in osmoregulation and its reported use as a nitrogen source for phytoplanakton growth.

Emmet [321] developed a colorimetric method involving chlorination of the urea with hypochlorite, followed by condensation with phenol. The limit of detection for this method was $0.2 \, \mu g \, l^{-1}$ as N. The method was easily adaptable to automatic analysis.

Most workers now use a colorimetric method based on the reaction of urea with biacetylmonoxime [322, 323]. The method has been adapted for automated analysis by De Marche, Curl and Coughenower [324].

(e) Hydroxylamine

Von Breymann, de Angelis and Gordon [325] have described a method for the determination of hydroxylamine in seawater based on gas chromatography with electron capture detection.

(f) Nitro-compounds

The gas chromatographic determination of isomers of dinitrotoluene in seawater has been described by Hashimoto and co-workers [326, 327]. These authors

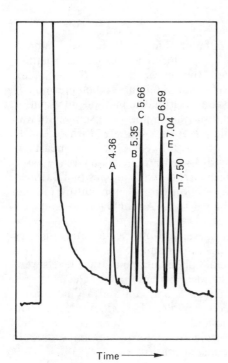

Time ⟶

Figure 7.15 Typical chromatogram of dinitrotoluenes with Apiezon L grease SCOT glass capillary column. Chromatographic peaks are: A, 2,6-DNT; B, 2,3-DDT; C, 2,5-DNT; D, 2,4-DNT; E, 3,4-DNT; and F, 3,5-DNT. Chromatographic conditions: detector, ^{63}Ni electron capture; carrier gas N_2, flow rate, 80 ml min^{-1}; column temperature 140°C; injection temperature 180°C and detector temperature 275°C. The numbers on the peaks are the retention times in minutes (Reproduced from Hashimoto, A., *et al.* (1980) *Analyst, London*, **105**, 509, Royal Society of Chemistry, by courtesy of authors and publishers.)

Table 7.17 Concentrations of 2,6- and 2,4-dinitrotoluene (μg l^{-1}) in seawater from Dokai Bay, Japan, from 4 April to 6 August 1979

Date	Sampling site	Sample*	Isomer 2,6-	Isomer 2,4-	2,6- to 2,4-isomer ratio
4.4.79	A	s	ND	1.19	0
		b	0.18	5.40	0.033
	B	s	ND	1.74	0
		b	ND	0.25	0
15.5.79	A	s	1.40	16.0	0.088
		b	0.22	3.58	0.061
	B	s	ND	0.78	0
		b	ND	0.59	0
5.6.79	A	s	0.57	14.1	0.040
		b	0.12	3.08	0.039
	B	s	ND	1.11	0
		b	ND	1.03	0
4.7.79	A	s	2.10	28.3	0.074
		b	0.44	2.66	0.165
	B	s	0.23	3.00	0.077
		b	ND	0.13	0
6.8.79	A	s	0.26	4.80	0.054
		b	0.65	5.91	0.110
	B	s	0.08	2.26	0.035
		b	ND	0.97	0

*s = Surface water sample at high tide; b = bottom water sample at high tide.
ND = Not detected.
Reproduced from Hashimoto, A., *et al.* (1980) *Analyst, London*, **105**, 787, Royal Society of Chemistry, by courtesy of authors and publishers.

describe the complete separation of six dinitrotoluene isomers using gas chromatography with support-coated open tubular glass capillary columns and electron-capture detection. The method was applied to the qualitative and quantitative analyses of trace levels of isomers in seawater and the results were found to be satisfactory with no need for further clean-up procedures.

The separation achieved on a Apiezon L grease SCOT glass capillary column is shown in Figure 7.15. Complete separation of six dinitrobenzene isomers took 8 minutes.

The method was used for routine monitoring of dinitrotoluene concentrations in seawater from Dokai Bay, Japan. The results are shown in Table 7.17. Both 2,6- and 2,4-dinitrotoluene were detected. Concentrations of 2,4-dinitrotoluene in surface water samples were higher than those in bottom water samples in 8 out of 10 samples.

(g) Azaarenes

Shinohara *et al.* [328] have described a procedure based on gas chromatography for the determination of traces of two, three and five ring azaarenes in seawater.

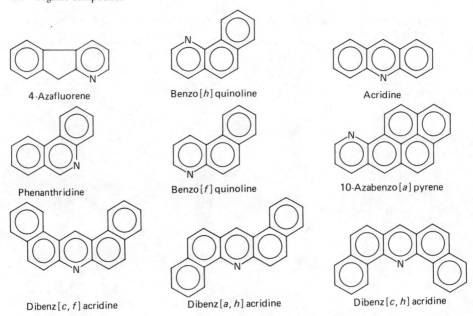

4-Azafluorene   Benzo[h]quinoline   Acridine

Phenanthridine   Benzo[f]quinoline   10-Azabenzo[a]pyrene

Dibenz[c, f]acridine   Dibenz[a, h]acridine   Dibenz[c, h]acridine

Chemical structures of azaarenes

The procedure is based on the concentration of the compounds on Amberlite XAD-2 resin, separation by solvent partition [329] and determination by gas chromatography using a flame thermionic detector and gas chromatography–mass spectrometry with a selective ion monitor. Detection limits by the flame thermionic detector were 0.5–3.0 ng and those by gas chromatography–mass spectrometry were in the range 0.02–0.5 ng. The preferred solvent for elution from the resin was dichloromethane and the recoveries were mainly in the range 89–94%. The method has been used for the analysis of seawater samples in Japan.

The recovery of azaarenes added to water is shown in Table 7.18.

Table 7.18 Recovery of azaarenes from water using an XAD-2 resin column

Compound	Recovery (%)	Coefficient of variation (%)
Quinoline	92.4	1.1
2-Methylquinoline	91.7	1.2
4-Methylquinoline	95.0	0.7
6-Methylquinoline	96.0	2.2
1-Methylisoquinoline	79.9	1.0
2,4-Dimethylquinoline	83.5	0.8
2,6-Dimethylquinoline	95.1	1.8
4-Azafluorene	92.9	3.3
Benzo[h]quinoline	93.8	3.7
Acridine	90.1	2.5
Phenanthridine	93.9	2.1
Benzo[f]quinoline	92.6	2.0
10-Azabenzo[a]pyrene	89.6	3.1
Dibenz[a,j]acridine	73.4	4.1
Dibenz[a,h]acridine	89.3	1.2
Dibenz[c,h]acridine	88.9	2.9

Reproduced from Shinohara, R., et al. (1983) Journal of Chromatography, **256**, 81, Elsevier Science Publishers, by courtesy of authors and publishers.

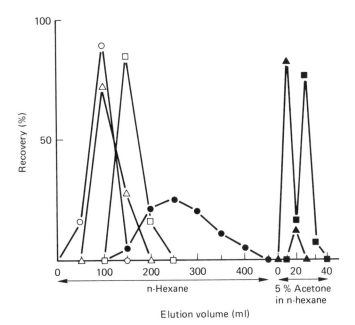

Figure 7.16 Alumina column chromatograms of azaarenes and polyaromatic hydrocarbons. ○, Phenanthrene; △, pyrene; □ chrysene; ● benzo(*a*)pyrene; ▲, dibenz(*c.h*)acridine; ■, dibenz(*a,j*)acridine (Reproduced from Shinohara, R., *et al.* (1983) *Journal of Chromatography*, **256**, 81, Elsevier Science Publishers, by courtesy of authors and publishers.)

Polyaromatic hydrocarbons of lower molecular weight than benzo(a)pyrene were almost completely eluted with 450 ml of *n*-hexane, the dibenzacridines subsequently being quantitatively recovered with 50 ml of *n*-hexane solution containing 5% of acetone (Figure 7.16).

This procedure was applied to a sample of seawater collected from Dohkai Bay, Japan. At first, the two- and three-ring azaarenes (quinolines and benzoquinolines) and the five-ring azaarenes (dibenzacridines and 10-azabenzo(*a*)pyrene) were analysed by gas chromatograms–flame thermionic detector and gas chromatograms thus obtained are shown in Figures 7.17 and 7.18. Although many peaks with retention times corresponding to the standard two- and three-ring azaarenes were observed, positive identification was difficult. Gas chromatograms using a flame thermionic detector which is selective but not a specific detector, should be applied to the routine analysis and sample screening, because there are many geometric isomers of azaarenes due to the type of ring fusion and the position of the aza-nitrogen in the molecules and their alkylated isomers. Consequently gas chromatography–mass spectrometry–selective ion monitoring was used for the determination of trace azaarenes in seawater.

The molecular ion peak of each azaarene which is not alkylated always becomes a base peak, as with polyaromatic hydrocarbons. As shown in Figures 7.19 and 7.20, gas chromatography–mass spectrometry–selective ion monitoring using the m/z value of each molecular ion did not give interfering peaks on the selective ion monitoring chromatograms obtained from a sample.

310

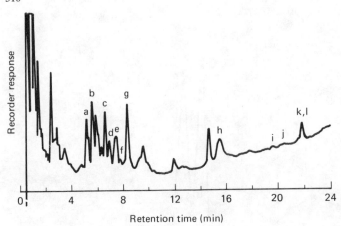

Figure 7.17 FTD gas chromatograms of two- and three-ring azaarene fraction separated from seawater (Reproduced from Shinohara, R., *et al.* (1983) *Journal of Chromatography*, **256**, 81, Elsevier Science Publishers, by courtesy of authors and publishers.)

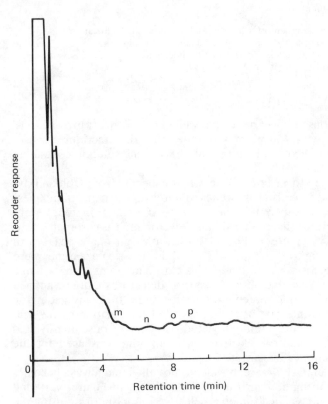

Figure 7.18 FTD gas chromatograms of five-ring azaarene fraction separated from seawater; m–p correspond to compounds shown in Figure 7.16 (Reproduced from Shinohara, R., *et al.* (1983) *Journal of Chromatography*, **256**, 81, Elsevier Science Publishers, by courtesy of authors and publishers.)

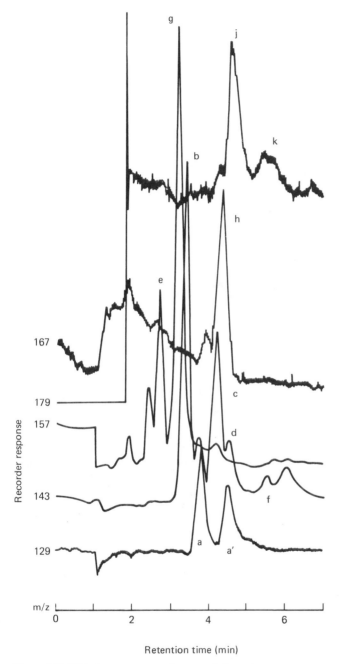

g

j

k

b

h

e

167

179

157

d

c

143

f

129

a

a'

m/z

0   2   4   6

Retention time (min)

Figure 7.19 SIM chromatograms of two- and three-ring azaarenes separated from seawater, a = Quinoline; a' = isoquinoline; b = 2-methylquinoline; c = L-methylisoquinoline; d = 6-methylquinoline; e = 2,6-dimethylquinoline; f = 4-methylquinoline; g = 2,4-dimethylquinoline; h = 4-azafluorene; j = acridine; k = phenanthridine and benzol(f) quinoline. Gas chromatography conditions as described in Experimental (Reproduced from Shinohara, R., *et al.* (1983) *Journal of Chromatography*, **256**, 81, Elsevier Science Publishers, by courtesy of authors and publishers.)

Recorder response

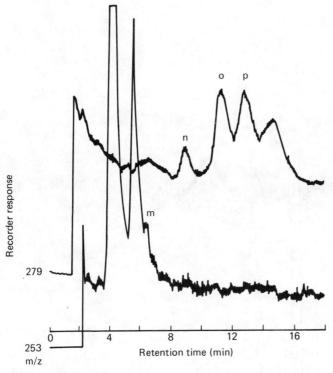

Figure 7.20 SIM chromatograms of five-ring azaarenes separated from seawater. m = 10-Azabenzo(*a*) pyrene; n = dibenz(*c,h*)acridine; o = dibenz(*a,h*)acridine; p = dibenz (*a,j*)acridine. Gas chromatography conditions as described in text (Reproduced from Shinohara, R., *et al.* (1983) *Journal of Chromatography*, **256**, 81, Elsevier Science Publishers, by courtesy of authors and publishers.)

Table 7.19 Detection limits of azaarenes and their concentration in seawater from Dohkai Bay, Japan

Compound	Detection limit (ng l^{-1})*	Concentration (ng l^{-1})
Quinoline	0.5	22
Isoquinoline	0.5	13
2-Methylquinoline	0.2	46
1-Methylisoquinoline	0.4	43
6-Methylquinoline	0.5	4.1
4-Methylquinoline	0.4	3.2
2,6-Dimethylquinoline	0.2	16
2,4-Dimethylquinoline	0.2	55
4-Azafluorene	0.3	5.8
Benzo[*h*]quinoline	0.3	ND
Acridine	0.5	9.5
Phenanthridine or/and benzo[*f*]quinoline	0.3	2.4
10-Azabenzo[*a*]pyrene	2.0	ND
Dibenz[*c,h*]acridine	0.4	0.66
Dibenz[*a,h*]acridine	0.4	3.1
Dibenz[*a,j*]acridine	0.5	4.1

*The volume of the seawater sample was 20 litres and the volume of the sample solution submitted to GC and the volume injected into the GC system were 0.5–1 ml and 5 µl, respectively.
ND = Not detected.
Reproduced from Shinohara, R., *et al.* (1983) *Journal of Chromatography*, **256**, 81, Elsevier Science Publishers, by courtesy of authors and publishers.

The detection limits of azaarenes in seawater and their estimated concentrations in the water sample collected from Donkai Bay are listed in Table 7.19.

(h) Nucleic acids

Pillai and Ganguly [330] have concentrated the nucleic acids from seawater by adsorption on homogeneously precipitated barium sulphate, then hydrolysed with 0.02 M hydrochloric acid and analysed for the constituents.

Hicks and Riley [331] have described a method for determining the natural levels of nucleic acids in lake and seawaters, which involves pre-concentration by adsorption on to a hydroxyapatite, elution of the nucleic acids and then photometric determination of the ribose obtained from them by hydrolysis.

Enzyme activity
Enzyme activity has been sought in seawater; Strickland and Solorzano [332] looked for photomonoesterase activity, while Maeda and Taga [333] used a fluorimetric method for assaying deoxyribonuclease activity in both seawater and sediment samples.

7.15 Sulphur compounds

Dimethyl sulphide and dimethyl disulphide have been measured in seawater and in the atmosphere by gas chromatography [334] and by gas chromatography–mass spectrometry [335]. Since these components are easily vaporized, the main effort in such analyses must be in the accumulation of enough sample. Some variety of cryogenic trapping is often used.

Andreae [336] has described a gas chromatographic method for the determination of nanogram quantities of dimethyl sulphoxide in natural waters, seawater and phytoplankton culture waters. The method uses chemical reduction with sodium borohydride to dimethyl sulphide which is then determined gas chromatographically using a flame photometric detector.

Andreae [336] investigated two different apparatus configurations. One consisted of a reaction/trapping apparatus connected by a six-way valve to a gas chromatograph equipped with a flame ionization detector, the other apparatus combined the trapping and separation functions in one column, which was attached to a flame photometric detector. The gas chromatographic flame ionization detector system was identical with that described by Andreae [337] for the analysis of methylarsenicals, with the exception that a reaction vessel which allowed the injection of solid sodium borohydride pellets was used. The flame photometric system is modified after a design by Braman, Ammions and Bricker [338].

7.16 Lipids

Chromarod thin-layer chromatography with Iatroscan flame ionization detection (TLCFED) (i.e. thin layer chromatography followed by strong heating of separated compounds and detection of combustion products by gas chromatography with a flame ionization detector) has been used to analyse lipid classes from diverse sources.

Delmas, Parrish and Ackman [339] studied various qualitative and quantitative aspects of this system to assess its usefulness in the analysis of lipid classes present in seawater. Marine lipids are important biological energy sources and have been used as tracers in food-web studies [340–343]. Some lipids, however, are pollutants [344, 345] and all lipids can potentially act as solvents, transporters or sinks for pollutants [344, 346–348].

The rapidity with which the Iatroscan THIO MK II Iatron Laboratories, Tokyo, TLCFID system provides synoptic lipid class data from small samples suggested that it would be useful for screening seawater samples prior to performing more detailed chromatographic analyses. The lipid class concentrations obtained by TLCFID provide reference values for the concentrations of the individual components obtained by other techniques. In particular, shipboard TLCFID analysis would help in deciding sampling strategies for more detailed investigations. In addition, the TLCFID technique alone could provide an overall picture of spatial or temporal variations in the distribution of a complete range of lipid classes.

Standards used for the construction of calibration curves had alkyl chain lengths in the range C-16 to C-19. Table 7.20 gives the principal compounds used as

Table 7.20 Seawater lipid classes and standards used for their identification and calibration

Class	Abbreviation	Standards and suppliers
Aliphatic hydrocarbon	HC	Nonadecane (Polyscience)
Wax and sterol esters	WE	Hexadecyl palmitate (Analabs)
3-Ketone (internal stand)	KET	Hexadecan-3-one (K & K Labs)
Free fatty acid	FFA	Palmitic acid (Supelco)
Triglyceride	TG	Tripalmitin (Supelco)
Free alcohol	ALC	Hexadecan-1-ol (Polyscience)
Free sterol	ST	Cholesterol (Supelco)
Polar lipid	PL	Dihexadecanoyl lecithin (Supelco)

Reproduced from Delmas, R.P. (1984) *Analytical Chemistry*, **56**, 1272, American Chemical Society, by courtesy of authors and publishers.

representatives of each class. Lipids were separated on the thin-layer plate with solvents of increasing polarity.

This technique was used by Delmas, Parrish and Ackman [339] to separate lipid extracts in seawater into various classes. Lipid classes that have been eluted away from the point of application may be burnt off the rod in a partial scan, allowing those lipids remaining near the origin to be developed into the place that has just been simultaneously scanned and reactivated. By analysis of complex mixtures of neutral lipids in this stepwise manner it is possible to be more selective about lipid class separations as well as to be more confident about assigning identities to peaks obtained from a seawater sample. In addition, this approach also reduces the possibility of peak contamination by impurities which would normally coelute with marine lipid classes (e.g. phthalate esters [349]).

The separation scheme used by Delmas, Parrish and Ackman [339] is an elaboration of a two-step procedure proposed for the TLCFID analysis of neutral lipid classes. In the first development (40 min in hexane-diethyl ether-formic acid, 98:2:0.1) the less polar classes (HC, WE, KET, FFA; Figure 7.21(a)) are moved

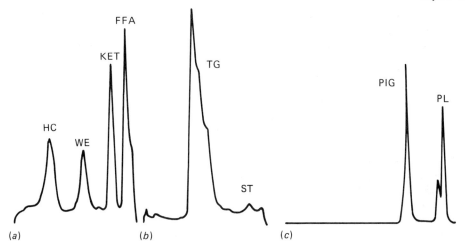

Figure 7.21 Consecutive dissolved lipid chromatograms from the same Chromarod; developing direction is from right to left, while scanning direction is from left to right. (*a*) First partial scan. (*b*) Second partial scan. (*c*) Final complete scan at a higher attenuation (Reproduced from Delmas, R.P., *et al.* (1984) *Analytical Chemistry*, **56**, 1272, American Chemical Society, by courtesy of authors and publishers.)

away from the point of application. These are scanned and the scan is stopped manually at the lowest point on the tail of the FFA peak (Figure 7.21(a)).

The low polarity solvent system used for the first separation was chosen because it separates ethyl ketones from other lipid classes and from other ketones [349]. e-Hexadecanone was used as an internal standard.

The remaining neutral lipid classes (TG, ALC, ST) are separated in the second development (40 min in hexane-diethyl ether-formic acid 80:20:0.1). Again a partial scan is performed (Figure 7.21(b)) and this is followed by a short development in acetone (3 cm above the origin). This development separates glycolipids from phospholipids, which are immobile in acetone and it results in two polar lipid peaks which are quantified in the final scan (Figure 7.21(c)). Since the acetone-mobile lipids visibly include pigments, the chromatographic peak associated with this material has been designated PIG (Figure 7.21(c)).

The fatty acid and the fatty acid esters used for calibrations (FFA, WE, TG, PL; Table 7.20, Figure 7.22) had C-16 fatty acid chain lengths. This and the use of nonadecane and cholesterol as standards (HC, ST) was thought to best represent what is known as the composition of lipids in seawater.

Coefficients of variation of between 5 and 10% were obtained by this procedure in replicate determinations on a sample.

After a single development, hydrocarbons give peaks with broad bases and small areas (Figures 7.22(a) and 7.22(b)). It was possible to reverse this process by performing a second development to refocus the material into a narrower band on the Chromarod. After a double development, the hydrocarbon peak is recorded as a narrower peak of greater area (Figure 7.23(b)).

Figure 7.23 shows some selected profiles for dissolved and particulate marine lipid classes. For the dissolved lipids each data point is the mean of three or four

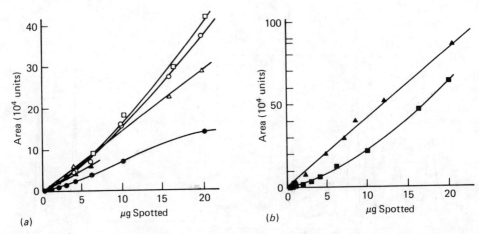

Figure 7.22 Calibration curves. (*a*) Low-response compounds: (●) HC, (▲) KET, (△) TG, (○) WE, and (□) FFA. (*b*) High-response compounds: (■) ST and (▲) PL (Reproduced from Delmas, R.P. (1984) *Analytical Chemistry*, **56**, 1272, American Chemical Society, by courtesy of author and publishers.)

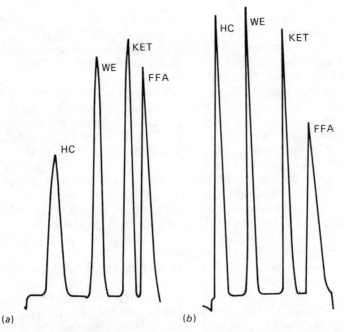

Figure 7.23 Single (*a*) and double (*b*) developments of 6 µg each lipid class on the same rod; partial scans to the lowest point on the tail of FFA. After 40 min in hexane-diethyl ether-formic acid (98:2:0.1) and subsequent drying. (*a*) was scanned and (*b*) was redeveloped for a further 30 min in the same solvent system (Reproduced from Delmas, R.P., *et al.* (1984) *Analytical Chemistry*, **56**, 1272, American Chemical Society, by courtesy of authors and publishers.)

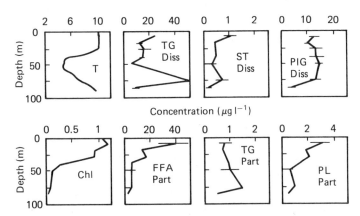

Concentration (μg l^{-1})

Concentration (μg l^{-1})

Figure 7.24 Vertical profiles of temperature (T), chlorophyll a (Chl) and dissolved (Diss) and particulate (Part) lipid classes from the top 100 m of the water colum near the edge of the Scotian Shelf; Cruise SS XXV, Station 56 (Reproduced from Delmas, R.P., *et al.* (1984) *Analytical Chemistry*, **56**, 1272, American Chemical Society, by courtesy of authors and publishers.)

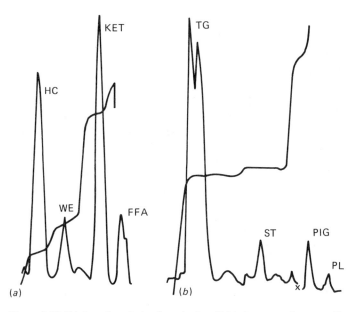

Figure 7.25 Shipboard analysis of particulate lipid classes on the same Chromarod. (*a*) Non-polar neutral lipids and internal standard, partial scan. (*b*) Remaining neutral lipids and polar lipids, full scan, increased attenuation at X (Reproduced from Delmas, R.P., *et al.*, (1984) *Analytical Chemistry*, **56**, 1272, American Chemical Society, by courtesy of authors and publishers.)

determinations. The lipid class profiles (Figure 7.24) show a high degree of variability which is likely to be related to the distribution of physical and biological parameters (Figure 7.24). The seven lipid classes detected in the dissolved and particulate samples can be divided into three groups; those showing a distinct subsurface maximum as exemplified in Figure 7.24, by dissolved and particulate TG, those showing a near surface maximum as exemplified by particulate FFA and PL, and those showing little variability with depth (at least by comparison with the size of the error bars; dissolved ST and PIG).

Figure 7.25 shows a more detailed presentation of the type of results obtained by this procedure (for key see Table 7.20).

7.17 Sterols

The sterols differ from the other compounds in that no class reaction has been proposed for the measurement of total sterols. Instead, various fractionation methods, usually derived from the biochemical literature, have been adapted to the concentrated materials collected from seawater. Certain of the more important sterols, particularly those used in the evaluation of water quality, have been determined by the use of a compound-specific reaction, after concentration from solution. Thus Walker and Litsky [350, 352] measured coprostanol, a faecal sterol in seawater, after collection and separation, by extraction using liquid–liquid partitioning or extraction on a column of Amerlite XAD-2 resin.

When solvent extraction was performed, the procedure used by Dutka, Chau and Coburn [351] was followed with a slight modification: 2 ml of concentrated hydrochloric acid and 5 ml of 20% (w/v) sodium chloride were added to each litre of water sample. The sample was extracted with vigorous mixing three times with 100 ml each of hexane for 30 min. The combined extract was washed with two 50 ml portions of acetonitrile (saturated with hexane) followed by two 50 ml portions of 70% ethanol. The hexane was then brought to dryness on a rotary evaporator under reduced pressure. The sample was re-dissolved in 100–200 μl of carbon disulphide and 1–5 μl of the solution was used for gas chromatographic analysis.

^{14}C-labelled cholesterol was used to test the recovery of 5–100 μg of faecal sterols from seawater (labelled coprostanol not being available). The radioactivity of the samples and eluates was measured by a two-channel liquid scintillation counter. Percentage recovery was calculated on the basis of the amount of labelled material recovered in the acetone eluant. The results (Table 7.21) indicate that column extraction efficiency is not adversely affected by the salinity of the water samples.

Table 7.21 Extraction of radioactive cholesterol from seawater

Water sample (1 litre)	Cholesterol			
	Added		Acetone solution	
	μg l^{-1}	dpm	dpm	% recovery
Synthetic seawater	5	1.6×10^6	1.52×10^6	95
	100	1.16×10^6	1.09×10^6	95
Natural seawater	50	1.48×10^6	1.43×10^6	97

Column: 15 mm × 20 cm, packed with XAD-2 (60–120) resin. Flow rate: 18 ml min^{-1}. Eluent: acetone, 100 ml at 3 ml min^{-1}.
Reproduced from Wun, C.K., *et al.* (1976) *Water Research*, **10**, 955, Pergamon Publications, by courtesy of authors and publishers.

Most often the sterols have been collected by liquid-liquid extraction, using petroleum ether and ethyl acetate [353], chloroform and methanol [354], *n*-hexane [355, 356] or chloroform [357, 358]. After concentration, gas chromatography was generally used for the final separation and determination, although thin-layer chromatography has also been employed. The extra sensitivity of the electron capture detector could be used by reacting the concentrated sterols with bromomethyldimethylchlorosilane (BMDS) before separation and measurement [359].

An analysis performed by gas chromatography can usually also be performed by coupled gas chromatography–mass spectrometry, often with some increase in sensitivity, and with considerably greater certainty in the identification of the compounds. This technique has been applied to seawater [360] and to marine sediments [361].

While most of the interest in sterols has been in the materials in solution, Kanazawa and Teshima [362] have investigated the compounds present in suspension. The suspended matter was fractionated by filtration through a graded series of filters, the sterols removed by extraction with an organic solvent and the final separation and determination was made by flame ionization gas chromatography.

Gas chromatography appears to give adequate separation and measurement of the various sterols to be found in the marine environment; where it is less than satisfactory is in the identification of the substances being measured. With compounds whose structures can be so similar, only gas chromatography–mass spectrometry can be expected to provide reasonable identifications.

7.18 Chelators

While a great many compounds that have been or might be found in the marine environment have been accused of chelation, this section deals only with the non-specific measurements of chelation, with what has been called 'chelation capacity'. In general, this capacity is measured by spiking the solution with a transition metal, preferably one that is easily measured, and then determining either how much is complexed or how much is left over. While the principle of all of the methods is the same, the details are different, and often quite ingenious.

The metals used are usually copper or cobalt. A good example of a relatively simple approach is the paper by Hanck and Dillard [363]. They add an excess of Co^{II}, which forms strong but labile complexes with the organic materials present. After complex formation, dilute hydrogen peroxide is added, to oxidize the complex to the Co^{III} form, which is much less labile. The excess hydrogen peroxide is then destroyed and the unreacted Co^{II} measured by differential pulse polarography.

Another relatively simple approach is that of Stolzberg and Rosin [364]. They spike the sample with an excess of copper, then pass it through a Chelax 100 column. The column retains the free copper ion, but passes the copper associated with strong ligands. The chelated copper eluted from the column is measured by plasma emission spectrometry.

A kind of 'standard additions' approach can also be used for the measurement of apparent complexing capacity. In this technique, labile copper is measured by differential pulse anodic stripping voltammetry after each of a number of spikes of ionic copper have been added to the sample [365].

In terms of solution chemistry, 'apparent' capacities derived from the extremely dilute and diverse natural samples, determined at natural pH values, are very crude. To get results that are more exact, or tell more about the possible nature of the complexing materials, it is necessary to concentrate, and sometimes to separate out, the organic chelators. This approach has been used by Buffle *et al.* [366]. Working with river water, they were able to concentrate and clean up their samples to the point where they were able to treat the chelating material almost as reagents, and to determine mean molecular weights for the ligands, stability constants for the complexes and the pH dependency of the stability of the complexes. This work would be considerably more difficult in seawater, but if the same organic materials were to be found in both fresh and seawater, it might be possible to determine the various values for fresh water and then to determine the effect of ionic strength on these quantities.

When the chelators are actually known, as in the case of industrial materials injected into the environment, it is possible to derive much more information from the analyses. Thus high pressure liquid chromatography has been used to separate the copper chelates of EDTA, NTA, EGTA and CDTA with the final measurement of copper being made by atomic absorption [367, 368].

A spectrophotometric method based on the light absorption of the coloured Co^{III} complexes has been used to determine EDTA and NTA in fresh water [369]. In these few cases, actual well-defined compounds were present at concentrations high enough so that the individual compound could be isolated, identified and measured. This is seldom the case for the chelators in seawater; we are usually measuring an attribute, not a compound, with little idea of the actual identity of the compounds.

7.19 Humic materials and plant pigments

Humic materials cover three main classes of compounds which are discussed below under separate headings. They include:

1. The fluorescent yellow materials known collectively as 'Gelbstoff' (yellow material).
2. The lignins and lignin sulphonates.
3. The humic and fulvic acids.

While in a strict sense only the last class of compounds should fall into the category of humic materials, there has been much dispute concerning the origin of the marine humic materials; both Gelbstoff and the lignins have been mentioned as probable starting compounds for the marine humic and fulvic acids.

Chlorophyll is also discussed in this section.

(a) Gelbstoff

Under this heading are discussed both the naturally occurring fluorescent material and the yellow substances which give coastal waters their generally greenish colour. It is usually considered that these two categories are the same, or at least overlap almost entirely.

An *in situ* technique for measuring fluorescence in the ocean has been developed by Egan [370]. His sensors, set to measure separately both chlorophyll and Gelbstoff fluorescence, can be lowered to 600 m and operate unattended.

Shapiro [371] developed concentration and analysis methods for the yellow acids in lake waters. He used low-temperature reduced pressure evaporation to remove much of the water, solvent extraction to concentrate and separate the yellow materials and paper chromatography for the final separations. Gel filtration with Sephadex was used by Ghassemi and Christman [372] to make separations by molecular size on water samples concentrated by vacuum evaporation. Fluorescence was also used as one method for following the fractionation. Molecular size was also used by Gjessing [373] but with pressure dialysis as the method of separation. A similar method of concentration and separation was used by Brown [374] to follow the dispersion of these materials as fresh and salt water mixed in the Baltic Sea.

A method of concentration by adsorption on aluminium oxide has been proposed. It achieved almost complete adsorption.

The argument has been made by several investigators (see the review by Wangersky [375]) that the humic materials in seawater are not derived from terrestrial sources, but result from light-induced condensation reactions of phenolic material and proteins released by fixed algae in the coastal regions. These materials can be collected on nylon-packed chromatographic columns [376]. Methods of collection and fractionation, as well as studies on the chemical and physical properties of these compounds, were discussed by the authors [377]. These compounds seem to be relatively easy to collect, to fractionate and to follow through the various chemical procedures.

(b) Lignins

Lignins and lignin sulphonates are considered to be important tracers of man's activities. The major source of these compounds is the pulp and paper industry; lignin is not a marine product, and any large accumulation of these compounds suggests a local dominance of terrestrial materials.

Pocklington and Hardstaff [378] react sediment samples with 1,3,5-trihydroxybenzene in alcoholic hydrochloric acid to produce a colour in the particulate lignins, facilitating their identification under the microscope. Samples high in lignins can then be subjected to the normal methods of analysis. This is an excellent screening technique (semi-quantitative).

A concentration step is often used in order to bring the sample within the concentration range of the method. Thus Stoeber and Eberle [379] extracted the lignosulphonic acids with trioctylamine in chloroform before final determination and Revina and Kriul'kov [380] employed spectrophotometric methods. Extraction is also used to remove interferences in spectrophotometric methods [381].

Direct spectrophotometric methods have been proposed for both particulate and dissolved lignosulphonic materials. Kloster [382] used the Folin Ciocalteu method, which actually measures hydroxylated aromatic compounds. A general review of spectrophotometric methods was published by Bilikova [383].

Certain derivatives of the lignin sulphonic acids can be determined directly in water. The nitroso derivatives, which are easily formed in solution, can be determined by differential pulse polarography [384]. Vantillin can be formed by alkaline hydrolysis [385] or alkaline nitrobenzene oxidation [386], extracted into an organic solvent and determined by gas chromatography.

As a general rule, fluorimetric methods are considerably more sensitive than spectrophotometric methods, although standardization is more difficult. Direct fluorimetric procedures for lignin and lignin sulphonates have been described [387–389].

(c) Humic and fulvic acids

The designation of certain classes of organic materials as humic and fulvic acids unfortunately implies a certainty and regularity of structure which is far from true. These terms, derived from soil science, indicate only the products resulting from a particular sequence of fractionation; humic acids are those compounds extracted from soils by alkaline solutions and precipitated upon acidification, while fulvic acids are those extracted by alkaline solutions but left in solution on subsequent acidification [390]. While these terms were originally specific to soil science, and referred only to materials isolated from soils, the usage was gradually extended to marine sediments and then by further extension to materials isolated from solution by similar techniques.

Evidence is accumulating which shows that materials isolated from the marine environment are quite unlike those isolated from soils, at least in $^{13}C/^{12}C$ ratios and probably in structure.

The methods for isolating humic and fulvic acids from marine samples can be found in King [391], Rashid and King [392] and Pierce and Felbeck [393]. These techniques, which derive from soil chemistry, may be used when the sample is a marine sediment. However, when it is the material in solution that is being separated, the application of the methods is not so straightforward. Several workers have used macroreticular resins, usually XAD-2, to collect high molecular weight materials from seawater [394–396]. It has been demonstrated that this resin will, in fact, collect the humic and fulvic acids separated from soils by the usual methods [396]. It has therefore been assumed that the materials collected by this resin from seawater are the humic and fulvic acids, although the characteristics of the fractions so collected are different from those collected from terrestrial soils and marine sediments [394, 397].

Applications of NMR, ESR, thermal analysis, spectrophotometry, gas chromatography and gas chromatography–mass spectrometry to humic and fulvic acid analysis have been reviewed by Schnitzer [398].

Salfeld [399] has shown that qualitative information on the nature of the compounds present can be found by means of derivative spectrophotometry. The concentration of fulvic acids in natural waters can be calculated from the ultraviolet absorption at 400 and 340 nm, after removal of the humic acids [400]. If a somewhat greater degree of uncertainty as to the actual composition of the materials is acceptable, sample preparation can almost be eliminated, and the sensitivity of the method increased, by using fluorescence rather than absorption as the measure of the humic materials present [387, 389]. However, it must always be remembered that in this method of analysis we are equating humic materials with all of the soluble compounds fluorescing in the proper region; the correspondence is not necessarily exact. In addition, the fluorescence of humic and fulvic acids is rapidly decreased as sea surface sunlight intensities. The fluorescence may therefore behave as a highly non-conservative indicator for these materials in surface waters.

Many of the methods used for lignins can also be applied to humic acids. Thus the use of pulse polarography after the formation of nitroso derivatives is possible with humic acids as well as lignins [379, 384]. If the fulvic acids are removed on activated charcoal, then eluted with acid, they can be determined by a.c. polarography [401].

Separation of the humic acids into molecular size classifications by gel permeation chromatography has been used largely to determine the distribution of sizes within this group of compounds [402]. The information coming from such fractionation cannot be taken at face value, however, since the elution of the humic materials is strongly influenced by the functional groups present, the pH, the ionic strength. Such fractionation can profitably be used as a first step in the gas chromatography–mass spectrometry determination of the compounds present. For those compounds too large or too complex to be determined directly by gas chromatography–mass spectrometry, oxidation [403] or hydrolysis [404] can be useful preliminary treatments.

For a rough estimation of the relative amounts of humic and fulvic acids, the fluorimetric method has much to recommend it, not the least being its simplicity. Also, the ability of the method to measure short-range variability might be quite valuable. It is not possible to be certain of the value of any more detailed or more exact information until a great deal more is known about the nature of the compounds in each of these fractions.

(d) Chlorophylls

While the spectrophotometric determination of plant pigments, particularly the chlorophylls, has become a routine procedure [405, 406], refinements aiming at either greater accuracy (usually by the separation and determination of the separate pigments) or greater sensitivity continue to appear in the literature. A review of the various methods, both for total pigments and for each separate pigment, was written by Rai [407]. A method using separation of the pigments by TLC and determination by densitometry was described by Messiha-Hanna [408], while Quirry et al. [409] discussed a laser absorption spectrometry method with a sensitivity a hundred times greater than the usual spectrophotometric method.

Bjarnborg [410] has studied shipboard methods of continuous recording in vivo chlorophyll fluorescence and extraction based chlorophyll determinations in Swedish archipeligo waters.

Boto and Bunt [411] used thin-layer chromatography for the preliminary separation of chlorophylls and phaeophytins from seawater and combined this with selective excitation fluorimetry for the determination of the separated chlorophylls a, b and c and their corresponding phaeophytin components. An advantage of the latter technique is that appropriate selection of excitation and emission wavelengths reduces the overlap between the emission spectra of each pigment to a greater extent than is possible with broad band excitation and the use of relatively broad band filters for emission.

The fluorescence studies were performed on 90% acetone solutions with an Aminco-Bowman spectrofluorimeter (J4-8203G Model).

Figure 7.26 shows an excitation and emission spectra for chlorophylls a, b and c and their respective phaeophytins. The excitation spectra were obtained by holding the emission wavelength at the emission maximum and slowly scanning the

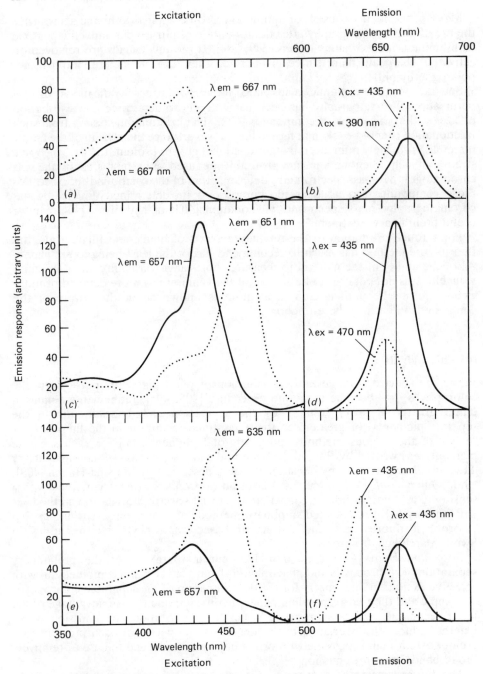

Figure 7.26 Excitation and emission spectra. (*a* and *b*) Chlorophyll a (- - -) and phaeophytin a (——); concentration of both = 0.134 µg ml⁻¹. (*c* and *d*) chlorophyll b (- - -) and phaeophytin b (——); concentration of both = 0.172 µg ml⁻¹. (*e* and *f*) chlorophyll c (- - -) and phaeophytin c (——); concentration of both 0.042 µg ml⁻¹. Note that the phaeophytin c curves are shown on a ten-fold higher sensitivity scale to the others. All solutions are in 90% acetone (Reproduced from Boto, K.G., *et al.* (1978) *Analytical Chemistry*, **50**, 392, American Chemical Society. by courtesy of authors and publishers.)

excitation wavelength over the required range. The emission spectra shown have not been corrected for changes in photomultiplier response with wavelength. These spectra show that with suitable choice of excitation wavelengths, good selectivity can be achieved. Assuming that a given mixture contains all three chlorophylls and their phaeophytins, a set of simultaneous equations can be derived for the emission responses at the usual emission maxima of each pigment when the mixture is excited at various chosen wavelengths. After acidification of the mixture to phaeophytins alone, further information can be obtained as indicated below (solutions of the phaeophytins were prepared from the corresponding chlorophylls by adding 2 drops of 0.1 M hydrochloric acid per 100 ml solution).

Although the equations are complex, in most cases, the minor terms are such that many can be neglected. Further simplification is often possible when one considers that, in most offshore seawater samples, the only primary pigments found are chlorophylls a and c along with phaeophytin a. Chlorophyll b and phaeophytins b and c are either absent or present in minor amounts. Phaeophorbides also are commonly found. However, these would probably have very similar fluorimetric properties to the phaeophytins and, hence, could not be separately determined by any fluorimetric method.

Taking into account the most likely pigment compositions of the phytoplankton in most seawater samples, a simplified set of equations was obtained. As the quantum efficiency of chlorophyll c appears to be very high its emission spectrum is very little affected by interference from the emissions of other pigments.

Table 7.22 Results of analyses using the fluorimetric method for artificial mixtures and actual seawater extracts (in 90% acetone)

Sample type	Actual concentrations*			Measured concentrations (errors in parentheses)				
	C_a	C_b	C_c	C_a	C_b	C^1_{pa}	C^1_{pb}	C^1_{pc}
Prepared mixture	0.021	0.007	0.00774	0.020 (−3.8%)	0.0053 (−2.4%)	0.0076 (−1.8%)		
Prepared mixture	0.115		0.563	0.120 (+4.3%)	0.006	0.057 (+1.2%)	0.110 (−4.3%)	0.0003	0.050 (−11.1%)
Seawater extract†				0.059	0.001	0.0087	0.0086	0.0018	0.0327
Similar collected near same spot†				0.046	0.0002	0.0088	0.090	0.0012	0.0458

*Concentrations in μg ml ml⁻¹ actual extract.
†Results shown here are averages of duplicates for each seawater sample.
Reproduced from Boto, K.G., et al. (1978) Analytical Chemistry, **50**, 392, American Chemical Society, by courtesy of authors and publishers.

Therefore, its concentration can be quite accurately calculated and this value then used to give accurate estimates of the others.

Typical results obtained by this procedure are given in Table 7.22.

Murray et al. [412] carried out an intercomparison of the determination of chlorophyll in marine waters by methods based on high performance liquid chromatography, spectrophotometry and fluorimetry. Good agreement was obtained between these methods for chlorophyll a and chlorophyll b.

7.20 Vitamins

Only one analytical method has been widely applied to the measurement of vitamins in seawater. That method, bioassay, is not really within the realm of the analytical chemist, since it requires the maintenance of cultures of test organisms. The tests also usually require a minimum of 4 days before results are available.

The older methods for vitamin B_{12} used the optical density of the culture medium as a measure of growth rate of the assay organism [413, 414]. The sensitivity of the method could be increased somewhat by following ^{14}C uptake as a measure of growth [415, 416]. These methods are sensitive to 0.1 ng well below the amounts of the vitamin which could be measured by any available chemical technique.

Thiamine has also been measured by bioassay, a marine yeast being used as the assay organism [417, 418]. Marine bacteria [419, 420], marine yeasts [418] and dinoflagellates [421] have been used for the assay of biotin.

There are many uncertainties in the use of bioassay methods, not the least of these being the genetic stability of the assay organisms. Stock cultures cannot be treated like chemical reagents, to be put back on the shelf and forgotten between analytical runs. This is an area of analysis best left to the microbiologists; if the information is absolutely necessary, the maintenance of cultures and the actual assays should not be left solely in the hands of the analytical chemists.

References

1. Blumer, M., Blokker, P.C., Cowell, E.B. and Duckworth, E.F. In *A Guide to Marine Pollution*, (ed. E.D. Goldberg) Gordon and Breach Science Publishers, New York, pp. 19–40 (1972)
2. Farrington, J.W. *Analytical Techniques for the Determination of Petroleum Contamination in Marine Organisms*, A.D. Report No. 766792/6. US National Technical Information Service (1973)
3. Andelman, J.B. and Snodgrass, J.E. *Critical Review of Environmental Control*, **4**, 69 (1974)
4. Institute of Petroleum *Marine Pollution by Oil: Characterization of Pollutants, Sampling, Analysis and Interpretation.* Applied Science Publications, Barkins (1974)
5. McGlynn, J.A. A review of techniques for the characterization and identification of oil spillages. In *Examination of Waters, Evaluation of Methods for Selected Characteristics*, Resource Council Technical Paper, no. 8, pp. 85–89 (1974)
6. Whitham, B.T., Duckworth, D.F., Harvey, A.A.B., Jeffrey, P.G. and Perry, S.G. Characterization of pollutants, sampling, analysis and interpretation. In *Marine Pollution by Oil* (ed. B.T. Whitham), Applied Science Publishers, Essex (1974)
7. Erskine, R.L. and Whitehead, E.V. *Iranian Journal of Science and Technology*, **3**, 221 (1975)
8. IOC. *Report of the ICG for GIPME Task Team for the Evaluation of Recommendation No. IV of the Joint IOC/WMO Task Team II on the Marine Pollution Monitoring Project*, UNESCO, Paris (1975)
9. Moulder, D.S. and Varley, A. *A Bibliography on Marine and Estuarine Oil Pollution, Supplement 1*, Marine Biology Association (1975)
10. National Academy of Science, USA. *Petroleum in the Marine Environment.* A workshop on inputs, fates and the effects of petroleum in the marine environment held in Airlie on 21st May 1973 (1975)
    crude oils, heating oils and marine tissues. In *1975 Conference on Prevention and Control of Oil Pollution*, American Chemical Society, San Francisco, CA, pp. 103–113 (1975)
12. Morris, B.F. *Science*, **173**, 430 (1971)
13. Brown, R.A., Searl, T.D., Elliott, J.J., *et al.* Distribution of heavy hydrocarbons in some Atlantic Waters. In *Proceedings of The Conference on Prevention and Control of Oil spills.* American Petroleum, Washington, pp. 509–519 (1973)
14. Hites, R.A. and Biemann, K. *Science*, **178**, 158 (1972)
15. Hites, R.A. and Bieman, W.G. Identification of specific organic compounds in a highly anoxic

sediment by gas chromatographic–mass spectrometry and high resolution mass spectrometry. In *Analytical Methods in Oceanography*, (ed. R.P. Gibb, Jr.), American Chemical Society, Washington, pp. 188–201 (1975)

16. Mengerhauser, J.V. *Crystal Microbalance for Determination of Oil in Water*, US National Technical Information Service A.D. Report, No. 762107 (1973)

17. Bieri, R.H., Walerk, A.L., Lewis, B.W., *et al. Identification of Hydrocarbons in an Extract from Estuarine Water Accommodated No. 2 Fuel Oil*. National Bureau of Standards, US Special Publication, No. 409, 149 (1974)

18. Hertz, H.S., Chesler, S.N., May, W.E., *et al. Methods for Trace Organic Analysis in Sediments and Marine Organisms*, National Bureau of Standards US, Special Publication, No. 409, 197 (1974)

19. Cretney, W.J., Johnson, W.K. and Wong, C.S. *Trace Analysis of Oil in Sea Water by Fluorescence Spectroscopy*. Pacific Marine Science Report, unpublished manuscript, 77–75 (1977)

20. Barbier, M., Joly, D., Saliot, A. and Tourres, D. *Deep Sea Research*, **20**, 305 (1973)

21. Zobova, N.A. *Okhr. Ikh. Resur.*, **4**, 45 (1973)

22. Matsushima, H. and Hanya, T. *Bunseki Kagaku*, **24**, 505 (1975)

23. Maxmanidi, N.D., Kovaleva, G.I. and Zobova, N.S. *Okeanologiya*, **15**, 453 (1974)

24. Hellman, H. *Zeitschrif für Analytische Chemie*, **278**, 263 (1976)

25. Hughes, D.R., Belcher, R.S. and O'Brien, E.J. *Bulletin of Environmental Contamination and Toxicology*, **10**, 170 (1973)

26. Majori, L., Petronio, F. and Nedoclan, G. *Ig. Mod.*, **66**, 150 (1973)

27. Ahmed, S.M., Beasley, M.D., Efromson, A.C. and Hites, R.A. *Analytical Chemistry*, **46**, 1858 (1974)

28. Brown, R. Al., Elliott, J.J. and Searl, T.D. *Measurement and Characterisation of Nonvolatile Hydrocarbons in Ocean Water*. National Bureau Standards US Special Publication, No. **409**, p. 131 (1974)

29. Mallevialle, J. *Water Research*, **8**, 1071 (1974)

30. Brown, R.A., Elliott, J.J., Kelliher, J.M. and Searl, T.D. Sampling and analysis of nonvolatile hydrocarbons in ocean water. In *Analytical Methods in Oceanography* (ed. R.P. Gibb, Jr.), American Chemical Society, Washington DC, pp. 172–187 (1975)

31. LeRoux, J.H. *Instrument for Detecting and Measuring Trace Quantities of Hydrocarbons in Water*, S. African, Patent No. 02,379 (Cl. BOID. GOIF) (1975)

32. Zsolnay, A. *Chemosphere*, **2**, 253 (1973)

33. Zsolnay, A. *Determination of Aromatic and Total Hydrocarbon Content in Submicrogram and Microgram Quantities in Aqueous Systems by Means of High Performance Liquid Chromatography*, National Bureau of Standards (US) Special Publications, No. **409**, p. 119 (1974)

34. Zsolnay, A. *Journal of Chromatography*, **90**, 79 (1974)

35. Tokuev, Yu. S. and Oradovskii, S.G. *Trudy gos Okeanogr. Inst.* **54**, 113 (1972)

36. Zsolnay, A. *Deep Sea Research*, **20**, 923 (1973)

37. Sutton, C. and Calder, J.A. *Environmental Science and Technology*, **8**, 654 (1974)

38. Hargrave, B.T. and Phillips, G.A. *Environmental Pollution*, **8**, 193 (1975)

39. Walker, J.D., Colwell, R.R., Hamming, M.C. and Ford, H.T. *Bulletin of Environmental Contamination and Toxicology*, **13**, 245 (1975)

40. Ellis, D.W. *Analysis of Aromatic Compounds in Water using Fluorescence and Phosphorescence*. US National Technical Information Service, PB Report No. 212268 (1979)

41. Mackie, P.R., Whittle, K.J. and Hardy, R. *Marine Science*, **2**, 359 (1974)

42. Simoneit, B.R., Smith, D.H., Eglinton, G. and Burlingame, A.L. *Archives of Environmental Contamination and Toxicology*, **1**, 193 (1973)

43. Blaylock, J.W., Bean, R.M. and Wildung, R.E. *Determination of Extractable Organic Material and Analysis of Hydrocarbon Types in Lake and Coastal Sediments*, National Bureau Standards (US) Special Publication, No. **409**, p. 217 (1974)

44. Jeffrey, L.M., Rasby, B.F., Stevenson, B. and Hood, D.W. Lipids of ocean water. In *Advances in Organic Geochemistry* (eds. U. Colombo and G.D. Hobson), Pergamon Press, Oxford, pp. 175–197 (1964)

45. Zafiricu, O.C. *Estuary and Coastal Marine Science*, **1**, 81 (1973)

46. Anderson, J.W., Clark, R.C. and Stegemen, J. Petroleum hydrocarbons. In *Proceedings of a Workshop on Marine Bioassays*. Marine Technology Society, Washington, pp. 36–75 (1974)

47. Frankenfeld, J.W. and Schulz, W. *Identification of Weathered Oil Films Found in the Marine Environment*. US NTIS, AD Report AD/A015883 (1974)

48. Mommessin, P.R. and Raia, J.C. *Chemical and Physical Characterization of Tar Samples from the Marine Environment*. Paper No. CG-D20-75 (1974)

49. Giger, W. and Blumer, M. *Analytical Chemistry*, **46**, 1663 (1974)

50. Kawahara, F.E. *Laboratory Guide for the Identification of Petroleum Products*, Department of the Interior Federal Water Pollution Central Administration Analytical Quality Central Laboratory, 1014 Broadway, Cincinnati, Ohio, USA (1969)
51. Institute of Petroleum Standardization Committee. *Journal of the Institute of Petroleum*, **56**, 107 (1970)
52. Ramsdale, S.J. and Wilkinson, R.E. *Journal of the Institute of Petroleum*, **54**, 326 (1968)
53. Brunnock, J.V. *Journal of the Institute of Petroleum*, **54**, 310 (1968)
54. Duckworth, D.F. Aspects of petroleum pollutant analysis. In *Water Pollution in Oil* (ed. P. Hepple), London, Institute of Petroleum, p. 165 (1971)
55. Blumer, M. *Marine Biology*, **5**, 196 (1970)
56. Adams, I.M. *Process Biochemistry*, **2**, 33 (1967)
57. Freegarde, M., Hatchard, C.G. and Parker, C.A. *Laboratory Practice*, **20**, 35 (1971)
58. Erhardt, M. and Blumer, M. *Environmental Pollution*, **3**, 179 (1972)
59. Zafiron, O.C., Myers, J. and Freestone, F. *Marine Pollution Bulletin*, **4**, 87 (1973)
60. Boylan, D.B. and Tripp, B.N. *Nature (London)*, **230**, 44 (1971)
61. Bulten, N.J., Morris, B.F., Sas, J. and Bermuda, J. *Biological Research*, Special Publication No. 10, (1973)
62. Zafiron, O.C. and Oliver, C. *Analytical Chemistry*, **45**, 952 (1973)
63. Hertz, H.S., May, W.E., Chester, S.N. and Gump, B.H. *Environmental Science and Technology*, **10**, 900 (1976)
64. Rasmussen, D.V. *Analytical Chemistry*, **48**, 1562 (1976)
65. Wilson, C.A., Ferreto, E.P. and Coloman, H.J. *American Chemical Society Division of Chemistry Reports*, **20**, 613 (1975)
66. Garza, M.E. and Huth, J. *Environmental Science and Technology*, **8**, 249 (1974)
67. Adland, E.R., Creaser, L.F. and Matthews, P.H. *Analytical Chemistry*, **44**, 297 (1972)
68. Blumer, M. and Sass, J. *Science*, **176**, 1120 (1972)
69. Ramsdale, S.T. and Wilkinson, R.E. *Journal of the Institute of Petroleum*, **54**, 326 (1968)
70. Adlard, E.R., Creaser, L.F. and Matthews, P.H.D. *Analytical Chemistry*, **46**, 64 (1972)
71. Brunnock, J.V., Duckworth, D.F. and Stephens, G.G. *Journal of the Institute of Petroleum*, **54**, 310 (1968)
72. McKay, T.R. In *Proceedings of the Ninth International Symposium on Gas Chromatography, Montreux, Switzerland*, pp. 33–38, October (1972)
73. Adlard, E.R. and Matthews, P.H.D. *Natural Physical Science*, **233**, 83 (1971)
74. Garra, M.E. and Muth, J. *Science and Technology*, **8**, 249 (1974)
75. Wasik, S.P. and Brown, R.L. Determination of hydrocarbon solubility in seawater and the analysis of hydrocarbons in water-extracts. In *Proceedings of the Conference on Prevention and Control of Oil Spills*, American Petroleum Institute, Washington, pp. 223–237 (1973)
76. May, W.E., Chester, S.N., Cram, S.P., *et al. Journal of Chromatographic Science*, **13**, 555 (1975)
77. Brooks, J.M. and Sackett, W.M. *Journal of Geophysical Research*, **78**, 5248 (1973)
78. Atkinson, L.P. and Richard, F.A. *Deep Sea Research*, **14**, 673 (1967)
79. Perras, J.C. *Portable Gas Chromatographic Technique to Measure Dissolved Hydrocarbons in Seawater*, USNTIS, AD Report No. 786583/5GA (1973)
80. Swinnerton, J.W. and Linnenbom, V.J. *Science*, **156**, 1119 (1967)
81. Bellar, T.A. and Lichtenberg, J.J. *The Determination of Volatile Organic Compounds at the μg/ l Level in Water by Gas Chromatography*, US National Technical Information Service Report No. 237973/3GA (1975)
82. Albaiges, J. and Albrecht, P. *International Journal of Environmental Analytical Chemistry*, **6**, 171 (1979)
83. Ramsdale, S.J. and Wilkinson, R.E. *Journal of the Institute of Petroleum*, **54**, 326 (1968)
84. Brunnock, J.V., Duckworth, D.F. and Stephens, G.G. *Journal of the Institute of Petroleum*, **54**, 310 (1968)
85. Erhardt, M. and Blumer, M. *Environmental Pollution*, **179**, 17–22 (1972)
86. Adlard, E.R., Creaser, L.F. and Matthews, P.H.D. *Analytical Chemistry*, **44**, 64 (1972)
87. Boylan, D.B. and Tripp, B.W. *Nature (London)*, **230**, 44 (1971)
88. Hertz, H.S., May, W.E., Chesler, S.N. and Gump, B.H. *Environmental Science and Technology*, **10**, 900 (1978)
89. Krstulovic, A.M., Rosie, D.M. and Brown, P.R. *Analytical Chemistry*, **48**, 1383 (1976)
90. Done, J.M. and Reid, W.K. *Separation Science*, **5**, 825 (1970)
91. Zurina, L.F., Stradomskaya, A.G. and Semenov, A.D. *Gidrokhim. Mater.*, **57**, 141 (1973)
92. Dawson, R. and Ehrhardt, M. Determination of aromatic hydrocarbons in seawater. In *Methods of Seawater Analysis*, (ed. K. Grasshoff), Verlag Chemie, New York, pp. 227–234 (1976)

93. Van Buuren, B.L. *Analytical Chemistry*, **32**, 1436 (1960)
94. Keizer, P.D. and Gordon, D.C., Jr. *Journal of the Fisheries Research Board of Canada*, **30**, 1039 (1973)
95. Cretney, W.J. and Wong, C.S. National Bureau of Standards (US) Special Publication, No. **409**, p.175 (1974)
96. Hellman, H. and Sehle, H. *Fresenius' Zeitschrift für Analytishe Chemie*, **265**, 245 (1973)
97. Hellman, H. *Fresenius' Zeitschrift für Analytishe Chemie*, **275**, 109 (1975)
98. Schwarz, E.P. and Wasik, S.P. *Analytical Chemistry*, **48**, 524 (1976)
99. Freegarde, M. *Laboratory Practice*, **20**, 35 (1971)
100. Thruston, A.D. and Knight, H.W. *Environmental Science and Technology*, **5**, 64 (1971)
101. Lloyd, J.B.F. *Journal of the Forensic Science Society*, **11**, 83 (1971)
102. Lloyd, J.B.F. *Journal of the Forensic Science Society*, **11**, 153 (1971)
103. Lloyd, J.B.F. *Journal of the Forensic Science Society*, **11**, 235 (1971)
104. Parker, C.A. and Barnes, W.J. *Analyst (London)*, **85**, 3 (1960)
105. Maher, W.A. *Bulletin of Environmental Contamination and Toxicology*, **30**, 413 (1983)
106. Red'Kin, Yu, R., Voitenko, A.M. and Teplyakcv, P.A. *Okeanologiya*, **13**, 908 (1973)
107. Lynch, P.F. and Brown, C.W. *Environmental Science and Technology*, **1**, 1123 (1973)
108. Ahmadjian, M., Baer, C.D., Lynch, P.F. and Brown, C.W. *Environmental Science Technology*, **10**, 777 (1971)
109. Mattson, J.S., Mattson, C.S., Spencer, M.J. and Starks, S.A. *Analytical Chemistry*, **49**, 297 (1977)
110. Wasik, S.P. *Journal of Chromatographic Science*, **12**, 845 (1974)
111. Zsolnay, A. and Kiel, W.J. *Journal of Chromatography*, **90**, 79 (1974)
112. Schuldiner, J.A. *Analytical Chemistry*, **23**, 1676 (1951)
113. Brown, R.A. and Huffman, H.L. *Science*, **101**, 847 (1976)
114. Institute of Petroleum. *Journal of the Institute of Petroleum*, **56**, 107 (1970)
115. Millard, J.P. and Arveson, J.C. *Applied Optics*, **11**, 102 (1972)
116. Freegarde, M., Hatchard, O.G. and Parker, C.A. *Chemical Abstracts*, **20**, 35 (1971)
117. Wilson, A.J. and Forester, J. *US Environmental Protection Agency Report No. EPA-600-7-74*, *108* (1974)
118. Wilson, A.J., Forester, J. and Knight, J. *US Fish Wildlife Service Circular 355, 18–20*, Centre for Estuaries and Research, Gulf Breeze, Florida, USA (1969)
119. *US Environmental Protection Agency Analytical Quality Control Laboratory*, US Government Printing Office, Washington D.C. (1972)
120. Wilson, A.J. and Forester, J.J. *Association of the Officers in Analytical Chemistry*, **54**, 525 (1971)
121. Picer, I.V., Picer, M. and Strohal, P. *Bulletin of Environmental Contamination and Toxicology*, **14**, 565 (1975)
122. Wilson, A.J. *Bulletin of Environmental Contamination and Toxicology*, **15**, 515 (1976)
123. Weber, K. *Water Research*, **10**, 237 (1976)
124. Aspila, K.I., Carron, J.H. and Chau, A.S.Y. *Journal of the Association Official Analytical Chemists*, **60**, 1097 (1977)
125. *Analytical Methods Manual*. Inland Water Directorate, Water Quality Branch, Ottawa, Ontario, Canada (1974)
126. Miles, J.R., Tu, C.M. and Harris, O.R. *Journal of Economics and Entomology*, **62**, 1334 (1969)
127. Chau, A.S.Y., Rosen, J.D. and Cochrane, W.B. *Bulletin of Environmental Contamination and Toxicology*, **6**, 225 (1971)
128. Chau, A.S.Y. *Journal of the Americal Oil and Colours Association*, **57**, 585 (1974)
129. Thompson, J.E. *Analysis of Pesticide Residues in Human and Environmental Samples*, US Environmental Protection Agency, Research Triangle Park, US Environmental N.C. Sec. 10A, DG 12/2/74 (1974)
130. Eichellberger, J.W. and Lichtenberg, J.L. *Environmental Science and Technology*, **5**, 541 (1971)
131. McFarrer, E.F., Lishka, R.I. and Parker, H.J. *Analytical Chemistry*, **42**, 358 (1970)
132. Zitko, V. and Choi, P. *PCB and Other Industrial Halogenated Hydrocarbons in the Environment*. Fish Research Board, Ottawa, Canada, Technical Report No. 272, Biological Station, St. Andrews, N.B. (1971)
133. Oller, W.L. and Cramer, M.F. *Journal of Chromatographic Science*, **13**, 296 (1975)
134. Edwards, R. *Chemistry and Industry (London)*, 1340 (1970)
135. Fishbein, L. *Journal of Chromatography*, **68**, 345 (1972)
136. Schulte, E., Thier, H.P. and Acker, L. *Deut. Lebensmittel Rundschau*, **72**, 229 (1976)
137. Goke, G. *Deut. Lebensim-Rundschau*, **71**, 309 (1975)
138. Ashling, B. and Jensen, S. *Analytical Chemistry*, **42**, 1483 (1970)
139. Musty, P.R. and Nickless, G. *Journal of Chromatography*, **89**, 185 (1974)
140. Girenko, D.B., Klisenko, M.A. and Dishhcholka, Y.K. *Hydrobiological Journal*, **11**, 60 (1975)

141. Elder, D. *Marine Pollution Bulletin*, **7**, 63 (1976)
142. Harvey, G.R. *Report of the US Environment Protection Agency*, EDA-R2-73-177 (1973)
143. Picer, N. and Picer, M. *Journal of Chromatography*, **193**, 357 (1980)
144. Holden, A.V. and Marsden, K. *Journal of Chromatography*, **44**, 481 (1969)
145. Snyder, D.E. and Reinert, R.E. *Bulletin of Environmental Contamination and Toxicology*, **6**, 385 (1971)
146. Picer, M. and Abel, M. *Journal of Chromatography*, **150**, 1191 (1978)
147. Dawson, R., Riley, J.P. and Tennant, R.H. *Marine Chemistry*, **4**, 83 (1976)
148. Lovelock, J.E., Maggs, R.J. and Wade, R.L. *Nature (London)*, **241**, 194 (1973)
149. Lovelock, J.E. *Nature (London)*, **256**, 193 (1975)
150. Murray, A.J. and Riley, J.P. *Nature (London)*, **242**, 37 (1973)
151. Murray, A.J. and Riley, J.P. *Analytica Chimica Acta*, **65**, 261 (1973)
152. Eklund, G., Josefsson, B. and Roos, C. *Journal of High Resolution Chromatography*, **1**, 34 (1978)
153. Eklund, G., Josefsson, B. and Roos, C. *Journal of Chromatography*, **142**, 575 (1977)
154. Grob, K. and Grob, G. *Journal of Chromatography*, **125**, 471 (1970)
155. Grob, K. *Chromatographie*, **10**, 181 (1977)
156. Minear, R.A. *Environmental Science and Technology*, **6**, 431 (1972)
157. Minear, R.A. and Walanski, K.A. *Investigation of the Chemical Identity of Soluble Organophosphorus Compounds found in Natural Waters*, Research Report UILU-WRC-74-0086 (1974)
158. McElvroy, W.D. and Strehler, B.L. *Archives of Biochemistry*, **22**, 420 (1949)
159. Wildish, D.J. *Determination of Adenosine 5'-triphosphate in Estuarine Water and Sediments by Firefly Bioluminescence Assay*, Technical Report of the Fisheries Maritime Service Research Division, No. 649, (1976)
160. Hofer-Siegrist, L. and Schweiz, A. *Hydrology*, **38**, 49 (1976)
161. Michel, H.O., Gordon, E.C. and Epstein, J. *Environmental Science and Technology*, **7**, 1045 (1973)
162. Grift, N. and Lockhart, W.L. *Journal of the Association of Official Analytical Chemists*, **57**, 1282 (1974)
163. Daughton, C.G., Crosby, D.C., Garnas, R.L. and Hsieh, D.P.H. *Journal of Agriculture and Food Chemistry*, **24**, 236 (1976)
164. Cambella, A.D. and Antia, N.J. *Marine Chemistry*, **19**, 205 (1986)
165. Lores, E.M., Moore, J.C., Knight, J. *et al. Journal of Chromatographic Science*, **23**, 124 (1985)
166. Formaro, L. and Trasatti, S. *Analytical Chemistry*, **40**, 1060 (1968)
167. Zvonaric, T., Kutic, V. and Branica, M. *Thalassia Jugosi*, **9**, 65 (1973)
168. Kozarac, Z., Cosovic, B. and Branica, M. *Marine Science Communications*, **1**, 147 (1975)
169. Hagihara, K. *Kogai Bunseki Shiskin*, **5**, 68 (1972)
170. Wang, L.K., Yang, J.Y. and Wang, M.H. *Engineering Bulletin Purdue University Engineering External Series*, **142**, 76 (1973)
171. Wang, L.K., Yang, J.Y. and Wang, M.H. *Journal of the American Water Works Association*, **67**, 6 (1975)
172. Takano, S., Takasake, S., Kunihiro, K. and Yamanaka, M. *Yukagaku*, **25**, 31 (1976)
173. Favretto, L. and Tunis, F. *Analyst (London)*, **101**, 198 (1976)
174. Crisp, P.T., McKert, J.M. and Gibson, N.A. *Analytica Chimica Acta*, **78**, 391 (1975)
175. Le Bihan, A. and Courtot-Coupez, J. *Bulletin of the Chemical Society of France*, **406** (1970)
176. Courtot-Coupez, J. and Le Bihan, A. *Analyst Letters*, **2**, 211 (1969)
177. Le Bihan, A. and Courtot Coupez, J. *Analyst Letters*, **10**, 759 (1977)
178. Gagnon, M.J. *Water Research*, **13**, 53 (1979)
179. Bhat, S.R., Eckert, J.M. and Gibson, N.A. *Analytica Chimica Acta*, **116**, 191 (1980)
180. Crisp, P.T., Eckert, J.M. and Gibson, N.A. *Analytica Chimica Acta*, **78**, 391 (1975)
181. Crisp, P.T., Ekert, J.M., Gibson, N.A. *et al. Analytica Chimica Acta*, **87**, 97 (1976)
182. Hon Nami, H. and Hanya, T. *Water Research*, **14**, 1251 (1980)
183. Hon Nami, H. and Hanya, T. *Journal of Chromatography*, **161**, 205 (1978)
184. American Public Health Association. *Standard Methods for the Examination of Waters and Wastewaters*. APHA (1975)
185. Oba, K., Mori, A. and Tomiyama, S. *Yukagaku*, **17**, 517 (1968)
186. Evans, H.C. *Journal of the Society of Chemical Industry (London)*, Supplement Issue No. 2, 76–80 (1950)
187. Courtot-Coupez, J. and Le Bihan, A. *Analytical Letters*, **2**, 567 (1969)
188. Courtot-Coupez, J. and Le Bihan, A. *Riv. Ital. Soztanze Grasse*, **20**, 672 (1971)
189. Grieff, O. *Analytical Abstracts*, **13**, 3650 (1966)
190. Crisp, P.T., Eckert, J.M. and Gibson, W.A. *Analytica Chimica Acta*, **104**, 93 (1979)

191. Harwood, J.E. and Huyser, D.J. *Water Research*, **2**, 631 (1968)
192. Treguer, P., Le Corre, P. and Courtot, P. *Journal of the Marine Biological Association of the UK*, **52**, 1045 (1972)
193. Dunges, W. *Analytical Chemistry*, **49**, 442 (1977)
194. Antia, N.J., McAllister, C.D., Parsons, T.R. *et al. Limnology and Oceanography*, **8**, 166 (1963)
195. Shah, N.M. and Fogg, G.E. *Journal of the Marine Biological Association of the UK*, **53**, 321 (1973)
196. Shah, N.M. and Wright, R.T. *Marine Biology*, **24**, 121 (1974)
197. Stradomskaya, A.G. and Goncharova, I.A. *Gidrokhim Mater*, **43**, 57 (1967)
198. Mueller, H.F., Larson, T.E. and Ferretti, M. *Analytical Chemistry*, **32**, 687 (1960)
199. Koyama, T. and Thompson, T.G. *Journal of the Oceanographic Society of Japan*, **20**, 209 (1964)
200. Maksimova, I.V. and Pimenova, M.N. *Mikrobiology*, **37**, 77 (1969)
201. Lamar, W.L. and Goerlitz, D.F. *Organic Acids in Naturally Colored Surface Waters*, Geological Survey Water-Supply Paper 1817-A: 1–17 (1966)
202. Horikawa, K. *Buseki Kagaku*, **21**, 806 (1972)
203. Goncharova, I.A. and Khomenko, A.N. *Gidrokhim Mater*, **53**, 36 Ref: Zhur Khim 199D (1970) Abstract No. 59339 (1971)
204. Borch, R.F. *Analytical Chemistry*, **47**, 2437 (1975)
205. Hoffman, N.W. and Ligo, J.C. *Analytical Chemistry*, **48**, 1104 (1976)
206. Garrett, W.D., Timmons, C.O., Jarvis, N.L. and Kagarise, R.E. *Constitution and Surface Chemical Properties of Sea Slicks*, part 1, Bay of Panama, Naval Research Laboratory, Report 5925 (1963)
207. Slowey, J.F., Jeffrey, L.M. and Hood, D.W. *Geochimica and Cosmochimica Acta*, **26**, 607 (1962)
208. Williams, P.M. *Journal of the Fisheries Research Board of Canada*, **22**, 1107 (1965)
209. Jeffrey, L.M. *Journal of the American Oil Chemists Society*, **43**, 211 (1966)
210. Ackman, R.G., Sipos, J.C. and Tocher, C.S. *Journal of the Fisheries Research Board of Canada*, **24**, 635 (1967)
211. Mahadevan, V. and Stenroos, L. *Analytical Chemistry*, **39**, 1652 (1967)
212. Blumer, M. Dissolved organic compounds in seawater. Saturated and olefinic hydrocarbons and singly branched fatty acids. In *Organic Matter in Natural Waters* (ed. D.W. Hood), Institute of Marine Science University of Alaska, Publication No. 1, pp. 153–167 (1970)
213. Stauffer, T.B. and MacIntyre, W.G. *Chesapeake Science*, **11**, 216 (1970)
214. Quinn, J.G. and Meyers, P.A. *Limnology and Oceanography*, **16**, 129 (1971)
215. Meyers, P.A. *Limnology and Oceanography*, **21**, 315 (1976)
216. Garrett, W.D. *Deep Sea Research*, **14**, 221 (1967)
217. Barger, W.R. and Garrett, W.D. *Journal of Geophysical Research*, **75**, 4561 (1970)
218. Duce, R.A., Quinn, J.G., Olney, C.E., *et al. Science*, **176**, 161 (1972)
219. Schultz, D.M. and Quinn, J.G. *Journal of Fisheries Research*, **29**, 1482 (1972)
220. Marty, J.C. and Saliot, A. *Journal Rech. Atmosphere*, **13**, 563 (1974)
221. Ackman, R.G. *Journal of Chromatographic Science*, **10**, 560 (1972)
222. Di Corcia, A. and Samperi, F. *Analytical Chemistry*, **46**, 140 (1974)
223. Bjork, R.G. *Analytical Biochemistry*, **63**, 80 (1975)
224. White, W.R. and Leenheer, J.A. *Journal of Chromatographic Science*, **13**, 386 (1975)
225. Sever, J. and Parker, P.L. *Science*, **164**, 1052 (1969)
226. Schultz, D.M. and Quinn, J.M. *Marine Biology*, **27**, 143 (1974)
227. White, R.H. and Miller, S.L. *Science*, **193**, 885 (1976)
228. Kamata, E. *Bulletin of the Chemical Society of Japan*, **39**, 1227 (1966)
229. Afghan, B.K., Kulkarni, A.U., Leung, R. and Ryan, J.F. *Environmental Letters*, **7**, 53 (1974)
230. Zika, R.G. *An Investigation in Marine Photochemistry*, PhD Thesis, Dalhousie University (1977)
231. Belman, S. *Analytica Chimica Acta*, **29**, 120 (1963)
232. Slawinska, D. and Slawinski, J. *Analytical Chemistry*, **47**, 2101 (1975)
233. Afghan, B.K., Kulkarni, A.U. and Ryan, J.F. *Analytical Chemistry*, **47**, 488 (1975)
234. MacKinnon, M.D. *The Analysis of the Total Organic Carbon in Seawater: a. Development of Methods for the Quantification of T.O.C. b. Measurement and Examination of the Volatile Fraction of the T.O.C.* PhD Thesis, Dalhousie University, pp. 1–183 (1977)
235. Eberhardt, M.A. and Sieburth, J.M. *Marine Chemistry*, **17**, 199 (1985)
236. Katase, T. *Kogyo Josui*, **1972**, 19 (1972)
237. Corcoran, E.G. *Environmental Health Perspectives*, **3**, 13 (1978)
238. Giam, C.S., Chan, H.S. and Neff, G.S. *Analytical Chemistry*, **13**, 2225 (1975)
239. Hites, R.A. *Environmental Health Perspectives*, **1973**, 17 (1973)
240. Lewis, G.J. and Rakestraw, N.W. *Journal of Marine Research*, **14**, 253 (1955)

332   Organic compounds

241. Josefsson, B. and Koroleff, F. Polyphenolic substances. In *Methods of Seawater Analysis*, (ed. K. Grasshoff), Verlag Chemie, New York, pp. 220–227 (1976)
242. Collier, A. *Limnology and Oceanography*, **3**, 33 (1958)
243. Josefsson, B., Uppstrom, L. and Ostling, G. *Deep Sea Research*, **19**, 385 (1972)
244. Mopper, K. and Gindler, E.M. *Analytical Chemistry*, **56**, 440 (1973)
245. Johnson, K.M. and Sieburth, J. McN. *Marine Chemistry*, **5**, 1 (1977)
246. Burney, C.M. and Sieburth, J. McN. *Marine Chemistry*, **5**, 15 (1977)
247. Hirayama, H. *Analytica Chimica Acta*, **70**, 141 (1974)
248. Williams, P.M. and Craigie, J.S. Microdetermination of uronic acids and related compounds in the marine environment. In *Organic Matter in Natural Waters* (ed. D.W. Hood), Institute of Marine Science, University of Alaska, Publication No. 1. pp. 509–519 (1970)
249. Mopper, K. *Marine Chemistry*, **5**, 585 (1977)
250. Hicks, S.E. and Carey, F.G. *Limnology and Oceanography*, **13**, 361 (1968)
251. Andrews, P. and Williams, P.J. Leb. *Journal of the Marine Biological Association*, **51**, 11 (1971)
252. Whittaker, J.R. and Vallentyne, J.R. *Limnology and Oceanography*, **2**, 98 (1957)
253. Degens, E.T., Reuter, J.H. and Shaw, K.N.F. *Geochimica and Cosmochimka Acta*, **28**, 45 (1964)
254. Starikova, N.D. and Yablokova, O.G. *Okeanologiya*, **12**, 431 (1972a)
255. Starikova, N.D. and Yablokova, O.G. *Okeanologiya*, **12**, 898 (1972b)
256. Geller, A. *Archives Hydrobiology*, **47**, 295 (1975)
257. Modzeleski, J.E., Laurie, W.A. and Nagy, B. *Geochimica and Cosmochimica Acta*, **34**, 825 (1971)
258. Tesarik, K. *Journal of Chromatography*, **65**, 295 (1972)
259. Eklund, G., Josefsson, B. and Roos, C.U. *Journal of Chromatography*, **142**, 575 (1977)
260. Josefsson, B.P. *Analytica Chimica Acta*, **52**, 65 (1970)
261. Larsson, L.T. and Samuelson, O. *Microchimica Acta*, **2**, 328 (1967)
262. Mopper, K. and Degens, E.G. *Analytical Biochemistry*, **45**, 147 (1972)
263. Williams, P.J. Le B. *Limnology and Oceanography*, **9**, 138 (1964)
264. Alekseeva, L.M. *Trudy gos Oceanogr. Inst.*, **60**, 17 (1972)
265. Goulden, P.D., Brooksbank, P. and Day, M.B. *Analytical Chemistry*, **45**, 2430 (1973)
266. Schlungbaum, G. and Behling, A. *Acta Hydrochimica Hydrobiologica*, **2**, 423 (1974)
267. Stilinovic, L., Mujko, Il. and Vukic, B. *Arh. Hig. Rada Tokskko*, **25**, 247 (1974)
268. Gales, M.E., Jr. *Analýst (London)*, **100**, 841 (1975)
269. Wexler, A.S. *Analytical Chemistry*, **35**, 1936 (1963)
270. Fountaine, J.E., Joshipura, P.B., Keliher, P.N. and Johnson, J.D. *Analytical Chemistry*, **46**, 62 (1975)
271. Williamson, J.A. Rapid determination of phenol content of water. In *Water Resource Instruments Proceedings of an International Seminar* (eds. R.J. Krizek and E.F. Mosonyi), Ann Arbor Science, Ann Arbor, MI, pp. 175–186 (1975)
272. Sinel'nikov, V.E. and Breusov, N.G. *Trudy Inst. Biol. V nutr. Vod, Akad. Nauk. SSSR*, 212 (1973)
273. Afghan, B.K., Belliveau, P.E., Larose, R.H. and Ryan, J.F. *Analytica Chimica Acta*, **71**, 355 (1974)
274. Guilbault, G.G., Kramer, D.N. and Hackley, E. *Analytical Chemistry*, **38**, 1897 (1966)
275. Goncharova, I.A., Khomenko, A.N. and Shuk, I.P. *Gidrokhim. Mater*, **61**, 100 (1974)
276. Chriswell, C.D., Kissinger, L.D. and Fritz, J.S. *Analytical Chemistry*, **47**, 1325 (1976)
277. Kunte, H. and Slemrova, J. *Zeitung Wasserabwasser Forschung*, **8**, 176 (1975)
278. Kleverlaan, N.T.M. *Chemishe Weekblad*, **71**, 13 (1975)
279. Cassidy, R.M., LeGay, D.S. and Frei, R.W. *Journal of Chromatographic Science*, **12**, 85 (1974)
280. Park, K., Williams, W.T., Prescott, J.M. and Hood, D.W. *Science*, **138**, 531 (1962)
281. Riley, J.P. and Segar, D.A. *Journal of the Marine Biological Association*, **50**, 713 (1970)
282. Hosaku, K. and Maita, Y. *Journal of the Oceanographic Society of Japan*, **27**, 27 (1971)
283. Starikova, N.D. and Korzhikova, R.I. *Oceanogiya*, **9**, 509 (1969)
284. Lee, C. and Bada, J.L. *Earth Planetary Science Letters*, **26**, 61 (1975)
285. Zlobin, V.S., Perlyuk, M.F. and Orlova, T.A. *Okeanologiya*, **15**, 643 (1975)
286. Debeika, E.V., Fyabov, A.K. and Nabivanets, B.I. *Gidrobiol. Zhur*, **9**, 109 (1973)
287. Udeufriend, S., Stein, S., Bohlen, P., Dairman, W., Leimgruber, W. and Weigele, M. *Science*, **178**, 871 (1972)
288. North, B.B. *Limnology and Oceanography*, **20**, 20 (1975)
289. Parkard, T.T. and Dortch, Q. *Marine Biology*, **33**, 347 (1975)
290. Baier, R.M., Goupil, D.W., Perimutter, S. and King, R. *Journal Rech. Atmosphere*, **8**, 1 (1974)
291. Semenov, A.D., Ivleva, I.N. and Datsko, V.G. *Trudy Komiss, Anal. Khim. Akad. Nauk. SSSR*, **13**, 62 (1964)

292. Coughenower, D.D. and Curl, H.C., Jr. *Limnology and Oceanography*, **20**, 128 (1975)
293. Froehlich, P.M. and Murphy, L.D. *Analytical Chemistry*, **49**, 1606 (1977)
294. Whitledge, T.E. and Dugdale, R.C. *Limnology and Oceanography*, **17**, 309 (1972)
295. Litchfield, C.D. and Prescott, J.M. *Limnology and Oceanography*, **15**, 250 (1970)
296. Clark, M.E., Jackson, G.A. and North, W.J. *Limnology and Oceanography*, **17**, 749 (1972)
297. Palmork, K.H. *Acta Chimica Scandanavica*, **17**, 1456 (1963)
298. Siegal, A. and Degens, E.T. *Science*, **151**, 1098 (1966)
299. Webb, K.L. and Wood, L. Improved techniques for analysis of free amino acids in seawater. In *Automation in Analytical Chemistry (Technicon Symposium, 1966)* (eds N.B. Scova,*et al.*) New York Mediad, Incorp., Vol. 1, pp. 440–444 (1967)
300. Hobbie, J.E., Crawford, C.C. and Webb, K.L. *Science*, **159**, 1463 (1968)
301. Brockman, U.H., Eberlein, K., Junge, H.D., *et al. Entwicklung naturlicher Plankton-populationen in einem outdoor-tank mit nahrstoffarmen Meerwasser. II. Konsentrationsverander-ungen von gelosten neutralen Kohlen Meeresforschung*. Vol. 6. Springer, Berlin (1974)
302. Williams, P.J., LeB. Berman, T. and Holm-Hansen, O. *Marine Biology*, **35**, 41 (1976)
303. Bohling, H. *Marine Biology*, **6**, 213 (1970)
304. Bohling, H. *Marine Biology*, **16**, 281 (1972)
305. Garrasi, C. and Degens, E.T. *Analytische Methoden zur saulenchromatographischen bestimmung von Aminosauren und Suckern im Meereswasser und Sediment. Berichte aus dem Projekt DFG-De 74/e: 'Litoralforschung-Abwasser in Kustennahe'*, DFG-Abschlubkolloquim, Bremerhaven (1976)
306. Tetsumota, M., Williams, W.T., Prescott, J.M. and Hood, D.W. *Journal of Marine Research*, **19**, 89 (1961)
307. Roth, M. *Analytical Chemistry*, **43**, 880 (1971)
308. Bayer, E., Grom, E., Kaltenegger, B. and Uhmann, R. *Analytical Chemistry*, **48**, 1106 (1976)
309. Roth, M. and Hampai, A. *Journal of Chromatography*, **83**, 353 (1973)
310. Felix, A.M. and Terkelsen, G. *Archives of Biochemistry and Biophysics*, **157**, 177 (1973)
311. Gardner, W.S. *Marine Chemistry*, **6**, 15 (1975)
312. Dawson, R., and Pritchard, R.G. *Marine Chemistry*, **6**, 27 (1978)
313. Mopper, K. *Science of the Total Environment*, **49**, 115 (1986)
314. Pocklington, R. *Analytical Biochemistry*, **45**, 403 (1972)
315. Gehrke, C.W., Lamkin, W.M., Stalling, D.L. and Shahrokhi, F. *Biochemistry and Biophysics*, **19**, 328 (1965)
316. Gardner, W.S. and Lee, G.F. *Environmental Science and Technology*, **7**, 719 (1973)
317. Batley, G.E., Florence, T.M. and Kennedy, J.R. *Talanta*, **20**, 987 (1973)
318. Di Corcia, A. and Samperi, R. *Analytical Chemistry*, **46**, 977 (1974)
319. El-Dib, M.A., Abdel-Rahman, M.O. and Aly, O.A. *Water Research*, **9**, 515 (1975)
320. Varney, M.S. and Preston, M.R. *Journal of Chromatography*, **348**, 265 (1985)
321. Emmet, R.T. *Spectrophotometric Determination of Urea in Natural Waters with Hypochlorite and Phenol*. NAVSHIPRANDLAB Annapolis Report 2663 (1969)
322. Newell, B.S., Morgan, B. and Cundy, J. *Journal of Marine Research*, **25**, 201 (1967)
323. Muravskaya, Z.A. *Gidrobiol. Zhur.* **9**, 127 (1973)
324. DeMarche, J.M., Curl, H. Jr. and Coughenower, D.D. *Limnology and Oceanography*, **18**, 686 (1973)
325. Von Breymann, M.T., de Angelis, M.A. and Gordon, L.I. *Analytical Chemistry*, **54**, 1209 (1982)
326. Hashimoto, A., Sakinot, H., Yamagami, E. and Tateishi, S. *Analyst (London)*, **105**, 787 (1980)
327. Hashimoto, A., Kozima, T., Shakino, H. and Arikey, T. *Water Research*, **13**, 509 (1979)
328. Shinohara, R., Kido, A., Okomoto, Y. and Takeshita, R. *Journal of Chromatography*, **256**, 81 (1983)
329. Juuk, G.A., Richard, J.J., Greeser, M.D., *et al. Journal of Chromatography*, **99**, 745 (1974)
330. Pillai, T.N.V. and Ganguly, A.K. *Current Science*, **39**, 501 (1970)
331. Hicks, E. and Riley, J.P. *Analytica Chimica Acta*, **116**, 137 (1980)
332. Strickland, J.D.H. and Solorzano, L. *Determination of Monoesterase Hydrolysable Phosphate and Phosphomonesterase Activity in Seawater. Some Continued Studies in Marine Science*, pp. 665–674 Springer, Berlin (1966)
333. Maeda, M. and Taga, N. *Marine Biology*, **20**, 58 (1973)
334. Lovelock, J.E., Maggs, R.J. and Rasmussen, R.A. *Nature (London)*, **237**, 452 (1972)
335. Yamaoka, Y. and Tanimota, T. *Journal of The Agricultural Chemical Society*, **50**, 147 (1976)
336. Andreae, M.O. *Analytical Chemistry*, **52**, 150 (1980)
337. Andreae, M.O. *Analytical Chemistry*, **49**, 820 (1977)
338. Braman, R.S., Ammions, J.M. and Bricker, J.L. *Analytical Chemistry*, **50**, 992 (1978)

339. Delmas, R.P., Parrish, C. and Ackman, R.G. *Analytical Chemistry*, **56**, 1272 (1984)
340. Paradis, M. and Ackman, R.G. *Journal of the Fisheries Research Board, Canada*, **34**, 2156 (1977)
341. Lewis, R.W. *Limnology and Oceanography*, **14**, 35 (1969)
342. Lee, R.F., Nevenzei, J.C. and Pattenhofer, G.A. *Marine Biology*, **9**, 99 (1971)
343. Volimirov, B. *Marine Ecology*, **3**, 97 (1982)
344. Olsen, C.R., Cutshall, N.H. and Larsen, T.L. *Marine Chemistry*, **11**, 501 (1982)
345. Whittle, K.J., Hardy, R., Mackie, P.R. and McGill, A.S. *Philosophical Transactions of the Research Society, London, Series B*, **297**, 193 (1982)
346. Sullivan, K.R., Atlas, E.L. and Glam, C.S. *Environmental Science and Technology*, **16**, 428 (1982)
347. Simkiss, K. *Journal of the Marine Biological Association, UK*, **63**, 1 (1983)
348. Florence, T.M., Lumsden, B.G. and Fardy, J. *Analytica Chimica Acta*, **151**, 281 (1983)
349. Parrish, C.C. and Ackman, R.G. *Journal of Chromatography*, **262**, 103 (1983)
350. Wun, C.K., Walker, R.W. and Litsky, W. *Water Research*, **10**, 955 (1976)
351. Dutka, B.T., Chau, A.S.Y. and Coburn, J. *Water Research*, **8**, 1047 (1974)
352. Wun, C.K., Walker, R.W. and Litsky, W. *Health Laboratory Science*, **15**, 67 (1978)
353. Adams, D.D. and Richards, R.A. *Deep Sea Research*, **15**, 471 (1960)
354. Belyaeva, A.N. *Okeanologiya*, **14**, 1101 (1974)
355. Matthews, S.W. and Smith, L.L. *Lipids*, **3**, 239 (1968)
356. Smith, L.L. and Gouron, R.E. *Water Research*, **3**, 141 (1969)
357. Saliot, A. and Barbier, M. *Deep Sea Research*, **20**, 1077 (1973a)
358. Saliot, A. and Barbier, M. *Advances in Organic Geochemical Procedures Institute Meeting* in *Proc. Mtg. US Geochemical Institute*, 607–617 (1973b)
359. Matsushima, H., Wakimoto, T. and Tatsukawa, R. *Japan Analyst*, **24**, 342 (1975)
360. Gagosian, R.B. *Geochimica and Cosmochimica Acta*, **39**, 1443 (1975)
361. Attaway, D. and Parker, P.L. *Science*, **169**, 674 (1970)
362. Kanazawa, A. and Teshima, S.I. *Journal of the Oceanographic Society of Japan*, **27**, 207 (1971)
363. Hanck, K.W. and Dillard, J.W. *Analytical Chemistry*, **49**, 404 (1977)
364. Stolzberg, R.J. and Rosin, D. *Analytical Chemistry*, **49**, 226 (1977)
365. Chau, Y.K., Gachter, R. and Lum-Shue-Chan, K. *Journal of the Fisheries Research Board of Canada*, **31**, 1515 (1974)
366. Buffle, J., Greter, France-Line and Haerdi, W. *Analytical Chemistry*, **49**, 216 (1977)
367. Jones, D.R. and Mahahan, S.E. *Analytical Letters*, **8**, 421 (1975)
368. Jones, D.R. and Mahahan, S.E. *Analytical Chemistry*, **48**, 502 (1976)
369. Kaiser, K.L.E. *Water Research*, **7**, 1465 (1973)
370. Egan, W.G. *Journal of the Marine Technical Society*, **8**, 40 (1974)
371. Shapiro, J. *Limnology and Oceanography*, **2**, 161 (1957)
372. Ghassemi, M. and Christman, R.F. *Limnology and Oceanography*, **13**, 583 (1968)
373. Gjessing, E.T. *Environmental Science and Technology*, **4**, 437 (1970)
374. Brown, W. *Marine Chemistry*, **3**, 253 (1975)
375. Wangersky, P.J. Production of dissolved organic matter. In *Marine Ecology* (ed. O. Kinne), Vol. 4, John Wiley, New York (1978)
376. Sieburth, J. McN. and Jensen, A. *Journal of Marine Biology and Ecology*, **2**, 174 (1968)
377. Sieburth, J. McN. and Jensen, A. *Journal of Experimental Marine Biology and Ecology*, **3**, 275 (1969)
378. Pocklington, R. and Hardstaff, W.R. *Journal of the Conservation of International Mercantile Exploration*, **36**, 92 (1974)
379. Stoeber, H. and Eberle, S.H. *Organic Acids in the Rhine and Some of its Tributaries*. Kernforschungszentrum Karlsrune, KFK 1969 UF. pp. 58–81 (1974)
380. Revina, S.K. and Kriul'Kov, V.A. *Trudy gos Oceanographic Institute*, No. 113, 42 (1972)
381. Bilikova, A. *Vod. Hospod, B*, **23**, 44 (1973)
382. Kloster, M.B. *Journal of the American Water Works Association*, **66**, 44 (1975)
383. Bilikova, A. *Vod. Hospod. B*, **24**, 20 (1974)
384. Eberle, S.H., Hoesle, C. and Krueckeberg, C. *New Method for Simultaneously Determining Lignosulphonic Acid and Humic Acid in Water, Using Differential Pulse Polarography*. Kernforschungs-zentrum Karlsruhe, KFK 1969 UF pp. 44–57 (1974)
385. Paluch, J. and Stangret, S. *Fortschr. Wasserchem. Ihrer Grenzgeg*, **14**, 89 (1972)
386. Pocklington, R. and MacGregor, C.D. *International Journal of Environmental Analytical Chemistry*, **3**, 81 (1973)
387. Almgren, T. and Josefsson, B. *Sveridge Papperstidn.*, **76**, 19 (1973)
388. Wilander, A., Kvarnas, H. and Lindell, T. *Water Research*, **8**, 1037 (1974)
389. Almgren, T., Josefsson, B. and Nyquist, G. *Analytica Chimica Acta*, **78**, 411 (1975)

390. Kononova, M.M. In *Soil Organic Matter*, Pergamon Press, New York, (1966)
391. King, L.H. *Isolation and Characterization of Organic Matter from Glacial-marine Sediments on the Scotian Shelf*. Report B.I.O. 67-4: 1–18 (1967)
392. Rashid, M.A. and King, L.H. *Geochimica and Cosmochimica Acta*, **33**, 147 (1969)
393. Pierce, R.H.Jr. and Felbeck, G.T.Jr. A comparison of three methods of extracting organic matter from soils and marine sediments. In *Humic Substances*, (eds D. Povoledo and H.L. Golterman), Center Agriculture Publication Document, Wageningen, pp. 217–232 (1972)
394. Stuermer, D.H. and Harvey, G.R. *Nature (London)*, **250**, 480 (1974)
395. Stuermer, D.H. and Harvey, G.R. *Deep Sea Research*, **24**, 303 (1977)
396. Mantoura, R.F.C. and Riley, J.P. *Analytica Chimica Acta*, **76**, 97 (1975)
397. Stuermer, D.H. and Payne, J.R. *Geochimica and Cosmochimica Acta*, **40**, 1109 (1976)
398. Schnitzer, M. Chemical, spectroscopic, and thermal methods for the classification and characterization of humic substances. In *Humic Substances* (eds. D. Povoledo and H.L. Golterman), Center for Agriculture Publication and Documentation, Wageningen, pp. 293–310 (1972)
399. Salfeld, J. Chr., Ultraviolet and visible absorption spectra of humic systems. In *Humic Substances* (eds D. Povoledo and H.L. Golterman), Center for Agriculture Publications and Documentation, Wageningen, pp. 269–280 (1972)
400. Generalova, V.A. *Gidrokhim. Mater*, **60**, 186 (1974)
401. Sil'chenko, S.A. Rubinshtein, R.N. and Vasileva, L.N. Polarographic determination of fulvic acids. *Zhur Analytical Khim.*, **27**, 2465 (1972)
402. Manka, J., Rebhun, M., Mandelbaum, A. and Bortinger, A. *Environmental Science and Technology*, **8**, 1017 (1974)
403. Schnitzer, M. and Skinner, S.I.M. *Canadian Journal of Chemistry*, **52**, 1072 (1974)
404. Skinner, S.I.M. and Schnitzer, M. *Analytica Chimica Acta*, **75**, 207 (1975)
405. Strickland, J.D.H. and Parsons, T.R. *A Practical Handbook of Seawater Analysis*, 2nd edn. (Bulletin of the Fisheries Research Board of Canada, No. 067) (1972)
406. Strickland, J.D.H. and Parsons, T.R. Determination of soluble organic carbon. In *A Guide to Marine Pollution* (ed. E.D. Goldberg) Gordon and Breach Science Publishers, New York, pp. 70–76 (1972)
407. Rai, H. *Theoretical Angerolimnology* **18**, 1864 (1973)
408. Messiha-Hanna, R.G. *Okeanologiya*, **13**, 704 (1973)
409. Quirry, M.R., Holland, W.E., Waring, R.C. and Hale, G.M. *High Sensitivity Laser Absorption Spectroscopy of Laboratory Aqueous Solutions and of Natural Missouri Waters*. US National Technical Information Service, PB Report No. 243123 (1975)
410. Bjarnborg, B. *Report No. 1239, P22ABJA*. Swedish Environmental Protection Board, Solna, Sweden. (1979)
411. Boto, K.G. and Bunt, T.S. *Analytical Chemistry*, **50**, 392 (1978)
412. Murray, A.P., Gibbs, C.F., Longmore, A.R. and Flett, D.J. *Marine Chemistry*, **19**, 211 (1986)
413. Ryther, J.H. and Guillard, R.R.L. *Canadian Journal of Microbiology*, **8**, 437 (1962)
414. Ohwada, K., Otshhata, M. and Taga, N. *Bulletin of the Japanese Society of Fish Science*, **28**, 817 (1972)
415. Gold, K. *Limnology and Oceanography*, **9**, 343 (1964)
416. Carlucci, A.F. and Silbernagel, S.B. *Canadian Journal of Microbiology*, **12**, 175 (1966)
417. Natarajan, K.V. and Dugdale, R.C. *Limnology and Oceanography*, **11**, 621 (1966)
418. Natarajan, K.N. *Applied Microbiology*, **16**, 366 (1968)
419. Litchfield, C.D. and Hood, D.W. *Applied Microbiology*, **13**, 886 (1965)
420. Ohwada, K. *Marine Biology*, **14**, 10 (1972)
421. Carlucci, A.F. and Silbernagel, S.B. *Canadian Journal of Microbiology*, **13**, 979 (1967)

Chapter 8

Organometallic compounds

8.1 Organomercury compounds

The interest in mercury contamination, and particularly in the organic mercury compounds, is a direct reflection of the toxicity of these compounds to man. Some idea of the proliferation of work can be derived from the reviews of Krenkel [1]; Robinson and Scott [2] (460 references) and Uthe and Armstrong [3] (283 references).

All forms of mercury are potentially harmful to biota, but monomethyl and dimethyl mercury are particularly neurotoxic. The liphophilic nature of the latter compounds allows them to be concentrated in higher trophic levels and the effects of this biomagnification can be catastrophic [4]. Certain species of micro-organisms in contact with inorganic mercury produce methylmercury compounds [5]. Environmental factors influence the net amount of methylmercury in an ecosystem by shifting the equilibrium of the opposing methylation and demethylation processes. Methylation is the result of mercuric ion (Hg^{2+}) interference with biochemical C-1 transfer reactions [6]. Demethylation is brought about by non-specific hydrolytic and reductive enzyme processes [7–9]. The biotic and abiotic influences that govern the rates at which these processes occur are not completely understood.

Although much of the early work on the cycling of mercury pollutants has been performed in fresh-water environments, estuaries are also subject to anthropogenic mercury pollution [10]. A strong negative correlation exists between the salinity of anaerobic sediments and their ability to form methylmercury from Hg^{2+}. As an explanation for this negative correlation the theory was advanced that sulphide, derived by microbial reduction of sea salt sulphate, interferes with Hg^{2+} methylation by forming mercuric sulphide which is not readily methylated [11–14]. There are several reports in the literature concerning the methylation of Hg^{2+} by methylcobalamin [6, 15, 16].

Loss of mercury on storage

Mercury, more than most metals, is subject to loss on storage. Methylmercury and some of the other organomercury compounds are quite volatile and easily lost from apparently sealed bottles. The problems of storage have been reviewed by Mahan and Mahan [17] and by Olson [18]. Olson felt that the greatest losses occurred from the inorganic pool while Mahan and Mahan were more concerned with loss from samples with a high organic content. There have also been problems associated with the kinds of bottles used for storage [19].

Stoeppler and Matthes [20] have made a detailed study of the storage behaviour of methylmercury and mercuric chloride in seawater. They recommended that samples spiked with inorganic and/or methylmercury chloride were stored in carefully cleaned glass containers acidified with hydrochloric acid to pH 2.5. Brown glass bottles were preferred. Storage of methylmercury chloride should not exceed 10 days.

From the experience of the workers in this field, it would seem that the best method of handling seawater samples when analysing the mercury is, if possible, to make the analyses on shipboard as soon after collection as possible.

May et al. [21] used radiochemical studies to ascertain the behaviour of methylmercury chloride and mercuric chloride in seawater under different storage conditions.

The application of ^{203}Hg unambiguously revealed that the loss of mercury observed upon storage of unacidified seawater samples in polyethylene bottles was due to adsorption and to the diffusion of metallic $Hg(Hg^0)$ through the container wall.

For the chemical speciation of mercury compounds the storage time and the kind of storage are of paramount imortance. After a three-day storage in brown glass bottles 47% of ^{203}HgCl$_2$ added to the seawater became reduced to Hg^0, but a complete reduction of Hg^{II} in the samples was not achievable. The Hg^{II} species not reduced by stannous chloride may be to a distinct extent iodides and sulphides. Within 35 days of storage CH_3 ^{203}HgCl decomposed into Hg^0 and Hg^{II}; 35% could be identified as Hg^0 and 27% as reactive mercury. Strong solar radiation does not influence the transformation of ^{203}Hg^{II} into Hg^0, but after a strong 3 day solar radiation of a sample containing CH_3^{203}HgCl, Hg^0, reactive Hg^{II} and $Hg(CH_3)_2$ could be observed. Thus cooling and darkness during storage are important prerequisites for the subsequent differentiation of mercury species. Experiments with sodium borohydride as an alternative reducing agent showed in comparison with stannous chloride a quantitative reduction of all Hg^{II} species in seawater including CH_3^{203}Hg which gave a yield of nearly 80%.

Forms of mercury

Most of the methods of analysis for mercury actually measure inorganic mercury; to measure either organic or total mercury by such methods, it is necessary to decompose any organic mercury compounds present. This decomposition can be effected by ultraviolet irradiation of the samples. Systems of this sort have been described [22–24]. Since as much as 50% of the mercury may be present in organic form [22] the differentiation between inorganic and organic mercury can be of considerable importance.

Various workers have reported on the levels of total mercury in seawater. Generally, the levels are less than $0.2 \, \mu g \, l^{-1}$ with the exception of some parts of the Mediterranean where additional contributions due to man-made pollution are found [25–28].

Fitzgerald [29] used a cold trap to concentrate mercury from large volumes of seawater. Using this technique, he could achieve a detection limit of 0.2 ng Hg, and a coefficient of variation of 15% at the $25 \, ng \, l^{-1}$ level. Most oceanic samples contained less than $10 \, ng \, l^{-1}$ but coastal samples could approach $50 \, ng \, l^{-1}$.

With so many variations in methodology in the literature, some sort of standardization of analytical methods is necessary. One study of standardization was produced by Krenkel, Burrows and Dickinson [30]. They produced a simplified extraction and clean-up procedure and recommended gas chromatography as the method of measurement.

Atomic absorption spectrometry used either by direct aspiration (to determine total mercury) or as an element specific detector for gas chromatography (to determine organically bound mercury) are now discussed.

Atomic absorption spectrometry

Agemian and Chau [31] have described an automated method for the determination of total dissolved mercury in fresh and saline waters by ultraviolet digestion and cold vapour atomic absorption spectroscopy. A flow-through ultraviolet digester is used to carry out photo-oxidation in the automated cold vapour atomic adsorption spectrometric system. This removes the chloride interference. Work was carried out to check the ability of the technique to degrade seven particular organomercury compounds. The precision of the method at levels of $0.07 \,\mu g \, l^{-1}$, $0.28 \,\mu g \, l^{-1}$ and $0.55 \,\mu g \, l^{-1}$ Hg was $\pm \, 6.0\%$, $\pm \, 3.8\%$ and $\pm \, 1.00\%$ respectively. The detection limit of the system is $0.02 \,\mu g \, l^{-1}$.

Many of the standard methods for the reduction of organomercury compounds to metallic mercury such as acidic potassium permanganate and potassium periodate are inapplicable to saline waters as the large amount of chloride ion present reduces all of the oxidant used in the system, thus interfering with the oxidation of organomercurials. In addition, large amounts of chlorine are produced which unless reduced back to chloride, would absorb at the 253.7 nm line causing a positive interference. Because of these problems, high chloride samples could not be analysed in an automated system and required a manual predigestion step.

Agemian and Chau [31] showed that any organomercurials could be decomposed by ultraviolet radiation and that the rate of decomposition of organomercurials increased rapidly in the presence of sulphuric acid and with increased surface area of the ultraviolet irradiation. They developed a flow-through ultraviolet digestor which had a delay time of 3 min which was used to carry out the photo-oxidation in the automated system. The ultraviolet radiation has no effect on chloride. The method can be applied to both fresh and saline waters without the chloride interference.

The equipment used for the analysis is illustrated in Figures 8.1 and 8.2.

Jenne and Avotins [32] have pointed out the requirement for a strong oxidizing agent together with a strong acid for the preservation of low levels of mercury. Both potassium permanganate and dichromate have been used as the oxidants. The former is inadequate for samples with high chloride levels since it would be readily consumed by chloride. Carron and Agemian [33] showed that 1% sulphuric acid containing 0.05% potassium dichromate make a very effective preservative for sub-ppb levels of mercury in water, for extended periods of time especially when coupled with glass at the container. Agemian and Chau [31] preserved samples by adding 1 ml concentrated sulphuric acid and 1 ml of 5% potassium dichromate in glass containers at the start of the dilutions or sampling.

The system has a detection limit of $0.02 \,\mu g \, l^{-1}$ and is linear up to about $5 \,\mu g \, l^{-1}$. A rate of 30 samples per hour was found to be the practical limit for the system.

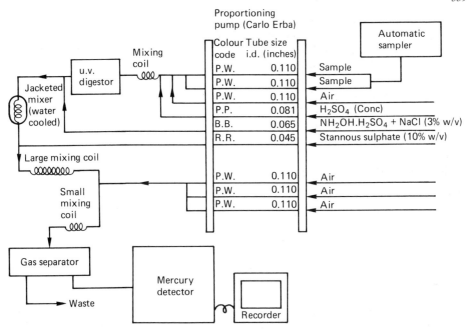

Figure 8.1 Mercury manifold No. 1 (Reproduced from Agemian, H. and Chau, A.S.Y. (1978) *Analytical Chemistry*, **50**, 13, American Chemical Society, by courtesy of authors and publishers.)

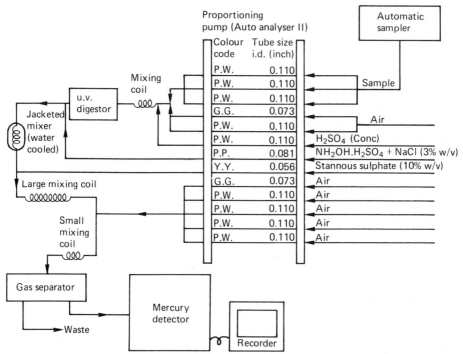

Figure 8.2 Mercury manifold No. 2 (Reproduced from Agemian, H. and Chau, A.S.Y. (1978) *Analytical Chemistry*, **50**, 13, American Chemical Society, by courtesy of authors and publishers.)

Table 8.1 Recoveries of organomercury compounds for different oxidation methods in the automated system

5 µg l^{-1} Hg as organic compound	Method (% recovery ± 5)				
	1	2	3	4	5
Phenylmercuric acetate	74	80	95	61	102
Phenylmercuric nitrate	73	75	95	71	98
Diphenylmercury	65	92	84	100	91
Methylmercuric chloride	41	46	89	46	98
Ethylmercuric chloride	81	88	88	98	95
Methoxyethylmercuric chloride	70	75	94	96	93
Ethoxyethylmercuric chloride	85	91	91	85	95

1. H_2SO_4. 2. H_2SO_4 + 4% (w/v) $K_2Cr_2O_7$. 3. H_2SO_4 + 0.5% (w/v) $KMnO_4$ + 0.5% (w/v) $K_2S_2O_8$. 4. Ultraviolet oxidation. 5. H_2SO_4 + ultraviolet oxidation.
Reproduced from Agemian, H. and Chau, A.S.Y. (1978) *Analytical Chemistry*, **50**, 13, American Chemical Society, by courtesy of authors and publishers.

Table 8.1 (column 5) shows that complete recovery of seven organomercurials is obtained by using the ultraviolet digestor described above in the presence of sulphuric acid. The recoveries (Table 8.1) are for $5 \mu g l^{-1}$ Hg solutions which is much higher than mercury levels in most natural waters.

Workers at the Department of the Environment UK have described a method for the determination of organic and inorganic mercury in seawater [34]. This method is suitable for determining dissolved inorganic mercury, and those organomercury compounds that form dithizonates, in saline sea and estuary water. In this method inorganic mercury is extracted from the acidified saline water as its dithizonate into carbon tetrachloride, but not all these compounds form dithizonates and those that do not may not be determined by this method. In general, organomercury compounds of the type RHg-X, in which X is a simple anion, form dithizonates, whereas the type R_1-Hg-R_2 does not. Monomethylmercury ion is extracted though it only appears to have a transient existence in aerobic saline water. The dithizonates are decomposed by the addition of hydrochloric acid and sodium nitrite and the mercury or organomercury compound returned to the aqueous phase. Some organomercury compounds may not be completely re-extracted into the aqueous phase. The mercury in this aqueous phase is determined by the stannous chloride reduction–atomic absorption spectroscopic technique.

The performance characteristics of this method are outlined in Table 8.2.

Fitzgerald and Lyons [35] have described flameless atomic absorption methods for determining organic mercury compounds respectively in coastal and seawaters, using ultraviolet light in the presence of nitric acid to decompose the organomercury compounds. In this method two sets of 100 ml samples of natural water are collected in glass bottles and then adjusted to pH 1.0 with nitric acid. One set of samples is analysed directly to give inorganically bound mercury, the other set is photo-oxidized by means of ultraviolet radiation for the destruction of organic material and then analysed to give total mercury. The element is determined by a flameless atomic absorption technique, after having been collected on a column of 1.5% of OV-17 and 1.95% of QF.1 on Chromosorb WHP (80–100 mesh). The precision of analysis is 15%. It was found that up to about 50% of the mercury

Table 8.2 Department of the Environment (UK) conditions for determination of organomercury compounds

Range of application	Up to 100 ng l^{-1}	
Calibration curve	Linear to 250 ng l^{-1}	
Standard deviation	Mercury concentration	Standard deviation
	(ng l^{-1})	(ng l^{-1})
	0.0	1.30
	50.0	1.15
	(each with 9 degrees of freedom)	
Limit of detection	4 ng l^{-1} (with 9 degrees of freedom)	
Sensitivity	100 ng l^{-1} is equivalent to an absorbance of approximately 0.1	
Bias	None detected	
Interferences	The combined effect of the commonly presently ions in estuarine and seawaters at the concentration normally encountered in these waters is less than 1 ng l^{-1} at a mercury concentration of 30 ng l^{-1}	
Time required for analysis	For six samples the total analytical and operator times are approximately 140 minutes and 60 minutes respectively	

Reproduced from *Department of the Environment Report* (1978), 22-Ab ENV, H.M. Stationary Office, London, by courtesy of the author and publishers.

present in river and coastal waters is organically bound or associated with organic matter.

Millward and Bihan [36] studied the effect of humic material on the determination of mercury by flameless atomic absorption spectrometry. In both fresh and seawater association between inorganic and organic entities takes place within 90 min at pH values of 7 or above, and the organically bound mercury was not detected by an analytical method designed for inorganic mercury. The amount of detectable mercury was related to the amount of humic material added to the solutions. However, total mercury could be measured after exposure to ultraviolet radiation under strongly acid conditions.

Yamamoto, Kaneda and Hisaka [37] determined picogram quantities of methyl mercury and total mercury in seawater by gold amalgamation and atomic absorption spectrometry.

Methylmercury was extracted with benzene and concentrated by a succession of three partitions between benzene and cysteine solution. Total mercury was extracted by wet combustion of the sample with sulphuric acid and potassium permanganate. The proportion of methylmercury to total mercury in the coastal seawater sampled was around 1%.

Graphite furnace atomic absorpiton spectrophotometry has been used for the determination down to 5 ng l^{-1} inorganic and organic mercury in seawater [38]. The method used a preliminary pre-concentration of mercury using the ammonium pyrrolidine dithiocarbamate-chloroform system. 85–86% of recovery of mercury was reproducibly obtained in the first chloroform extract and consequently it was possible to calibrate the method on this basis. A standard deviation of 2.6% was obtained on a seawater sample containing 529 ng l^{-1} mercury.

Typical recorder tracings are shown in Figure 8.3 for chloroform extracts introduced directly into the graphite furnace. The relative standard deviation of ten repeated determinations of 500 ml distilled water containing 10 ng mercury (II) chloride was 17.4%.

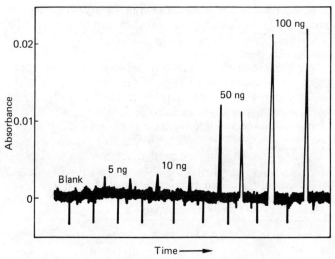

Figure 8.3 Recorder tracings obtained from chloroform extracts with various mercury concentrations (20 μl aliquot in the furnace) (Reproduced from Filipelli, A. (1984) *Analyst*, **109**, 515, Royal Society of Chemistry, by courtesy of author and publishers.)

Gas chromatography

The high sensitivity and selectivity of some gas chromatographic detectors are used to advantage in the measurement of organic mercury compounds. In the simplest approach, methyl mercury is extracted from seawater and converted to the iodide for electron capture gas chromatography [39].

Ealy [40] also used conversion to alkyl mercury iodides for the gas chromatographic determination of organomercury compounds in benzene extracts of water. The iodides were then determined by gas chromatograph of the benzene extract on a glass column packed with 5% of cyclohexane-succinate on Anakron ABS (70–80 mesh) and operated at 200°C with nitrogen (56 ml min⁻¹) as carrier gas and electron capture detection. Good separation of chromatographic peaks was obtained for the mercury compounds as either chlorides, bromides or iodides. The extraction recoveries were monitored by the use of alkylmercury compounds labelled with ^{203}Hg.

An increase in specificity, permitting a lesser concern with clean-up procedures, can be gained by using a microwave emission spectrometer as the gas chromatography detector [41]. A similar method, using extraction of methylmercury compounds into benzene, followed by back-extraction into L-cysteine, conversion to the chloride, back-extraction into benzene and analysis by electron capture gas chromatography was devised by Chau and Saitch [42].

The importance of selectivity and sensitivity in the detection system can be seen by examining the concentration and clean-up procedures used by Hobo *et al.* [43]. They used foam separation, with the addition of a surface-active compound, to collect the organic mercury compounds, followed by extraction of the foam with an organic solvent and removal of interfering heavy metals by column chromatography. Cappon and Smith [44] used a double extraction, in which the organic mercury compounds were first removed and the remaining inorganic mercury was converted to methylmercury by reaction with tetramethyltin and then extracted. The two organic fractions were then cleaned up by standard procedures and analysed by gas chromatography.

Davies, Graham and Pirie [45] described a method for the determination of methylmercury in seawater at the $0.06 \mu g l^{-1}$ level. Mussels from a clean environment were suspended in cages at several locations in the Firth of Forth. A small number were removed periodically homogenized and analysed for methylmercury by solvent extraction–gas chromatography, as described by Westoo [46]. The rate of accumulation of methylmercury was determined and, by dividing this by mussel filtration rate, the total concentration of methylmercury in the seawater was calculated.

The methylmercury concentration in caged mussels increased from low levels (less than $0.01 \mu g g^{-1}$) to 0.06–$0.08 \mu g g^{-1}$ in 150 days (Figure 8.4), giving a mean

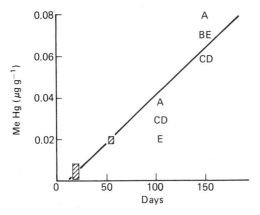

Figure 8.4 The increase with time of methylmercury concentration in cage mussels at positions A–E. Methylmercury was not detectable ($0.01 \mu g g^{-1}$) after 20 days and animals from all five positions contained $0.02 \mu g g^{-1}$ after 55 days' exposure, as shown by the shaded rectangles. The case at position B was not sampled at 106 days' exposure (Reproduced from Davies, I.M., *et al.* (1979) *Marine Chemistry*, 7, 111, Elsevier Science Publishers, by courtesy of authors and publishers.)

uptake rate of $0.4 ng g^{-1}$ daily, i.e. a 10 g mussel accumulated 4 ng daily. The average percentage of total mercury in the form of methylmercury increased from less than 10% after 20 days to 33% after 150 days. This may be compared with analyses of natural intertidal mussels from the area in which the proportion of methylmercury was higher in mussels of lower (less than $10 \mu g g^{-1}$) than of higher total mercury concentrations.

Davies, Graham and Pirie [45] calculated the total methylmercury concentration in the seawater as $0.06 \mu g l^{-1}$ i.e. 0.1–0.3% of the total mercury concentration as opposed to less than 5–$32 ng l^{-1}$ methylmercury found in Minamata Bay, Japan. These workers pointed out that a potentially valuable consequence of this type of bioassay is estimates of the relative abundance of methylmercury may be obtained at different sites by the exposure of 'standardized' mussels as used in their experiment, in cages for controlled periods of time, and by the comparison of the resultant accumulations of methylmercury.

Compeau and Bartha [47] have discussed the abiotic methylation of mercuric ion and mercuric ion sea salt anion complexes to methylmercury by methylcyanocobalamin in seawater and saline sediments under aerobic and anaerobic conditions.

The reaction of 30 μmol l^{-1} methylcobalamin and 60 nmol l^{-1} mercuric chloride to produce methyl mercury was measured by two different methods. A spectrophotometer was used to measure changes in absorbance at 351 nm and 380 nm respectively. These are changes characteristic of the transition of methylcobalamin to aquocobalamin [48].

Methylmercury formed in the reaction was also measured gas chromatographically in benzene extracts of the aqueous phase [49].

An increase in absorbance at 351 nm and a concomitant decrease in absorbance at 380 nm in the ultraviolet visible spectrum of methylcobalamin during the abiotic transfer of the methyl group to Hg^{2+} are characteristic for the loss of the methyl group and formation of aquocobalamin. In experiments monitored by both analytical techniques, gas chromatographic measurements of methylmercury formation were in good agreement with the spectrophotometric measurement of aquocobalamin formation from methylcobalamin at 351 nm. Aerobic versus anaerobic reaction conditions had no measurable effect on either the methyl transfer rates, the stability of the reactants, or on the reaction products.

A number of analytical methods for the separation of organic mercury compounds use an initial extraction of the organic materials with an organic solvent. Klisenko and Shmigidina [50] then converted both the inorganic mercury held in the aqueous fraction and the organic mercury in the chloroform extract, to dithizonate, separated the components on chromatographic columns and determined the concentration of the various fractions by comparison with reference standards. This method is semi-quantitative, at best.

Miscellaneous

Differentiation of inorganic and organic mercury can be achieved in a number of different ways, many of which depend upon the reduction and vaporization of the inorganic mercury, followed by reduction [51] or oxidation [52, 53] of the organic mercury compounds and a final measurement by atomic absorption or mass spectrometry. Similar methods of separation of the inorganic and organic components are used in the pre-treatment of samples where the final analysis for mercury is to be made by neutron activation analysis [54, 55].

Sipos et al. [56] used subtractive differential pulse voltammetry at a twin gold electrode to determine total mercury levels in seawater samples taken from the North Sea.

Ke and Thibert [57] have described a kinetic microdetermination of down to 0.05 μg ml^{-1} of inorganic and organic mercury in river water and seawater. Mercury is determined by use of the iodide-catalysed reaction between Ce^{IV} and As^{III} which is followed spectrophotometrically at 273 nm.

8.2 Organolead compounds

Work on lead has consisted almost entirely of the measurement of lead; the total organic fraction has largely been ignored. If we look at the inter-comparison studies that have been made on the measurement of lead in seawater [58] two points become obvious: first, that almost no one can measure lead accurately in seawater; and second, that there is much less lead present than anyone had estimated. With so little lead present in any form, identification of the organic moiety or an organolead compound becomes a major problem.

We are aided in this by our knowledge of the form in which the major anthropogenic addition is put into the system, i.e. as the tetraethyl lead additive in gasoline. Methods have been developed for the analysis of tetraalkyl lead compounds, both the original compound and those resulting from biological conversions, in the air [59]. The method for atmospheric materials using cold trapping ($-80°C$) on a column of UV-1, followed by separation by gas chromatography and analysis by atomic absorption. The method for biological materials uses benzene extraction in the presence of EDTA, followed by digestion to free the lead and determination by atomic absorption. While neither of these methods is specifically designed for seawater, it should be possible to adapt them to aqueous samples.

At least one of the organolead compounds has been shown to be a normal constituent of the aqueous environment. Tetramethyl lead has been found as a product of anaerobic bacterial metabolism in lake sediments [60]. It should perhaps be sought in the air over coastal mud flats.

Bond et al. [61] examined interferences in the stripping voltammetric method determination of trimethyl lead in seawater using polarography and mercury 199 and lead 207 NMF. It was shown that Hg^{II} reacts with $((CH_3) Pb)^+$ in seawater. Consequently, anodic stripping voltammetric methods for determining $((CH_3)_3Pb)^{2+}$ and inorganic Pb^{II} may be unreliable.

8.3 Organotin compounds

Several investigators have recently reported ng–μg l^{-1} concentrations of organotin compounds in both fresh water and marine samples. Inorganic tin, methyltins and butyltins have been detected in marine and fresh water environmental samples [62–65]. The presence of inorganic tin, butyltin and methyltin species has been reported in Canadian lakes, rivers and harbours [66, 67]. Both organotins and inorganic tin were reported to be highly concentrated by factors of up to 10^4 in the surface microlayer relative to subsurface water [66, 67]. Inorganic tin, mono-, di- and tri-methyltins have been detected at ng l^{-1} levels in saline, estuarine and fresh water samples [64, 68]. Methylation of tin compounds by biotic as well as abiotic processes has been proposed [69, 70].

Possible anthropogenic sources of organotins have recently been suggested. Both polyvinylchloride (PVC) and chlorinated polyvinylchloride (CPVC) have been shown to leach methyltin and dibutyltin compounds, respectively, into the environment [71]. Monobutyltin has been measured in marine sediments collected in areas associated with boating and shipping. Butyltin was not detected in areas free of exposure to maritime activity [72]. The use of organotin antifouling coatings in particular has stimulated interest in their environmental impact.

The techniques used for the investigation of organotin compounds in seawater are atomic absorption spectrometry, gas chromatography or gas chromatography using atomic absorption spectrometry as a detector.

Atomic absorption spectrometry

Hodge, Seidel and Goldberg [73] have described an atomic absorption spectroscopic method for the determination of butyltin chlorides and inorganic tin in natural waters, coastal sediments and macro algae in amounts down to 0.4 ng.

Valkirs *et al.* [74] determined the range of butyl tin compounds in samples collected within San Diego Bay. Results suggested that in certain areas of the bay the use of tin containing antifouling paints on ships was increasing. Water extracted with methylisobutyl ketone and analysed by graphite furnace atomic absorption spectrophotometry consistently yielded higher tin concentrations than the same samples analysed by a hydride reduction/flame atomic absorption method [76, 77] suggesting a non-hydride reducible tin fraction was present. Gas chromatography–mass spectrometry confirmed the presence of tributyltin in an environmental marine water sample following derivatization to the hydride species.

Surface and bottom water samples were collected in 500 ml, 1 or 4 litre polycarbonate bottles. Polycarbonate bottles have been shown to retain 97% of an initial spike of bis(tri-*n*-butyltin) oxide in seawater at a concentration of $0.5 \, \text{mg} \, l^{-1}$ over a weeklong period [75]. Samples were analysed immediately after collection and transported to the laboratory, or were stored frozen at $-20°C$ and analysed at a later date. Frozen storage has been shown to be effective in preserving sample stability with respect to monobutyltin, dibutyltin and tributyltin concentrations for a period of at least 100 days.

An estimate of the accuracy of both analytical methods was performed on bis(tri-*n*-butyltin) oxide and tri-*n*-butyltin chloride solutions ($8.9–35.6 \, \mu g \, l^{-1}$) prepared in filtered (0.45 μm) near-shore seawater free of detectable organotins. Average mean recoveries of 92.8% by both methods were determined for tributyltin standard solutions. Some typical analyses obtained by the hydride reduction technique are shown in Table 8.3.

Organotin concentrations measured by the hydride reduction method were consistently low when compared with concentrations determined by the graphite

Table 8.3 Inorganic and butyltin concentrations (ng l⁻¹) in water samples collected in San Diego Bay

Station	Date	Inorganic Sn^{IV}	Monobutyltin	Dibutyltin	Tributyltin
1					
surface	3 Jan 83	0.001	0.01	ND	ND
bottom		0.005	0.03	ND	0.01
2-1					
surface	3 Jan 83	0.009	0.01	0.13	0.06
bottom		0.003	0.04	0.13	0.10
surface	20 Jun 84	NM	0.06	0.27	0.27
surface	7 Mar 85	NM	ND	0.19 (3)	0.29 (3)
				SD ± 0.03	SD ± 0.03
surface	25 Sep 85	NM	0.02	0.46 (3)	0.78 (3)
				SD ± 0.05	SD ± 0.03

NM = Not measured.
ND = Not detectable; detection limit was <0.01 μg l⁻¹ for butyltin species. All butyltin species are reported as the mono-, di-, or tributyltin chloride compounds.
Numbers in parentheses represent the number of replicate samples analysed for the reported value. All other values are reported on single measurements.
Reproduced from Valkirs, A.O., *et al.* (1986) *Marine Bulletin*, **17**, 319, by courtesy of authors and publishers.

Table 8.4 A comparison of hydride detection and atomic absorption (HDAA) analysis and methyl isobutyl ketone (MIBK) extraction at pH 1.0 with graphite furnace atomic absorption (GFAA) analysis of subsurface samples from San Diego Bay, Jan 3, 1983 (results given as µg Sn per litre)

Station	HDAA	GFAA	GFAA:HDAA
1	0.004	0.03	7.5
2–1	0.077	0.52	6.8
2–3	0.015	0.25	16.7
3–1	0.086	1.20	14.0
3–4	0.004	0.13	32.5
5	0.008	0.16	20.0

Reproduced from Valkirs, A.O., *et al.* (1986) *Marine pollution Bulletin*, **17**, 319, by courtesy of authors and publishers.

furnace atomic absorption spectrometry technique (Table 8.4). The large differences between total extractable tin analysed by graphite furnace atomic absorption spectrometry and the alkyltin determined by hydride reduction methods were possibly caused by non-hydride reducible tin compounds other than butyltins associated with the methylisobutyltin ketone extractable fraction. Standard additions to samples did not reveal significant differences in the butyltin concentrations measured by hydride reduction suggesting that matrix interferences were not responsible for the differences in tin measured by the two methods.

Valkirs *et al.* [78] have conducted an interlaboratory comparison on determinations of di- and tributyltin species in marine and estuarine waters using hydride generation with atomic absorption detection and hydride generation with flame photometric detection.

Gas chromatography

Studies by Braman and Tompkins [79] have shown that non-volatile methyltin species $Me_nSn_{aq}^{(4-n)+}$ (n=1–3), are ubiquitous at $ng\,l^{-1}$ concentrations in natural waters including both marine and fresh water sources. Their work, however, failed to establish whether tetramethyltin was present in natural waters because of the inability of the methods used to trap this compound effectively during the combined pre-concentration purge and reductive derivatization steps employed to generate volatile organotin hydrides necessary for tin specific detection.

Tin compounds are converted to the corresponding volatile hydride (SnH_4, CH_3SnH_3, $(CH_3)_2SnH_2$, and $(CH_3)_3SnH$) by reaction with sodium borohydride at pH 6.5 followed by separation of the hydrides and then atomic absorption spectroscopy using a hydrogen-rich hydrogen–air flame emission type detector (Sn-H band). The apparatus used is shown in Figures 8.5 and 8.6.

The technique described has a detection limit of 0.01 ng as tin and hence parts per trillion of organotin species can be determined in water samples.

Braman and Tompkins [79] found that stannane (SnH_4) and methylstannanes (CH_3SnH_3, $(CH_3)_2SnH_2$, and $(CH_3)_3SnH$) could be separated very well on a column comprising silicone oil OV-3 (20% w/w) supported on Chromosorb W. A typical separation achieved on a water sample is shown in Figure 8.7.

The generation and recovery of stannane, methyl- dimethyl- and trimethyl-stannane were studied in seawater. Average tin recoveries ranged from 96 to 109% for six samples analysed to which were added 0.4–1.6 µg methyltin compounds and

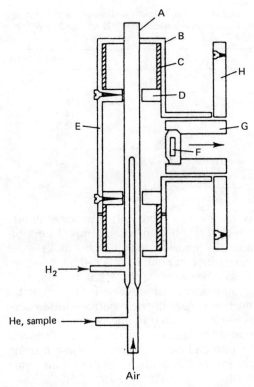

Figure 8.5 Quartz burner and housing. A = Quartz burner; B = PVC cap; C = PVC tubing; D = mounting ring; E = PVC T-joint, 1.25 inch; F = filter and holder; G = PVC coupling; H = connection to photomultiplier housing (PM) (Reproduced from Braman, R.S. and Tomkins, M.A. (1979) *Analytical Chemistry*, **51**, 12, American Chemical Society, by courtesy of authors and publishers.)

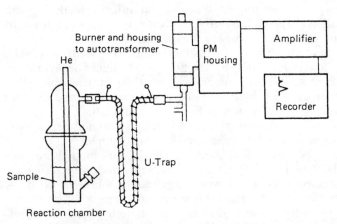

Figure 8.6 Apparatus arrangement for tin analysis (Reproduced from Braman, R.S. and Tomkins, M.A. (1979) *Analytical Chemistry*, **51**, 12, by courtesy of authors and publishers.)

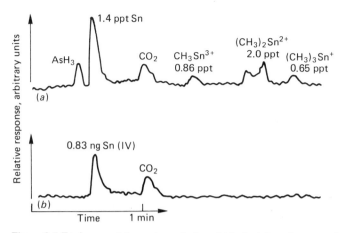

Figure 8.7 Environmental sample analysis and blank; (*a*) environmental analysis, Old Tampa Bay, (*b*) typical blank. (Reproduced from Braman, R.S. and Tomkins, M.A. (1979) *Analytical Chemistry*, **51**, 12, American Chemical Society, by courtesy of authors and publishers.)

Table 8.5 Analysis of saline and estuarine water samples. Data are average of duplicates (given in ng l^{-1}; results in parentheses are percentages of total tin)

Sample	TIN(IV)		Methyltin		Dimethyltin		Trimethyltin		Total tin
Saline waters									
Gulf of Mexico, Sarasota	62	(73)	15	(18)	7.0	(8.3)	0.98	(1.2)	85
Gulf of Mexico, Fort Desoto	2.2	(6.0)	ND		0.74	(2.0)	0.71	(20)	3.6
Gulf of Mexico, St Petersburg	4.5	(54)	0.62	(7.4)	3.2	39	ND		8.3
Old Tampa Bay, Oldsmar	0.3	9.7	0.86	33	0.88	34	0.61	24	2.6
Old Tampa Bay, Safety Harbour	1.4	29	0.86	17	2.0	40	0.65	13	5.0
Old Tampa Bay, Philipee Park	0.8	32	1.1	44	0.60	24	ND		2.5
Old Tampa Bay, Davis Municipal	ND		0.98	35	0.91	32	0.95	34	2.8
Old Tampa Bay Courtney Campbell	2.7	54	ND		1.7	34	0.61	12	5.0
Average	1.7	40	0.63	15	1.4	33	0.50	12	4.2
Estuarine surface waters									
Sarasota Bay	5.7	47	3.3	27	2.0	16	1.1	9.1	12
Tampa Bay	3.3	27	8.0	66	0.79	6.5	ND		12
McKay Bay	20	88	ND		2.2	9.6	0.45	2.0	23
Hillsborough Bay	ND		ND		1.8	71	0.71	29	2.5
Hillsborough Bay, Seddon Channel North	12	86	0.74	5.3	0.91	6.6	0.35	2.5	14
Hillsborough Bay, Seddon Channel South	13	83	ND		2.4	15	0.31	1.9	16
Manatee River	4.8	61	1.4	1.7	1.1	14	0.65	8.2	7.9
Alafia River	3.4	73	ND		0.75	16	0.55	12	4.7
Palm River*	567	98	ND		4.6	0.80	4.0	0.69	576
Bowes' Creek	8.6	42	8.5	42	3.3	16	ND		20
Average	7.9	63	2.4	19	1.7	14	0.46	3.7	12

ND less than 0.01 ng l^{-1} for methyltin compounds and 0.3 ng l^{-1} for inorganic tin.
*This set of values was not used in computing the average.
Reproduced from Braman, R.S. and Tomkins, M.A. (1979) *Analytical Chemistry*, **51**, 1256, American Chemical Society, by courtesy of authors and publishers.

3 ng inorganic tin. Reanalysis of analysed samples shows that all methyltin and inorganic tin is removed in one analysis procedure.

A number of natural waters, from in and around the Tampa Bay, Florida, area were analysed for tin content. All samples were analysed without pre-treatment. Samples that were not analysed immediately were frozen until analysis was possible. Polyethylene bottles, 500 ml volume, were used for sample acquisition and storage. The results of the analyses appear in Table 8.5. The average total tin content of fresh, saline and estuarine waters are 9.1, 4.2 and 12 ng l^{-1} respectively. Approximately 17–60% of the total tin present was found to be in the methylated forms. The saline waters appear to have the highest percentage of methylated tin compounds, 60% of the total tin present was found to be in the methylated forms, the dimethyltin form contributes approximately half of this value.

The above procedure, although valuable in itself, is incomplete in that any monobutyltin species present escape detection. Excellent recoveries of monobutyltin species are achieved with tropolone.

Meinima, Burger and Wiersina [81] studied the effect of combinations of various solvents with 0.05% tropolone on the recoveries of mono-, di- and tributyltin species either individually or simultaneously present in aqueous solutions. The results obtained by gas chromatography–mass spectrometry after methylation show that Bu_3Sn and Bu_2Sn recoveries appear to be almost quantitative both for neutral and from hydrobromic acid-acidified aqueous solutions. Bu_1Sn recovery appears to be influenced by the presence of hydrobromic acid in that, in general, recovery rates are higher from solutions acidified with hydrobromic acid than from non-acidified solutions. Bu_3Sn and Bu_2Sn recoveries remain fairly constant with ageing of an aqueous solution of these species over a period of several weeks. Bu_1Sn recoveries, however, do decrease with time to a notable extent (20–40%) most likely as a result of adsorption/deposition of Bu_1Sn species to the glass wall of the vessel. Addition of hydrobromic acid obviously affects the desorption of these species as recovery of Bu_1Sn species increases to almost the same values as obtained from hydrobromic acid-acidified freshly prepared aqueous solutions of Bu_1Sn species.

Jackson et al. [82] devised trace speciation methods capable of ensuring detection of tin species along with appropriate pre-concentration and derivatization without loss, decomposition, or alteration of their basic molecular features.

They describe the development of a system employing a Tenax GC filled purge and trap sampler, which collects and concentrates volatile organotins from water samples (and species volatilized by hydrodization with sodium borohydride), coupled automatically to a gas chromatograph equipped with a commercial flame photometric detector modified for tin-specific detection [83–85].

The system was applied to the analysis of a series of water samples obtained from the Cheapeake Bay at both industrially polluted and relatively pristine sites.

Table 8.6 tabulates concentrations of various organotin compounds found by this procedure in marine harbour waters.

The volatile species shown in the table under mode I were determined by the purge and trap gas chromatographic flame photometric detector method without the sodium borohydride reduction step. Non-volatile species (mode II) were volatilized, trapped and detected with the volatile species present in the samples. Generally no changes in the concentration of the volatile species were observed with the hydride reduction step. An increase in the amount detected, as with Me_2SnH_2 and $BuSnH_3$, indicated that the solvated cations were also present in the

Table 8.6 Three-week Baltimore Harbor sampling for organotin compounds

Site‡	Collection date	SnH₄ (0.88 ± 0.07)* 0.92 ± 0.01†		Me₂SnH₂ (1.57±0.04)* Me₂S (1.71 ± 0.03) 1.75 ± 0.04†		Me₃SnH (2.16±0.03)* 2.23 ± 0.13†		Me₄Sn (2.86 ± 0.11)* 2.99 ± 0.05†		BuSnH₃ (3.98 ± 0.08)* 4.16 ± 0.06†	
		Mode I§	Mode II‖,¶	Mode I**,¶	Mode II**,¶	Mode I	Mode II¶	Mode I¶	Mode II¶	Mode I¶	Mode II¶
JFS	09/30/80		10–20	<0.01	<0.01		<0.01			0.05–0.1	0.2–0.3
JFS	10/07/80		0.3–0.5	<0.01	<0.01			0.2–0.3	0.2–0.3		
JFS	10/15/80		3–4								
JFB	09/30/80		0.6–0.7	<0.01	<0.01					0.05–0.1	0.05–0.1
JFB	10/07/80		0.2–0.3	0.02–0.05	0.05–0.1		0.01–0.02	0.05–0.1	0.05–0.1		
JFB	10/15/80		0.7–0.8	<0.01	<0.01						
CCS	09/30/80			<0.01	0.5–0.1		<0.01	0.01–0.05	0.01–0.05	0.05–0.1	0.05–0.1
CCS	10/07/80			<0.02	<0.02		<0.01				
CCS	10/15/80			<0.005	<0.005		<0.005	<0.01	<0.01		
CCB	09/30/80			<0.02	<0.02			0.15–0.3	0.15–0.3	0.05–0.1	0.05–0.1
CCB	10/07/80			<0.02	<0.02						
CCB	10/15/80			<0.005	<0.005						

*Identified species with mean $t_r \pm$ SE (minutes) as determined with standards.
†Mean $t_r \pm$ SE (minutes) determined from samples over 3-week period.
‡S = surface samples; B = bottom samples; JF = Jones Falls; and CC = Colgate Creek.
§Mode I: samples analysed for volatile species without addition of $NaBH_4$ for reduction.
‖Mode II: samples with $NaBH_4$ added; gives simultaneous analysis for volatile and nonvolatile species.
¶Quantities expressed in concentration of µg l⁻¹.
**Assuming species to be Me_2SnH_2.
Reproduced from Jackson, J.A., et al. (1982) *Environmental Science and Technology*, p.111, American Chemical Society, by courtesy of authors and publishers.

water samples along with the respective free stannane. The prevailing gas chromatographic peak at 1.75 ± 0.04 min has not been completely resolved or identified. The retention time is within the range of both dimethyl sulphide (Me₂S) and Me₂SnH₂. Dimethyl sulphide is known to be a microbial metabolite present in environmental water; therefore, this is a possible interference.

Miscellaneous methods

Luskima and Syavtsillo [86] have described a spectrophotometric procedure utilizing phenylfluorone for the determination of organotin compounds in water.

A method for the determination of organotin compounds by anodic stripping polarography has been published [87]. It has yet to be applied to seawater. Since the sensitivity permits the measurement of 0.01 ppm of the tin compounds, it is likely to be not quite sensitive enough for seawater and would require a pre-concentration step.

Laughlin, Guard and Coleman [88] analysed chloroform extracts of tributyltin dissolved in seawater using nuclear magnetic resonance spectroscopy. It was shown that an equilibrium mixture occurs which contains tributyltin chloride, tributyl tin hydroxide, the aquo complex and a tributyltin carbonate species. Fluorimetry has been used to determine triphenyltin compounds in seawater [84]. Triphenyltin compounds in water at concentrations of $0.004-2 \, pmg \, l^{-1}$ are readily extracted into toluene and can be determined by spectrofluorimetric measurements of the triphenyltin-3-hydroxyflavone complex.

The observed excitation and emission maxima for the compounds used by Bluden and Chapman [89] are given in Table 8.7 (Perkin Elmer Model 1000 fluorescence spectrometer).

Tri-, di-, and monobutyl and di- and monoethyltin compounds did not fluoresce under the conditions used for the determination of triphenyltin. However, trimethyltin compounds react in a similar manner with 3-hydroxylflavone, and although the emission maximum is at approximately 510 nm, this is not sufficiently different from the emission maximum of triphenyltin compounds (approximately 495 nm) for these compounds to be determined in the presence of each other.

Spiking recoveries by the above procedure carried out on standard solutions of triphenyltin chloride in various types of water ranged from 74% at the $4 \, \mu g \, l^{-1}$ tin level (relative SD 8.9%) to 93.6% at the $2 \, mg \, l^{-1}$ tin level (relative SD 4.2%).

Table 8.7 Excitation and emission maxima in fluorescence spectrometry for organotin compounds

Compound	Approximate wavelength (nm)	
	Excitation	Emission
$(C_6H_5)_3SnCl$	415	495
$(C_6H_5)_3SnOCOCH_3$	415	495
$(C_6H_5)_2SnCl_2$	415	450
$(C_6H_5)SnCl_3$	415	450
$(C_6H_3)_3SnCl$	415	510
$SnCl_4$	395	450
3-Hydroxyflavone	380	525

Reproduced from Bluden, S.J. and Chapman, A.H. (1978) *Analyst*, **103**, 1266, Royal Society of Chemistry, by courtesy of authors and publishers.

8.4 Organoarsenic compounds

Large amounts of arsenic enter the environment each year because of the use of arsenic compounds in agriculture and industry as pesticides and wood preservatives. The main amount is used as inorganic arsenic (arsenite, arsenate), and about 30% as organoarsenicals such as monomethylarsinate and dimethylarsinate. Arsenic is known to be relatively easily transformed between organic and inorganic forms in different oxidation states by biological and chemical action [90, 91]. Until recently, most of the analytical work has been concerned only with the total content of arsenic. But as the toxicity and biological activity of the different species vary considerably, information about the chemical form is of great importance in environmental analysis.

Arsenic being present in inorganic form and as a variety of organic compounds implies the use of analytical strategies which use physical and chemical separations, as well as general or specific detectors. Since the lower organic arsenic compounds, as well as some inorganic arsenic compounds, are easily vaporized, both gas chromatography and direct vapour generation atomic absorption are favoured analytical methods.

A review of the analytical chemistry of arsenic in the sea, including occurrence, analytical methods, and the establishment of analytical standards, has been published [92].

Earlier investigations have shown that the major known organic arsenic compounds in the environment is dimethyl arsinate. For that reason, specific and sensitive methods for the determination of this compound are needed. Several methods exist for this determination often together with other arsenic species. Hydride generation and selective vaporization of cold-trapped arsines in combination with various detection systems seem to be the methods most frequently used [93–95]. These methods are applicable to the species As^{III}, As^{V}, monomethylarsinates, dimethyl arsinate and trimethylarsine oxide. The optimum conditions for generation of the respective arsine are, however, different for all these species with regard to the pH of the generating solution [96, 97]. These methods also suffer from interferences from numerous inorganic ions [98]. Pre-concentration in a toluene cold-trap following arsine generation was successful for monomethyl arsinate and dimethyl arsinate but non-quantitative recoveries of dimethyl arsinate were reported, probably because of the problem mentioned above with the arsine generation step. Molecular rearrangements occurring during arsine generations are, however, reported to be minimized if sodium tetrahydroborate is introduced as a pellet [99]. To simplify the determination of the different species, improved separation of the arsines using gas chromatography may be necessary.

The simplest analytical method is direct measurement of arsenic in volatile methylated arsenicals by atomic absorption [100]. A slightly more complicate system, but one that permits differentiation of the various forms of arsenic, uses reduction of the arsenic compounds to their respective arsines by treatment with sodium borohydride. The arsines are collected in a cold trap (liquid nitrogen) then vaporized separately by slow warming, and the arsenic measured by monitoring the intensity of an arsenic spectral line, as produced by a direct current electrical discharge [101, 102, 104]. Essentially the same method was proposed by Talmi and Bostick [103] except that they collected the arsines in cold toluene ($-5°C$), separated them on a gas chromatography column and used a mass spectrometer as the detector. Their method had a sensitivity of $0.25\,\mu g\,l^{-1}$ for water samples.

Another variation on the method [105] with slightly higher sensitivity (several nanograms per litre) used the liquid nitrogen cold trap and gas chromatography separation, but used the standard gas chromatography detectors or atomic absorption, for the final measurement. These workers found four arsenic species in natural waters.

Atomic absorption spectrometry

Persson and Irgum [106] fulfilled a requirement to determine sub ppm levels of dimethyl arsinate by preconcentrating the organoarsenic compound on a strong cation-exchange resin (Dowex AG 50 W-XB). By optimizing the elution parameters, dimethyl arsinate can be separated from other arsenicals and sample components, such as group I and II metals, which can interfere in the final determination. Graphite-furnace atomic absorption spectrometry was used as a sensitive and specific detector for arsenic. The described technique allows dimethyl arsinate to be determined in a sample (20 ml) containing a 10^5 fold excess of inorganic arsenic with a detection limit of 0.02 ng As per ml.

A Perkin-Elmer 372 atomic absorption spectrometer provided with background correction was used. The instrument was equipped with a HGA74 graphite furnace which was connected to a power supply with facilities for temperature-controlled heating of the graphite tube [107]. The instrumental parameters are given in Table 8.8.

Figure 8.8 shows elution profiles for dimethyl arsinate for two flow rates. The total peak volume is of the same size in both cases but a sharper profile is obtained

Table 8.8 Instrumental parameters in atomic absorption spectrometry for organoarsenic compounds

	Time (s)	*Temperature* (°C)	*Heating rate* (°C s^{-1})
Drying	60	110	4
Ashing	40	800	45
Atomization	5	2400	1800
Cleaning	4	Max.	
Graphite tube	Non-pyrolitic	Background correction	Used
Wavelength	193.7 nm	Argon flow	
Spectral band width	2.0 nm	Internal (litres min^{-1})	0.4*
Lamp source	EDL 9 W	External (litres min^{-1})	1.5

*Gas stop during atomization.
Reproduced from Persson, J. and Irgum, K. (1982) *Analytica Chimica Acta*, **138**, 111, Elsevier Science Publishers, by courtesy of authors and publishers.

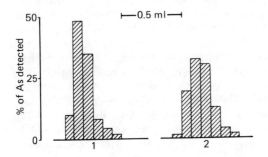

Figure 8.8 Elution profile for 10 ng DMA at two flow rates; (1) 0.36 ml min^{-1}; (2) 0.86 ml min^{-1}. The volume of each fraction taken was 80 μl (Reproduced from Persson, J. and Irgum, K. (1982) *Analytica Chimica Acta*, **138**, 111, Elsevier Science Publishers, by courtesy of authors and publishers.)

for the lower flow rate; 80% of the eluted dimethyl arsinate could be recovered within 160 μl when a flow rate of 0.36 ml min^{-1} was used. For higher flow rates than those shown here, larger peak volumes were obtained. Fraction size was set to 10 drops, giving a volume of 0.4 ml. This allowed for duplicate measurements and standard additions, with 100 μl injections, in the final stage. Normally more than 95% of the dimethyl arsinate was recovered within two 0.41 ml fractions, depending on how elution and fraction collecting were synchronized.

Table 8.9 Recovery of added dimethyl arsinate (DMA) from water samples based on a standard curve

Sample	Salinity (‰ NaCl)	Sample volume (ml)	DMA (ng As ml^{-1})			Recovery (%)
			Sample	Added	Total found	
Artificial seawater	27.0	4	ND	4.8	4.7 ± 0.31	98 ± 6.6
Artificial seawater	2.7	20	ND	0.050	0.048 ± 0.0044	96 ± 9.2
Brackish water I	4.3	20	0.37 ± 0.027	0.50	0.78 ± 0.048	82 ± 18
Brackish water II	4.3	6	0.20 ± 0.054	1.0	0.92 ± 0.018	72 ± 10
Seawater I	17.3	4	0.245 ± 0.006	1.08	1.16 ± 0.011	85 ± 1.9
Seawater II	17.3	3.5	0.210 ± 0.043	1.0	0.95 ± 0.061	74 ± 14

ND = Not detected.
Reproduced from Persson, J. and Irgum, K. (1982) *Analytica Chimica Acta*, **138**, 111, Elsevier Science Publishers, by courtesy of authors and publishers.

The results given in Table 8.9 show that good recoveries were obtained from the artificial seawaters, even at the 0.05 μg l^{-1} level, but for natural seawater samples the recoveries were lower (74–85%). This effect could be attributed to organic sample components that eluted from the column together with dimethyl arsinate.

Miscellaneous methods

It is also possible to separate the organic arsenic compounds by column chromatography first, and then reduce them to arsines later [108].

Haywood and Riley [109] showed that arsenic in the form of tetraphenylarsonium chloride 1-(o-arsonophenylazo)-2-naphthol-3,6-disulphuric acid and o-arsonophenylazo-p-dimethylaminobenzene are quantitatively decomposed in seawater by ultraviolet radiation.

Quantitative recoveries (98%) of arsenic were obtained from tetraphenylarsonium chloride, 1-(o-arsonophenylazo)-2-naphthol-3,6-disulphuric acid and o-arsonophenylazo-p-dimethylaminobenzene.

Organic arsenic species, these can be rendered reactive either by photolysis with ultraviolet radiation or by oxidation with potassium permanganate or a mixture of nitric acid and sulphuric acids. Arsenic(V) can be determined separately from total inorganic arsenic after extracting As^{III} as its pyrrolidine dithiocarbamate into chloroform.

In the method for inorganic arsenic the sample is treated with sodium borohydride added at a controlled rate (Figure 8.9). The arsine evolved is absorbed in a solution of iodine and the resultant arsenate ion is determined photometrically by a molybdenum blue method. For seawater the range, standard deviation, and detection limit are 1–4 μg l^{-1}, 1.4% and 0.14 μg l^{-1} respectively; for potable waters they are 0–800 μg l^{-1}, about 1% (at 2 μg l^{-1} level) and 0.5 μg l^{-1} respectively. Silver and copper cause serious interference at concentrations of a few tens of mg l^{-1};

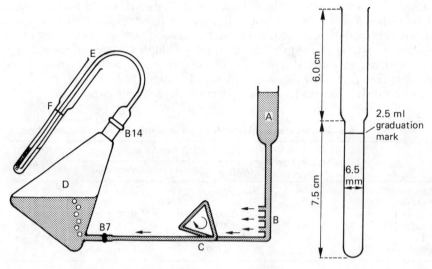

Figure 8.9 Apparatus for the evolution and trapping of arsenic. The dimensions of the absorption tube F are given in the inset (Reproduced from Haywood, M.G. and Riley, J.P. (1976) *Analytica Chimica Acta*, **85**, 219, Elsevier Science Publishers, by courtesy of authors and publishers.)

however, these elements can be removed either by preliminary extraction with a solution of dithizone in chloroform or by ion-exchange.

The precision of the method was tested by carrying out replicate analyses (10) on 150 ml aliquots of two seawater samples from the Irish Sea. Mean ($\pm$SD) arsenic concentrations of 2.63 ± 0.05 and $2.49 \pm 0.05\,\mu g\,l^{-1}$ were found. The recovery of arsenic was checked by analysing 150 ml aliquots of arsenic-free seawater which had been spiked with known amounts of arsenic (V). The results of these experiments (Table 8.10) shows that there is a linear relationship between absorbance and arsenic concentration and that arsenic could be recovered from seawater with an average efficiency of 98.0% at levels of $1.3–6.6\,\mu g\,l^{-1}$. Analagous experiments in which arsenic (III) was used gave similar recoveries.

Although purely thermodynamic considerations suggest that arsenic should exist in oxic seawaters practically entirely in the pentavalent state, equilibrium rarely appears to be attained, probably because of the existence of biologically mediated reduction processes. For this reason, the arsenic in most of these waters exists to an appreciable extent in the trivalent state; and $As^{III}{:}As^{V}$ ratios as high as 1:1 have been found in a number of instances.

Table 8.10 Recovery of arsenic (V) from spiked samples of arsenic-free seawater (150 ml)

As added (μg)	0.200	0.300	0.400	0.600	0.800	1.000
Absorbance per $\mu g\,l^{-1}$	0.063	0.063	0.064	0.053	0.063	0.063
As found (μg)	0.196	0.294	0.396	0.586	0.778	0.985
Recovery (%)	98.1	97.8*	99.1	97.9	97.2	98.5

*Average for 10 determinations; range 96.0–99.2%, standard deviation $\pm$ 1.35%.
Reproduced from Haywood, M.G. and Riley, J.P. (1976) *Analytica Chimica Acta*, **85**, 219, Elsevier Science Publishers, by courtesy of authors and publishers.

Table 8.11 Determination of arsenic (V) in seawater

As^{3+} added (μg)	1.000	1.000	0.000	0.000	1.000	1.000	1.000	1.000
As^{5+} added (μg)	0.000	0.000	1.000	1.000	1.000	1.000	1.000	1.000
As found (μg)	0.018	0.015	0.979	0.985	0.990	0.980	0.976	1.032

*150 ml aliquots extracted with APDC in chloroform before the evolution and determination of arsenic.
Reproduced from Haywood, M.G. and Riley, J.P. (1976) *Analytica Chimica Acta*, **85**, 219, Elsevier Science Publishers, by courtesy of authors and publishers.

Haywood and Riley [109] found that arsenic(III) can be separated from arsenic(V) even at levels of $2\,\mu g\,l^{-1}$ by extracting it as the pyrrolidine dithiocarbamate complex with chloroform. They applied this technique to samples of seawater spiked with AS^V and As^{III} and found (Table 8.11) that arsenic(V) could be satisfactorily determined in the presence of arsenic(III).

References

1. Krenkel, P.A. *International Critical Review of Environmental Control*, **3**, 303 (1973)
2. Robinson, S. and Scott, W.B. *A Selected Bibliography on Mercury in the Environment, with Subject Listing*. Life Science Miscellaneous Publication, Royal Ontario Museum, p. 54 (1974)
3. Uthe, J.F. and Armstrong, F.A.J. *Toxicological Environment Chemistry Review*, **2**, 45 (1974)
4. D'Itri, P.A. and D'Itri, F.M. *Environmental Management*, **2**, 3 (1978)
5. Jensen, S. and Jernelov, A. *Nature (London)*, **753**, 223 (1969)
6. Wood, J.M. *Science*, **183**, 1049 (1974)
7. Furakawa, K. and Tonomura, K. *Agricultural and Biological Chemistry*, **35**, 604 (1971)
8. Furakawa, K. and Tonomura, K. *Agricultural and Biological Chemistry*, **36**, 217 (1972a)
9. Furakawa, K. and Tonomura, K. *Agricultural and Biological Chemistry*, **36**, 2441 (1972b)
10. Brinckman, F.E. and Iverson, W.P. In *Marine Chemistry in the Coastal Environment* (ed. T. Church), American Chemical Society Symposium 18. American Chemical Society, Washington, DC (1975)
11. Fagerstrom, T. and Jernelov, A. *Water Research*, **5**, 121 (1971)
12. Yamada, M. and Tonomura, K. *Journal of Fermentation Technology*, **50**, 159 (1972a)
13. Yamada, M. and Tonomura, K. *Journal of Fermentation Technology*, **50**, 893 (1972b)
14. Yamada, M. and Tonomura, K. *Journal of Fermentation Technology*, **50**, 901 (1972c)
15. Bertilsson, L. and Neujahr, H.Y. *Biochemistry*, **10**, 2805 (1971)
16. Imura, N.E., Sukegawa, S.K., Pan, K., *et al. Science*, **172**, 1248 (1971)
17. Mahan, K.I. and Mahan, S.E. *Analytical Chemistry*, **49**, 662 (1977)
18. Olson, K.R. *Analytical Chemistry*, **49**, 23 (1977)
19. Heiden, R.W. and Aikens, D.A. *Analytical Chemistry*, **49**, 668 (1977)
20. Stoeppler, M. and Matthes, W. *Analytica Chimica Acta*, **98**, 389 (1978)
21. May, K., Reisinger, K., Rlucht, R. and Stoeppler, M. *Sendeedenck Vom Wasser*, **55**, 63 (1980)
22. Fitzgerald, W.F. and Lyons, W.B. *Nature (London)*, **242**, 452 (1973)
23. Frimmel, F. and Winkler, H.A. *Zeitung Wasser-Abwasser-Forschung*, **8**, 67 (1975)
24. Kiemeneij, A.M. and Kloosterboer, J.G. *Analytical Chemistry*, **48**, 575 (1976)
25. Thibaud, Y. *Science et Peche, Bulletin Instrumental Peche Maritime*, **209**, 1 (1971)
26. Cumont, G., Viallex, G., Lelievre, H. and Babenricth, P. *Revue International Oceanographie Mediteranean*, **26**, 95 (1972)
27. Renzoni, A. and Baldi, F. *Accua and Aria*, 597 (1975)
28. Stoeppler, M., Backhaus, F., Matthes, W., *et al. Proc. Verb. XXVth Congress and Plenary Assembly of ICSEM Split*, (1976)
29. Fitzgerald, W.F. Mercury analyses in seawater using cold trap pre-concentration and gas phase detection. In *Analytical Methods in Oceanography*, (ed. T.R.P. Gibb, Jr.), American Chemical Society, Washington, pp. 99–109, 147 (1975)
30. Krenkel, P.A., Burrow, W. and Dickinson, W. *Standardization of Methylmercury Analysis*, Rep. COM-75-10673 (1975)

31. Agemian, H. and Chau, A.S.Y. *Analytical Chemistry*, **50**, 13 (1978)
32. Jenne, E.A. and Avotins, P. *Journal of Environmental Quality*, **4**, 427 (1975)
33. Carron, J. and Agemian, H. *Analytica Chimica Acta*, **92**, 61 (1977)
34. Department of the Environment and National Water Council (UK) Mercury in Waters, Effluents, Soils and Sediments etc., Additional Methods (Pt. 22-AG ENV) HMSO, London (1985)
35. Fitzgerald, W.F. and Lyons, W.B. *Nature (London)*, **242**, 452 (1973)
36. Millward, G.E. and Bihan, A.I. *Water Research*, **12**, 979 (1978)
37. Yamamoto, J., Kaneda, Y. and Hisaka, Y. *International Journal of Environmental Analytical Chemistry*, **16**, 1 (1983)
38. Filipelli, M. *Analyst (London)*, **109**, 515 (1984)
39. Longbottom, J.E., Dressman, R.C. and Lichtenberg, J.J. *Journal of the Association of Official Analytical Chemists*, **56**, 1297 (1973)
40. Ealy, J., Schulz, W.D. and Dean, D.A. *Analytica Chimica Acta*, **64**, 235 (1974)
41. Talmi, U. *Analytica Chimica Acta*, **74**, 107 (1975)
42. Chau, Y.K. and Saitoh, H. *International Journal of Environmental Analytical Chemistry*, **3**, 133 (1973)
43. Hobo, T., Ogura, T., Suzuki, S. and Araki, S. *Bunseki Kagaku*, **24**, 288 (1975)
44. Cappon, C.J. and Smith, J.C. *Analytical Chemistry*, **49**, 365 (1977)
45. Davies, I.M., Graham, W.C. and Pirie, S.M. *Marine Chemistry*, **7**, 11 (1979)
46. Westoo, G. *Acta Chemica Scandinavia*, **22**, 2277 (1978)
47. Compeau, G. and Bartha, R. *Bulletin of Environmental Contamination and Toxicology*, **31**, 486 (1983)
48. Dolphin, D. In *Methods in Enzymology*, (eds. D.B. McCormick and L.D. Wright), Vol. XVIII, Part C Academic Press, New York (1971)
49. Blum, J. and Bartha, R. *Bulletin of Environmental Contamination and Toxicology*, **25**, 404 (1980)
50. Klisenko, M.A. and Shmigidina, A.M. *Gig. Sahit.*, **1974**, 64 (1974)
51. Kamada, T., Hayashi, Y., Kumamaru, T. and Yamamoto, Y. *Bunseki Kagaku*, **22**, 1481 (1973)
52. Baltisberger, R.J. and Knudson, C.L. *Analytica Chimica Acta*, **73**, 265 (1974)
53. El-Awady, A.A., Miller, R.B. and Clark, M.J. *Analytical Chemistry*, **48**, 110 (1976)
54. Sloot, H.A. Van der and Das, H.A. *Analytica Chimica Acta*, **73**, 235 (1974)
55. De Jong, E.G. and Wiles, D.R. *Journal of the Fisheries Research Board of Canada*, **33**, 1324 (1976)
56. Sipos, L., Nurenberg, H.W., Valenta, P. and Branicia, M. *Analytica Chimica Acta*, **115**, 25 (1980)
57. Ke, P.J. and Thibert, R.T. *Mikrochimica Acta*, **3**, 417 (1973)
58. Patterson, C. and Settle, D. *Marine Chemistry*. **4**, 389 (1976)
59. Chau, Y.K., Wong, P.T.S. and Saitoh, H. *Journal of Chromatographic Science*, **14**, 162 (1976)
60. Wong, P.T.S., Chau, Y.K. and Luxon, P.L. *Nature (London)*, **253**, 263 (1975)
61. Bond, A.M., Bradbury, J.R., Hanna, P.J., *et al. Analytical Chemistry*, **56**, 2392 (1984)
62. Hodge, V.F., Seidel, S.L. and Goldberg, E.D. *Analytical Chemistry*, **51**, 1256 (1979)
63. Waldock, M.J. and Miller, D. *ICES Paper CM 1983 E:12 (mimeograph)* International Council for the Exploration of the Sea, Copenhagen (1983)
64. Tugrul, S., Balkas, T.L. and Goldberg, E.D. *Marine Pollution Bulletin*, **14**, 297 (1983)
65. Mueller, M.D. *Fresenius Zeitschrift für Analytische Chemie*, **317**, 32 (1984)
66. Maguire, R.J., Chau, Y.K., Bengert, G.A., *et al. Environmental Science and Technology*, **16**, 698 (1982)
67. Maguire, R.J. *Environmental Science and Technology*, **18**, 291 (1984)
68. Braman, R.S. and Tompkins, M.A. *Analytical Chemistry*, **51**, 12 (1979)
69. Ridley, W.P., Dizikes, L.J. and Wood, J.M. *Science*, **197**, 329 (1977)
70. Guard, H.E., Cobet, A.B. and Coleman, W.M. *Science*, **213**, 710 (1981)
71. Boettner, E.A., Ball, G.L., Hollingsworth, Z. and Aquino, R. USEPA Health Effects Research Laboratory, Cincinnati, Report OH EPA-600/SL-81-062 (1982)
72. Seidel, L.S., Hodge, V.F. and Goldberg, E.D. *Thalassia Jugoslavia*, **16**, 209 (1980)
73. Hodge, V.F., Seidel, S.L. and Goldberg, E.D. *Analytical Chemistry*, **51**, 1256 (1979)
74. Valkirs, A.O., Seligman, P.F., Stang, P.M., *et al. Marine Pollution Bulletin*, **17**, 319 (1986)
75. Dooley, C.A. and Homer, V. *Organotin Compounds in the Marine Environment: Uptake and Sorption Behaviour*. Naval Ocean Systems Center Technical Report No. 917, San Diego, CA (1983)
76. Braman, R.S. and Tompkins, M.A. *Analytical Chemistry*, **51**, 12 (1979)
77. Hodge, V.F., Seidel, S.L. and Goldberg, E.D. *Analytical Chemistry*, **51**, 1256 (1979)
78. Valkirs, A.O., Seligman, P.F., Olsen, G.J., *et al. Analyst (London)*, **112**, 17 (1987)
79. Braman, R.S. and Tompkins, M.A. *Analytical Chemistry*, **51**, 12 (1979)
80. Hodge, V.F., Seidel, S.L. and Goldberg, E.D. *Analytical Chemistry*, **51**, 1256 (1979)

81. Meinema, H.A., Burger, N. and Wiersina, T. *Environmental Science and Technology*, **12**, 288 (1978)
82. Jackson, J.A., Blair, W.R., Brinckman, F.E. and Iveson, W.P. *Environmental Science and Technology*, **16**, 111 (1982)
83. Aue, W.A. and Flinn, G.C. *Journal of Chromatography*, **142**, 145 (1977)
84. Huey, C., Brinckman, F.E., Grim, S. and Iveson, W.P. In *Proceedings of the International Conference on the Transport of Persistent Chemicals in Aquatic Ecosystems* (eds A.S.W. Freitas, D.J. Kushner and D.S.U. Quadri), National Research Council of Canada, Ottawa, Canada, pp. II-73–II-78 (1974)
85. Nelson, J.D., Blair, W., Brinckman, F.E., *et al. Applied Microbiology*, **26**, 321 (1973)
86. Luskina, B.M. and Syavtsillo, S.V. *Nov. Obl. Prom. Sanit. Khim.*, 186 (1969)
87. Woggon, H. and Jehle, D. *Nahrung*, **19**, 271 (1975)
88. Laughlin, R.B., Guard, H.E. and Coleman, W.M. *Environmental Science and Technology*, **20**, 201 (1986)
89. Bluden, S.J. and Chapman, A.H. *Analyst (London)*, **103**, 1266 (1978)
90. Braman, R.S. and Foreback, C.C. *Science*, **182**, 1247 (1973)
91. Woolson, E.A. *Arsenical Pesticides*, ACS Symposium Series, No. 7 American Chemical Society, Washington (1975)
92. Penrose, W.R. *Critical Reviews of Environmental Control*, **4**, 465 (1974)
93. Andraea, M.O. *Analytical Chemistry*, **49**, 820 (1977)
94. Braman, R.S., Johnson, D.L., Foreback, C.C., *et al. Analytical Chemistry*, **49**, 621 (1977)
95. Howard, A.G. and Arbab-Zavar, M.H. *Analyst (London)*, **106**, 213 (1981)
96. Arbab-Zavar, M.H. and Howard, A.G. *Analyst (London)*, **105**, 744 (1980)
97. Hinners, T.A. *Analyst (London)*, **105**, 751 (1980)
98. Pierce, F.D. and Brown, H.R. *Analytical Chemistry*, **49**, 1417 (1977)
99. Talmi, Y. and Bostick, D.T. *Analytical Chemistry*, **47**, 2145 (1975)
100. Edmonds, J.S. and Francesconi, K.A. *Analytical Chemistry*, **48**, 2019 (1976)
101. Johnson, D.L. and Braman, R.S. *Deep Sea Research*, **22**, 503 (1975)
102. Braman, R.S., Johnson, D.L., Foreback, C.C. *et al. Analytical Chemistry*, **49**, 621 (1977)
103. Talmi, Y. and Bostick, D.T. *Analytical Chemistry*, **47**, 2145 (1975)
104. Braman, R.S. and Foreback, G.C. *Science*, **182**, 1247 (1973)
105. Andraea, M.O. *Analytical Chemistry*, **49**, 820 (1977)
106. Persson, J. and Irgum, K. *Analytica Chimica Acta*, **138**, 111 (1982)
107. Lundgren, G., Lundwork, L. and Johansson, G. *Analytical Chemistry*, **46**, 1028 (1974)
108. Yamamoto, M. *Soil Science Society American Proceedings*, **39**, 859 (1975)
109. Haywood, M.G. and Riley, J.P. *Analytica Chimica Acta*, **85**, 219 (1976)

Chapter 9

Elemental analysis

9.1 Carbon

The quantity of dissolved organic carbon in the oceans has been estimated to be about 10^{18} g and constitutes one of the major reservoirs of organic carbon. Although large in total mass, the concentration of organic carbon in seawater is low (typically 0.5–1.5 mg C per litre).

The determination of organic carbon has always presented analytical difficulties. The content of inorganic carbon present in seawater is thirty or more times as great as that of organic carbon. To measure the organic carbon, either the organic or the inorganic carbon must be removed from solution. The retention of even a small percentage of the inorganic carbon could easily double or triple the apparent organic carbon content of a sample. While this observation might seem obvious, it is a fact that several workers have described methods in which the organic carbon was determined as the difference between a total and an inorganic carbon measurement, and where the measurement of inorganic carbon had been incomplete. The resulting dissolved organic carbon values were too high by a factor of up to 10.

Even if removal of inorganic carbon is complete, the analysis of the remaining carbon is difficult. At a concentration of 1 ppm, a normal value for surface water in the open ocean, a 1 ml sample will contain 1 µg carbon. If we are interested in differences between samples, we must strive for a precision of ± 5% or ± 0.05 µg C per ml sample. This requirement places severe constraints on the sensitivity and precision of instrumentation.

Since any attempt to measure the productivity of a marine ecosystem must eventually require measurements of organic carbon in the various reservoirs of the system, an extensive literature exists on methods for the measurement of organic carbon. The methods usually distinguish between total organic carbon (TOC), particulate organic carbon (POC), dissolved organic carbon (DOC) and volatile organic carbon (VOC) and these are discussed below under separate headings. A review of the available methods for all of these components up to 1975 was published by Wangersky [1].

Both biochemical oxygen demand and chemical oxygen demand are measurements frequently made in water laboratories. Both of these methods are subject to interference when applied to saline samples as is discussed in the concluding section in this chapter.

(a) Total and dissolved organic carbon

Because they are so intimately related, for the purposes of this discussion, the categories of 'total' and 'dissolved' organic carbon are combined in this section. Also it is doubtful in any case that anyone ever measures a true 'total' organic carbon in seawater. The total organic carbon should include the dissolved, particulate and volatile organic fractions, in the process of removing the inorganic carbon, usually by acidification and bubbling, some portion of the volatile organic carbon must be removed. Thus the total organic carbon as measured by direct determination must always be the total minus the volatile fraction.

The determination of dissolved organic carbon by oxidation methods in water comprises three analytical steps; the removal of inorganic carbon from the sample oxidation of the organic compounds to carbon dioxide, and the quantitative determination of the resulting carbon dioxide. The methods of oxidation can be classified into three major groups:

1. Wet chemical oxidation methods, using oxidants such as persulphate, are widely used in oceanographic and limnologic work [2, 3]. The main drawbacks of these methods are the manual and cumbersome techniques and incomplete oxidation of some organic compounds [4].
2. High temperature combustion, or dry oxidation methods developed for fresh and wastewater analysis, led to a whole range of discontinuous [5–7] and continuous [8] commercial instruments. The discontinuous injection method lacks in sensitivity and shows high blank values [9] whereas while the continuous combustion methods, with high sensitivity, do not tolerate the high ionic strength of seawater, even if designed for analysis of highly buffered high pressure liquid chromatography affluents [10].
3. The photo-oxidation of dissolved organic carbon in batch procedures [11, 12] requires long reaction times, because of the thickness of the sample. Automated, continuous procedures, where the sample is irradiated by a 'medium pressure' mercury lamp [13, 14] brought great improvement and are utilized in several commercial instruments.

The merits and limitations of wet chemical oxidation, high temperature combustion and photo-oxidation methods for seawater analysis were summarized by Gerskey *et al.* [15]. The specifications for the future development of a sensitive, accurate and automated method were suggested by Wangersky [16].

The various methods used for the determination of total and dissolved organic carbon are now discussed under separate headings.

Ultraviolet absorption
Armstrong and Boalch [17] have examined the ultraviolet absorption of seawater particularly in the wavelengths between 250 and 300 nm, where the absorption is considered to result from the presence of aromatic compounds. Light absorption is a particularly useful measure, if it can be made to work, since it is not too difficult to construct an *in situ* colorimeter which can produce continuous profiles of dissolved organic carbon with distance or depth [18].

Ogura and Hanya [20] investigated the components of the ultraviolet absorption, in an attempt to devise a useful method for oceanic dissolved organic carbon measurements [19–21]. They concluded that while the method might have limited application in coastal waters, most of the absorption in oceanic waters was due to the inorganic components, principally nitrate and bromide ions.

The method has since been revived for use in fresh water [22], where comparisons with total organic carbon determinations by direct injection showed the ultraviolet method to give results that were slightly (3.8%) low, with a fairly large scatter (SD=19% of the mean). Mattson et al. [23] applied a variant of the method to coastal water, again calibrating against a direct injection total organic carbon method and claimed reasonable results. However, Wheeler [24] found that correlations between total organic carbon and ultraviolet absorption, while quite strong over limited geographical regions, could switch from positive to negative in adjacent regions.

Kulkarni [25] described an ultraviolet absorption method for the measurement of the organic carbon level in seawater which compares the intensities at two wavelengths, one less than 350 nm and the other greater than 400 nm.

Biggs et al. [26] compared various instrumental methods for monitoring organic pollution in water. These included total organic carbon, total oxygen demand, chemical oxygen demand and biochemical oxygen demand, and an ultraviolet method based on measurement of the ratio of the light transmitted by the sample at 254 nm in the ultraviolet to that at 510 nm in the visible region. The results obtained by the ultraviolet method were claimed to be largely independent of the presence of inert suspended solids in the sample. They also showed that reasonably good correlations were obtained between ultraviolet absorbance at 254 nm on the one hand and total organic carbon, or biochemical oxygen demand or chemical oxygen demand on the other (Figures 9.1–9.3). Although these investigations were performed only on sewage effluents it could be used as a rapid monitoring procedure for total organic carbon in seawater.

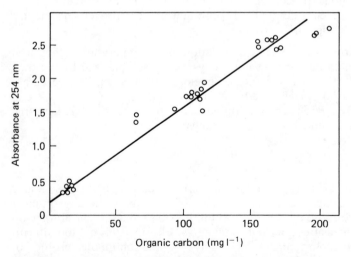

Figure 9.1 Relation between ultraviolet absorbance and total organic carbon of samples of sewage at various stages of treatment, obtained over 4 months. Total organic carbon determination accurate to ± 60% lower range and ± 12½% in mid and upper ranges. Absorbance precision approximately ± 5% full scale (Reproduced from Biggs, R., et al. (1976) *Water Pollution Control*, **47**, 2 , Water Pollution Control Federation, by courtesy of authors and publishers.)

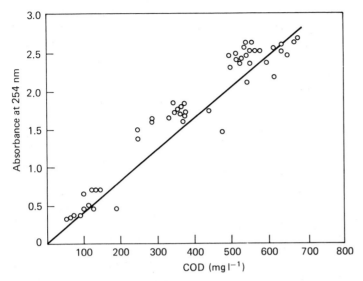

Figure 9.2 Relation between ultraviolet absorbance and the chemical oxygen demand of samples of sewage at various stages of treatment, obtained over 4 months. Chemical oxygen demand determination accurate to ± 10%, absorbance to ± 5% (Reproduced from Biggs, R., *et al.* (1976) *Water Pollution Control*, **47**, 2, Water Pollution Federation, by courtesy of authors and publishers.)

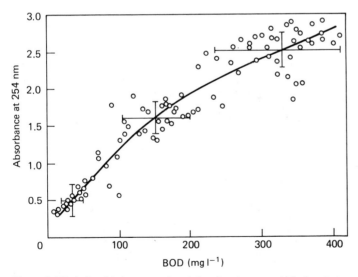

Figure 9.3 Relationship between ultraviolet absorbance and biochemical oxygen demand of samples of sewage at various stages of treatment obtained over a 4-month period. Biochemical oxygen demand precision estimated at approximately ± 30%. Absorbance precision approximately ± 5% of full scale (Reproduced from Biggs, R., *et al.* (1976) *Water Pollution Control*, **47**, 2, Water Pollution Control Federation, by courtesy of authors and publishers.)

Wet oxidation methods

Another group of attractive techniques for determining total organic carbon are those using wet oxidation of the organic carbon to carbon dioxide. In these methods, the inorganic carbon is first removed by acidification and gas purging to remove carbon dioxide produced by the decomposition of metallic carbonates. The wet oxidation of organic carbon is then carried out by the addition of the preferred oxidant, usually with heating, and the carbon dioxide resulting from the oxidation is measured by various methods.

Several different oxidants have been used in this work. The trend has been to stronger oxidants, in the hope that more complete oxidation would result. Among those used have been potassium peroxide [26] dichromate in sulphuric acid [27–30], silver-catalysed potassium dichromate [31], potassium persulphate [32–34] and silver-catalysed potassium persulphate [35, 36]. Similarly, many methods for measuring the evolved carbon dioxide have been devised: titration of the oxidant consumed by coulometric titration (Duursma [31]), conductometry and gas chromatography [29]. By far the most popular, however, has been the non-dispersive infra-red gas analyser [36, 37].

Low temperature ultraviolet promoted chemical oxidation has provided more reliable and precise data in the determination of low $\mu g\, l^{-1}$ levels of total organic carbon [6, 38].

These instruments employ a continuous flow of persulphate solution to promote oxidation prior to ultraviolet irradiation, and have a low system blank and low detection limit. Since all reactions take place in the liquid phase, problems suffered by combustion techniques, such as catalyst poisoning, reactor corrosion and high temperature element burnouts, are obviated. However, the ultraviolet promoted chemical oxidation technique is not designed to handle particulate-containing samples and tends to give incomplete oxidation for certain types of compounds such as cyanuric acid.

While the precision of these methods of the order of ± 0.1 mg C per litre is not high enough for purposes of maintaining budgets on various fractions of the total organic carbon it is certainly good enough to provide some idea of world-wide distributions. The great disadvantage of all wet oxidation methods is the uncertainty of completeness of oxidation. Each change to a more powerful oxidant has been made with the implicit assumption that this particular oxidant would finally work on all of the compounds and give total organic carbon values equivalent to total combustion values. In the absence of a recognized referee method, the only checks on the completeness of oxidation have come from the oxidation of known pure compounds and from the oxidation of ^{14}C-labelled material derived from plankton cultures. Both of these techniques seem to show essentially total oxidation by the high temperature persulphate method [2].

There is considerable controversy concerning the completeness of the wet oxidation. Some of this controversy has centred on the claculation of the blank value for the method. One method for the preparation of organic-free water consists of the oxidation of the residual organic matter with phosphoric acid and persulphate in the process of distillation. This procedure should produce water that is immune to further oxidation with the same reagents and therefore produces zero blanks. However, many workers using total combustion techniques have found that small, but measurable amounts of organic carbon are resistant to this, or to any other, wet chemical oxidation. This is not surprising, since there are a number of organic compounds that are not completely oxidized even under much more stringent conditions, such as boiling with nitric and perchloric acids for periods of up to 45 minutes and at temperatures up to 203°C [39].

While there have been very few intercalibration studies done between the various kinds of methods, the few available comparisons suggest that both the ultraviolet irradiation method and the various total combustion methods discussed in later sections find more organic carbon than do the chemical wet oxidation methods.

However, the work of Menzel and Vaccaro [2] and others [40, 41] suggests that wet oxidation is useful for determining the dissolved organic content of seawater samples, and the oxidation has been shown to be essentially complete [42]. Seawater is first freed of inorganic carbon by treatment with a small volume of 3% phosphoric acid and the organic carbon is then oxidized in sealed glass ampules in an autoclave at 130°C using potassium persulphate as an oxidant. The resulting carbon dioxide is passed through a non-dispersive infra-red analyser whose signals are related to milligrams of carbon in the sample.

Methods based on chemical oxidation are still being developed as is evidenced by the recent (1983) method of Hannaker and Buchanan [42], based on wet oxidation with potassium persulphate [43], for the determination of dissolved organic in concentrated brines following the removal of inorganic carbonates with phosphoric acid. The method involves wet oxidation with potassium persulphate at 130°C followed by a hot copper oxidation and gravimetric measurement of the carbon dioxide produced. The technique overcomes difficulties of calibration curvature, catalytic clogging and instrument fouling often encountered with instrumental methods.

Low volatility natural organic material such as polysaccharides and higher molecular weight proteins sometimes produced low results. In the Hanniker and Buchanon [42] method these problems are overcome by using a solution-phase oxidant and enclosing the system in a sealed tube. In this way all of the constituents are fully contained and exposed to oxidation and, moreover, oxidation of the organic matter to carbon dioxide is complete for the greater majority of compounds.

Hannaker and Buchanan [42] differ from Menzel and Vaccaro [2] in that they use carbosorb or soda asbestos tubes to estimate the carbon dioxide produced instead of the non-dispersive infra-red analyser used by the latter workers.

The equipment was arranged as shown in Figure 9.4. The apparatus allows finely

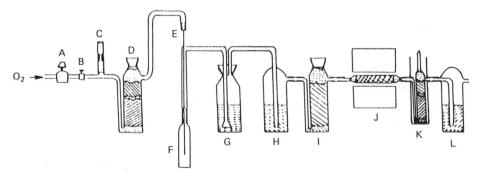

Figure 9.4 Arrangement of analytical equipment. The items are expanded in the text (Reproduced from Hannaker, P. and Bucharan, A.S. (1983) *Analytical Chemistry*, **55**, 1922, American Chemical Society, by courtesy of authors and publishers.)

regulated oxygen (A, B, C) to be purified by passing it through a tube of Carbosorb and magnesium perchlorate. The gas is then passed through a hypodermic needle (E) connected to a glass ampoule (F) using silicone tubing for all connections. The needle is inserted through silicone tubing and should reach to the bottom of the ampoule. The silicone tubing should be clear to allow for visible manipulation of the needle. The silicone tubing is connected to a potassium iodide–sulphuric acid scrubber (30 g potassium iodide dissolved in 75 cm^3 of 10% (v/v) sulphuric acid) to complex any chlorine produced by the wet oxidation of the brine sample. This scrubber bottle (G) is connected to a silver nitrate–nitric acid scrubber bottle (H) to trap any I_3^- carried over from (G) and to prevent clogging of the magnesium perchlorate water removal bottle (I). The drying bottle (I) is connected to a silica glass tube containing cupric oxide maintained at a temperature of 600°C (J) to ensure conversion of carbon products to carbon dioxide. This tube is connected to the Carbosorb-magnesium perchlorate absorption tube (K) inserted to trap the resulting carbon dioxide produced from the oxidation procedure. The system is sealed from the atmosphere by allowing the exit gas to bubble through concentrated sulphuric acid (L).

Using this method, a series of organic species were measured at various concentrations. The results presented in Table 9.1 show that very close agreement

Table 9.1 Analysis of a series of organic species at various concentration levels, both with and without NaCl present

Organic compound tested	Range tested, ppm	NaCl range, ppm	No. of samples tested	Percentage mean recovery
Mannitol	54–432	0–300	10	98.4
Acetic acid	30–720	0–300	11	97.9
D-tartaric acid	76–604	0–300	9	102.7
Oxalic acid	57–1140	0–300	7	100.9
Malic acid	36–716	0–300	9	99.0
Propan-2-ol	45–180	0–300	4	94.4
Acetylacetone	232–465	0–300	4	99.8
Ethanol	208–417	0–300	4	98.1
Glycine	10–382	0–300	10	95.5
Glycerol	10–392	0–300	9	98.7
Trisodium citrate	123–490	0–300	7	96.0
Salicylic acid	42–406	0–300	9	98.3

Reproduced from Hannaker, P. and Buchanan, A.S. (1983) *Analytical Chemistry*, **55**, 1922, American Chemical Society, by courtesy of authors and publishers.

was obtained for a variety of organic compounds both with and without sodium chloride present.

Table 9.2 presents data for a number of natural brine solutions measured by using the same apparatus arrangements.

Each sample was centrifuged and split into two. The first portion was analysed without filtration, the second with filtration. The results show an approximately constant higher value owing to contamination from the filtration medium, indicating that centrifugal removal only of particulates is desirable.

Samples with phosphorus present provide a nutrient source for micro-organisms in solution. Therefore, if analysis cannot be completed at once, brine samples

Table 9.2 Analysis of natural brine solutions from a solar salt field illustrating the possible high results due to soluble organic species from the untreated filter paper

Sample	Soluble organic carbon, ppm	
	Unfiltered	Filtered
Pond A	26	29
Pond B	40	42
Pond C	46	49
Pond D	47	49
Pond E	59	50
Pond F	44	48
Pond G	66	67
Pond H	70	78
Pond I	8	11
Seawater inlet	23	26

Reproduced from Hannaker, P. and Buchanan, A.S. (1983) *Analytical Chemistry*, **55**, 1922, American Chemical Society, by courtesy of authors and publishers.

Table 9.3 The effect of storage on the soluble organic material in solution

Sample	Month of analysis	Soluble organic carbon (ppm) on 4 aliquots		
		High	Low	Mean
Brine 1	February	133	123	128
	April	131	121	125
	June	126	124	126
	August	128	122	124
	October	130	120	123
Brine 2	February	94	86	90
	April	90	84	88
	June	98	89	91
	August	89	85	87
	October	90	84	88

Reproduced from Hannaker, P. and Buchanan, A.S. (1983) *Analytical Chemistry*, **55**, 1922, American Chemical Society, by courtesy of authors and publishers.

should be stored in the dark. Strickland and Parsons [3], however, have suggested that quick deep-freezing stabilizes samples for many months.

Table 9.3 provides measurements of the effect of storage on the soluble organic material in brines. For these two samples micro-organism activity is not affecting the results to any marked degree over a period of 9 months. The samples were stored in brown glass bottles in the dark.

Detection limits are dependent upon the volume of sample measured. With a five or six decimal place balance, levels down to approximately $5\,\mu g\,l^{-1}$ may be determined. Improved accuracy at lower levels may be obtained by increasing the volume of brine measured and the quantity of potassium persulphate used.

Ultraviolet irradiation methods
Another promising wet oxidation method uses high-intensity ultraviolet light in the presence of an oxidizing agent such as hydrogen peroxide or potassium

persulphate. This method was proposed by Beattie, Bricker and Garvin [11] who used a mass spectrometer as a detector for the carbon dioxide produced. Armstrong and his co-workers adapted the method to seawater and showed that the method gave essentially the same results as the persulphate oxidation [44, 45]. They noted the incomplete oxidation of urea, using this technique.

Williams [46, 47] found that a high-energy ultraviolet oxidation method gave dissolved organic carbon results for seawater samples that were higher than those obtained by the wet persulphate oxidation method described by Menzel and Vaccaro [2]. Other variations on the method have been published [12, 48].

One of the great advantages of this method is that it is easily adapted for automatic analysis [13, 14, 36]. Ehrhardt [13] detected generated carbon dioxide by a conductiometric method, Collins and Williams [14] used an infra-red analyser and Armstrong and Tibbets [45] and Goulden and Brooksbank [36] used a continuous photoelectric method. A disadvantage of these automatic methods is that owing to long irradiation times, there is a considerable lag between the intake of the first sample and the issuance of the first result.

An example of an automated colorimetric method oxidation by ultraviolet radiation for the determination of dissolved organic carbon in the presence of potassium persulphate [50] is that of Schreurs [49]. The method uses a Technicon dissolved organic carbon analyser, is fast and precise, and may be used for measuring dissolved organic carbon in both seawater and fresh water over a range of 0.1–10 mg C per litre (Figure 9.5).

Figure 9.6 shows the flow diagram and the sequestrial steps in the oxidization process and carbon dioxide measurements. The sample is acidified and the inorganic carbon is removed with nitrogen. An aliquot is resampled for analyses. Buffered persulphate is added and the sample is irradiated in the ultraviolet destructor for about 9 minutes. The hydroxylamine is added and the sample stream passes into the dialysis system. The carbon dioxide generated diffuses through the gas-permeable silicone membrane. A weakly buffered phenolphthalein indicator solution is used as the recipient stream and the colour intensity of this solution decreases proportionately to the change in pH caused by the absorbed carbon dioxide gas. The measurement of the colour intensity is done in a 15 mm flow cell at 550 nm.

A plot of the concentration found against the concentration calculated for potassium hydrogen phthalate standards is shown in Figure 9.7. Linear response to within 0.1% of the theoretical, least squared fit line is obtained with the intercept passing through zero.

Table 9.4 shows recoveries obtained for a range of organic compounds representing various organic classes and containing a variety of functional groups. These organic compounds with a concentration of 2 and 4 mg C per litre were dissolved in standard seawater. Recovery was measured against potassium hydrogen phthalate standards. As Table 9.4 shows the average recovery is 98.9%.

The development of methods based on ultraviolet oxidation continues as is evidenced by the recent work (1983) of Mueller and Bandaranayake [51] on an automated method for the determination of dissolved organic carbon in seawater using continuous thin film ultraviolet oxidation in a quartz spinal ultraviolet 'low pressure' mercury lamp. Tests with a range of organic compounds showed that the instrument is capable of rapid, continuous and reliable dissolved organic carbon analyses in seawater and is suitable for use aboard a research vessel.

Mueller and Bandaranayake [51] modified for seawater analysis the instrument

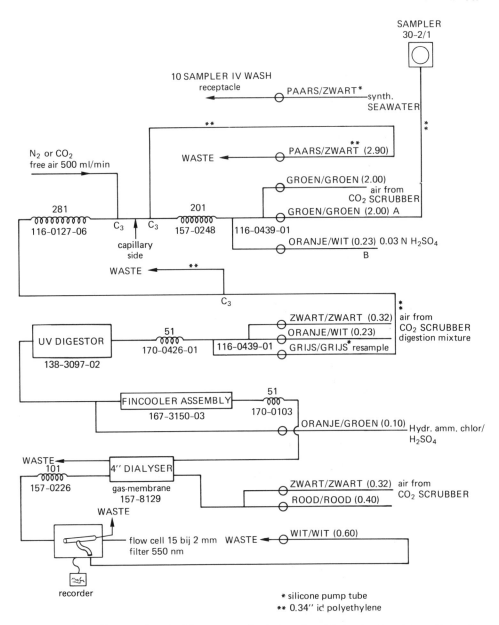

Figure 9.5 Block diagram of method for automated colorimetric oxidation by ultraviolet radiation for determination of dissolved organic carbon in the presence of potassium persulphate (Reproduced from Schreurs, W. *Hydrobiology Bulletin*, **12**, 137 (1978).)

described by Graentzel [52] for the determination of dissolved organic carbon in fresh water (Figure 9.8).

The photo-oxidation reactor consists of two vertically placed concentric, high purity grade, quartz glass cylinders. The outer cylinder, which has inlets for the

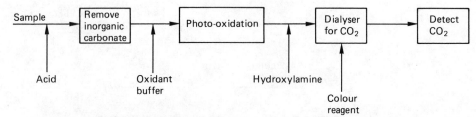

Figure 9.6 Flow diagram of the method described in Figure 9.5 (Reproduced from Schreurs, W. *Hydrobiology Bulletin*, **12**, 137 (1978).)

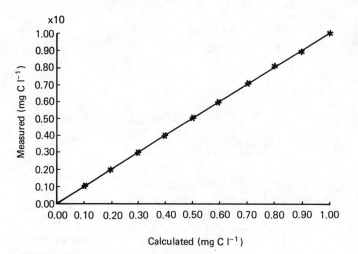

Figure 9.7 Linear regression of measured dissolved organic carbon values against known standard potassium hydrogen phthalate solutions (Reproduced from Schreurs, W. *Hydrobiology Bulletin*, **12**, 137 (1978).)

Table 9.4 Recovery for selected organic compounds

Compound	Theoretical carbon concentration (mg C/litre)	Carbon concentration measures (mg C/litre)	Recovery
Nicotinic acid	4	4.02	100.5
Glutamine	4	3.91	97.75
Gluconic acid	4	4.03	100.75
Sodium citrate	4	4.01	100.25
Thio urea	4	3.90	97.5
Sodium oxalate	4	3.99	99.75
Sucrose	4	3.99	99.75
Disodium-EDTA	4	4.02	100.5
ATP	4	3.74	93.5

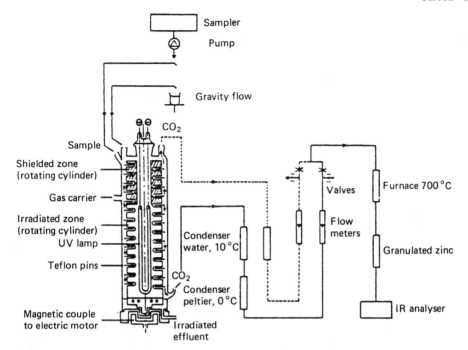

Figure 9.8 Schematic flow diagram of the Graentzel dissolved organic carbon analyser. ——— = Organic carrier; ---- = inorganic carrier (Reproduced from Mueller, H. and Bandaronayake, W.M. (1983) *Marine Chemistry*, **12**, 59, Elsevier Scientific Publishers, by courtesy of authors and publishers.)

sample and carrier gas and outlets for carbon dioxide is mounted, and the hollow inner cylinder is magnetically coupled to an electric motor. This cylinder carries Teflon pins, loosely arranged in indentations. The ultraviolet source is a 100 W, 250 mA 'low pressure' mercury lamp and it is placed in the centre of the inner cylinder. Ultraviolet radiation is emitted at wavelengths of 185 nm (18%) and 245 nm (80%) with about 80% of the oxidation of organic compounds being effected by the 185 nm emission band. The inner cylinder consists of two zones. In the upper zone, which is shielded from radiation, the inorganic carbonates of the acidified sample are completely stripped and removed. The organic compounds are oxidized to carbon dioxide in the lower irradiated zone. Rapid and efficient oxidation of the sample is facilitated by the high penetration depth of ultraviolet radiation into the thin sample film and by the large surface area for efficient gas exchange.

The photo-oxidation method was statistically investigated using the method of Gottschalk [53]. Accordingly, 24 standards (potassium phthalate in 3.5% sodium chloride solution, six concentrations) in a working range of 0.2–2.0 mg C per litre and 24 blanks were analysed. In Table 9.5 the results are compared with previously obtained data [54] on the performance of some commercial instruments using the same procedure. The method compares well with the ultraviolet photo-oxidation method of Collins and Williams [14] but is superior in sensitivity to all other instruments investigated.

Table 9.5 Statistical comparison of commercial organic carbon analysers with data from Collins and Williams [14] and calculated according to the method of Mueller [54], included for comparison

Instrument oxidation/ detector	Oxidation method	Optimal fit calibration	95% confidence limits (mg C/litre)	Determination limits (mg C/litre)
Beckman 915/IR 215	High temp.	$Y = cE + dE^2$ 0.23		0.30
Beckman 915/IR 865	High temp.	$Y = cE$ 0.19		0.24
Beckman 91/Horiba IR 2000	High temp.	$Y = cE + dE^2$ 0.14		0.20
Tocsin II/FID	High temp. continuous	$Y = cE$ 0.20		0.28
Carlo Erba TCM 420/FID	High temp.	$Y = cE$ 0.42		0.57
Dohrmann DC 54/FID	UV-rad.	$Y = cE + dE^2$ 0.14		0.19
Unor Maihak/IR Maihak	High temp. continuous	$Y = cE + dE^2$ 0.09		0.13
Graentzel/IR 865*	UV-rad. continuous	$Y = cE + dE^2$ 0.04		0.05
Collins and Williams (1977)	UV-rad. continuous	? 0.03		0.3**

*Used by Mueller and Bandaronayake [51] by running and re-running the same standards through the apparatus up to three successive times.
**Due to high blank of the procedure.
Reproduced from Mueller, H. and Bandaronayake, W.M. (1983) *Marine Chemistry*, **12**, 59, Elsevier Scientific Publishers, by courtesy of authors and publishers.

Mueller and Bandaranayake [51] were able to show that more than 95% of the following compounds were oxidized in the first run, when present in the water sample at the 5 mg C per litre level; oxalic acid, potassium phthalate, humic acid, glucose, sucrose, ascorbic acid, glycine and phenol; only sulphur compounds gave incomplete recoveries [15, 55].

A pH of 3 was optimal for the complete removal of inorganic carbon and most efficient oxidation of nitrogen-free organic compounds, while a pH of 2.5 was optimal for nitrogenous compounds (Figure 9.9).

These workers found that the efficiency of oxidation was a function of the residence time of the sample in the reactor and the flow rate of the carrier gas. A high precision of carbon dioxide determination was achieved at a sample flow rate of 50 ml h^{-1} and a carrier gas flow rate of 62 litres h^{-1} of which 40 litres h^{-1} passes through the shielded zone.

Using this procedure analysis can be completed in 5–10 minutes.

Evaporation–dry combustion methods
It seems that in principle the simplest of all possible total organic carbon methods would use acidification of the sample, followed by evaporation and dry high temperature combustion of the resulting sea salts. This method has been developed both for fresh water [56] and seawater [57]. The data for oceanic total organic carbon resulting from the latter method were considerably higher, by a factor of two or three, than those coming from the wet oxidation methods. This difference was later ascribed to contamination of the samples by the apparatus and impurities in room air.

A method using vaporation at low temperature in an atmosphere of organic-free gas, an extension of the method of Skopintsev and Timofeyeva [57], was developed by MacKinnon [59]. Contamination from organic vapours in the laboratory air was the major problem; as soon as this was understood, and sufficient care was given to safeguarding the evaporation process, both accuracy and precision reached

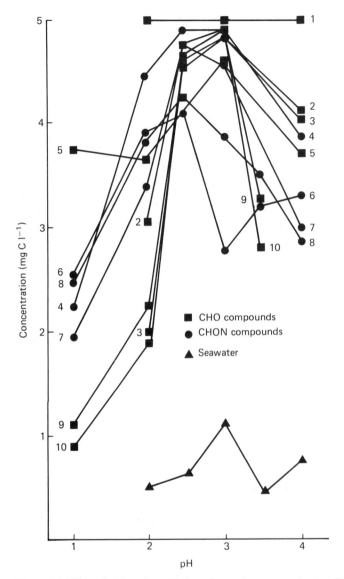

Figure 9.9 Effect of pH on the oxidation of organic compounds. 1 = Oxalic acid; 2 = potassium phthalate; 3 = absorbic acid; 4 = humic acid; 5 = phenol; 6 = guanidine; 7 = glycine; 8 = urea; 9 = glucose; 10 = sucrose. Nitrogen-free compounds peak at pH 3, nitrogen containing compounds at pH 2.5. The seawater sample was collected from Myrmidon Reef, 80 km north of Townsville (Reproduced from Mueller, H. and Bandaronayake, W.M. (1983) *Marine Chemistry*, **12**, 59, Elsevier Scientific Publishers, by courtesy of authors and publishers.)

reasonable levels. Comparisons were run between this dry combustion method and the Sharp [60] version of the persulphate wet oxidation method. The dry combustion method gave results consistently 15–25% higher than persulphate wet oxidation. On water from the same parts of the ocean, the values were well below those found by Skopintsev and Timofeyeva [57] and Gordon and Sutcliffe [58].

McKinnon [61] has described a high-temperature oxidation method for the accurate and precise determination of the total organic carbon in seawater. Problems of contamination in sample storage, preparation and oxidation which are evident in previous dry oxidation methods were controlled. The total organic carbon results from different areas are determined and compared directly with the results obtained by the persulphate oxidation method. A high correlation between the two methods was obtained. In this method, the organic matter is oxidized to carbon dioxide, in an oxidation similar to that of Skopintsev [62], after a sample preparation similar to that of Gordon and Sutcliffe [58]. The oxidation products are determined with a non-dispersive infra-red detector. Values obtained by this procedure are similar to the lower values obtained by Sharp [60] with his high-temperature combustion method. In the method by Gordon and Sutcliffe [58] the aqueous sample is evaporated to dryness by freeze drying and the resulting salt is oxidized in a high temperature furnace.

MacKinnon [61] carried out a very detailed study in which sample collection and storage, sample preparation, the dry oxidation procedure, water corrections, adsorption effects and precision and accuracy are all discussed in detail. Figure 9.10

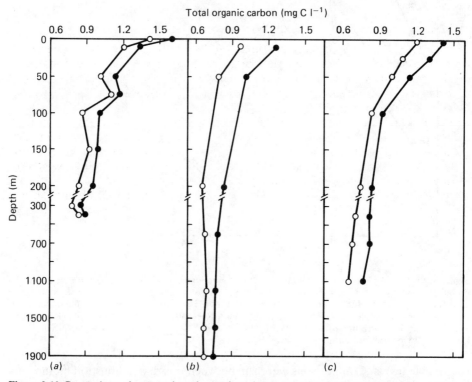

Figure 9.10 Comparison of averaged total organic carbon results for duplicate samples from different areas analysed by the dry oxidation (●) and persulfate oxidation (○) methods. (*a*) Gulf of St. Lawrence (5,6/75). (*b*) Scotian Shelf and Slope (8/75). (*c*) Central Eastern Atlantic (off Senegal) (2,3/76) (Reproduced from MacKinnon, M.D. (1978) *Marine Chemistry*, **7**, 17, Elsevier Science Publishers, by courtesy of author and publishers.)

compares total organic carbon values obtained by the dry oxidation procedure and the persulphate oxidation procedure for a range of seawater samples. The values obtained by dry oxidation are seen to be significantly higher (10–25%) than those obtained by persulphate oxidation.

Comparison of results obtained by evaporation–dry combustion, ultraviolet photo-oxidation
Initially, results reported for dry-combustion methods were found to be higher than wet-oxidation methods based on persulphate by factors of two or more. This discrepancy has steadily decreased as methodologies have improved. Contamination problems of dry methods have been reduced and the oxidation efficiency of the wet methods has been improved. While the differences between approaches have been discussed [60, 61, 63, 64], there is still uncertainty whether the remaining difference between the two techniques is real; a result of incomplete oxidation, incorrect estimation of blanks or a combination of both.

There is a need to resolve this question and to standardize on the methods that will meet practical requirements of oceanographers as well as satisfying the purist regarding accuracy.

Collins and Williams [14] have recently described a modification of Ehrhardt's earlier photochemical method [13] which offers the practical advantages of speed, convenience and the potential for 'real time' analyses. However, until the accuracy of the results is established, the method will not receive general acceptance. Collins and Williams [14] examined the completeness of oxidation of their photo-oxidation system using three independent methods but pointed out that while essentially complete oxidation was indicated, definitive proof was lacking. A more satisfactory solution to the problem might be found through comparison of results of the photo-oxidation method with the dry-combustion method which most analysts are willing to accept as complete [65].

In this connection, Gershay *et al.* [15] in a detailed study have compared results obtained for dissolved organic carbon in seawater using the evaporation dry combustion method of MacKinnon [61], the ultraviolet photochemical method of Collins and Williams [14] (slightly modified) and the persulphate oxidation method of Menzel and Vaccaro [2] and Sharp [60].

The samples were selected to provide a range of dissolved organic carbon concentrations (Figure 9.11) covering the normal range of values encountered in

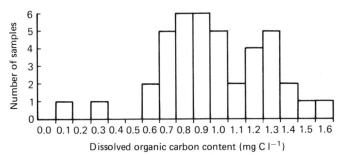

Figure 9.11 Frequency distribution of the organic content of the samples used in an intercomparison study. All illustrated values obtained by the dry-combustion method (Reproduced from Garshay, R.M., *et al.* (1979) *Marine Chemistry*, **7**, 289, Elsevier Science Publishers, by courtesy of authors and publishers.)

Table 9.6 Summary of results of carbon analysis for filtered samples

Reference	n	Range (mg C/litre)	Relative mean values			Ratio of mean values	Mean of ratios of paired analyses of individual samples	Regression analysis		
			Dry	Photo	Wet			R	B	A
Williams [46]	24	0.35–1.22	–	100	90	1.11	1.12 ± 0.029	0.92	1.03 ± 0.09	0.049 ± 0.063
Sharp [60]	68	0.57–1.74	100	–	78	–	–	0.50	–	–
Goulden and Brooksbank [36]*	–	1.49–3.08	100	101.9	97	–	–	Insufficient data		
MacKinnon [61]	20	0.74–2.44	100	–	86 }	1.16	1.16 ± 0.011	0.99	1.17 ± 0.027	0.0014 ± 0.026
Present: dry-persulphate	38	0.13–1.63	100	–	86 }					
dry-photo	40	0.13–1.63	100	95	–	1.06	1.05 ± 0.014	0.98	1.06 ± 0.033	0.005 ± 0.033

n = Number of samples in intercomparison; range, given for dry combustion analysis where available; dry, dry combustion; photo, UV-photo-oxidation; wet, persulphate oxidation; R, correlation coefficient; B, slope of regression line ± standard error; A, intercept of regression line ± standard error.
*Only selection of data available.
Reproduced from Gershay, R.M., et al. (1979) Marine Chemistry, 7, 289, Elsevier Science Publishers, by courtesy of authors and publishers.

Table 9.7 Performance of dissolved organic carbon methods during Gershay [47] and other intercomparisons

		Dry combustion method	Photo-oxidation method	Persulphate oxidation method
Gershay *et al.* [15]	Blank (as mg C per litre)	<0.1 ± 0.02	<0.02	≈ 0.2
	Mean deviation of replicate analyses (mg C per litre)	± 0.031	± 0.030	± 0.044
	Mean coefficient of variation of replicate analyses (%)	2.65	2.9	5.4
Sharp [60]	Coefficient of variation (%)	4.2	–	5.5
Goulden and Brooksbank [36]	Coefficient of variation (%)	3.4	1.2	0.9
MacKinnon [60]	Coefficient of variation (%)	2.5	–	3.1

Reproduced from Gershay, R.M., *et al.* (1979) *Marine Chemistry*, 7, 289, Elsevier Science Publishers, by courtesy of authors and publishers.

the sea (0.5–1.5 mg C per litre). The samples were collected from the continental shelf off Nova Scotia and from Halifax Harbour. Samples were filtered through precombusted (450°C) glass-fibre filters (Whatman GF/C). Each sample was subdivided into three aliquots. The samples were acidified to pH 2.5 with phosphoric acid and stored at 5°C or frozen until analysis. A standard solution of dextrose (1.0 mg C per ml) was used to calibrate all three methods.

The detailed results of comparisons for filtered samples are given in Table 9.6 and are displayed graphically in Figures 9.12 and 9.13.

The mean coefficient of variation of replicates obtained by each of the methods used in the comparison is given in Table 9.7. A few unfiltered samples were analysed and these results are included in Figures 9.12 and 9.13 as well as in Table 9.8. The performance of the photo-oxidation system with these unfiltered samples was significantly poorer from that seen with the filtered samples. The performance of the persulphate oxidation did not appear to be affected by the presence of particles (Table 9.8) and Figure 9.13. A problem with the dry combustion method used by Gershay *et al.* [15] is that volatile organic components of a sample will not be measured. This can be seen with the samples in Table 9.8 which were known to be contaminated with ethylalcohol.

Gershay *et al.* [15] have shown that the agreement between the three techniques for analysis of dissolved organic carbon in seawater is generally good, considering the problems inherent in this type of trace analysis. Taking their results and those of Goulden and Brooksband [36] together, one can conclude that the difference between the results of the dry combustion and the photo-oxidation methods is small (see Table 9.6) and of little practical consequence.

The versions of the persulphate oxidation methods used at present persistently yield results that are lower than those obtained using the dry combustion or photo-oxidation techniques (Table 9.9). Close agreement between the persulphate and other methods is obtained when the analyses are carried out on freshwater rather than seawater samples. If the persulphate oxidation procedure is to be retained as a method for seawater analysis, serious consideration should be given to abandoning the present batchwise procedure in favour of an automated procedure.

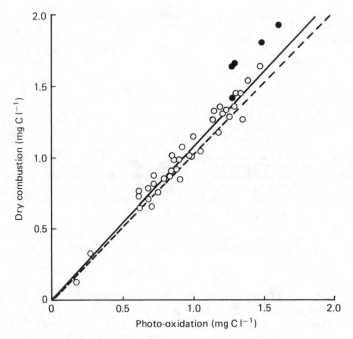

Figure 9.12 Regression analysis of the results of the dry-combustion and photo-oxidation methods. Dashed line indicates ideal relationship. Solid line fitted by least-squares analysis. ○ = Filtered samples, dissolved organic carbon; ● = unfiltered samples, total organic carbon (Reproduced from Gershay, R.M., *et al.* (1979) *Marine Chemistry*, **7**, 289 Elsevier Science Publishers, by courtesy of authors and publishers.)

Such a system has been demonstrated by Goulden and Brooksbank [36] to yield reliable results with excellent precision.

Gershay *et al.* [15] concluded that a continuous and automated photo-oxidation procedure of the type described by Collins and Williams [14] with the reported modifications, will probably satisfy most of the needs of the oceanographer concerning measurement of dissolved organic carbon. The convenience and rapidity of the method opens up a new area of research; the study of the small-scale temporal and spatial variations of the dissolved organic carbon content of the oceans.

Dry combustion direct injection
These procedures differ from the evaporation–dry combustion procedures described in the previous section as follows.

Evaporation–dry combustion.   Inorganic carbonate is removed by acid addition, evaporated at fairly low temperature to a dry sea salt residue, and the residue combusted to carbon dioxide in a high temperature furnace.

Dry combustion–direct injection.   Inorganic carbonate is removed by acid addition, and the acidified liquid sample injected directly into a high temperature furnace.

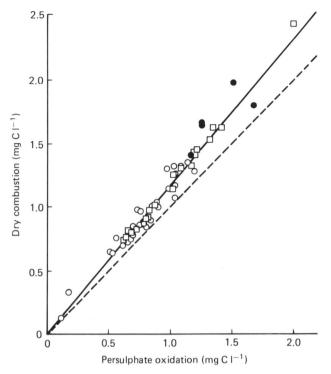

Figure 9.13 Regression analysis of the results of the dry-combustion and persulphate oxidation methods. Dashed line indicates ideal relationship. Solid line fitted by least squares analysis. ● = Filtered samples (present study); ○ = unfiltered samples (present study); □ = unfiltered samples (Reproduced from Gershay, R.M., *et al.* (1979) *Marine Chemistry*, **7**, 289, Elsevier Science Publishers, by courtesy of authors and publishers.)

The dry combustion–direct injection technique provides many advantages over other methods, such as quick response and complete oxidation for determining the carbon content of water. Its primary shortcoming is the need for rapid discrete sample injection into a high temperature combustion tube. When an aqueous sample is injected into the furnace, it is instantaneously vaporized at 900°C and a 5000-fold volume increase can be expected. Such a sudden change in volume causes so-called system blank and limits the maximum volume of injectable water sample, which in turn limits the sensitivity [75, 76].

A dry combustion-direct injection apparatus was applied to water samples by Van Hall, Safranko and Stenger [7]. The carbon dioxide was measured with a non-dispersive infra-red gas analyser. Later developments included a total carbon analyser [66], a diffusion unit for the elimination of carbonates [67] and finally a dual tube unit which measured total carbon by combustion through one pathway and carbonate carbon through another. Total organic carbon was then calculated as the difference between the two measurements [68].

Van Hall, Safrenko and Stenger [69] inject a 20 litre sample into a high-temperature furnace at 950°C containing a catalyst to promote oxidation of carbon compounds to carbon dioxide which is then passed into a non-dispersive infra-red

Table 9.8 Results of intercomparison of oxidation methods for the analysis of the organic matter in seawater

Area	Depth (m)	Salinity (‰)	Concentration of dissolved organic carbon (mg C per litre)		
			Photo-oxidation	Dry oxidation	Persulphate oxidation
Scotian shelf	0	32.11	0.85 ± 0.03	0.86 ± 0.01	0.75 ± 0.07
	10	32.12	0.93 ± 0.04	0.84 ± 0.03	0.81 ± 0.02
	25	32.12	0.88 ± 0.05	0.98 ± 0.06	0.73 ± 0.02
	50	32.40	0.79 ± 0.09	0.87 ± 0.05	0.79 ± 0.04
	0	31.88	–	0.95 ± 0.05	0.90 ± 0.03
	25	31.93	0.86 ± 0.02	0.90 ± 0.02	0.85 ± 0.02
	50	33.19	0.86 ± 0.02	0.90 ± 0.01	0.83 ± 0.04
	100	34.56	0.72 ± 0.05	0.78 ± 0.01	0.71 ± 0.07
	200	34.99	0.63 ± 0.01	0.72 ± 0.01	0.66 ± 0.06
	0	31.34	1.07 ± 0.01	1.04 ± 0.02	0.90 ± 0.03
	25	31.25	1.02 ± 0.04	1.14 ± 0.03	1.01 ± 0.03
	50	33.50	0.70 ± 0.02	0.70 ± 0.01	0.62 ± 0.04
	0	31.96	0.99 ± 0.02	1.01 ± 0.02	–
	25	31.93	0.87 ± 0.02	1.01 ± 0.03	0.85 ± 0.06
	50	34.95	0.91 ± 0.03	0.97 ± 0.04	0.76 ± 0.03
	100	34.56	0.81 ± 0.02	0.85 ± 0.01	0.70 ± 0.08
	400	35.07	0.72 ± 0.06	0.65 ± 0.01	0.51 ± 0.05
	0	31.34	0.94 ± 0.01	1.07 ± 0.01	1.04 ± 0.04
	25	31.98	1.01 ± 0.03	1.00 ± 0.01	0.91 ± 0.08
	100	33.74	0.90 ± 0.04	0.91 ± 0.02	0.80 ± 0.05
	150	34.42	0.74 ± 0.03	0.81 ± 0.01	0.71 ± 0.04
	0	31.96	0.92 ± 0.02	0.98 ± 0.00	0.84 ± 0.04
	50	34.95	0.81 ± 0.02	0.85 ± 0.01	0.73 ± 0.01
	100	34.99	0.77 ± 0.01	0.75 ± 0.01	0.69 ± 0.04
	200	35.06	0.63 ± 0.01	0.76 ± 0.03	0.56 ± 0.07
	500	35.08	0.64 ± 0.02	0.64 ± 0.02	0.53 ± 0.05
Halifax Harbour samples (a) filtered (DOC) sample					
1	1	30.24	1.28 ± 0.07	1.28 ± 0.03	1.20 ± 0.10
2	1	30.69	1.27 ± 0.06	1.29 ± 0.10	1.19 ± 0.04
3	1	30.92	1.20 ± 0.06	1.17 ± 0.06	1.04 ± 0.04
4	1	31.00	1.26 ± 0.01	1.33	1.08 ± 0.07
	10	30.89	1.17 ± 0.02	1.32 ± 0.08	1.03 ± 0.02
(b) unfiltered (TOC)					
1	1	30.24	1.51 ± 0.03	1.80 ± 0.06	1.68 ± 0.08
2	1	30.69	1.63 ± 0.14	1.90 ± 0.14	1.52 ± 0.10
3	1	30.92	1.32 ± 0.06	1.66 ± 0.05	1.27 ± 0.02
4	1	31.00	1.30 ± 0.02	1.64 ± 0.02	1.27 ± 0.03
	10	30.89	1.30 ± 0.02	1.41 ± 0.02	1.18 ± 0.01
Tap seawater spiked with an algal extract sample					
1		30.50	1.16 ± 0.02	1.26 ± 0.05	1.04 ± 0.02
2		30.50	1.21 ± 0.03	1.35 ± 0.02	1.14 ± 0.01
3		30.50	1.33 ± 0.05	1.44 ± 0.01	1.19 ± 0.04
4		30.50	1.41 ± 0.05	1.53 ± 0.03	1.32 ± 0.03
5		30.50	1.50 ± 0.02	1.63 ± 0.10	1.40 ± 0.05
Miscellaneous samples Combusted Salts prepared in deionized water			0.28 ± 0.02 0.18 ± 0.01	0.33 ± 0.04 0.13 ± 0.05	0.18 0.12 ± 0.01

Area	Depth (m)	Salinity (‰)	Concentration of dissolved organic carbon (mg C per litre)		
			Photo-oxidation	Dry oxidation	Persulphate oxidation
Loss of volatiles during DOC analysis					
Samples not contaminated:			1.36 ± 0.05	1.41 ± 0.01	1.21 ± 0.08
			1.31 ± 0.02	1.35 ± 0.01	1.21 ± 0.05
Contaminated with ethanol:			*39.3 ± 1.2	*1.27 ± 0.06	*13.05 ± 1.5
			*39.0	*1.30 ± 0.05	*11.58 ± 0.33

DOC = Dissolved organic carbon; TOC = total organic carbon.
*Not used in comparison.
Reproduced from Gershay, R.M., *et al.* (1979) *Marine Chemistry*, **7**, 289, Elsevier Science Publishers, by courtesy of authors and publishers.

Table 9.9 Comparison of the different oxidation methods for analysis of organic carbon in filtered and unfiltered seawater

Area	n	Averaged organic carbon concentration (mg C per litre)			Difference by paired *t*-test at 95% confidence level
		Dry oxidation	Photo-oxidation	Persulphate oxidation	
Scotian shelf	25	–	0.85 ± 0.14	0.77 ± 0.13	s
(0–500 m)	26	0.88 ± 0.13	0.85 ± 0.14	–	ns
(31–35‰)	25	0.88 ± 0.13	–	0.77 ± 0.13	s
Coastal area					
(a) filtered (DOC)	9	–	1.26 ± 0.06	1.12 ± 0.10	s
	9	1.31 ± 0.08	1.26 ± 0.06	–	ns
	9	1.31 ± 0.08	–	1.12 ± 0.10	s
(b) unfiltered (TOC)	5	–	1.41 ± 0.15	1.38 ± 0.21	ns
	5	1.68 ± 0.19	1.41 ± 0.15	–	s
	5	1.68 ± 0.19	–	1.38 ± 0.21	s

s = Significant; ns = not significant.
DOC = Dissolved organic carbon; TOC = total organic carbon.
Reproduced from Gershay, R.M., *et al.* (1979) *Marine Chemistry*, **7**, 289, Elsevier Science Publishers, by courtesy of authors and publishers.

analyser. The carbonate interference can be determined by passing an acidified portion of the sample through a low-temperature furnace [70–72].

One of the best known commercial instruments developed for organic carbon determinations is the Beckman total carbon analyser which utilizes an analysis scheme developed by Van Hall and co-workers [7, 68]. This instrument works reasonably well in fresh water. It has become a standard instrument in pollution control and water treatment [72]. The Beckman instrument has not worked as satisfactorily for seawater because of its high carbonate and low organic content.

Another instrument developed by the Precision Scientific Co. was based upon the work of Stenger and Van Hall [68, 73]. Experience has shown that application of the Beckman and Precision Scientific Co. instruments to concentrated or saturated brine solutions leads to erratic and unreliable results. There are several possible reasons for this: (1) the catalyst will rapidly become loaded with sodium chloride, (2) oxidation of Cl^- to chlorine will occur, and (3) volatile organics may not all be trapped by the solid catalyst.

Van Hall, Safranko and Stenger [7] have also pointed out that strong brines interfere with the method by producing 'fogs' which may be counted as carbon dioxide while in cases where the flame ionization detector is being used, large spikes appear in the recorded curve [74].

Sharp [60] has described a dry combustion–direct injection system built for oceanographic analyses. This unit used 100 μl samples, injected into a 900°C oven in an atmosphere of oxygen. The output from a non-dispersive infra-red carbon dioxide analyser was linearized and integrated.

Ton and Takahashi [77] describe a dry combustion–direct injection total organic carbon analyser, the DC-90 (produced by Dohrmann Division, Xertex Corporation; Figure 9.14) which, it is claimed, overcomes many of the problems encountered with earlier versions of this equipment. In this analyser, carrier water is continuously pumped into a ceramic combustion tube where a constant flow of oxygen is provided. The combustion tube is designed so that the carrier water is

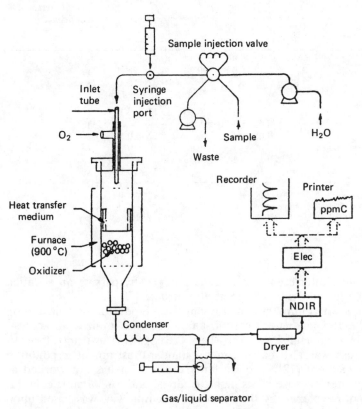

Figure 9.14 Flow diagram of DC-90 (Reproduced from Ton, N. and Takahari, Y. (1985) *International Laboratory*, **September**, p. 143, by courtesy of authors and publishers.)

dispersed and instantaneously evaporated. A sample is introduced by either a rotary injection valve (via a sample loop) or by syringe into the carrier water stream. Thus gas and steam are maintained at constant flow. Carbonaceous materials in the sample are completely oxidized to carbon dioxide in the presence of oxygen, an oxidation catalyst, and a heat transfer device. After the steam is condensed, carbon dioxide in oxygen is subsequently measured by a non-dispersive infra-red spectroscopy. The carbon dioxide generated by complete combustion is directly proportional to the total carbon in the stream. Since a large sample volume can be injected without disturbing flow conditions, the system blank is negligible; thus high precision and sensitivity can be achieved. This continuous water carrier flow–high temperature combustion technique works quite well for determining total organic carbon in samples such as seawater and brine solutions. The constant water flow prevents salt build-up, and deterioration of the combustion tube is eliminated by the use of a ceramic tube. Inorganic carbon can also be directly determined by injecting a sample into a gas/liquid separator (Figure 9.14). Total organic carbon can be determined either by externally acidifying and sparging the sample before analysis or calculated by the difference:

Total organic carbon = total carbon − inorganic carbon

Table 9.10 compares calculated total organic carbon values of various carbon-containing standards with results obtained with the high temperature oxidation method described by Ton and Takahashi [77]. This experiment was conducted at 900°C with 0.6 ml min^{-1} water and 100 ml min^{-1} oxygen; a 200 μl sample volume (sample loop and syringe) was used. The system was calibrated with 400 mg l^{-1} carbon potassium hydrogen phthalate standard. In all cases, recovery was well within 2% of the calculated values.

Ton and Takahashi [77] investigated several methods for obtaining carbon-free blank water. They found that ultraviolet treated deionized water served this purpose best. Acidified deionized water is pumped into a glass chamber which houses an ultraviolet light. With a constant flow of oxygen, carbonaceous matter in the water is oxidized. Resultant carbon dioxide is sparged out and carbon-free water is achieved. Since this treated water is continuously pumped directly to the furnace and does not come in contact with the atmosphere, contamination and carbon dioxide absorption are avoided.

Table 9.11 shows the results for several seawater samples injected from a 1 ml loop. The results agree with previous studies, which found the total organic carbon level of seawater near the surface to be about 1 mg l^{-1}. To study sample recovery in the presence of salt, 3% sodium chloride was added to known concentrations of methyl alcohol and sucrose, and the solutions were then analysed. As Table 9.11 shows, the change in recovery due to the addition of salt was statistically insignificant.

Shimadzu supply a fully automated total organic carbon analyser, Model TOC.500, based on dry combustion direct injection followed by non-dispersive infra-red spectroscopy. The system incorporates a microcomputer and has an optional automatic sample injector. The method is applicable to all waters including seawater and has a range of 1–3000 ppm. It is also applicable to the determination of volatile organic carbon. In this instrument (Figure 9.15) flow-controlled and humidified carrier gas (highly pure air) is allowed to flow through the TTCU (total carbon) combustion tube which is kept at 680°C. Then the gas flows through the drain separator where its moisture is removed and through the

Table 9.10 Comparison of calculated total organic carbon results of various standards with results obtained by high temperature oxidation method [77]

Standards	Calculated concentration	Obtained concentration (ppm C)	
		200 µl loop	200 µl syringe
KHP	400.0	399.7 ± 3.5	399.0 ± 1.2
MeOH	400.0	402.0 ± 2.1	405.0 ± 1.0
Sucrose	400.0	401.8 ± 1.2	400.3 ± 1.2
Nicotinic acid	401.3	395.0 ± 3.0	404.0 ± 1.7
Acetonitrile	368.6	362.0 ± 5.3	379.0 ± 2.8
Cyanuric acid	405.0	407.0 ± 0.6	404.0 ± 1.7
Proline	347.4	348.0 ± 3.2	
Aniline	410.1	413.0 ± 1.8	
Pyridine	441.0	442.0 ± 3.5	
Chloroform	423.5	426.0 ± 1.9	
Acetone	496.3	504.0 ± 4.3	507.0 ± 4.0
Acetic acid	428.0	423.0 ± 3.7	
Sodium lauryl sulfate	404.8	399.0 ± 2.1	
Urea	406.0	398.0 ± 4.4	
Humic acid	1000 mg/l		497.0 ± 4.7

Reproduced from Ton, N. and Takahari, Y. (1985) *International Laboratory*, **September**, 49, by courtesy of authors and publishers.

Table 9.11 Recovery of total organic carbon in salt-containing samples

Sample	Mean (±SD) concentration (ppm)
Sargasso Sea	0.85 ± 0.02
Eel Pond (MA)	1.41 ± 0.04
Vinyard Sound	1.07 ± 0.07
Santa Cruz	0.99 ± 0.07
Norwegian	1.00 ± 0.04
MeOH	10.07 ± 0.05
Before salt	9.98 ± 0.06
Sucrose	9.96 ± 0.02
Before salt	9.92 ± 0.02

Reproduced from Ton, N. and Takahari, Y. (1985) *International Laboratory*, **September**, 49, by courtesy of authors and publishers.

inorganic carbon reaction tube kept at 150°C, and through the electronic dehumidifier where its moisture is removed again. Finally, the gas flows through the cell of the non-dispersive infra-red analyser which measures the carbon dioxide concentration in the gas.

When a sample is injected into the total carbon injection port with a microlitre syringe, the carbon atoms in the organic and inorganic compounds in the sample are oxidized into carbon dioxide. The NDIR analyser generates a peaked signal with an area proportional to the carbon dioxide concentration. The built-in microcomputer measures the peak area and converts it into total carbon concentration value by means of the calibrating equation predetermined through measurement of standard solution samples.

When a sample is injected into the inorganic carbon injection port, the sample enters the inorganic carbon reaction tube, where only the inorganic carbon is turned into carbon dioxide. The carbon dioxide concentration is converted into

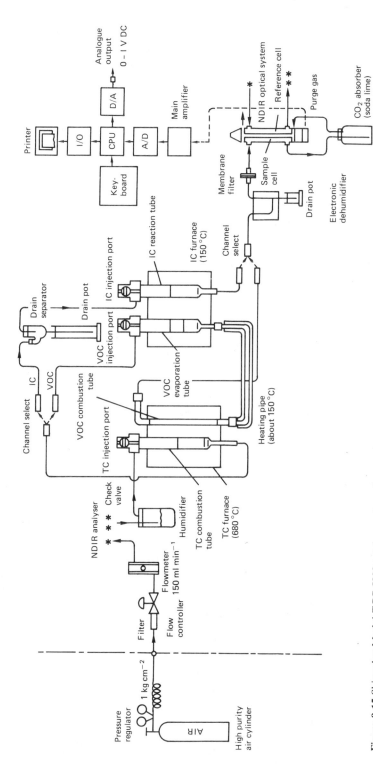

Figure 9.15 Shimadzu Model TCC 5000 total organic carbon analyser

inorganic carbon concentration value in the same way as total carbon. The total organic carbon concentration is obtained by subtracting inorganic carbon concentration from total carbon concentration and printed out.

With the optional volatile organic carbon measuring channel added, volatile organic carbon concentration is measured in the same way.

When the volatile organic carbon measuring channel is selected, the gas coming out of the total carbon combustion tube enters the volatile organic carbon evaporation tube, which is packed with volatile organic compound evaporation reagent and inorganic carbon absorbing reagent and is kept at 150°C. Then the gas flows through the heated tube into the volatile organic carbon combustion tube kept at 680°C and reaches the non-dispersive infra-red analyser via the electronic dehumidifier.

When a sample is injected into the volatile organic carbon injection port, only the volatile organic components are vaporized. Some inorganic carbon components generate carbon dioxide which is absorbed by the inorganic carbon absorbing reagent.

Volatile organic components are oxidized in the volatile organic carbon combustion tube into carbon dioxide. The carbon dioxide concentration is converted in the same way as inorganic carbon and printed out.

Dry combustion–direct injection techniques using gas chromatography as the method of measurement have been devised for fresh water and low salinity water but not for seawater. Nelson and Lysyj [78] used pyrolysis of the organic compounds without oxidation, with flame ionization as the detection method. The technique was further developed by Eggertsen and Stross [74]. Since the response of this detector varies according to the type of organic compound present, the accuracy of the method cannot match its precision. This problem was overcome [79–81] by injecting the sample into a high temperature furnace containing cupric oxide so that the organic components are oxidized to carbon dioxide then reducing the carbon dioxide to methane which is measured with a flame ionization detector.

A more recent development is the Dohrmann DC-54 Ultra Low Level total organic carbon analysis. This equipment is capable of determining total organic carbon down to $10 \, \mu g \, l^{-1}$ and pungeable organics down to $1 \, \mu g \, l^{-1}$. It is applicable to seawater. The principle employed is ultraviolet promoted chemical oxidation of organic carbon to carbon dioxide, followed by conversion to methane which is determined by flame ionization gas chromatography. Analysis time is 8–9 minutes and the range of application $0–10\,000 \, \mu g \, l^{-1}$.

The precision for total organic carbon is $\pm \, 1 \, \mu g \, l^{-1}$.

Preservation and storage of samples for the determination of dissolved organic carbon

As was expressed by Wangersky and Zika [65], sampling and storage of water samples for the determination of organic compounds can be an important source of errors. Duursma [82] stored his samples in 500 ml glass bottles, preserving with sulphuric acid. Wangersky and Zika [65] claim that addition of acid to samples can change organic compounds.

Merks and Vlasbom [83] filtered 20 ml sample through a glass fibre filter and then transferred the solution into a 20 ml glass ampoule. Immediately after that, the ampoule was closed by heating it in a flame. The filters were pre-treated by heating at 330°C and the ampoules by heating at 610°C, both for 2 hours. After closing, the ampoules were deep frozen at −20°C. Just before the determination of dissolved organic carbon the ampoule was thawed again and opened. Results of several compared analyses carried out by Olrichs [84] showed very good agreements using different analytical systems for measuring total organic carbon. However, Elgershuizen [85] did not find a good agreement between determinations of dissolved organic carbon which were carried out in two laboratories. As the sampling took place at the same time and in the same way, the cause of the bad results can only be a matter of preservation and storage.

In fact, the conditions of storage were very different. Merks and Vlasbom [83] used glass ampoules stored at −20°C whereas Elgershuizen [85] stored the samples in PVC bottles at 4°C acidified to pH 2.

Merks and Vlasbom [83] caried out some comparative experiments on standard seawater with potassium hydrogen phthalate and water from Eastern Scheidt and Western Scheidt. These three types of samples were stored in three different ways and in three types of storage bottles.

The storage occurred in a deep-freezer at −20°C, in a refrigerator at 4°C and at room temperature at about 20°C in the laboratory. The bottles used in this investigation were glass ampoules, hard plastic bottles and polystyrene sample cups as used in auto-analysers. The pre-treatment of those bottles was as follows: the glass ampoules were heated at 610°C for 2 hours, the plastic bottles were flushed several times with deionized water from a Millipore Milli-Q outfit and dried afterwards. The polystyrene cups were not pretreated at all. All samples were filtered over preheated glass fibre filters just before the three types of bottles were filled. Before filling with the sample, the plastic bottles and the polystyrene cups were rinsed twice with the filtered sample, while the glass ampoules were filled without rinsing.

From the results presented in Figure 9.16 it can be concluded that the decrease in dissolved organic carbon with storage time is always lowest when the samples come deep frozen; deep freezing in a glass ampoule resulted in the lowest decline. Even under these conditions the storage time should be as low as possible.

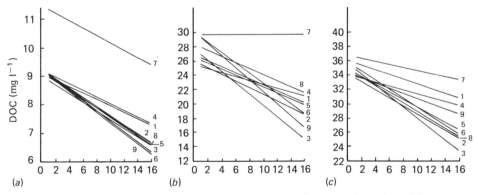

Figure 9.16 Regression lines of dissolved organic carbon concentrations and storage time of three types of water when stored at nine types of bottle–temperature combinations (1–9). (a) KHP in standard seawater. (b) Eastern Scheldt water. (:c) Western Scheldt water. 1 = glass ampoule, stored at − 20°C. 2 = glass ampoule, stored at 4°C. 3 = glass ampoule, stored at room temperature. 4 = hard plastic bottle, stored at − 20 °C. 5 = hard plastic bottle, stored at 4°C. 6 = hard plastic bottle, stored at room temperature. 7 = polystyrene cup, stored at − 20°C. 8 = polystyrene cup, stored at 4°C. 9 = polystyrene cup, stored at room temperature (Reproduced from Merks, A.G.A. and Ulassam, A.C. *Hydrobiological Bulletin*, **17**, 53 (1983).)

Miscellaneous methods for measurement of dissolved organic carbon
Methods that have been applied only to fresh water samples show some promise
for seawater analysis. One such system [86] measures inorganic, volatile organic,
and non-volatile organic carbon fractions on the same sample. The different
fractions are evolved separately, the inorganic carbon by acidification with
phosphoric acid, the volatile fraction by purging with oxygen and high-temperature
combustion of the vapours and the non-volatile fraction by ultraviolet photo-
oxidation. In each case, the carbon dioxide evolved is purified by cold trapping.

Another technique depends upon the difference in rates of volatilization to
differentiate between organic and inorganic carbon, and upon plasma emission
spectrography for the final measurement [87]. The sample is dispensed into a
platinum boat, dried, and then covered with an oxidant, in this case V_2O_5. The
sample is then moved into an oven held at 850°C. At this temperature the organic
carbon is volatilized a few seconds before the carbonate carbon. Light from a
microwave-excited argon plasma, into which the volatile material is passed, is
dispersed by a monochromator, and the 193 nm atomic carbon line is measured
with a photo-multiplier. An advantage of this technique is that oxidation to carbon
dioxide need not be complete. When comparisons were run between this method
and persulphate oxidation, no significant differences were obtained. The obvious
difficulty to be expected in using this technique for seawater analysis stems from
the vastly greater concentration of carbonates. The peaks for organic and inorganic
carbon are not completely separated. Even a small overlap between the peaks
would prove disastrous.

(b) Particulate organic carbon

The simplest of the components of the organic carbon system to measure is the
particulate phase, since the solid material can be isolated by filtration. Once the
sampling conditions for filtration have been defined, the analytical procedures are
straightforward. The major problem is that of sensitivity; depending upon the filter
used, the particulate organic carbon content of surface seawater will run between
25 and 200 µg C per litre, while the deep water will give values of 3–15 µg C per
litre. Almost without exception, the modern methods for particulate organic
carbon use a high-temperature combustion of the filter and its organic load. Many
early workers employed a wet oxidation with chromic and sulphuric acids, with the
actual measurement being a titration with ferrous ammonium sulphate [88]. A
rough conversion from wet oxidation to dry combustion values can be achieved by
multiplication of the wet oxidation values by a factor of 1.09 is commonly used for
this purpose.

The various combustion methods differ primarily in the method of measuring
the carbon dioxide generated from the organic carbon. The first really sensitive
carbon dioxide detector and the one still most used is the non-dispersive infra-red
gas analyser. The detecting element senses the difference in absorption of infra-
red energy between a standard cell, filled with a gas with no absorption in the infra-
red and a sample cell. Water vapour is the only serious interference, hence the
carbon dioxide must be dried before any measurements are made.

The carbon dioxide resulting from combustion of the particulate organic carbon
can be passed through the analyser, producing a single spike on a recorder, or it
can be pumped in a loop through the analyser, until an equilibrium concentration

is reached in the loop. Since the output of the analyser is non-linear, the latter technique has been favoured by some investigators [89].

Carbon analysers using the non-dispersive infra-red analysers have been described by Kuck *et al.* [90] and by Ernst [91] among others.

Electrometric methods have also been used for the final measurement of carbon dioxide; Szekielda and Krey [92] devised a conductometric method. By far the most popular methods, however, have been commercial CHN analysers for the final measurement [93–96]. The actual measurement of carbon dioxide with these instruments is made by gas chromatography; Perkin Elmer supply a CHN analyser.

Of course, since these analysers measure all of the carbon dioxide generated, any carbonate present in the sample will also be measured as organic carbon. Many investigators remove any carbonate carbon by fuming the filters with hydrogen chloride vapour, or by treating them with dilute hydrochloric acid. Differential combustion has also been suggested as a method for distinguishing between carbonate carbon and organic carbon; baking at 500°C for 4 h was used to remove organic carbon from carbonate-rich particulate matter [97]

An alternative method for the determination of particulate organic carbon in marine sediments is based on oxidation with potassium persulphate followed by measurement of carbon dioxide by a Carlo Erba non-dispersive infra-red analyser [98, 99]. This procedure has been applied to estuarine and high carbonate oceanic sediments and results compared with those obtained by a high-temperature combustion method.

Approximately 10 g sediment was dried at 110°C for 4 h then ground to a fine powder. The percentage moisture was determined by the measured weight loss before and after drying. The sediment sample was dried for 4 h before organic carbon analysis. In some cases sediment samples were acid-pretreated before analysis by the persulphate method. For these sediments 25 ml 4% hydrochloric acid (0.5 M) was added to the sample. After standing for at least 4 h, the mixture was filtered through a precombusted (450°C, 4 h) Whatman GF/C glass fibre filter, rinsed with distilled water and dried for 4 h.

A comparison of results of organic carbon analysis by persulphate oxidation and high-temperature combustion using the Carlo Erba elemental analyser are given in Table 9.12. There appears to be no systematic bias between the methods, and a t-test for a difference of means from zero showed no significant difference at the 95% confidence level.

The average relative standard deviation calculated for the 24 values obtained by the persulphate method presented in Table 9.12 is 11.4%. This is higher than the

Table 9.12 Comparison of potassium persulphate oxidation and high-temperature oxidation for measurements of organic carbon in sediments. (Results are given as mean organic carbon (mg/g dry weight) $\pm$ SD)

Sediment sample	$K_2S_2O_8$ oxidation	High-temperature oxidation	Number of replicates*
Providence River *A*	37.0 $\pm$ 2.7	30.0 $\pm$ 1.3	6
Providence River *B*	63.5 $\pm$ 10.9	65.6 $\pm$ 5.3	6
Providence River *C*	24.2 $\pm$ 2.1	16.4 $\pm$ 0.5	3
Narragansett Bay *B*	10.9 $\pm$ 1.8	13.4 $\pm$ 1.1	6
Open ocean *E*	2.51 $\pm$ 0.07	2.41 $\pm$ 0.05	3

*Number of replicates for each method.

Reproduced from Quinn, G.L. (1979) *Chemical Geology*, **25**, 155, by courtesy of author and publishers.

value of 5.1% calculated for the results from the high-temperature oxidation method. The overall precision of the persulphate method, including estuarine, coastal and oceanic sediments, is 6.4%.

More recent methods for the determination of particulate organic carbon have been described [100, 101].

Weliby *et al.* [101] described a procedure for the determination of both organic and inorganic carbon in a single sample of a marine deposit. Carbonate carbon is determined from the carbon dioxide evolved by treatment of the sample with phosphoric acid, the residue is then treated with a concentrated solution of dichromate and sulphuric acid to release carbon dioxide from the organic matter.

The carbon dioxide produced at the two stages of the analysis is estimated using a carbon analyser based on the thermal conductivity principle. In addition, total carbon content is determined on another subsample using the dry combustion furnace. This provides a check on the values determined by the phosphoric acid/dichromate technique.

The apparatus used is illustrated in Figure 9.17.

Sediment samples and seawater particulate matter from a variety of environments were analysed by two methods of carbon phase partitioning, the phosphoric acid/dichromate technique and the measurement of total carbon on the combustion furnace before and after removal of organic matter by burning at 500°C. In the latter case, C-CO$_3$ is determined in the sediment after burning. C-total is determined on an untreated subsample, and C-org is calculated by difference.

The comparison here serves to emphasize the problem with accurate carbon partitioning rather than that of carbon detection. Within a wide range of carbon

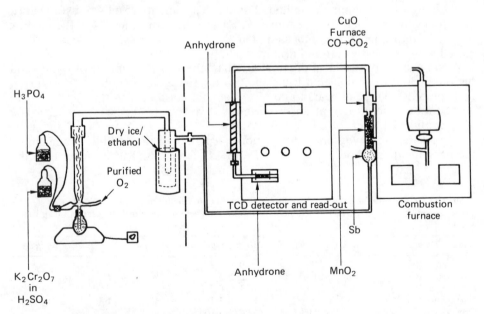

Figure 9.17 Schematic illustration of the modified LECO apparatus used in the H$_3$PO$_4$/dichromate method or the LECO dry combustion technique (Reproduced from Weliby, K., *et al.* (1983) *Limnology and Oceanography*, **28**, 1252, Springer Verlag, by courtesy of authors and publishers.)

Table 9.13 Comparison of carbon values determined by the H₃PO₄/dichromate method and furnace combustion (all data reported in wt%). C-sum and C-total values agree within the error of the two techniques; however, there are large differences between C-org and C-CO₃ determined by each method

Sample	H_3PO_4/dichromate			Combustion			%Mn in sediment
	%C-org	%C-CO₃	%C-sum	%C-total	%C-CO₃	%C-org	
EA 20–29 cm	0.30	8.32	8.62	8.50	7.82	0.68	–
110–120 cm	1.62	7.54	9.16	9.24	7.32	1.92	–
803–813 cm	0.34	7.53	7.89	7.89	7.30	0.59	–
CP 0–2 cm	0.24	8.71	8.95	8.88	7.73	1.15	–
2–4 cm	0.19	8.81	9.00	8.98	8.34	0.64	–
4–6 cm	0.19	9.09	9.28	9.18	8.44	0.74	–
R 0–4 cm	0.15	0.07	0.22	0.26	0.00	0.26	0.34
(0.2–0.5 μm)	0.27	0.04	0.31	0.31	–	–	0.53
NC 15–18 cm	0.70	0.05	0.75	0.81	–	–	1.96
(0.2–0.5 μm)	0.92	0.05	0.97	1.09	–	–	1.64
AA 121–126 cm	0.92	0.07	0.99	1.02	–	–	0.07
(0.2–0.5 μm)	1.04	0.04	1.08	1.09	–	–	0.08
OM 0–3 cm	2.36	0.15	2.51	2.51	–	–	0.04
(0.2–0.5 μm)	4.75	0.00	4.75	5.00	–	–	0.04
CR surface	0.61	0.00	0.61	0.63	–	–	0.06
H 0–2 cm	0.62	0.06	0.68	0.73	0.00	0.73	3.46
6–8 cm	0.70	0.20	0.90	0.86	0.00	0.86	4.44
12–14 cm	0.73	0.85	1.59	1.62	0.07	1.54	2.07
16–18 cm	0.70	1.36	2.07	2.13	0.13	2.00	1.77
20–22 cm	0.58	1.39	1.98	1.97	0.20	1.77	1.33
30–32 cm	0.46	0.83	1.29	1.29	0.02	1.26	1.12
ST(M-1)	7.54	4.20	11.74	11.77	1.61	10.16	0.50
ST(M-2)	7.21	4.41	11.62	11.81	2.14	9.67	0.50
ST(H-1)	5.65	7.32	12.97	12.91	5.74	7.17	0.20
ST(H-2)	5.55	7.25	12.80	12.84	5.87	6.97	0.17

Reproduced from Welisky, K., et al. (1983) *Limnology and Oceanography*, **28**, 1252, Springer Verlag, by courtesy of authors and publishers.

values and sediment types, excellent agreement is achieved between C-total, determined by combustion, and C-sum, determined by the phosphonic acid/dichromate technique (Table 9.13). Both estimates agree within 2% (or 0.02% carbon if the carbon value is ⌇⌇⌇ r 1%); this is within the error of each technique.

The importance of develc ⌇⌇g a reliable method for partitioning total carbon between the inorganic and ⌇ganic phases is exemplified in Figure 9.18. The two profiles shown are of the ⌇ ⌇ie sediment core, Vulcan I 62 BC. The values of CCO₃ and C-org in profile (⌇ were determined by the phosphoric acid/dichromate technique described by Welinky et al. [102]. The values in profile (b) were based on C-total and CCO₃ determinations by the combustion furnace method with selective removal of C-org by burning at 500°C. It is evident that the apparent C-org profile obtained from the combustion furnace values is merely a reflection of the calcium carbonate content of the sediment. Since the organic carbon values in Figure 9.18(b) were calculated by difference, the high C-org values apparently include carbonate carbon that was released to carbon dioxide during the attempt to remove organic carbon by burning. Therefore, correct carbon phase partitioning is crucial to the interpretation of C-org and CCO₃ trends in sediment and water column profiles.

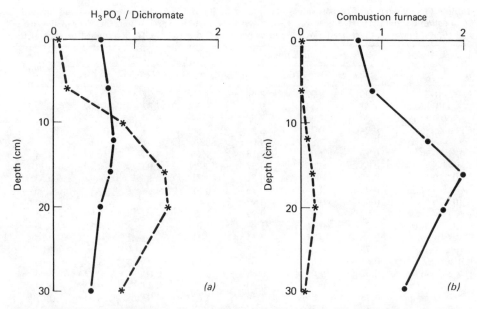

Figure 9.18 Differences in depth profiles obtained for % C-org (●) and % C-CO$_3$(×) by the H$_3$PO$_4$/ dichromate technique (*a*) and the combustion furnace method (*b*) where organic carbon is first selectively removed by burning at 500°C. Samples were taken from core Vulcan I 62 BC (Reproduced from Weliby, K., *et al.* (1983) *Limnology and Oceanography*, **28**, 1252, Springer Verlag, by courtesy of authors and publishers.)

(c) Volatile organic carbon

Many volatile organic compounds (hydrocarbons, alcohols, aldehydes, acids, esters, ketones, amines, etc.) have been identified in marine systems [102, 103]. These volatile materials may have an important role in the cycling of organic matter in natural waters. Volatile or low molecular organic materials may be produced *in situ* by biological [104–107] and chemical reactions [108–110] or can be introduced into the marine systems by human activity [111, 112], or through fluvial and atmospheric transport.

Direct methods of analysis such as distillation [104, 113, 114], liquid–liquid extraction [105, 115], headspace analysis [116–118], dynamic headspace analysis [103, 119–125] and direct injection [126] have been used mainly for specific volatile components (see Chapter 7).

As for all of the fractions of organic material in seawater, the volatile organic carbon fraction is defined by the method by which it is collected. In one of the earliest estimates, Skopintsev [127] defined the volatile fraction as the difference between total organic carbon values, as measured by evaporation and dry combustion, when the evaporations were carried out at room temperature and at 60°C. Thus Skopintsev's 'volatile fraction' consists of those compounds that are volatile from acidified solution taken to dryness at 60°C but not at 20°C. This fraction was found to be between 10 and 15% of the total organic carbon. He also noted a 15% difference in measured organic carbon with his dry combustion method when samples were dried at different temperatures and concluded that this difference was due to the loss of volatiles.

The volatile fraction as defined by the various wet oxidation methods and most of the direct injection methods would be that fraction removed by acidification and purging with inert gas at room temperature. In the freeze drying method of Gordon and Sutcliffe [58] the volatile fraction is that fraction lost by sublimation *in vacuo*. There have been no actual determinations of these losses, and for the most part Skopintsev's numbers were accepted as valid for all of these methods, largely because they are the only numbers available.

MacKinnon [59] and Wangersky [128] have made direct determinations of the volatile fractions from a variety of depths and stations in the North Atlantic. The volatile fractions as defined by MacKinnon's method is that fraction which can be removed from solution by purging with an inert gas at 80°C and a pH of 8 for 10–12 hours, then at 65°C for a further 10–12 hours.

The inert gas stream is flushed through an ice-packed condenser to remove water, then into a trap packed with Tenax GC followed by a U-shaped stainless steel cold trap held at −78°C. After 5 hours, the flow rate is reduced, the Tenax trap replaced and the cold trap taken out of line and analysed immediately. The extraction is continued for another 5–7 hours, then the Tenax trap is again replaced, the temperature reduced to 60°C and the extraction continued for another 10–12 hours. The volatile organic carbon is taken as the sum of the volatile organic carbon in the three Tenax traps and the cold trap.

In the actual analysis, material is desorbed from the Tenax columns into a cold trap (−78°C) by heating 175°C) in an atmosphere of nitrogen. When the desorption is complete, the cold trap is heated and the organic matter flushed through a furnace held at 950°C in an atmosphere of oxygen and the carbon dioxide resulting is measured with a non-dispersive infra-red gas analyser. With this system of collection and analysis, values ranged from 35 to 40 µg C per litre for surface waters and from 25 to 35 µg C per litre at depth. The volatile material collected in this manner may be used for qualitative and quantitative determinations. The fraction defined in this manner is remarkably constant, making up only 2–6% of the total organic carbon with the higher values occurring in the surface waters.

The relationship between Skopintsev's and MacKinnon's definitions of the volatile organic carbon are not easy to see. Skopintsev's values should certainly be the higher of the two, since his samples are taken to dryness. Any compound with a moderately high vapour pressure at the temperature of the evaporation will be removed. The volatile nitrogen compounds should be retained and the low molecular weight acids should be lost. MacKinnon's values may be either high or low, depending upon the definition chosen.

While several possible methods of fractionation and determination might be suggested, perhaps the most elegant is coupled gas chromatography–mass spectrometry which has shown the presence of very complex mixtures of volatile organics in seawater.

Gershay *et al.* [15] have pointed out that persulphate and photo-oxidation procedures will determine only that portion of the volatile organics not lost during the removal of inorganic carbonate [14, 37, 129, 130]. Loss of the volatile fraction may be reduced by use of a modified decarbonation procedure such as one based on diffusion [67]. Dry combustion techniques that use freeze-drying or evaporation, will result in the complete loss of the volatile fraction [29, 37, 127, 130].

The loss of volatile organics will only be a problem in areas where the volatile component is high. In open ocean areas, volatiles should be a small fraction (2–6%) of the total organic carbon [131]. Under strongly reducing conditions, many of the end-products of microbial metabolism are volatile in nature. Deuser [132] noted that his wet oxidation analysis of the dissolved organic carbon of water from the reducing part of the Black Sea gave systematically higher values than those of dry combustion. He concluded that it was due to substantial amounts of volatile material in the water produced as a consequence of anaerobic microbial metabolism.

In a more recent extension to the work discussed above, MacKinnon [133] has discussed in detail a method for the measurement of the volatile fraction of total organic carbon in seawater.

In this method volatile organic matter in seawater is concentrated on a Tenax GC solid adsorbent trap and dry-ice trap in series. The trapped organic material is then desorbed and oxidized to carbon dioxide which is measured with a non-dispersive infra-red analyser. A dynamic headspace method was used for the extraction with the assistance of nitrogen purging. Dynamic headspace analysis [134] is an efficient extraction procedure. The efficiency of extraction can be increased by using a purging gas and increasing the temperature of the sample to 60–80°C [119]. The efficiency and properties of Tenax GC have been examined [124, 136, 137].

Samples were taken with Niskin bottles fitted with stainless steel rather than rubber closing mechanisms. Since the samples were lowered open, there was a possibility of sample contamination while passing through the surface film. Samples were withdrawn immediately from the Niskin bottle with a glass delivery tube into amber glass bottles (650 ml). Care was taken to prevent the formation of bubbles during the transfer. After 0.5 ml mercuric chloride (3%) had been added to the sample, the bottle was sealed with an airtight metal cap. The sample was frozen until analysis.

A dynamic headspace method [119, 134, 135] was used for extraction of the volatile organic matter from seawater. The rate of extraction of the volatile organics was enhanced by purging the sample with an organic-free gas and by increasing the temperature of the water sample (Table 9.14). Although under these conditions, thermal decomposition of some of the more labile higher molecular weight compounds is possible, this did not appear to be significant. This was indicated by the decrease in the rate of extraction of volatile organic material during extended extractions (Figure 9.19). With elevated temperatures (60–80°C) for the extraction, a water condenser must be placed in line to prevent water vapour from clogging the traps.

After the sample had thawed, the sample bottle was decapped and fitted with a condenser. A sparger for the purging gas (20–40 ml min^{-1} purified nitrogen) was positioned in the sample (Figure 9.20). For seawater samples, a sample volume of 500–600 ml was used. A system blank for the extraction procedure and traps was measured to be about 4–5 µg C per litre. Extractions were carried out at a pH of about 8.1–8.3.

The organic materials stripped from water samples were concentrated with a combination of traps (Figure 9.20). The first trap was a 0.25–0.30 m stainless steel column (6 mm o.d.) packed with 2 ml Tenax GV (35/60 mesh) followed in line by a dry-ice (−78°C) trap (0.5 stainless steel U-shaped column of 6 mm o.d.) packed with quartz wool. The dry-ice trap is an efficient trap for volatile organics, but its

Table 9.14 Effect of temperature on the rate of extraction of the volatile organic matter from seawater

Sample	Temperature of sample (°C)	Time of extraction (h)	Organic matter measured (µg C)	Rate of extraction (µg C l⁻¹ h⁻¹)
A	30	11.5	6.09	0.53
		11.5	7.00	0.61
	60	11.5	10.10	0.88
		11.5	8.90	0.77
	90	11.5	14.90	1.30
		11.5	15.80	1.37
B	30	9.0	4.3	0.48
		9.0	4.7	0.52
	80	9.0	9.2	1.02
		9.0	11.2	1.24
C	25	9.0	7.39	0.82
	80	9.0	23.11	2.57
D	25	9.0	18.74	2.08
		9.0	20.91	2.32
	80	9.0	44.16	4.91
		9.0	51.20	5.69

Reproduced from MacKinnon, M.D. (1979) *Marine Chemistry*, **8**, 143, Elsevier Science Publishers, by courtesy of author and publishers.

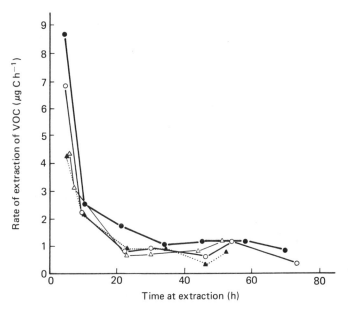

Figure 9.19 Effect of time on the rate of extraction (µg C per hour) of volatile organic carbon from seawater samples. ○ and ● = Seawater from Tower Tank facility at Dalhousie University. Biological bloom underway. △ and ▲ = seawater from tapped seawater system of Dalhousie University (Reproduced from MacKinnon, M.D. (1979) *Marine Chemistry*, **8**, 143, Elsevier Science Publishers, by courtesy of author and publishers.)

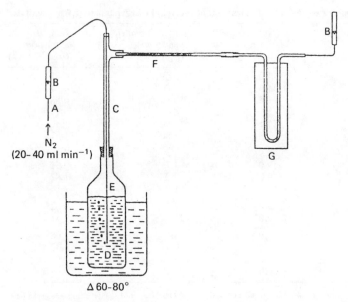

Figure 9.20 Apparatus for extraction of volatile organic matter from water samples. A = Nitrogen purging gas; B = flowmeters; C = air condenser; D = sample (500 ml), container in water bath; E = sparger; F = stainless steel trap (0.25 m × 4 mm i.d.) packed with 2 ml of Tenax G.C. (35/60); G = stainless steel U trap (0.5 m × 4 mm i.d.) in dry-ice ($-78°C$) (Reproduced from MacKinnon, M.D. (1979) *Marine Chemistry*, **8**, 143, Elsevier Science Publishers, by courtesy of author and publishers.)

efficiency is reduced because of problems of clogging by water vapour evolved during extraction. The Tenax trap is easy to handle and has a minimum adsorption of water and carbon dioxide.

After conditioning at 250–300°C under a flow of nitrogen for several hours, the cleaned Tenax trap was connected to the extraction system (Figure 9.20). The flow rate of the purging gas used for the extraction was limited since the Tenax GC column displays low adsorbent efficiencies at flow rates greater than $20–25\,ml\,min^{-1}$. During the initial extraction period (4–6 hours), the water sample was heated to 70–80°C and a flow rate of about $40\,ml\,min^{-1}$ of the stripping gas (nitrogen) was maintained. At this higher flow rate, a U-shaped stainless stell trap in dry-ice ($-78°C$) was connected in series so that organic components not retained by the Tenax trap were concentrated. After the initial extraction period, the traps were removed. The dry ice trap was analysed immediately while the Tenax trap was stored at 5°C until analysis (usually within 24 h). The Tenax trap was replaced and extraction was continued for a further 5–7 hours with a reduced nitrogen flow rate ($20\,ml\,min^{-1}$). Finally, a third Tenax trap was inserted and extraction continued under these conditions for a further 10–12 h. The volatile organic carbon concentration was calculated as the total organic material trapped by three Tenax traps and the dry-ice trap during this 24 h extraction.

The efficiency of extraction and analysis was determined for different classes of volatile organic materials (Table 9.15). Acetone was used as a test material to evaluate efficiency for the extraction, concentration and analysis procedures. While acetone has a high vapour pressure, it is difficult to extract efficiently from water

Table 9.15 Efficiency of recovery of volatile organic compounds added to pre-stripped seawater samples

Compound	MW	bp (°C)	Concentration (μg C l^{-1})	Mean ($\pm SD$) recovery (%)
Butanal	72.10	75.7	14–27	114 ± 10
2-Propanone	58.08	56.6	10–123	105 ± 2
2-Butanone	72.1	79.6	32–214	106 ± 4
Methanol	32.04	64.7	30–59	70 ± 10
2-Propanol	60.09	82.5	11–112	95 ± 20
Diethyl ether	74.12	34.6	23–116	112 ± 15
Ethyl acetate	88.12	77.1	30–60	100 ± 10
Acetonitrile	41.05	81.6	22–44	105 ± 5
Heptane	100.21	98.4	129–248	120 ± 10
Benzene	78.12	80.1	33–88	101 ± 3
Propanoic acid	74.08	140.99	121	27 ± 10
1,2-Ethylene diamine	60.11	116.5	90	44 ± 10

Reproduced from MacKinnon, M.D. (1979) *Marine Chemistry*, **8**, 143, Elsevier Science Publishers, by courtesy of author and publishers.

because of its high solubility. Almost complete recovery of acetone from seawater was obtained. Over the concentration range of 10–200 μg C per litre, high recoveries (100 ± 20%) of acetone added to pre-extracted seawater samples were calculated. For other compounds tested, high recoveries were noted for materials of low solubility and high vapour pressure (hydrocarbons, ethers, esters, aldehydes, ketones, nitriles) while poorer recoveries were observed for the more polar organic compounds (alcohols, amines, acids).

A schematic diagram of the system used for the measurement of extracted volatile organic matter is shown in Figure 9.21. The Tenax trap (C) was heated to 160–200°C in an oven (D). The adsorbed volatile organics were desorbed from the trap under a flow of organic-free nitrogen (40 min at 160–200°C with 30 ml min^{-1} purge nitrogen) into a dry-ice (−78°C) trap (F) where the organic components were trapped while inorganic carbon dioxide (usually about 10% of volatile organic carbon) which may have been entrained in the Tenax trap was not retained. After completion of the transfer, the dry-ice was removed and this trap was quickly heated with a Bunsen burner.

The organic components were flushed into the oxidation zone (900–950°C) of the combustion tube (1 m × 15 mm i.d. packed with CuO) and were oxidized to carbon dioxide. The combustion tube was Y-shaped (I); through one arm the volatile organics were carried with a stream of nitrogen while through the other a slow flow of organic free nitrogen (20–25 ml min^{-1}) was maintained. After the volatile components had been flushed into the combustion furnace, the flow rate of the oxygen was increased to about 350 ml min^{-1}. The carbon dioxide produced by the oxidation of the organic matter was dried of water vapour with a condenser (o) and drying traps (P) before being measured with the non-dispersive infra-red (Q). The time required for analysis of each trap was about 40 min for the initial desorption from the Tenax trap, about 6 min for flushing the dry-ice trap into the combustion furnace and 2 min for sweeping the oxidation products to the detector.

Over the range of values expected for natural samples (0–100 μg C per litre) the response of the analyser was linear. A calibration line was calculated by analysis of direct injections of an acetone standard solution into the dry-ice trap (F) and the treatment of this by the described procedure.

The dry-ice trap (Figure 9.20 (G) was analysed in a similar system except that the desorption step was shorter. The volatiles were flushed into the combustion furnace as soon as the signal from any inorganic carbon dioxide entrained in the dry-ice trap had passed through the infra-red detector.

The concentration of volatile organic matter was calculated by measuring the carbon dioxide produced by oxidation of the organic matter. Therefore, any interferences from inorganic carbon dioxide had to be controlled. The concentration of inorganic carbon dioxide measured in the Tenax traps was small (usually less than 10% of measured volatile organic carbon). During desorption of the traps (Figures 9.20 (F) and 9.20 (G)) into the dry-ice trap (Figure 9.21 (F)) inorganic carbon dioxide was not retained and was recorded by the infra-red detector before the volatile organic material was oxidized in the combustion furnace. Water that may have been trapped during the extraction steps did not affect the efficiency of oxidation. However, water vapour would interfere with the non-dispersive infra-red detector, so it was removed with a condenser and drying column (magnesium perchlorate) before entering the analyser.

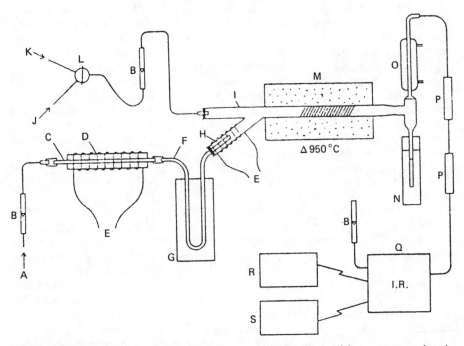

Figure 9.21 System for measurement of volatile organic matter extracted from water samples. A = Nitrogen (30 ml min^{-1}); B = flowmeters; C = Tenax trap or dry-ice trap; D = tube oven (175°C); E = variac; F = stainless steel U-tube; G = dry-ice; H = heated inlet arm of combustion tube; I = y-shaped quartz combustion tube (0.95 m × 13 mm i.d.) packed with CuO catalyst; J = slow flow of O$_2$ (20–25 ml/min^{-1}); K = fast flow of O$_2$ (250–300 ml min^{-1}); L = switching valve; M = furnace (950°C); N = water collector; O = condenser; P = drying tubes (Mg(ClO$_4$)$_2$); Q = non-dispersive infra-red analyser; R = recorder; S = integrator (Reproduced from MacKinnon, M.D. (1979) *Marine Chemistry*, **8**, 143, Elsevier Science Publishers, by courtesy of author and publishers.)

Table 9.16 Averaged volatile organic carbon (VOC) concentrations and VOC:total organic carbon (VOC/TOC) ratios from different areas

Area (time of sampling)	Depth	n	Average VOC concentrations (μg C l^{-1})	Range (μg C l^{-1})	Average VOC/TOC ratio (%)	Range (%)
Gulf of	0–10	9	31.8 ± 8.4	21.7–51.7	2.7 ± 0.9	1.9–4.8
St. Lawrence	10–50	6	36.0 ± 8.9	22.3–47.0	3.0 ± 0.6	2.0–3.7
(11/75)	50–100	5	29.8 ± 1.7	27.5–32.2	2.7 ± 0.4	2.2–3.1
	100–250	9	30.1 ± 6.5	22.8–38.8	3.1 ± 0.7	2.2–4.1
Scotian Shelf	0–10	26	40.9 ± 9.1	23.9–60.6	3.3 ± 0.9	1.8–5.9
& Slope	10–25	16	37.8 ± 8.1	24.0–53.0	3.4 ± 0.8	1.9–5.2
(5/74)	25–100	17	33.9 ± 9.9	24.0–68.5	3.5 ± 0.9	1.8–6.2
(8/75)	100–250	17	35.5 ± 3.7	28.3–41.2	4.1 ± 0.5	3.5–4.9
(3/76)	250–750	12	33.0 ± 6.4	22.4–46.4	4.5 ± 0.9	3.0–6.1
	750–1500	5	26.2 ± 2.4	23.2–29.7	3.7 ± 0.3	3.4–4.0
Central &	0–10	31	30.4 ± 9.7	17.6–52.4	3.1 ± 1.1	1.7–5.5
Northwestern	10–25	14	27.8 ± 6.2	18.8–36.7	2.8 ± 0.7	1.7–3.2
Atlantic	25–100	15	31.9 ± 9.8	22.6–52.2	3.4 ± 1.1	2.3–5.8
(10/74)	100–250	35	27.7 ± 6.5	17.0–52.3	3.3 ± 0.9	2.0–5.8
(2/75)	250–750	22	25.1 ± 4.4	19.6–34.0	3.5 ± 0.7	2.5–5.1
	750–1500	15	26.1 ± 4.6	20.2–34.7	3.6 ± 0.7	2.7–4.7
	1500–3000	15	24.4 ± 2.5	19.2–29.3	3.4 ± 0.4	2.5–4.3
	3000–5000	5	26.5 ± 1.7	25.0–29.1	3.8 ± 0.5	3.4–4.7
North West Arm Halifax Harbour (4/76)	0–10	37	32.6 ± 7.8	20.4–49.4	1.9 ± 0.4	1.2–2.9
St. Margaret's Bay, Nova Scotia (4/76)	0–40	28	30.9 ± 6.2	19.8–45.5	2.3 ± 0.5	1.4–3.4

Reproduced from MacKinnon, M.D. (1979) *Marine Chemistry*, **8**, 143, Elsevier Science Publishers, by courtesy of author and publishers.

The precision of this procedure on seawater samples was of the order of 5–10% coefficient of variation. Accuracy was difficult to assess.

Table 9.16 presents some typical results obtained by this procedure on seawater samples. MacKinnon [130] concluded that since the volatile organic carbon contents of normal (i.e. unpolluted) seawaters are small, the effect of complete or partial loss of volatile organic components during the determination of total organic carbon in most ocean areas (except highly reducing environments) with either the wet and direct injection methods or dry oxidation methods should be small (about 5%) and within the precision of these methods.

(d) Chemical oxygen demand

The chemical method for the determination of the chemical oxygen demand of non-saline waters uses oxidation of the organic matter with an excess standard acidic potassium dichromate in the presence of silver sulphate catalyst followed by estimation of unused dichromate by titration with ferrous ammonium sulphate. Unfortunately, in this method, the high concentrations of sodium chloride present in seawater react with potassium dichromate producing chlorine:

$$6Cl^- + Cr_2O_7^{2+} + 14H^+ - 2Cr^{3+} + 3Cl_2 + 7H_2O$$

Consequently the consumption of dichromate is many times higher than that due to organic material in the sample. To complicate matters any amines in the sample consume and release chlorine in a cyclic process leading to high chemical oxygen demands:

$$RNH \rightarrow NH_2 + CO_2 + H_2O$$
$$NH_4^+ + 3Cl_2 \rightarrow NCl_3 + 3Cl^- + 2H^-$$
$$2NCl_3 \rightarrow N_2 + 2Cl_2$$

Also, the addition of silver (I) sulphate causes precipitation of silver chloride which in the presence of organic compounds, is neither completely nor reproducibly oxidized.

This method, whilst being applicable to estuarine waters of relatively low chloride content, would present difficulties when applied to highly saline seawaters of low organic content. Zeitz [138] gives details of a method for the determination of chemical and oxygen demand, by potassium dichromate. Its advantages over previously published methods are that greater accuracy is achieved in recording even very small chemical oxygen demand measurements when there is a high chloride concentration, and the use of mercury compounds as buffering agents is obviated.

It is claimed that the process must be carried out in a closed vacuum flask. The concentration of chlorine in the gaseous phase is so slight that it can be disregarded when chemical oxygen demand values are being determined.

Wagner and Ruck [139, 140] propose various methods for overcoming interference by chloride in the determination of chemical oxygen demand. These include the quantitative oxidation of chloride ions to chlorine and a corresponding correction of the result of the chemical oxygen demand determination; use of mercuric sulphate as a sequestering agent for chloride ions, for which the limits of validity are discussed; elimination of chloride ions prior to the oxidation step by conversion to hydrogen chloride and its diffusion into a closed chamber to reduce losses of volatile constituents.

Because of the invalidity of the classical procedure, several workers have attempted to devise a method that is free from interference by chloride. Chloride interference can be eliminated by preventing the concurrent oxidation of organic material and chloride. This can be effected in two ways: either (1) by leaving the chloride in the test mixture but preventing its oxidation, or (2) by removing the chloride prior to the chemical oxygen demand test.

Both ways have been used in previously reported attempts to remove chloride interference. Three methods have been used in attempts to prevent chloride oxidation: masking with mercury (II) [141, 142]; precipitation of chloride using silver (I) [138]; or altering oxidation conditions [143].

Two methods have been used for removal of chloride; the chloride being removed as chlorine [144, 145] or as hydrogen chloride [146].

Baumann [147] collected the chlorine produced in the reaction in excess potassium iodide solution and back titrated against standard sodium thiosulphate to obtain the necessary correction for chloride.

Wagner and Ruck [140] carried out experiments to test the reliability of two standard procedures (DIN 38409 Parts 41/2 and 42) for the removal of chloride ions prior to chemical oxygen demand determinations on chloride containing samples. Acetic acid was used as a model substance on account of its volatility, and the use of special absorbers for hydrochloric acid vapour generated by the addition of sulphuric acid was tested under various conditions. In no case was there any appreciable loss of acetic acid but in some cases a yellow coloration was observed in the solution coupled with an increase in the measured chemical oxygen demand value. This was traced to the presence of nitrate ions which were reduced to nitrite ions by chloride in acid solution, the resulting nitrite ions being re-oxidized during the subsequent stages and hence contributing to an inflated value for the chemical oxygen demand. As a preventive measure in the case of samples containing nitrate ions the hydrogen chloride diffusion can be carried out using weaker sulphuric acid solutions and, if necessary, using a longer reaction time in the analysis.

Southway and Bark [148] reported their findings on their investigations into three processes for removing chloride from the test solution and suggested that the method (3) may be of some value:

1. Removal by precipitation as silver chloride with subsequent filtration, so that a negligible amount of organic matter is co-precipitated with the silver chloride.
2. Removal as chlorine in an oxidation stage, mild enough to prevent any significant oxidation of easily oxidized organic matter.
3. Removal by a ligand exchange process using the silver I or the mercury II form of polyvinyl pyridine [149].

Lloyd [150] has given details of a method for determining the chemical oxygen demand of saline samples which uses digestion of the sample at 150°C in a glass stoppered flask with silver nitrate to suppress interference by chloride. Within-batch relative standard deviation ranged from 2.2% at levels of $60 \, mg \, l^{-1}$ of chemical oxygen demand to 0.8% at $380 \, mg \, l^{-1}$ of chemical oxygen demand (4 degrees of freedom). Total relative standard deviation in analysis of a $300 \, mg \, l^{-1}$ chemical oxygen demand potassium hydrogen phthalate solution was 1.0% (8 degrees of freedom). These results did not differ significantly ($p = 0.5$) from data obtained using the standard procedure or the silver nitrate reflux chemical oxygen demand procedure.

Standard solutions of potassium hydrogen phthalate, spiked with chloride, were also analysed. These results are compared with those given for the standard procedure in Table 9.17.

Table 9.17 Analyses of chloride-spiked potassium hydrogen phthalate solutions by standard and sealed tube procedures (1 ml 25% m/v silver nitrate solution)

Expected COD (mg l⁻¹)	Observed COD at given chloride level (mg l⁻¹ Cl)				
	0	500	1000	1500	2000
0	–	15 (10)	19 (23)	21 (41)	32 (59)
100	102	112 (103)	115 (105)	121 (109)	128 (121)
200	199	203 (206)	209 (214)	221 (215)	217 (228)
300	295	300 (308)	297 (308)	311 (311)	316 (318)
400	391	401 (404)	403 (404)	405 (412)	407 (419)

*Results are the means of two determinations. Values in parentheses are means of results obtained using the standard sealed tube procedure.
Reproduced from Lloyd, A. (1982) *Analyst*, **107**, 1316, Royal Society of Chemistry, by courtesy of author and publishers.

Thompson *et al.* [151] have described a simple method for minimizing, but not eliminating, the effect of chloride on the determination of chemical oxygen demand without the use of mercury salts. This method minimized interference from the potassium dichromate oxidation of chloride to chlorine by adding a small amount of chromium (III) to the sample digest prior to heating. Chromium (III) was thought to complex any free chloride ion in the sample solution. The method was particularly useful at low chemical oxygen demand concentrations, chloride effects significantly decreasing with increases in chemical oxygen demand.

In the author's experience, no truly satisfactory method yet exists for the determination of chemical oxidation demand in highly saline samples such as seawater. With reservations, several of the above methods are, however, applicable to estuarine waters of low salinity.

(e) Biochemical oxygen demand (BOD)

It has been reported that whilst the dilution bottle method for biochemical oxygen demand yields satisfactory results, on fresh and low saline waters, a discrepancy exists when the test is performed on waters containing elevated levels of sodium chloride and other salts.

Kessick and Manchen [152], using a manometric BOD apparatus, found that the soluble waste portion in a salt water carrier was metabolized at the same rate by the microbial population in domestic sewage. However, the insoluble portion produced different rates, with that of domestic sewage in fresh water being three times greater than that of domestic sewage in a salt water carrier. Manchen [153] reported that only the degradation of suspended organic material was inhibited by salt and that the dissolved organic fraction of the BOD was unaffected. Davis, Petros and Powers [154] using hypersaline industrial wastes with salt tolerant bacteria species as seed, found that the BOD varied according to the salt concentrations. Increases in BOD resulted when the salt level was decreased. Ten-day BODs on salt water wastes were 3–6 times higher with standard dilution water than with a dilution water having the same salinity as the waste.

The rate of biological oxidation may vary considerably under different test conditions. Gotaas [155] reported that waste waters containing less than $10\,000\,mg\,l^{-1}$ chloride diluted with fresh water, had a rate of biological oxidation value greater than that of fresh water waste. The rate of biological oxidation increased to the point equivalent to 50% of that of seawater. When the 50% level was exceeded, the rate of biological oxidation decreased, until in 100% seawater, it was lower than in fresh water. This led to the hypothesis that a low salinity may stimulate microbial activity and result in increased BOD values. Seymour [156] also found that an increase in salinity resulted in a decrease of BOD values. Degradation rate decreased as the salinity level increased, yet only the particulate fraction seemed to be affected. Degradation of the soluble portion was not affected as adversely by salt content.

Lysis of bacteria will occur in a growth medium when sodium chloride is added. It will also occur when salt acclimated bacteria are placed in fresh water. If the change in salt concentration is less than $10\,000\,mg\,l^{-1}$ the degree of lysis is negligible, but if greater the lysis can be extensive. Cellular constituents released following lysis are metabolized by the remaining microbial population in preference to the substrate present, thus yielding erratic results [157, 158].

Davies *et al.* [159] set out to quantify the effects of fresh and saline dilution waters on the BOD of salt water wastes and to quantitatively establish the role of bacterial seed numbers and species associated with changes in BOD results. Sewage bacteria and salt-tolerant bacteria were evaluated to determine their capability to produce the same BOD values at various salt concentrations. When variations occurred in the BOD values Davies *et al.* [159] attempted to identify the reasons for such variations. They employed conventional and manometric BOD methods. Standard organic solutions and an industrial waste were tested with sewage seed and known species of salt-tolerant bacteria using standard and hypersaline dilution water at three salt concentrations. Significant BOD differences were found when saline wastes were diluted with standard (non-saline) BOD dilution water. Bacterial populations to genera were monitored and it was shown that equivalent numbers of bacteria did not have the same capability to degrade a given amount of waste with increases in salt concentrations to the 3% level. Seeding of hypersaline waste waters with known salt-tolerant species is recommended for consistent BOD results.

The BOD values obtained in sewage seed organic standards showed a significant trend, as salt concentration increased, BOD values decreased (Figures 9.22 and 9.23). Changes were significant at the 0.05 level. Dilution of sewage seeded saline organic standards with standard dilution water resulted in BOD values higher than those of corresponding non-saline organic standards due to increases in bacteria populations and increased organic removal (BOD values) in the presence of low levels of salt. Bacteria populations of sewage seed correlated with corresponding BOD values and concentrations of salt in organic standard solutions showed that the addition of 1% and 3% salt resulted in decreased initial bacteria population,

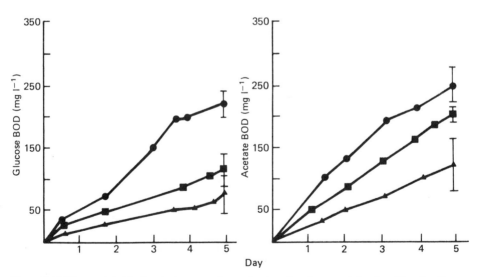

Figure 9.22 Manometric biochemical oxygen demand values for glucose–glutamic acid and sodium acetate standards with sewage seed at three salt concentrations. ● = 0%; ■ = 1%; ▲ = 3% (Reproduced from Davies, E.M., *et al.* (1978) *Water Research*, **12**, 917, Pergamon Publications, by courtesy of authors and publishers.)

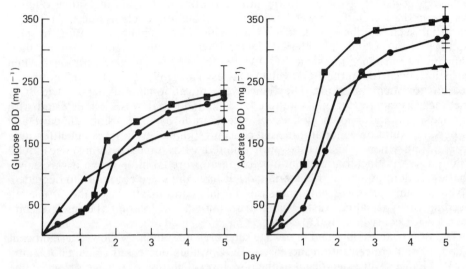

Figure 9.23 Manometric biochemical oxygen demand values for glucose–flutamic acid and sodium acetate standards with *Paracoccus halodenitrificans* as seed at three salt concentrations. ● = 0%; ■ = 1%; ▲ = 3% (Reproduced from Davies, E.M., *et al.* (1978) *Water Research*, **12**, 917, Pergamon Publications, by courtesy of authors and publishers.)

decreased growth rates and a decreased ability of an equivalent number of bacteria to degrade an equivalent amount of organic material.

Each salt-tolerant bacteria species yielded comparable BOD values for each organic standard conducted at three salt concentrations. Population increased and the rates of biological oxidation were similar at the three salt concentrations and when using either the standard dilution water or the saline dilution water.

Since sewage seed does not compare favourably with salt-tolerant bacteria in determining the BOD of saline waste waters, a salt-tolerant bacteria seed should be used to ensure an accurate and reproducible BOD value when hypersaline waste waters are being tested.

9.2 Organic nitrogen

The quantification of nitrogen in its soluble organic forms has interested marine chemists engaged in hydrographic, primary production and pollution studies [170–172]. The concentrations of particular dissolved organic nitrogen species in seawater may be low and while methods have been developed for some specific compounds (notably the amino acids) the bulk of the dissolved organic nitrogen remains uncharacterized. Whilst a large portion of the dissolved organic nitrogen in seawater is probably derived from soluble exudates from algae, it will also contain a wide range of other compounds including animal proteins, urea and other excretary products, nucleic acids, etc.

Dissolved organic nitrogen is commonly determined in four stages:

1. Removal of particulate organic nitrogen by filtering through a 0.45 μm pore membrane or equivalent glass-fibre filter.
2. Oxidation of soluble organic nitrogen to inorganic forms of nitrogen.
3. Determination of total inorganic nitrogen.
4. Correction for inorganic nitrogen species present before oxidation.

Until fairly recently, the only method available for the measurement of organic nitrogen in seawater was some version of the micro Kjeldahl method. This involves reducing all of the organic nitrogen to ammonia, then distilling the ammonia into an absorbing solution and determining it colorimetrically [162, 163, 173].

A comparative study has been carried out of the variations in final measurements of ammonia, by Astrani [160] and a Russian version which could also be used for the measurement of organic nitrogen in particulate matter [161].

Martens, Van Den Winkel and Massart [162] and Stevens [163] designed semi-automated versions of the micro Kjeldahl which avoided the distillation step altogether. In their versions, after the digestion step the digestion solution was diluted and the ammonia determined with an ammonia probe. The limitation on the sensitivity, then, is the sensitivity of the ammonia probe. This limits the method to the more productive oceanic waters.

Another approach to the organic nitrogen problem is to use persulphate wet oxidation to convert the nitrogen to nitrate or nitrite, in place of the reduction to ammonia [164, 165, 175, 176]. Results are fully comparable with those from the

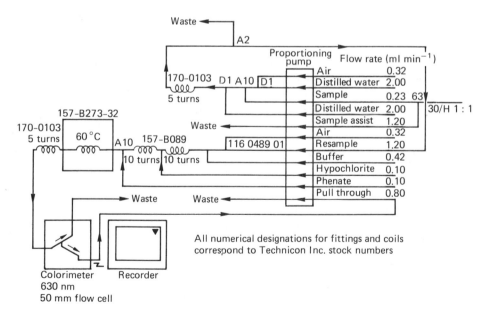

Figure 9.24 AutoAnalyzer II analysis scheme for the determination of Kjeldahl nitrogen in seawater via indophenol colorimetry (Reproduced from Adamski, J.M. (1976) *Analytical Chemistry*, **48**, 1194, American Chemical Society, by courtesy of author and publishers.)

Table 9.18 Effects due to extended digestion time on ammonium chloride standards prepared in distilled water

Sample	N concentration (mg l^{-1})		
	10 min digestion	20 min digestion	50 min digestion
1	1.51	1.44	1.06
2	1.48	1.58	1.75
3	1.50	1.47	1.46
4	1.51	1.47	1.91
5	1.48	1.55	0.90
6	–	–	1.61
SD (mg l^{-1})	±0.016	±0.06	±0.37
Relative SD (%)	±1.1	±4.0	±29

Reproduced from Adamski, J.M. (1976) *Analytical Chemistry*, **48**, 1194, American Chemical Society, by courtesy of author and publishers.

micro Kjeldahl but the technique is far simpler. The precision should also be higher, since the final step in the measurement, the colorimetric determination of nitrite, is much more precise than any of the ammonia methods.

The Adamski [164] procedure is semi-automated. It gives an accuracy of ± 8.1% and a precision of ± 8.2% for seawater samples spiked with 3.35 μg l^{-1} organic nitrogen. This procedure is based on the indophenal blue method and was employed using a Technicon-Auto Analyzer II system with the appropriate accessories, as shown in Figure 9.24.

The data in Table 9.18 for a series of replicate distilled water standards digested over various time intervals indicate the undesirable effects of prolonged digestion.

Various workers have described automated procedures for the determination of low levels of organic nitrogen in seawater [166, 167].

Ultraviolet photo-oxidation techniques can be used as a method for organic nitrogen. The organic nitrogen compounds are oxidized to nitrate and nitrite then determined by the standard seawater nitrate [168, 169, 174].

The oxidation of urea was not complete. The incomplete step in the photolysis of urea is much more likely to be the hydrolysis of urea to produce ammonia, since the oxidation of ammonia is known to occur under the conditions of the experiment.

If the freeze drying method for total organic carbon devised by Gordon and Sutcliffe [58] is used, with the final determination of carbon being run on a commercial CHN analyser, the analysis of total nitrogen is also obtained. If analyses for the inorganic nitrogen compounds have also been run on the sample, organic nitrogen can be calculated by difference.

Of the methods in common use, the two wet oxidation methods offer the best possibilities for further development. The photo-oxidation method in particular is well suited to automatic analysis. In the version of the total organic carbon method published by Collins and Williams [14] the effluent from the quartz photolysis coil could as easily be diverted to the nitrate analysis unit; if inorganic nitrogen were also measured, organic nitrogen could become a routine automatic method.

Shepherd and Davies [177] have described a semi-automated version of the alkaline peroxodisulphate procedure for the determination of total dissolved organic nitrogen in seawater. These workers carried out experiments on a range

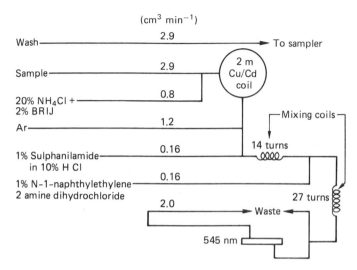

(cm³ min⁻¹)

Wash ——————— 2.9 —————————→ To sampler

Sample ——————— 2.9 ——————— [2 m Cu/Cd coil]

20% NH₄Cl + —————— 0.8 ———————
2% BRIJ

Ar ——————— 1.2 ———————

                                                ⌐Mixing coils⌐

1% Sulphanilamide —————— 0.16 ——————   14 turns
  in 10% H Cl

1% N-1-naphthylethylene —————— 0.16 ——————
2 amine dihydrochloride
                    2.0   27 turns
                            ——→ Waste ←——

                    545 nm

Figure 9.25 Manifold for the determination of nitrate in oxidized and natural seawater samples. (Reproduced from Shepherd, R.J. and Davies, I.M. (1981) *Analytica Chimica Acta*, **130**, 55, Elsevier Science Publishers, by courtesy of authors and publishers.)

Table 9.19 Recovery of inorganic nitrogen compounds added at the 10 µg N per dm³ level to natural seawater

	TDN (µg N per dm³)		*Recovery of added N (%)*	
Seawater	5.0	5.4	–	–
	5.0	5.4	–	–
Potassium nitrate	15.5	15.2	103	100
Potassium nitrite	16.2	16.2	110	110
Ammonium chloride	14.9	14.9	97	97
Urea	15.2	15.3	100	101
Glycine	14.3	14.0	91	88
2,2′-Bipyridine	16.2	16.5	110	113
Na₂-EDTA	15.5	15.0	103	98
Overall mean TDN	= 15.35, SD = 0.736, cv = 4.8%			
(excluding seawater)				
Overall mean recovery	= 101.5%, SD = 7.36%, cv = 7.2%			

TDN = Total dissolved nitrogen; CV = coefficient of variation.
Reproduced from Nydahl, F. (1978) Water Research, 12, 1123, Pergamon Publications, by courtesy of author and publishers.

of contaminated and clean seawaters using a version [178] of Koroleff's [175] alkaline peroxodisulphate oxidation technique.

Shepherd and Davies [177] made some modifications to the Nydahl [179] method to make it applicable to the automated routine analysis (Figure 9.25) of small samples of natural and polluted seawater and to samples of contaminated saline effluents.

Nydahl [179] published a list of 39 compounds which gave oxidation yields over 87% in distilled water compared with the Kjeldahl nitrogen determination. He also included eight compounds which gave poor (<83%) recoveries and concluded that yields from peroxodisulphate oxidation are low for compounds containing nitrogen–nitrogen bonds and HN=C groups. The determination of total dissolved nitrogen at the $10 \mu g$ N per dm^3 level for the compounds listed in Table 9.19 added as spikes to natural seawater gave recoveries of 88–113% at a total dissolved nitrogen concentration of $15.3 \mu g$ N per dm^3. The detection limit was $0.18 \mu g$ N per dm^3, i.e. 23 ng N per sample.

9.3 Organic phosphorus

The earlier methods for the measurement of organic phosphorus generally used fairly drastic measures to decompose the organic compounds present and free the phosphates. For example, Duurma [83] used the technique originally described by Harvey [180] which included autoclaving the seawater with sulphuric acid at 140°C for 6 hours, at a pressure of 3 atmospheres. These methods required great care and a good deal of time; as a result, organic phosphorus was never a routine determination.

Photo-oxidation was seen as a possible route to a total phosphorus method. Again, early work on the method was done by Armstrong, Williams and Strickland [166] and Armstrong and Tibbitts [45]. Grasshoff [181] adapted the method to continuous automatic analysis; a variation on this method is considered the 'standard' method for automatic analysis today [169]. Bikbulatov [182], on the other hand, feels that such important phosphorus compounds as ATP and DPN are not completely decomposed by ultraviolet irradiation and that persulphate oxidation gives better results.

Both the persulphate and ultraviolet oxidation methods have much to recommend them. Ultraviolet photo-oxidation methods have an advantage in that they are easily automated.

The simultaneous determination of dissolved organic carbon and phosphorus is feasible [51]. Phosphorglyceric acid (1 g equivalent P) was added to membrane filtered, pre-irradiated, seawater and the phosphate was measured before and after ultraviolet irradiation using an autoanalyser. The recovery of organic phosphorus was 100%.

Cembella, Antia and Taylor [188] have described a method for the determination of total phosphorus in seawater. The procedure used magnesium nitrate to oxidize organic compounds before standard molybdate colorimetric determination of ortho-phosphate. The method was applied to several pure organic phosphorus compounds and gave 93–100% recovery of phosphorus.

9.4 Total iodine

Schnepfe [183] has described a method for the determination of total iodine and iodate in seawater. One per cent aqueous sulphamic acid (1 ml) is added to seawater (10 ml) or to a solution of an evaporite sample (0.5 g in 10 ml water), then filtered, if necessary, and the pH adjusted to 2. After 15 minutes, 1 ml 0.1 M sodium hydroxide and 0.5 ml 0.1 M potassium permanganate are added and the steam bath heated for 1 h. The cooled solution is filtered, the residue washed, the filtrate plus washings is diluted to 16 ml and 1 ml of a 0.25 M phosphate solution (containing

0.3 µg iodine as IO_3^- per ml) added at 0°C then 0.7 ml 0.1 M Fe^{II} in 0.2% (v/v) sulphuric acid, 5 ml 10% aqueous sulphuric acid–phosphric acid (1:1) at 0°C and 2 ml starch–cadmium iodide reagent are added. The solution is diluted to 25 ml and after 10–15 min the extinction of the starch–iodine complex is measured in a 5 cm cell. For iodate the procedure is as described above except that the oxidation stage with sodium hydroxide–potassium permanganate is omitted and only 0.2 ml Fe^{II} solution is used; iodate standards were used in both procedures. The total iodine procedure is relatively free from interference by foreign ions but the iodate procedure is affected by bromate and sulphite. Down to 0.1 µg iodine can be determined in the presence of 500 mg chloride and 5 mg bromide by this procedure.

The classical method for the determination of iodide in seawater was described by Sugamara [184]. Various workers [185, 186] have improved the original method. Matthews and Riley [185] modified the method by concentrating iodide by means of co-precipitation with chloride using silver nitrate (0.23 g per 500 ml seawater). Treatment of the precipitate with aqueous bromine and ultrasonic agitation promote recovery of iodide as iodate which caused to react with excess of iodide ions under acid conditions, yielding I_3^-, which are determined either spectrophotometrically or by photometric titration with sodium thiosulphate. Photometric titration gave a recovery of 99.0 ± 0.4% and a coefficient of variation of ± 0.4% compared with 98.5 ± 0.6% and ± 0.8% respectively, for the spectrophotometric procedure.

Tsunogai [186] carried out a similar co-precipitation allowing a 20 hour standing period to ensure that iodide is fully recovered in the silver chloride co-precipitate. Again, the iodide is oxidized to iodate prior to spectrophotometric determination of the latter. This procedure also includes a step designed to prevent interference by bromine compounds.

9.5 Total sulphur

Taylor and Zeitlin [187] described an X-ray fluorescence procedure for the determination of total sulphur in seawater. They studied the matrix effects of sodium chloride, sodium tetraborate and lithium chloride and show that the X-ray fluorescence of sulphur in seawater experiences an enhancement by chloride and a suppression by sodium that fortuitously almost cancel out. The use of soft scattered radiation as an internal standard is ineffective in compensating for matrix effects but does diminish the effects of instrument variations and sample inhomogeneity.

References

1. Wangersky, P.J. Measurement of organic carbon in seawater. In *Analytical Methods in Oceanography* (ed. T.R.P. Gibb), American Chemical Society, Washington DC, pp. 148–162 (1975)
2. Menzel, D.W. and Vaccaro, R.F. *Limnology and Oceanography*, **9**, 138 (1964)
3. Strickland, J.D.H. and Parsons, T.R. *Bulletin Fisheries Research Board of Canada*, 167–311 (1968)
4. Sharp, J.H. *Marine Chemistry*, **1**, 211 (1973)
5. Skopintsev, B.A. *Marine Hydrophysic Institute*, **19**, 1 (1960)
6. Takahashi, Y. Ultra low level TOC analysis of potable waters. Presented at the *Water Quality Technology Conference, American Water Works Association, San Diego, December 5–8* (1976)

7. Van Hall, C.E., Safranko, J. and Stenger, H. *Analytical Chemistry*, **35**, 315 (1963)
8. Axt, G. *Vom Wasser*, **36**, 328 (1969)
9. Solomen, K. *Limnology and Oceanography*, **24**, 177 (1979)
10. Glor, R. and Leidner, H. *Analytical Chemistry*, **51**, 645 (1979)
11. Beattie, J., Bricker, C. and Garvin, D. *Analytical Chemistry*, **33**, 1890 (1961)
12. Woelfel, P. and Sontheimer, H. *Vom Wasser*, **43**, 315 (1974)
13. Ehrhardt, M. *Deep Sea Research*, **16**, 393 (1969)
14. Collins, K.J. and Williams, P.J. LeB. *Marine Chemistry*, **5**, 123 (1977)
15. Gershay, R.M., MacKinnon, M.D., Williams, P.J. LeB. and Moore, R.H. *Marine Chemistry*, **7**, 289 (1979)
16. Wangersky, P.J. *Trends in Analytical Chemistry*, **1**, 28 (1981)
17. Armstrong, F.A.J. and Boalch, G.T. *Journal of the Marine Biological Association*, **41**, 591 (1961)
18. Bauer, D., Brun-Cottan, J.C. and Saliot, A. *Canadian Oceanography*, **23**, 841 (1971)
19. Hanya, T. and Ogura, N. Application of ultraviolet spectroscopy to the examination of dissolved organic substances in water. In *Advances in Organic Geochemistry*, (eds U. Columbo and G.D. Hobson), Pergamon Press, Oxford, pp. 1–10 (1964)
20. Ogura, N. and Hanya, T. *Nature (London)*, **212**, 758 (1966)
21. Ogura, N. and Hanya, T. *Limnology and Oceanography*, **1**, 91 (1967)
22. Lewis, W.M. Jr. and Tyburczy, J.A. *Archives of Hydrobiology*, **74**, 8 (1974)
23. Mattson, J.S., Smith, C.A., Jones, T.T. and Gerchakov, S.M. *Limnology and Oceanography*, **19**, 530 (1974)
24. Wheeler, J.R. *Limnology and Oceanography*, **22**, 573 (1977)
25. Kulkarni, R.N. *Water, Air and Soil Pollution*, **5**, 231 (1975)
26. Biggs, R., Schofield, J.W. and Gorton, P.A.J. *Water Pollution Control Federation*, **47**, 2 (1976)
27. Maciolek, J.A. *USF and WS Research Report*, **60**, 1 (1962)
28. Camps, J.M. and Arias, E. *Invest. Pesq.*, **23**, 125 (1963)
29. Oppenheimer, C.H., Corcoran, E.F. and Van Arman, J. *Limnology and Oceanography*, **8**, 487 (1963)
30. Krey, J. and Szekielda, K.H. *Fresenius Zeitschrift fur Analytische Chemie*, **207**, 338 (1965)
31. Duursma, E.K. *Netherlands Journal of Marine Research*, **1**, 1 (1960)
32. Williams, P.J.LeB. *Limnology and Oceanography*, **14**, 292 (1969)
33. Bikbulatov, E.S. *Gidrokhim. Mater.*, **60**, 174 (1974)
34. Boehme, H. *Acta Hydrochimica et Hydrobiologica*, **3**, 327 (1975)
35. Baldwin, J.M. and McAtee, R.E. *Microchemical Journal*, **19**, 179 (1974)
36. Goulden, P.D. and Brooksbank, P. *Analytical Chemistry*, **47**, 1943 (1975)
37. Wilson, R.F. *Limnology and Oceanography*, **6**, 259 (1961)
38. Goulden, P.D. and Brooksbank, P. *Analytical Chemistry*, **47**, 1943 (1975)
39. Martinie, G.C. and Schilt, A.A. *Analytical Chemistry*, **48**, 70 (1976)
40. Armstrong, F.A.J., Williams, P.M. and Strickland, J.D.H. *Nature (London)*, **211**, 481 (1966)
41. Williams, P.H. *Limnology and Oceanography*, **14**, 297 (1969)
42. Hannaker, P. and Buchanan, S.S. *Analytical Chemistry*, **55**, 1922 (1983)
43. Gordon, M.A. *Journal of Physical Chemistry*, **18**, 55 (1914)
44. Armstrong, F.A.J., Williams, P.M. and Strickland, J.D.H. *Nature (London)*, **211**, 481 (1966)
45. Armstrong, F.A.J. and Tibbitts, S. *Journal of the Marine Biological Association*, **48**, 143 (1968)
46. Williams, P.M. *Limnology and Oceanography*, **14**, 297 (1969)
47. Williams, P.H. *Analytical Abstracts*, **17**, 3613 (1969)
48. Woelfel, P., Schuster, W. and Sontheimer, H. *Zeitung Wasser Abswasser Forsch*, **8**, 143 (1975)
49. Schreurs, W. *Hydrobiological Bulletin*, **12**, 137 (1978)
50. House, D.A. *Chemical Review*, **62**, 188 (1962)
51. Mueller, H. and Bondaranayake, W.M. *Marine Chemistry*, **12**, 59 (1983)
52. Graentzel, A. *Duennfilm, UV-DOC Me Beinrichtung Manual*, Physikalische Werkstaette, Karlsruhr, W., Germany (1980)
53. Gottschalk, G. *Freseninz Z. Analytische Chemie*, 275 (1975)
54. Mueller, H.R. *Evaluation of Organic Carbon Analysers*, EAWAG-Report Federal Institute of Water Resources and Water Pollution, Duehendorf, Switzerland (1978)
55. Moore, R.M. *Trace Metals Dissolved Organic Matter and their Association in Natural Waters*, PhD Thesis, University of Southampton (1977)
56. Montgomery, H.A.C. and Thom, N.S. *Analyst (London)*, **87**, 687 (1962)
57. Skopintsev, B.A. and Timofeyeva, S.N. *Marine Hydrophysics Institute*, **25**, 110 (1962)
58. Gordon, D.C. Jr. and Sutcliffe, W.H. Jr. *Marine Chemistry*, **1**, 231 (1973)

59. MacKinnon, M.D. *The Analysis of Total Organic Carbon in Seawater. (a) Development of Methods for the Quantification of TOC. (b) Measurement and Examination of the Volatile Fraction of the TOC Thesis*, Dalhousie University (1977)

60. Sharp, J.H. *Marine Chemistry*, **1**, 211 (1973)

61. MacKinnon, M.D. *Marine Chemistry*, **7**, 17 (1978)

62. Skopintsev, B.A. *Oceanography*, **6**, 361 (1966)

63. Williams, P.J. LeB. Determination of organic components. In *Chemical Oceanography* (eds J.P. Riley and G. Skirrow), Vol. 3, Academic Press, New York, pp. 443–477 (1975)

64. Skopintsev, B.A. *Oceanology*, **16**, 630 (1976)

65. Wangersky, P.J. and Zika, R.G. *The Analysis of Organic Compounds in Seawater*. Report 3, NRCC 16566, Marine Analytical Chemistry Standards Program (1978)

66. Van Hall, C.E. and Stenger, V.A. *Water Sewage*, **111**, 266 (1964)

67. Van Hall, C.E., Barth, D. and Stenger, V.A. *Analytical Chemistry*, **37**, 769 (1965)

68. Van Hall, C.E. and Stenger, V.A. *Analytical Chemistry*, **39**, 503 (1967)

69. Van Hall, C., Safranko, J. and Stenger, V. *Analytical Chemistry*, **35**, 319 (1963)

70. Golterman, H.L. *Methods for Chemical Analysis of Freshwater*, Blackwell, Oxford, pp. 133–145 (1969)

71. Water Pollution Research Laboratory. *Notes on Water Pollution*, No. 59, HMSO, London (1972)

72. Salvatella Battlori, N., Ribas Soler, F. and Oromi Durich, J. *Doc. Invest. Hidrol*, **17**, 303 (1974)

73. Stenger, V. and Van Hall, C.E. *Analytical Chemistry*, **39**, 206 (1969)

74. Eggertsen, F.T. and Stross, F.H. *Analytical Chemistry*, **44**, 709 (1972)

75. Bartz, A.M. *Transactions*, **7**, 279 (1968)

76. Hiser, L.L., Tarazi, D.S., Boldt, C.A. and Saenz, O. *American Laboratory*, **2**, 44 (1970)

77. Ton, N. and Takahashi, Y. *International Laboratory*, **September**, 49 (1985)

78. Nelson, K.H. and Lysyj, I. *Environmental Science and Technology*, **2**, 61 (1968)

79. Dobbs, R.A., Wise, R.H. and Dean, R.B. *Analytical Chemistry*, **39**, 1255 (1967)

80. Cropper, Heinekey, D.M. and Westwell, A. *Analyst (London)*, **92**, 436 and 443 (1967)

81. Croll, B.T. *Chemistry and Industry (London)*, 386 (1972)

82. Duursma, E.K. *Netherland Journal of Marine Research*, **1**, 148 (1960)

83. Merks, A.G.A. and Vlasbom, A.G. *Hydrobiological Bulletin*, **17**, 53 (1983)

84. Olrichs, S.H.S. *Projektur*, **71**, 211 (1982)

85. Elgershuizen, J.H.B.W. *Meitingen von DOC gehalten*, Notitie, DDMI 81–278 (1981)

86. Games, L.M. and Hayes, J.M. *Analytical Chemistry*, **48**, 130 (1976)

87. Mitchell, D.G., Aldous, K.M. and Canelli, E. *Analytical Chemistry*, **49**, 1235 (1977)

88. Fox, D.L., Isaacs, J.D. and Corcoran, E.F. *Journal of Marine Research*, **11**, 29 (1952)

89. Menzel, D.W. and Vaccaro, R.F. *Limnology and Oceanography*, **9**, 138 (1964)

90. Kuck, J.A., Berry, J.W., Andreatch, A.J. and Lentz, P.A. *Analytical Chemistry*, **34**, 403 (1962)

91. Ernst, W. *Inst. Meeresforsch. Bremerh.*, **15**, 269 (1975)

92. Szekielda, K.H. and Krey, J. *Mikrochimica Acta*, 149 (1965)

93. Hagell, G.T. and Pocklington, R. *A Seagoing System for the Measurement of Particulate Carbon, Hydrogen and Nitrogen* Report Ser./BI-R-73-14, 1–18 (1973)

94. Telek, G. and Marshall, N. *Marine Biology*, **24**, 219 (1974)

95. Ehrhardt, M. Determination of particulate organic carbon and nitrogen. In *Methods of Seawater Analysis* (ed. K. Grasshoff), Verlag Chemie, New York, pp. 215–220 (1976)

96. Ehrhardt, M. The automatic determination of dissolved organic carbon. In *Methods of Seawater Analysis* (ed. K. Grasshoff) Verlag Chemie, New York, pp. 289–297 (1976)

97. Hirota, J. and Szyper, J.P. Separation of total particulate carbon into inorganic and organic components. *Limnology and Oceanography*, **20**, 896 (1975)

98. Mills, J.G. and Quinn, G.L. *Chemical Geology*, **25**, 155 (1979)

99. Kerr, R.A. *The Isolation and Partial Characterization of Dissolved Organic Matter in Seawater*, PhD Thesis, University of Rhode Island, Kingston (1977)

100. Suzuki, J., Yokohama, Y., Unno, Y. and Suzuki, S. *Water Research*, **17**, 431 (1983)

101. Weliby, K., Suess, E., Ungerer, C.A., Muller, P.J. and Fischer, K. *Limnology and Oceanography*, **28**, 1252 (1983)

102. Giger, W. *Marine Chemistry*, **5**, 429 (1977)

103. Schwarzenbach, R.P., Bromund, R.H., Gschwend, P.M. and Zafiriou, O.C. *Organic Geochemistry*, **1**, 93 (1978)

104. Armstrong, F.A. and Boalch, G.T. *Nature (London)*, **185**, 761 (1960)

105. Koyama, T. *Journal of the Oceanographic Society of Japan*, **20**, 563 (1962)

106. Lovelock, J.E., Maggs, R.J. and Rasmussen, R.A. *Nature (London)*, **237**, 452 (1972)

412   Elemental analysis

107. Wangersky, P.J. *Deep Sea Research*, **23**, 457 (1976)
108. Wilson, D.F., Swinnerton, J.W. and Lamontagne, R.A. *Science*, **168**, 1577 (1970)
109. Dowty, B.L., Green, L.E. and Lasteter, J.L. *Analytical Chemistry*, **46**, 946 (1976)
110. Zika, R.G. *An Investigation in Marine Photochemistry*, PhD Thesis, Dalhousie University (1977)
111. Goldberg, E.D. Atmosphere transport. In *Impingement of Man on the Oceans* (ed. D.W. Hoo), Wiley-Interscience, New York, pp. 75–88 (1971)
112. Swinnerton, J.W. and Lamontagne, R.A. *Environmental Science and Technology*, **8**, 657 (1971)
113. Lamar, W.L. and Goerlitz, D.F. *Organic Acids in Naturally Colored Surface Waters*. Geological Survey Water Supply Paper 1817-A (1966)
114. Ryabov, A.K., Nabivanets, B.I. and Litvinenko, S.Z. *Hydrobiological Journal*, **8**, 97 (1972)
115. Kamata, E. *Bulletin of the Chemical Society of Japan*, **39**, 1227 (1966)
116. Bassette, R. and Ward, G. *Michrochemistry Journal*, **14**, 471 (1969)
117. Corwin, J.F. Volatile organic matter in seawater. In *Organic Matter in Natural Waters* (ed. D.W. Hood), Vol. 1, University of Alaska Press, pp. 169–180 (1970)
118. Hurst, R.E. *Analyst (London)*, **99**, 302 (1974)
119. Mieure, J.P. and Dietrich, M.W. *Journal of Chromatographic Science*, **11**, 559 (1973)
120. Novak, J., Zluticky, J., Kubelka, V. and Mostecky, J. *Journal of Chromatography*, **76**, 45 (1973)
121. Grob, K. *Journal of Chromatography*, **84**, 255 (1973)
122. Games, L.M. and Hayes, J.M. *Analytical Chemistry*, **48**, 130 (1976)
123. Zurcher, F. and Giger, W. *Vom Wasser*, **47**, 37 (1976)
124. Kuo, P.R.K., Chian, E.S.K., DeWalle, F.B. and Kim, J.H. *Analytical Chemistry*, **49**, 1023 (1977)
125. Sauer, T.C., Jr., Sackett, W.M. and Jeffrey, L.M. *Marine Chemistry*, **7**, 1 (1978)
126. Harris, L.E., Budde, W.L. and Eichelberger, J.W. *Analytical Chemistry*, **46**, 1912 (1974)
127. Skopintsev, V.A. *Okeanologiya*, **6**, 361 (1966)
128. Wangersky, P.J. Production of dissolved organic matter. In *Marine Ecology*, (ed. O. Kinne), Vol. 4., Wiley, New York (1978)
129. Duursma, E.K. *Journal of Sea Research*, **1**, 1 (1961)
130. MacKinnon, M.D. *Marine Chemistry*, **7**, 17 (1978)
131. MacKinnon, M.D. *The Analysis of Total Organic Carbon in Seawater. (a) Development of Methods for the Quantification of TOC. (b) Measurement and Examination of the Volatile Fraction of the TOC* Thesis, Dalhousie University (1977)
132. Deuser, W.S. *Deep Sea Research*, **18**, 995 (1971)
133. MacKinnon, M.D. *Marine Chemistry*, **8**, 143 (1979)
134. Grob, K., Grob, K., Jr. and Grob, G. *Journal of Chromatography*, **106**, 299 (1975)
135. Zlatkis, A., Lichtenstein, H.A. and Tichbee, A. *Chromatographia*, **6**, 67 (1973)
136. Russell, J.W. *Environmental Science and Technology*, **9**, 1175 (1975)
137. Daemen, J.M., Dankelman, H.W. and Hendriks, M.E. *Journal of Chromatographic Science*, **13**, 79 (1975)
138. Zeitz, U. *Gas-u-Wassertach (Wasser Abwasser)*, **117**, 181 (1976)
139. Wagner, R. and Ruck, W. *Zeitschrift für Wasser und Abwasser Forshung*, **14**, 145 (1981)
140. Wagner, R. and Ruck, W. *Zeitschrift für Wasser und Abwasser*, **15**, 287 (1982)
141. Dobbs, R.A. and Williams, R.T. *Analytical Chemistry*, **35**, 1064 (1963)
142. Cripps, J.M. and Jenkins, D. *Journal of the Water Pollution Control Federation*, **36**, 1240 (1964)
143. Ryding, S.O. and Forsberg, A. *Water Research*, **11**, 801 (1977)
144. Cameron, W.M. and Moore, T.B. *Analyst (London)*, **82**, 677 (1957)
145. Hoffmann, H.J. *Labor Prax (Warsberg)* **3**, 38 (1979)
146. Frankland, E. and Armstrong, H.E. *Journal of the Chemical Society*, **21**, 77 (1868)
147. Baumann, F.J. *Analytical Chemistry*, **46**, 1336 (1974)
148. Southway, C. and Bark, L.S. *Analytical Proceedings*, **18**, 7 (1981)
149. Heinerich, F. *Nature (London)*, **189**, 1001 (1961)
150. Lloyd, A. *Analyst (London)*, **107**, 1316 (1982)
151. Thompson, K.C., Mendham, D., Best, D. and De Casseres, K.E. *Analyst (London)*, **111**, 483 (1986)
152. Kessick, M.A. and Manchen, K.L. *Journal of the Water Pollution Control Federation*, **48**, 2131 (1976)
153. Manchen, L.K. *Unpublished Thesis*, Rice University (1974)
154. Davis, E.M., Petros, J.K. and Powers, E.L. *Industrial Wastes*, **23**, 22 (1977)
155. Gotaas, H.B. *Sewage Works Journal*, **21**, 818 (1949)
156. Seymour, M.P. *The Biodegradation of Particulate Organic Matter in Saltwater Media*, Thesis, Rice University (1977)
157. Kincannon, D.F. and Gaudy, A.F. *Biotechnical Engineering*, **8**, 371 (1966)

158. Kincannon, D.F. and Gaudy, A. *Journal of the Water Pollution Control Federation*, **38**, 1148 (1966)
159. Davies, E.M., Bishop, J.R., Gurhrie, R.K. and Forthafers, R. *Water Research*, **12**, 917 (1978)
160. Astrani, U. *Analysis of Waters for Nitrogen*, USNTIS PB Rep. No. 238056/6GA (1973)
161. Yablokava, O.G. *Okeanologiya*, **14**, 1113 (1974)
162. Mertens, J., Van Den Winkel, P. and Massart, D.L. *International Journal of Environmental Analytical Chemistry*, **47**, 25 (1975)
163. Stevens, R.J. *Water Research*, **10**, 171 (1976)
164. Adamski, J.M. *Analytical Chemistry*, **48**, 1194 (1976)
165. D'Elia, C.F., Steudler, P.A. and Corwin, N. *Limnology and Oceanography*, **22**, 760 (1977)
166. Armstrong, F.A.J., Williams, P.M. and Strickland, J.D.H. *Nature (London)*, **211**, 481 (1966)
167. Tenny, A.M. *Automated Analytical Chemistry*, **38**, 580 (1966)
168. Manny, B.A., Miller, M.C. and Wetzel, R.G. *Limnology and Oceanography*, **16**, 71 (1971)
169. Bruegmann, L. and Wilde, A. *Acta Hydrochimica Hydrobiologica*, **3**, 203 (1975)
170. Holm-Hansen, O., Strickland, J.D.H. and Williams, P.M. *Limnology and Oceanography*, **11**, 548 (1966)
171. Butler, E.I. *Estuarine Coastal Marine Science*, **8**, 195 (1979)
172. Butler, E.I., Knox, S. and Liddicoat, M.I. *Journal of the Marine Biological Association, UK*, **59**, 239 (1979)
173. Strickland, J.D.H. and Parsons, T.R. *A Practical Handbook of Seawater Analysis*, Bulletin 167, Fisheries Research Board Canada, p. 310 (1972)
174. Armstrong, F.A.J. and Tibbitts, S. *Journal of the Marine Biological Association, UK*, **48**, 143 (1968)
175. Koroleff, F. *Determination of Total Nitrogen in Natural Waters by Means of Digestion*, International Council for the exploration of the sea, Hydrography Committee, C.M. 1969/C:8 p. 4 (1969)
176. Koroleff, F. In *Methods of Seawater Analysis* (ed. K. Grasshoff), Verlag Chemie, New York, p. 116 (1976)
177. Shepherd, R.J. and Davies, I.M. *Analytica Chimica Acta*, **130**, 55 (1981)
178. D'Elia, C.F., Stendler, P.A. and Corwin, N. *Limnology and Oceanography*, **22**, 760 (1977)
179. Nydahl, F. *Water Research*, **12**, 1123 (1978)
180. Harvey, H.W. *Journal of the Marine Biological Association*, **27**, 337 (1948)
181. Grasshoff, K. *Fresenius Zeitschrift für Analytishe Chemie*, **220**, 89 (1966)
182. Bikbulatov, E.W. *Gidrokhim. Mater.*, **60**, 167 (1974)
183. Schnepfe, M.M. *Analytical Chimica Acta*, **58**, 83 (1972)
184. Sugawara, D. *Analytical Abstracts*, **93**, 1169 (1956)
185. Matthews, A.D. and Riley, J.P. *Analytica Chimica Acta*, **51**, 295 (1970)
186. Tsunogai, S. *Analytica Chimica Acta*, **55**, 444 (1971)
187. Taylor, D.L. and Zeitlin, H. *Analytica Chimica Acta*, **64**, 139 (1973)
188. Cembella, A.D., Antia, N.J. and Taylor, F.J.P. *Water Research*, **20**, 1197 (1986)

Index

The following abbreviations have been used in this index: aas = atomic absorption spectrometry; det of = determination of; glc = gas liquid chromatography; hplc = high performance liquid chromatography.